ELEMENTS OF
CHEMICAL PROCESS
ENGINEERING

ELEMENTS OF
CHEMICAL
PROCESS
ENGINEERING

D. S. J. Jones
Consultant Chemical Engineer, Calgary, Canada

JOHN WILEY & SONS
Chichester · New York · Brisbane · Toronto · Singapore

Copyright © 1996 by John Wiley & Sons Ltd,
                        Baffins Lane, Chichester,
                        West Sussex PO19 1UD, England

                  National   01243 779777
                  International (+44) 1243 779777

Other Wiley Editorial Offices

John Wiley & Sons, Inc., 605 Third Avenue,
New York, NY 10158-0012, USA

Jacaranda Wiley Ltd, 33 Park Road, Milton,
Queensland 4064, Australia

John Wiley & Sons (Canada) Ltd, 22 Worcester Road,
Rexdale, Ontario M9W 1L1, Canada

John Wiley & Sons (Asia) Pte Ltd, 2 Clementi Loop #02-01,
Jin Xing Distripark, Singapore 0512

Library of Congress Cataloging-in-Publication Data

Jones, D. S. J.
   Elements of chemical process engineering / D. S. J. Jones.
     p. cm.
   Includes bibliographical references and index.
   ISBN 0-471-96154-X (hardcover)
   1. Chemical engineering.   I. Title.
   TP155.J635   1996
   660–dc20   95-32241
                                                  CIP

British Library Cataloguing in Publication Data

A catalogue record for this book is available from the British Library

ISBN 0 471 96154 X

Typeset in 10/12pt Times by Keytec Typesetting Ltd, Bridport, Dorset
Printed and bound in Great Britain by Bookcraft (Bath) Ltd, Midsomer-Norton, Avon
This book is printed on acid-free paper responsibly manufactured from sustainable forestation,
for which at least two trees are planted for each one used for paper production.

CONTENTS

LIST OF FIGURES

LIST OF TABLES

PREFACE

It has always been my opinion that a considerable gap exists between the academic training necessary for chemical engineers and the role expected of them in industry. This gap, of course, is filled over a period of time by sheer experience gained in day-to-day work. In the meantime, I believe engineers experience some frustration and in many cases a sense of inadequacy in applying their acquired knowledge quickly and effectively, as is required of them in industry. I was one of the lucky ones. On joining the development department of a large petroleum company after graduation, I had the good fortune to work for a very experienced chemical engineer. He took it upon himself to teach me 'the tricks of the trade', and had the patience to guide me through the traumas of those early days. From him I learnt those rules of thumb so necessary in everyday engineering life, as well as the quick methods that could be used to arrive at acceptable problem solutions.

In this book I have compiled the more common and important of these techniques that I acquired during my early days as a process engineer and those I obtained later during some 20 years in practising this profession. I also demonstrate how these may be used in accomplishing process engineering activities in the chemical industry. These methods, with their rules of thumb, were used extensively in plant design, development, and troubleshooting, before the introduction of the excellent computer simulation packages of today, I believe that they still have a big part to play in providing the process engineer with an understanding of what is contained in the computer programs and the significance of the data the program requires as input.

The chapters of this book are sequenced to cover most aspects of the responsibilities of the process engineer in the chemical industry. They begin with a review of the most common unit operations, the build-up of material and heat balances, and the review of engineering diagrams. Chapter 2 introduces the reader to the common engineering systems met with in all process plants: the control system, safety system, and the utility and offsite systems. The latter two systems are important because they represent a large portion of the capital cost of any chemical complex.

Chapters 3–7 are devoted to an in-depth look at the major equipment used in this industry. All these chapters begin with a description of the various types and proceed with calculation techniques in sizing or defining them. Finally, most chapters give a description on how to develop the process specifications for the items. I believe this is extremely important. The process engineer must be able to tell the equipment manufacturer or specialist engineer exactly what duty the equipment is expected to perform and what the process constraints are. This is essential to ensure that the correct piece of equipment is supplied or the correct data obtained from the manufacturer or the specialist engineer.

Finally, the last two chapters of the book are devoted to the role the process engineer is usually expected to play in the management of a company or engineering project. Chapter 8 is concerned with the role of the process engineer in putting training and experience to work thinking up and evaluating new processing schemes that will meet a company's operating strategy. This chapter takes the process engineer through a sequence of events that finally offers for selection two or three process configurations for profit enhancement. The selection process is then defined in terms of a discounted cash flow type return on investment. The development of this type of return on investment is described in some detail with worked examples covering each of its parts.

The final chapter describes and discusses the process engineer's role in the installation of new facilities or the revamping of an existing one. This chapter covers the role of an engineer in the client's organization and that of an engineer in the engineering contractor's organization. It proceeds to cover the major activities of process engineers as they relate to the management of the project and the implementation of the engineering contract.

As you may have gathered, the purpose of this book is to offer process engineers in the chemical industry some of the techniques and experience I have gained during my professional career. Hopefully, this may, in part, help the newer process engineers to find a foothold in the industry they have chosen. It may also add a little to the experience and technical 'know-how' of the more experienced engineer.

D. Stan Jones
Calgary, Alberta
1995

1 BASIC PROCESS ENGINEERING PRINCIPLES

The role of the process engineer in the chemical industry may vary from company to company but certain basics are common to all chemical and associated companies. In the chemical manufacturing industry the process engineer will be responsible for:

- Setting parameters for the efficient operation of process units.
- Monitoring the process performance of operating units.
- Carrying out feasibility studies for new or upgraded process units. This will involve developing the process design, heat and material balance, and basic flowsheets. It will also include economic appraisals and analysis of one or more appropriate processes for management decision making.
- Setting the process specification for the engineering and construction of approved new process plants or a revamp of existing plants.
- Monitoring the performance and conformance to specification of the engineering contractor.
- Commissioning or participating in the commissioning of new or revamped plants. Preparing start-up reports, and participating in acceptance procedures.

In the engineering and construction industry the process engineer's duties will include:

- Initiating, the process engineering design from the client's specification.

Preparing flowsheets and equipment data sheets. Soliciting licensor participation where necessary.

• Taking a major role in the development of 'approved for construction' flow sheets, heat and material balances and the utilities balances.

• Participating with other disciplines in the control of cost and schedule for the project.

• Carrying out duties similar to the process engineer in the chemical manufacturing industry in monitoring performance of equipment vendors, subcontractors, and licensors.

• Initiating and developing process performance guarantees.

• Developing plant operating manuals and start-up procedures. Participating in plant commissioning and carrying out acceptance test runs when required.

The following chapters provide guidelines in the form of chemical engineering techniques and equipment characteristics to meet the duties required of the process engineer. While most of those calculations and correlations used by process engineers are available in some excellent computer packages it is still wise to know what is contained in these packages. It is also of benefit to be able to execute these calculations by hand should the need arise. This first chapter then describes and discusses the basic principles of a process engineering function. It begins with the material balance.

1.1   The Material Balance

The material balance of a process defines the flow of material within that process in terms of mass, mols and usually volume in unit time. The most important of these is the mass balance. This is the item that will usually be used to develop the volume and molal quantities which between them will be used to size the equipment and systems used in the process.

Before commencing the material balance the engineer must have some knowledge of the process itself. The sequence of flow from one item of equipment to another must be known and also the reason for the flow and the sequence. The nature of the feed and the product must also be known, together with the occurrence of intermediate products or byproducts. With this knowledge the task of putting quantities to them can commence. In the case of the mass and molal balances the total feed in must exactly equal the total products out in terms of weight per unit time and mols per unit time. A typical example of a material balance used for downstream process systems sizing is given in the following table.

The table opposite is a working material balance for the distillation tower of a crude oil vacuum unit. From the data in this table tower heat balances giving the internal flows of the tower can be calculated. These data are also the basis for the sizing of downstream equipment such as heat exchangers, pumps, and any surge vessels that may be required.

The format of the material balance describing the flows through the unit to establish downstream equipment sizing will be slightly different. This form of the

Kuwait crude vacuum unit material balance

	TBP cut range °F	% Volume		Volume				Weight			Moles	
		Cut	Cumulative	BPSD	GPH	SG at 60 °F	lb/gal	lb/h	% wt	MW	Moles/h	Mid-BP °F
Atmos residue	+690	100	100	13 800	24 150	0.957	8.16	197 085	100.0	547.6	359.9	
LVGO*	IBP – 750	24	24	3 312	5 796	0.894 (26.6% °API)	7.443	43 140	21.9	325 (1)	132.7	715
HVGO	985	43	67	5 934	10 384	0.930 (21% °API)	7.747	80 445	40.8	542	148.4	835
Bitumen	+985	33	100	4 554	7 970	1.090 (0% °API)	9.20	73 500	37.3	932	78.8	
Total products	–	100	–	13 800	24 150	0.957	8.16	197 085	100.0	547.6	359.9	
				Flash zone material balance								
Overflash	985 – 1025 °F	2	69.0	276	483	0.96 (16% °API)	7.99	3 859	2.0	552	7.0	Mid-BP 1015 °F
Product vapour	985	67.0	67.0	9 246	16 180	0.915 (23% °API)	7.64	123 585	62.7	439.6	281.1	
Total vapour	–	69.0	69.0	9 552	16 663	0.920 (22.5%)	7.65	127 444	64.7	442.4	288.1	
Liquid phase	+1025	31.0	31.0	4 278	7 487	1.10 (−4% °API)	9.30	69 641	35.3	970	71.8	

*The LVGO contains 1498 lb/h of overhead vapour.

material balance is usually part of the process flow diagram and this is described more fully in Section 1.11. A part of such a material balance format is given in the example below.

Stream no.	1	2	3	4	5
Stream name	Feed to vac. col.	Vac. col. residue	LT vac. Gasoil	HY vac. gasoil	Vac. col O/hd vapours
Normal flow lb/h	197 085	73 500	41 646*	80 445	1 494
BPSD at 60 °F	13 800	7 970	5 596	10 384	–
Cut point °F	+690	+985	750	985	–
SG at 60 °F	0.957	1.090	0.894	0.930	MW 218
°API	16.6	0	26.6	21.0	–
Max flow lb/h	226 648	84 525	47 893	92 512	1 718

*The overhead vapours are shown separately here.

The format of the material balance given here can be varied to meet the convenience of the process to which it applies. For example, in a chemical plant environment, barrels per stream day (bpsd) could be changed to gallons per hour and °API become lb/gal. A row giving mol weight or mol/h could be added if this helps to follow the flow diagram or assists in the downstream equipment and piping sizing calculations. The most important consideration in any material balance is that it does balance and that the mass per unit time of material in is exactly equal to that of the material out. These data will be used for all kinds of analysis and design, such as the heat balances within the unit. If the material balance is wrong then the other criteria cannot be correct.

CHANGE OF PHASE

In most chemical plants this generally means phase change from liquid to vapour or vice versa, and therefore only this phase change will be considered here. Compounds in a mixture of compounds have an unique composition in the vapour and liquid phase for a given condition of temperature and pressure. The quantities of the compounds that are present in the respective phases may be calculated using the *equilibrium constant K* for each compound or component in the mixture. The composition of the vapour and the liquid for a mixture at a condition of a temperature and a pressure is the *equilibrium flash vaporization* for the mixture. This is discussed in Section 1.4.

CALCULATING A COMPONENT BALANCE

To complete a material balance it is sometimes necessary to carry out a balance of all the components that make up the various streams. This is very much the case in establishing the material balance for multi-component fractionation units. This section then deals with estimating the component balance over such a fractionator. The technique uses the minimum number of trays concept at total reflux. This concept is dealt with in more detail in Section 1.5.

Minimum trays at total reflux is expressed by[1]:-

$$N_{M+1} = \log\left[\left(\frac{LT_{key}}{HY_{key}}\right)_D \cdot \left(\frac{HY_{key}}{LT_{key}}\right)_W\right] \div \log \phi$$

where

N_{M+1} = the minimum trays at total reflux plus 1
LT_{key} = moles of the light key component
HY_{key} = moles of the heavy key component
D = the distillate
W = the bottom product
ϕ = the relative volatility of the key components
= the equilibrium constant k of the light key divided by that of the heavy key. These values are taken at an estimated mean tower conditions.

To carry out the component distribution of components in the distillate stream and the bottom product stream of a fractionator the following steps are required:

- *Step 1* Establish the component analysis of the feed stream. Then select two adjacent components that will be in the distillate stream and the bottoms product stream. These are the 'key' components. The component with the lowest boiling point is the light key and the other the heavy key.
- *Step 2* Fix the amount of one component that is to be in either the distillate or the bottom product. This is usually fixed by a specification. Let x be the moles of the other key component in the second stream.
- *Step 3* Predict the number of trays for this type of separation. This can be determined from previous experience. For example:
Separation of C_2 from C_3—usually 20 to 30 trays.
Separation of C_3 from C_4—usually 30 to 40 trays.
Light paraffins from aromatics—30 to 40 trays.
C_5 and C_6 separation—20 to 30 trays.
- *Step 4* Calculate the number of theoretical trays from the selected figure for actual trays by multiplying actual trays with an efficiency factor of 60–65%. The estimated minimum theoretical trays may then be found by dividing the theoretical number of trays by 1.5.
- *Step 5* Calculate the relative volatility ϕ. Make a reasonable guess at the mean conditions. Accuracy for this is not a major requirement.
- *Step 6* Use the Fenske equation to solve for x and complete the material balance around the tower.

Example calculation

It is required to calculate the material balance over a debutanizer which has the following feed composition:

	Mole/h
H_2	1.3
C_1	4.0
C_2	14.8
C_3	38.5
iC_4	19.5
nC_4	31.7
iC_5	19.4
nC_5	22.4
C_6	140.1
C_7	130.5
C_8	87.0
Total	509.2

The specification for the distillate is such that the C_5 content be no more than 1.2% of the total C_4s in the distillate.

Solution

Set the C_5 content at 1.0% of the total C_4s in the feed. Then C_5s in the distillate product is 0.511 mole/h (as iC_5). Select the light key as nC_4 and the heavy key as iC_5.

Assume tower will have 30 trays and tray efficiency is 63%. Then number of theoretical trays = $30 \times 0.63 = 18.9$.

$$\text{Minimum number of trays at total reflux} = \frac{18.9}{1.5}$$

$$= 12.6$$

$$\text{and } N_{M+1} = 13.6$$

Let x be the mole/h of iC_4 in the bottom product. Then the Fenske equation is as follows:

$$13.6 = \text{Log}\left[\frac{(31.7 - x)}{0.511}\right]_D \cdot \left[\frac{(21.89)}{x}\right]_W \div \text{Log } 1.77$$

$$\phi = \frac{kiC_4}{knC_5} \text{ at estimated mean conditions of 309 °F and 239 psig}$$

$$x = 6.31 \text{ mole/h}$$

The component balance in mole/h can now be written as follows:

	Feed	O/Head dist.	Bottom prod.
H_2	1.3	1.3	
C_1	4.0	4.0	
C_2	14.8	14.8	
C_3	38.5	38.5	
iC_4	19.5	19.5	
nC_4	31.7	25.4	6.3
iC_5	19.4	0.5	18.9
nC_5	22.4		22.4
C_6	140.1		140.1
C_7	130.5		130.5
C_8	87.0		87.0
Total	509.2	104.0	405.2

Mass and volume balances can now be developed by applying the mol weight and SGs to the molar balance components.

1.2 Heat Balances

Heat balances are prepared to describe and define the flow of heat energy to and from equipment items and systems within the process configuration. As in the case of the mass balance, it follows the basic condition that the amount into a system must exactly equal the amount out of it. Process engineers use heat balances in their work to determine:

- The utilities required for the process
- The sizing of equipment items
- Equipment performance
- Process operating conditions

HEAT BALANCES TO PREDICT UTILITIES

The heat to a process or heat removed from a process is supplied or rejected by sources outside that process. That is, heat is supplied either directly or indirectly by a fuel such as fuel oil, fuel gas or coal. It is removed from a process by either air or water cooling. Thus the heat brought into a process by the feed streams and that removed from the process by the product streams are balanced by the fuel and the cooling medium utilized in the process. In processes where a chemical reaction takes place the heat of reaction must be considered as part of the heat brought into the process by the feed or taken out of it by the products. This also applies to changes of phase that may occur between the feed in and the product out.

Fuel input in this case includes steam or hot oil streams generated outside of the process (indirect fuel usage). Cooling of process streams includes that by air coolers which determines part of the electrical utilities consumption or by a cooling-water utility system generated outside the process. Refrigeration, where applicable, fits into this latter system.

USING HEAT BALANCES TO SIZE EQUIPMENT

Much of the data used to size some process equipment in the chemical industry are developed or calculated by using heat balances. Some common items that fall into this category are:

- Fractionation towers and tray diameter
- Heat exchanger duties
- Fired heater and steam boiler duties
- Basis for reactor design temperature control
- Temperature control and sizing equipment where recycle quench streams are required.

In all these instances heat balance around items of equipment or process systems provide the means to determine either temperature levels or fluid flows. The principle is always that the unknown quantity forms part of the equation

$$\text{Heat in} = \text{heat out}$$

where heat is measured by

$$\text{Mass} \times \text{specific heat} \times \text{temperature difference}$$

when there is no change of phase, and by

$$\text{Mass} \times \text{latent heat}$$

where there is a change of phase.

There are many sources in the literature where these equations are simplified by being able to read them from prepared curves in terms of enthalpy per unit mass versus temperature. These curves are usually based on either specific gravity of the stream, mol wt or average boiling points. Maxwell's *Data Book on Hydrocarbons* presents curves for many pure hydrocarbons such as butanes, propane, etc. and for hydrocarbon fractions based on average boiling points. The one thing to remember or note in using these curves is the datum temperature that is used. This is the temperature taken by the author where enthalpy is equal to zero. In developing heat balances enthalpy data used from sources with a different datum must be adjusted to the same datum base.

HEAT BALANCES USED FOR EVALUATING EQUIPMENT PERFORMANCE AND OPERATING CONDITIONS

Heat balances are frequently used to translate plant data into an analysis of the performance of equipment items. The principle used for this is the same as that described for sizing equipment. In this case, however, the parameters are data read from plant instruments. For example, in evaluating the performance of a heat exchanger the temperatures of the streams in and out of shell or tube side are read together with their respective mass flows to the exchanger. A simple heat balance of mass times enthalpy difference provides the actual duty of the exchanger in BTU/h, cal/h, etc. From the equipment data sheet the effective area of heat exchange or the overall heat transfer coefficient can be calculated using

$$Q = U \cdot A \cdot \Delta T_m$$

where

Q = duty in BTU/h
U = overall heat transfer coefficient in BTU/h.ft^2.°F
A = surface area (ft^2)
ΔT_m = log mean temperature difference (°F)

The comparison of the overall heat transfer coefficient with the design (from the data sheet) gives an indication of the item's performance.

Setting process operating conditions often uses this principle of heat balance to set temperature levels for process streams. The same heat balance calculations are used as described earlier. In this case, however, one or other of the temperatures making up the temperature difference becomes the unknown. The mass of the stream and the enthalpy of one or other side of the heat equation must be known. The unknown enthalpy is calculated and its corresponding temperature is read from the enthalpy curve or calculated using specific heat and/or latent heat criteria.

Example calculation

Among the most common use of heat balances is to establish the internal flows in fractionating columns. This example therefore concerns the flow of liquid and vapour in the top section of a debutanizer column. In this problem the overhead product contains a relatively high percentage of hydrogen and methane. It cannot therefore be completely condensed in an overhead condenser at practical conditions of temperature and pressure. The following conditions of temperature, pressure, and flows have already been calculated and are:

Product vapour = 2385 lb/h
Product distillate = 12 042 lb/h
External reflux = 27 564 lb/h

Tower top temperature = 158 °F
Tower top pressure = 220 psia

Reflux drum temperature = 95 °F
Reflux drum pressure = 215 psia

The heat balance diagram is as follows:

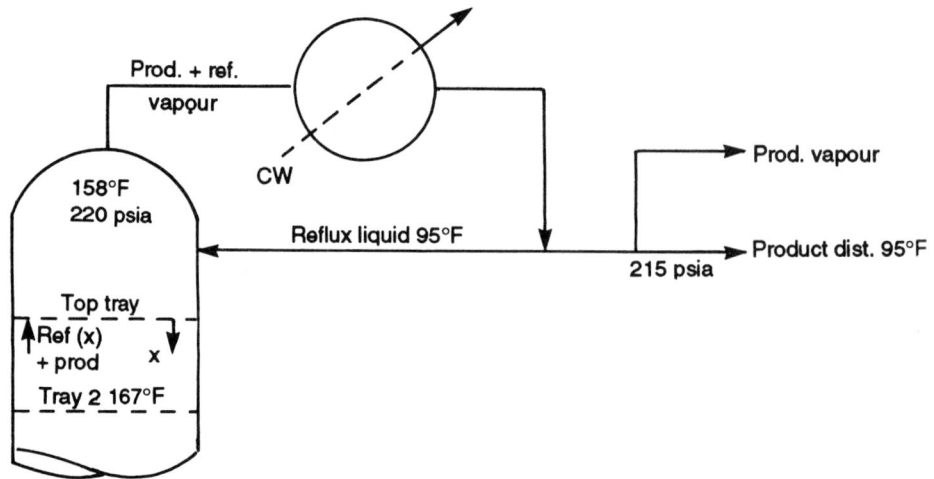

To calculate vapour and liquid load on top tray
(a) Calculate condenser duty

	Vapour or liquid	°API or (mol wt)	°F	lb/h	Enthalpy	
					BTU/lb	mmBTU/h
IN						
Product vapour	V	(46)	158	14 427	330	1.587
Reflux vapour	V	(48)	158	27 564	332	3.050
TOTAL IN				41 991		13.912
OUT						
Product vapour	V	(36)	95	2 385	320	0.763
Product distillate	L	(48)	95	12 042	170	2.047
Liquid reflux	L	(48)	95	27 564	170	4.686
Condenser		(By diff.)				6.416
TOTAL OUT				41 991		13.912

(b) Calculate vapour/liquid traffic

	V or L	°API or (MW)	°F	lb/h	Enthalpy	
					BTU/lb	BTU/h × 10⁶
IN						
Vapour ref. ex tray 2	V	(54)	167	X	332	332 X
Vapour prod. ex tray 2	V	(46)	167	14 427	335	4.833
TOTAL IN				14 427 + X		4.833 + 332 X
OUT						
Prod vapour.	V	(36)	95	2 385	320	0.763
Prod dist.	L	(48)	95	12 042	170	2.047
Internal reflux	L	(54)	162	X	200	200 X
Condenser duty						6.416
TOTAL OUT				14 427 + X		9.226 + 200 X

$$X = \frac{4\,393\,000}{132} = 33\,280\,\text{lb/h}$$

Traffic over the top tray

Vapour to tray	Liquid from tray
lb/h = 47 700	gph @ 60 = 119.7
mole/h = 929.7	Hot gph = 153.0
acfs = 7.83	Hot cfs = 0.339
lb/ft^2 ρ_V = 1.69	lb/h = 33 280
Temp. (°F) = 167	lb/ft^3 ρ_L = 27.3
Pressure (psia) = 220	Temp. (°F) = 162

1.3 Dew Points and Bubble Points

Dew points and bubble points are terms used to define the temperature and pressure condition at which a mixture of compounds begins to condense or boil. Thus in the case of hydrocarbons, for example,

* *Dew point* is the temperature and pressure at which a hydrocarbon vapour begins to condense.
* *Bubble point* is the temperature and pressure at which a hydrocarbon liquid begins to boil.

Both these conditions are calculated using the phase equilibria concept and examples of these calculations now follow.

The overheads from a fractionator has the following component:

Mole fraction composition:

C_2	0.008
C_3	0.054
iC_4	0.021
nC_4	0.084
C_5s	0.143
C_6	0.155
C_7	0.175
Comp. 1	0.124
Comp. 2	0.124
Comp. 3	0.075
Comp. 4	0.037
Total	1.000

Components 1 to 4 are pseudo-components which have properties similar to real components with the same boiling point. The mid-boiling points of these components (°F) are:

Comp. 1	260
Comp. 2	300
Comp. 3	340
Comp. 4	382

The reflux drum temperature will be fixed at 100 °F. The pressure will be calculated at the bubble point of this material at 100 °F.

Bubble point is defined as the sum of all y's = sum of all Kx's. Where x is the mole fraction of a component in the liquid phase and y is the mole fraction of the component in the vapour phase at equilibrium. At bubble point $\sum y = \sum Kx$. K is the equilibrium constant and can be read from the curves found in textbooks such as Maxwell's *Data book on Hydrocarbons* or can be considered (approximate and not to be used for definitive design) as:

$$\frac{\text{vapour pressure}}{\text{total pressure}} = K$$

This relationship will be used for this calculation. The calculation is iterative (trial and error) as follows:

At 100 °F

	Mol fract. x	First trial at 5 psig		Second trial at 12 psig	
		K	$Y = Kx$	K	$Y = Kx$
C_2	0.008	40.6	0.325	32.4	0.259
C_3	0.054	9.3	0.502	7.42	0.401
iC_4	0.021	3.55	0.075	0.38	0.05
nC_4	0.084	2.54	0.213	0.03	0.171
C_5	0.143	0.89	0.127	0.71	0.102
C_6	0.155	0.254	0.039	0.20	0.031
C_7	0.175	0.084	0.015	0.067	0.011
Comp. 1	0.124	0.023	0.003	0.020	0.002
Comp. 2	0.124	Neg.	Neg.		
Comp. 3	0.075	Neg.	Neg.		
Comp. 4	0.037	Neg.	Neg.		
	1.000		1.299		1.027

FOR SECOND TRIAL (ESTIMATE)

Take the K value of the highest fraction of y (in this case C_3) where $K = 9.3$. Take

$$\frac{K}{1.299} = 7.16 \,(\text{new } K)$$

Make the second trial with KC_3 at 7.16 which gives a systems pressure P as follows:

$$\frac{VP\ C_3}{P} = 7.1 \text{ where } VP\ C_3 \text{ at } 100\,°F \text{ is } 190 \text{ psia}$$

$$\text{Then } P = \frac{190}{7.1} = 26.5 \text{ psia}$$

Second trial pressure = 26.5 psia = 11.8 psig

Let us set it at 12 psig

The second trial gives sum of Y's $= 1.027$ and this is considered to be close enough. Then the drum will be operated at $100\,°F$ and at 12 psig.

DEW POINT CALCULATION

A similar calculation is made to establish the dew point for a hydrocarbon vapour. In this case, however, $\sum x = \sum y/K$ at the dew point condition of temperature and pressure of the system. Thus consider the dew point of the same hydrocarbon stream as used in the bubble point example. In this case the systems pressure is fixed at 8.3 psia and the dew point temperature will be calculated as follows.

Comp.	Mole Frac. Y	First trial at 220 K	First trial at 220 X = Y/K	Second trial at 225 °F K	Second trial at 225 °F X = Y/K	Mol wt	Weight factor	SG	Vol factor	Liquid prop.
C$_2$	0.008	–	NEG		NEG					
C$_3$	0.054	84.3	0.001	93.9	0.001	44	0.044	0.508	0.009	
iC$_4$	0.021	38.6	0.001	39.8	0.001	58	0.058	0.563	0.010	Mol wt = 130.7
nC$_4$	0.084	29.52	0.003	30.1	0.003	58	0.174	0.584	0.030	
C$_5$	0.143	12.53	0.011	12.65	0.011	72	0.792	0.629	0.126	SG = 0.766
C$_6$	0.155	4.70	0.033	4.94	0.031	85	2.635	0.675	0.390	
C7	0.175	2.17	0.081	2.19	0.080	100	8.00	0.721	1.110	°API = 5
Comp. 1	0.124	1.00	0.124	1.16	0.107	114	12.198	0.743	1.642	
Comp. 2	0.124	0.506	0.245	0.518	0.239	126	30.114	0.765	3.936	K = 12
Comp. 3	0.075	0.229	0.328	0.253	0.293	136	39.848	0.776	5.135	
Comp. 4	0.037	0.108	0.343	0.126	0.294	152	44.688	0.788	5.671	
Totals	1.000		1.170		1.06	130.7	138.551	0.767	18.059	

$$K = \frac{\text{Vapour pressure at selected temperature}}{\text{Total systems pressure}}$$

Second trial $= 0.108 \times 1.170$ (K for mid-BP 362 component)

New $K = 0.126$ then $VP = 8.3$ psia $\times 0.126 = 1.05$ psia $\equiv 225\,°F$
Second trial is close enough to actual temperature:
$0.126 \times 1.06 = 0.134$ $VP = 1.11 = 229\,°F$

Notes

(a) In estimating for the second trial and final temperature the K of the highest X component is multiplied by the total value of the X function. Then vapour pressure curves are used to give the component temperature corresponding to this new vapour pressure.

(b) The molal composition of the final X is the composition of the liquid in equilibrium with the product vapour.

1.4 Equilibrium Flash Vaporization (EFV)

When a mixture of compounds vaporize or condense there is an unique relationship between the composition of the mixture in the liquid phase and that in the corresponding vapour phase at any condition of temperature and pressure. This relationship is termed the equilibrium flash vaporization for the mixture. It can be calculated using the composition of the feed mixture and the equilibrium constant of the components in the mixture. This is expressed by

$$L = \frac{x_f}{1 + (V/L)K}$$

where

L = total mol/h of a component in the liquid phase
X_f = mol/h of the component in the feed
V/L = The ratio of total mol vapour to total mol liquid
K = the equilibrium constant for each component at the flash condition of temperature and pressure.

There are several publications giving values for K. Among these are the charts in Maxwell's 'Data on Hydrocarbons' which are based on fugacities. Others may be found in engineering data books such as those issued by the Gas Processors Suppliers Association which are based on convergence pressures. A rough and ready substitute for K factors is to use the vapour pressure of the component divided by the systems pressure. This, however, should not be used for any definitive design work nor in systems which have azeotropes or are near their critical conditions. A method for calculating equilibrium flash vaporization is given by the following steps:

- *Step 1* Prepare a table with the first column giving the components making up the feed. The second column will be the composition of the feed in mol/h. The third column is a listing of the equilibrium constant K for each component at the temperature and pressure of the flash condition. Allow up to three columns following for assumptions of V/L. Each of these three columns should be subdivided into two, the first giving the product of $(V/L)K$ and the second for listing the L for each component. Other columns may be added to calculate mol wt of vapour and SG of the liquid phase.
- *Step 2* Assume a value for V/L. This is a judged assumption but start with 1.0 or 0.1, whichever seems to be the more realistic. Calculate $(V/L)K$ for each component.
- *Step 3* Calculate L for each component from

$$L = \frac{X_f}{1 + (V/L)K}$$

- *Step 4* The calculated V/L is now obtained by adding the L column and subtracting this value from the total mols of feed in column 1. This subtraction is the vapour mols as calculated. Then the calculated V/L is arrived at by dividing the total V by the total of the L column.
- *Step 5* The calculation is complete when V/L calculated is equal to V/L assumed. An answer to within 5% is usually acceptable. If the calculated versus assumed is not within this limit make another assumption for V/L and

repeat steps 2–4. For this second assumption try 5, 0.5, or 0.05, whichever is more appropriate.

- *Step 6* If there is still no agreement between assumed and calculated V/L, plot the two trial points (assumed Vs calculated) on log log graph paper. Draw a straight line through these two points and note where on this line assumed V/L = calculated V/L. This value is the next assumed V/L. Repeat the calculation steps 2–4 using this value. This usually completes the calculation. If it does not, then check that the conditions for the flash are within the boiling point and condensing point for the feed.

Example calculation

In this example it is required to determine the amount of vapour and liquid and their composition in a feed to a fractionator at 112 psig and 300 °F.

Calculate feed flash

	Moles/h (°F)	127 psia K 300 °F	V/LK	1st trial $V/L = 0.3$ $L = \dfrac{X_F}{1 + V/LK}$	V/LK	2nd trial $V/L = 0.2$ $L = \dfrac{X_F}{1 + V/LK}$	V/LK	3rd trial $V/L = 0.1$ $L = \dfrac{X_F}{1 + V/LK}$
C_2	6.4	9.1	4.55	1.15	1.82	2.27	0.9	3.37
C_3	43.5	5.0	2.50	12.43	1.00	21.75	0.5	29.0
iC_4	16.9	3.3	1.65	44.79	0.66	10.18	0.33	12.71
nC_4	67.6	2.9	1.45	30.04	0.58	42.78	0.29	52.40
iC_5	80.5	1.8	0.90	42.37	0.26	59.19	0.18	68.22
nC_5	34.6	1.6	0.80	19.22	0.32	26.21	0.16	29.83
C_6	124.9	0.85	0.425	87.65	0.17	106.75	0.085	115.12
C_7	140.9	0.48	0.240	113.63	0.096	128.56	0.048	134.45
NP260	99.8	0.212	0.106	90.24	0.042	95.74	0.021	97.75
NP300	99.8	0.116	0.058	94.33	0.023	97.54	0.012	98.62
NP340	60.4	0.063	0.032	58.53	0.126	59.55	0.006	60.04
NP382	29.8	0.035	0.0175	29.29	0.007	29.60	0.004	29.80
Total	805.1			623.67		680.07		731.31
Calculated V/L				0.29		0.184		0.1

Liquid properties from 3rd trial

MW	Liquid lb/h	lb/gal	gph
30	101	2.97	34
44	1 276	4.69	301.75
58	737	4.69	157.1
58	3 039	4.87	624.0
72	4 912	5.21	942.8
72	2 148	5.26	408.4
86	9 900	5.54	1 787.0
100	13 445	5.74	2 342.3
114	11 144	6.18	1 803.2
126	12 426	6.37	1 950.7
136	8 165	6.46	1 263.9
152	4 530	6.56	690.5
Total 98.2	71 823	5.4	12 305.6 °API = 70

$$\text{Lb/h liquid} = 71\,823 \qquad \text{lb/gal} = 5.4$$

$$\text{Lb/h vapour} = 4998 \qquad \text{mol wt} = 67.7$$

$$\text{Lb/h feed} = 76\,821$$

Note that the components NP260 to NP382 are pseudo-components having mid-boiling points of 260 °F, 300 °F, 340 °F, 382 °F respectively. K for these components is based on their vapour pressure and system pressure relationship.

1.5   Fractionation

Next to heat transfer the most widely used unit operation in the chemical process industry is fractionation. This section therefore describes the basic principles of this operation and illustrates a short-cut method to arrive at the number of trays required for a unit.

Fractionation is used to separate components of a mixture by boiling range. Basically the principle of such a separation is by a series of flash vaporizations occurring on trays over a range of temperatures and pressures in a column. The compositions on individual trays are influenced by a vapour stream moving up the column bubbling through a liquid level on each tray. This liquid level is also influenced by a liquid reflux stream cascading down the column from tray to tray. The method of calculating fractionation by a tray-to-tray basis is a long and tedious exercise and today it is only performed using one of the many computer programs available. There is, however, a well-proven short-cut method to arrive at a reasonably good answer in calculating the number of trays in a fractionation column. Indeed it is recommended that this method be used prior to running the rigorous tray-to-tray program on the computer. This gives a first good guessed input to the program and therefore enhances the convergence of the programme.

The calculation employs the use of the Underwood equation[2] to calculate minimum reflux at infinite stages, then the Fenske[1] equation for minimum stages or trays at total reflux and a correlation of both to give theoretical trays required at a finite reflux.[3]

The first part using the Underwood equation is a rigorous trial-and-error calculation. The following steps describe this and the other procedures to arrive at the total number of theoretical trays.

- *Step 1* Obtain the mole fraction composition of the fractionator feed, top product, and the bottom product. This is obtained by a material balance over the tower (see Section 1.1).
- *Step 2* Identify one of the key components, that is, one of the components that appears in the top product that is to be separated and also in the bottom product. In the example below this will be either nC_4 or iC_5.
- *Step 3* Calculate the dew point conditions at the top of the tower and the bubble point temperature of the bottom product at the bottoms pressure (see Section 1.3). Calculate the mean temperature and pressure conditions in the tower.
- *Step 4* Divide the equilibrium constant (K) of each component at the mean

tower conditions by the K value for one or other of the key components. These will be the relative volatilities (ϕ) for each component.

- *Step 5* Construct the Underwood table for calculating the value of B in the equation $(X_f \cdot \phi)/(\phi - B)$ for each component. The sum of these must equal zero for a solution to the value of B.

- *Step 6* The columns in this table commence with the feed composition, then the relative volatilities for each component. This is followed by a column giving the product for each component of mole fraction multiplied by the relative volatilities ($X_f \cdot \phi$). The remaining columns are trial calculation for values of B that sum to zero.

- *Step 7* When the value for B is found this is used in the equation

$$\sum \frac{X_D - \phi}{(\phi - B)} = R_{M+1}$$

where X_D is mole fraction of each component in the distillate (or overhead product) and R_M is the minimum reflux at an infinite number of trays.

- *Step 8* Calculate minimum number of trays at total reflux using the Fenske equation. This has been described in Section 1.1.

- *Step 9* Using the Gilliland correlation[3] (Figure 1.1) read off a value for $(N - N_M)/(N + 1)$ where N is the theoretical number of trays. R is usually taken as $R_M \times 1.5$.

- *Step 10* Tray efficiency may be calculated by a method given in Maxwell's *Data Book on Hydrocarbons*, using average liquid fluidity (Figure 1.2). Usually rectifying trays (above the feed inlet) have an efficiency of 70–75% while stripping trays (below the feed tray) have an efficiency of 55–65%.

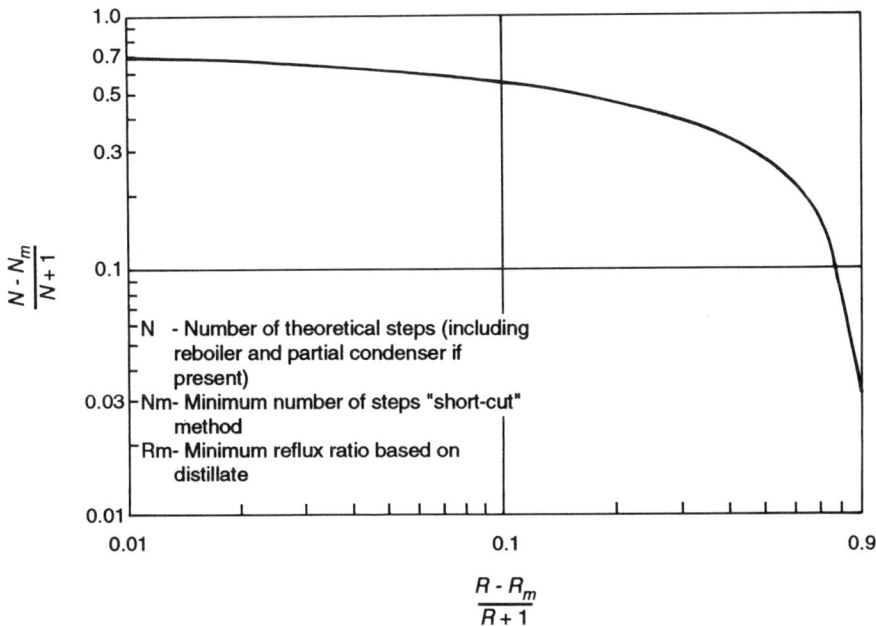

N - Number of theoretical steps (including reboiler and partial condenser if present)

Nm- Minimum number of steps "short-cut" method

Rm- Minimum reflux ratio based on distillate

Figure 1.1. The Gilliland correlation for calculating theoretical trays

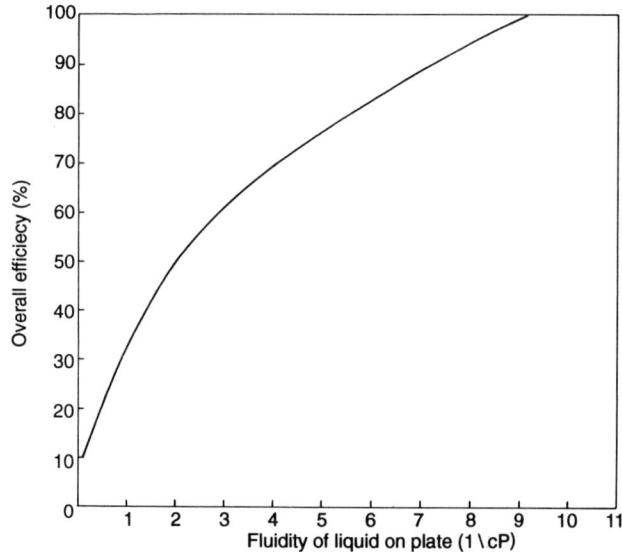

Figure 1.2. Overall plate efficiency versus fluidity of liquid on plates (Reproduced by permission of Kreiger Publishing Company, Malabar, Florida, 1950, Maxwell, Data Book on Hydrocarbons)

Example calculation

This calculation determines the number of trays required to fractionate an aromatic hydrocarbon feed to remove a light paraffin product overhead and leave a high content aromatic product as the tower bottoms. The material balance and tower conditions are as follows.

Material balance

	Feed					Overheads			Stabilized reformate		
	Moles/h	MW	lb/h	lb/gal	gph	Moles/h	lb/h	gph	moles/h	lb/h	gph
H_2	1.3	2	3	–		1.3	3	–			
C_1	4.0	16	64	–		4.0	64	–			
C_2	14.8	30	444	2.96	150	14.8	444	150			
C_3	38.5	44	1694	4.22	401	38.5	1694	401			
iC_4	19.5	58	1131	4.68	242	19.5	1131	242			
nC_4	31.7	58	1839	4.86	378	24.9	1444	297	6.8	395	81
iC_5	19.4	72	1397	5.20	269	0.4	29	6	19.0	1368	263
nC_5	22.4	72	1613	5.25	307				22.4	1613	307
Benz.	140.1	81	11348	6.83	1661				140.1	11348	1661
Tol.	130.5	102	13311	6.84	1946				130.5	13311	1946
Xyl.	87.0	128	11136	6.94	1605				87.0	11136	1605
Total	509.2	86.4	43980	6.32	6954	103.4	4809	1096	405.8	39171	5863

Calculating minimum reflux plates—using the Underwood equation: Calculating for B

Com.	Mol fract. in feed X_f	ϕ_{ave}	$X_f\phi_{ave}$	$\sum\dfrac{X_f\phi_{ave}}{\phi - B} = 0$		
				Trial 1 $B = 1.1$	Trial 2 $B = 1.09$	Trial 3 $B = 1.094$
H_2	0.0031	52	0.456	0.003	0.003	0.003
C_1	0.008	7.2	0.058	0.010	0.009	0.009
C_2	0.029	3.4	0.094	0.043	0.043	0.043
C_3	0.076	1.9	0.144	0.180	0.178	0.179
iC_4	0.038	1.17	0.045	0.643	0.563	0.592
nC_4 (key)	0.062	1.00	0.062	−0.620	−0.689	−0.660
iC_5	0.038	0.61	0.023	−0.047	−0.048	−0.048
nC_5	0.044	0.54	0.024	−0.043	−0.044	−0.043
Benz.	0.275	0.19	0.052	−0.057	−0.058	−0.058
Tol.	0.256	0.08	0.020	−0.020	−0.020	−0.020
Xyl.	0.171	0.03	0.005	−0.005	−0.005	−0.005
Total	1.00			0.087	−0.068	0.008

Trial 3 is within acceptable limits: $B = 1.094$

Use key as nC_4 at 309 °F and 254 psia: calculate for R_{M+1}

	Mol fract in distillate X_D	ϕ_{ave}	$X_D - \phi_{ave}$	$R_{M+1} = \dfrac{X_D\phi_{ave}}{(\phi_{ave} - B)}$
H_2	0.013	152	1.976	0.013
C_1	0.039	7.2	0.281	0.046
C_2	0.143	3.4	0.486	0.211
C_3	0.371	1.9	0.705	0.875
iC_4	0.189	1.17	0.221	0.908
nC_4	0.241	1.00	0.241	−2.560
iC_5	0.004	0.61	0.002	−0.004
Total	1.000			1.489

$R_{M+1} = 1.489$
$R_M = 0.489 \; R = 0.489 \times 1.5 = 0.7335$

Calculating minimum trays at total reflux

Using the Fenske equation (see also Section 1.1)

$$N_{M+1} = \text{Log}\left[\left(\frac{LT_{key}}{HY_{key}}\right)_D \cdot \left(\frac{HY_{key}}{LT_{key}}\right)_W\right] \div \text{Log } \phi$$

$$N_{M+1} = \text{Log}\left[\left(\frac{0.241}{0.004}\right) \cdot \left(\frac{0.047}{0.017}\right)\right] \div \text{Log } 1.77$$

$$N_{M+1} = 9 \text{ trays}$$

$$N_M = 8$$

Using the Gilliland correlation

$$\frac{R - R_M}{R + 1} = \frac{0.7335 - 0.489}{1.7335} = 0.141$$

From Figure 1.1

$$\frac{N - N_M}{N + 1} = 0.52$$

$$N - 8 = 0.52\,(N + 1)$$

$$N - 0.52\,N = 9$$

$$N = 19 \text{ trays}$$

Efficiency (say) 60%

$$\text{Then number of actual trays} = \frac{19}{0.6}$$

$$= 32$$

1.6   Absorption

The unit operation of absorption and stripping is used extensively in the chemical process industry to remove heavier gas components in a gas stream. This concept may be used to purify a light gas stream or to recover a valuable component or components from a bulk gas stream. A typical example of the former is the removal of 'heavy ends' from a natural gas stream to meet dew point requirements of the gas main. In the latter case absorption is used in many processes to recover saleable products such as propane and butanes from light gas process streams high in hydrogen and methane content.

As in the case of fractionating units there are many computer programs that can simulate absorption processes. It is worth knowing, however, how to calculate a material balance around an absorption system if only to apply the answer as a first guess for a computer program input. The following steps and the worked example which follows gives a method to evaluate the amount absorbed in a simple absorption process.

Consider the process which is required to remove the heavy ends from a gas stream to meet a dew point condition for the finished gas using a suitable liquid as the absorbent. Now the selection of this absorbent is important to the operation. Ideally the absorbent should possess an equilibrium constant as far distant as possible from the heaviest of the components to be absorbed. However, heavy liquids (that is, liquids with relatively high boiling points) usually have high viscosities that reduce tower tray efficiency. Selection therefore has to be a compromise between these properties and, of course, the availability of suitable streams.

In this calculation the number of trays and therefore the number of theoretical stages are assumed for the absorber tower. The rate of absorbent is also assumed.

Usually this will be dependent on the amount of the stream available. Alternatively, a molal ratio of absorbent to gas feed of between 0.7 to 1.0 will be set. The amount of each component absorbed and the composition of the dry product gas and the rich liquid absorbent is calculated using the following steps:

- *Step 1* Establish the tower top pressure. This will be such that the gas feed stream can enter the tower and proceed upward through the trays. Assume a tower top temperature and a tower bottom temperature. As a first guess let the tower top be at the temperature of the gas feed and the bottom temperature be 20 °F higher.
- *Step 2* Read off the K values for the gas components at the tower top conditions. Assume that the components with $K = 1.0$ or greater will be absorbed. Using this as a criterion, calculate the percentage of the gas stream that will be absorbed.
- *Step 3* Calculate values for L/V for the tower top and the tower bottom using the following expressions:

$$(L/V)_{top} = \frac{L_{n+1} + B_n}{V_t - B}$$

where

L_n = moles/h of absorbent
$B_n = 0.35 \times$ moles of gas absorbed/h
V_t = total moles/h of gas feed
B = moles/h of gas absorbed

$$(L/V)_{bot} = \frac{L_{n+1} + B}{V_t - B_n}$$

- *Step 4* Calculate the absorption factors and stripping factors for the tower from the expressions

$$\text{Absorption } A_e = \sqrt{A_{bot} \times (1 + A_{top}) + 0.25} - 0.5$$
$$\text{Stripping } S_e = \sqrt{S_{top} \times (1 - S_{bot}) + 0.25} - 0.5$$

where

$$A = \frac{(L/V)}{K}$$
$$S = \frac{(VK)}{L}$$

A and S are calculated with K values at the tower top conditions and the tower bottom conditions respectively for each component in these equations.

- *Step 5* Using the absorption factors from step 4 calculate the quantity absorbed from the expression

$$M = V_a(F_a) - l_{n+1}(F_s)$$

where

$$M = \text{moles/h absorbed}$$
$$n = \text{number of theoretical stages}$$
$$V_a = \text{total moles in wet gas (feed)}$$
$$l_{n+1} = \text{total moles in liquid entering top tray}$$
$$F_a \text{ and } F_s = \text{factors from the expressions given below:}$$

$$F_a = \frac{A_e^{n+1} - A_e}{A_e^{n+1} - 1}$$

$$F_s = \frac{S_e^{n+1} - S_e}{S_e^{n+1} - 1}$$

- *Step 6* From the quantity absorbed calculate the composition of the 'dry gas' (product gas) using the expression

$$m_g = V_a - M$$

for each component in the gas. Then calculate the composition of the 'rich liquid' leaving the bottom of the tower by difference.
- *Step 7* Carry out a dew point calculation on the dry gas and a bubble point calculation on the rich liquid to establish top and bottom temperatures (see Section 1.3). Compare with the assumed temperatures. If there is a large discrepancy, recalculate using the new temperatures. If these conditions are within 3–5 °F of those assumed, the compositions as calculated may be used to input a computer simulation program.
- *Step 8* This calculation can be used to determine the number of theoretical stages required for a specified gas composition or recovery of a component in the gas. In this case the assumed temperatures versus calculated should be within 2 °F to be acceptable. The final compositions should then be examined at another two or more iterations using different values for n (number of theoretical stages) to determine the best match to that specified. Clearly, this would be an exercise using a computer simulation program.

Example calculation

It is required to determine the composition and conditions (temperature and pressure) of a dry gas and rich liquid absorbent leaving a 20-tray absorber. The 'wet' feed gas to the tower enters at 175 psia and 95 °F and has the following molal composition:

	Mole/h	Mol fraction
H_2	95.04	0.467
C_1	38.60	0.190
C_2	12.11	0.059
H_2S	49.24	0.242
C_3	6.04	0.030
iC_4	1.32	0.006
nC_4	1.32	0.006
Total	203.67	1.000

The 'lean' absorbent is a heavy naphtha with properties similar to C_9 (nonane). The composition of this stream which has been stripped elsewhere and recycled is as follows:

	Mole/h
C_3	0.093
iC_4	0.120
nC_4	0.140
C_9	75.047
Total	75.400

The temperature of this stream entering the column is 90 °F. Consider the following diagram:

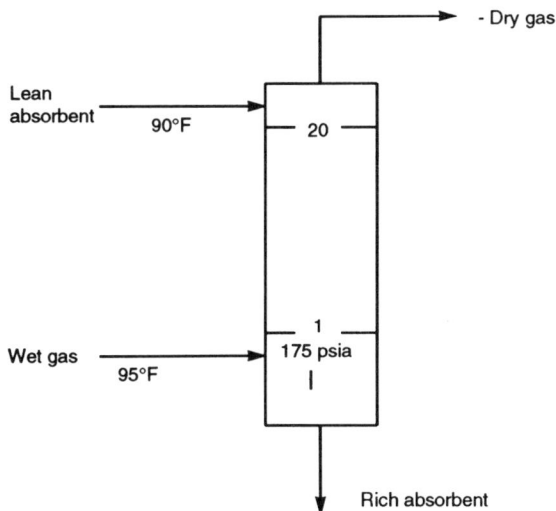

Read off K values for rich gas components at 175 psia and 95 °F.

	K
H_2	110
C_1	14
C_2	3.3
H_2S	2.4
C_3	1.0
iC_4	0.46
nC_4	0.34

Assume all components of $K = 1$ or less will be absorbed. Then moles absorbed are

$$
\begin{array}{ll}
 & \text{Moles/h} \\
C_3 & 6.04 \\
iC_4 & 1.32 \\
nC_4 & 1.32 \\
\text{Total} & 8.68
\end{array}
$$

$$\% \text{ absorbed} = \frac{8.68}{203.67} = 4.26\%$$

For a first assumption of temperatures out use the following:

$$T_{\text{dry gas}} = T_{\text{lean oil}} + 0.5 \,(\% \text{ absorbed})$$

$$T_{\text{rich oil}} = T_{\text{wet gas}} + \% \text{ absorbed}$$

Then

$$T_{\text{dry gas}} = 90 + 2.13$$

$$= 92\,°\text{F (this is assumed tower top temperature)}$$

$$T_{\text{rich oil}} = 95 + 4.26$$

$$= 99\,°\text{F}$$

To calculate L/V ratios

$$(L/V)_{\text{top}} = \frac{L_{n+1} + B_n}{V_t - B}$$

$$= \frac{76.47 + 0.35\,(8.68)}{203.67 - 8.68}$$

$$= 0.408$$

$$(L/V)_{\text{bot}} = \frac{L_{n+1} + B}{V_t - B_n}$$

$$= \frac{76.47 + 8.68}{203.67 - 0.35\,(8.68)}$$

$$= 0.424$$

To calculate absorption and stripping factors

Tower tray efficiency is taken as 20%. Then number of theoretical stages = $0.20 \times 20 = 4$.

Absorption factors

	K_{top}	K_{bot}	A_{top}	A_{bot}	A_{e}
H_2	115	110	0.004	0.0039	0.004
C_1	14	14	0.029	0.030	0.030
C_2	3.3	3.3	0.124	0.128	0.119
H_2S	2.6	2.5	0.157	0.170	0.168

C_3	1.0	1.0	0.408	0.424	0.420
iC_4	0.46	0.47	0.887	0.902	0.897
nC_4	0.34	0.35	1.200	1.211	1.207
C_9	0.0026	0.003	156.92	141.333	148.9

Stripping factors

$$(V/L)_{top} = \frac{1}{0.408} = 2.405$$

$$(V/L)_{bot} = \frac{1}{0.424} = 2.358$$

	K_{top}	K_{bot}	S_{top}	S_{bot}	S_e
C_3	1.0	1.0	2.405	2.358	2.385
iC_4	0.46	0.47	1.106	1.108	1.107
nC_4	0.34	0.35	0.818	0.825	0.820
C_9	0.0026	0.003	0.006	0.007	0.006

To calculate quantity absorbed

$$M = V_a(F_a) - l_{n+1}(F_s)$$

	Mole/h gas V_a	Moles/h liquid l_{n+1}	F_a	F_s	M moles abs
H2	95.04		0.004	–	0.380
C1	38.60		0.030	–	1.158
C2	12.11		0.119	–	1.441
H2S	49.24		0.168	–	8.270
C3	6.04	0.093	0.412	0.982	2.397
iC4	1.32	0.120	0.754	0.838	0.895
nC4	1.32	0.140	0.867	0.714	1.044
C9	–	75.047	1.000	0.006	−0.453
Total	203.67	75.400			15.132

To calculate the composition of the dry gas and rich oil

	← Moles/h →				Mol %	Mol %
	Wet gas V_a	Absorbed M	Dry gas M_d	Rich oil l	Dry gas	Rich oil
H_2	95.04	0.380	94.66	0.380	0.502	0.004
C_1	38.60	1.158	37.442	1.158	0.199	0.013
C_2	12.11	1.441	10.669	1.441	0.057	0.016
H_2S	49.24	8.270	40.97	8.370	0.217	0.091
C_3	6.04	2.397	3.643	2.490	0.019	0.028
iC_4	1.32	0.895	0.425	1.015	0.002	0.011
nC_4	1.32	1.044	0.276	1.184	0.001	0.013
C_9	–	−0.453	0.453	74.594	0.002	0.824
Total	203.67	15.132	188.538	90.532	1.000	1.000

To calculate the dew point temperature of the dry gas
Pressure at the bottom of the tower is 175 psia. Allow a pressure drop of 0.25 psi
per tray. Then tower top pressure is $175 - (20 \times 0.25) = 170$ psia. The top
temperature as the dew point of the dry gas will be calculated at 170 psia.

	Dry gas Mol. frac. y	1st trial 92 °F K	$x = \dfrac{y}{K}$
H_2	0.502	115	0.004
C_1	0.199	14	0.014
C_2	0.057	3.3	0.017
H_2S	0.218	2.5	0.087
C_3	0.019	1.0	0.019
iC_4	0.002	0.44	0.005
nC_4	0.001	0.33	0.003
C_9	0.002	0.0024	0.833
Total	1.000		0.982

Actual temperature $= 0.0024 \times 0.982 = 0.00235$

K C_9 at $0.00235 = 91.5$ (say 92 °F)

To calculate the bubble point temperature of rich oil
Bottom tower pressure $= 175$ psia
Bottom tower temperature will be bubble point of the rich oil.

	Rich oil Mol. frac. x	1st trial 99 °F K	$y = xK$
H_2	0.004	110	0.440
C_1	0.013	14.5	0.189
C_2	0.016	4.6	0.074
H_2S	0.091	2.5	0.228
C_3	0.028	1.0	0.028
iC_4	0.011	0.47	0.005
nC_4	0.013	0.36	0.005
C_9	0.824	0.0032	0.003
Total	1.000		0.972

Actual temperapure $\dfrac{110}{0.972} = 113$

K H_2 at 175 psia $= 113$. Temperature of this $K = 98$ °F. This is acceptable and no
further trials are required.

The assumed conditions of temperature and pressure in this case is acceptable.
Therefore the compositions can be used for input to a computer simulation
program without further iterations.

1.7   Calculating and Optimizing Heat Exchanger Configurations in a Process System

This section describes a calculation technique to develop the best configuration and sequence of heat exchangers in a preheater train. The objective is to use heat exchangers to supplement the duty and therefore minimize the operating costs associated with preheating a process feed. Consider the following diagram of a typical preheat of a fractionator feed. In the diagram the product streams 1 to 5 are heat exchanged in a series of heat exchangers against the heater feed which enters the system at 60 °F. It is required to fix the inlet temperature of the heater using the best sequence of the product stream exchange to the feed. This calculation follows these steps:

- *Step 1*   Develop the enthalpy curve for the feed in BTU/h versus tempera-ture. This is accomplished by selecting four or five temperatures for the feed. Then starting at the inlet temperature (60 °F in this case) multiply the mass of the feed in lb/h by the enthalpy at the temperatures selected. Plot the result against the selected temperatures on linear graph paper. The result is the enthalpy of the *liquid* feed. (Note: the feed will be maintained as a liquid in this system until it passes through the heater control valves into the heater.)
- *Step 2*   Calculate the heat balance over the fractionator. Now this can only be completed when the fractionation calculations have been done to establish the degree of vaporization required for the feed and the product draw-off temperatures. For the purpose of this example these calculations have been completed and the overall heat balance over the tower can be developed. It is important when developing a heat balance to use a method system. The author prefers to tabulate the heat balance data as shown in the worked example. However, engineers may use any other format. The important criterion is that the mass used for 'heat in' balances the mass used for the 'heat out' and that the enthalpy applied is for the correct phase of the material (i.e. either vapour or liquid or a mixture of both).
- *Step 3*   Using the data in the heat balance, develop enthalpy curves for the

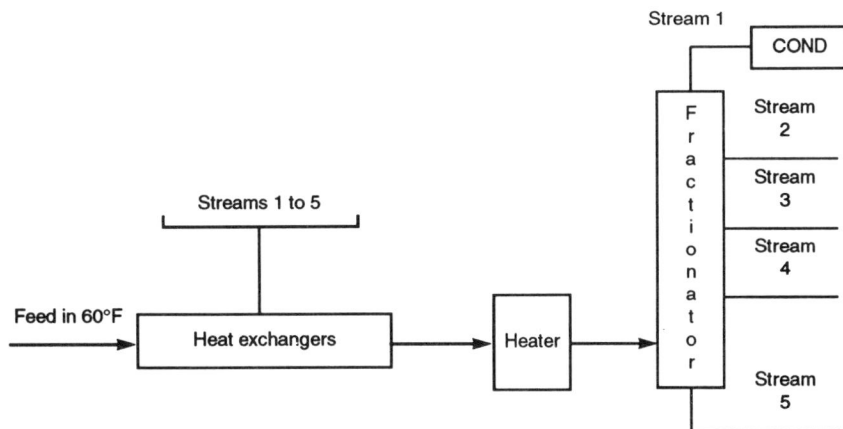

product streams. For stream 1 there is a change of phase in this example. Therefore a flash curve at three temperatures must be calculated and the enthalpy data for the liquid and vapour portion at these temperatures be added to arrive at the overall enthalpy curve.

- *Step 4* For the remaining product streams, which are all liquid, it is necessary only to select a temperature lower than the draw-off. Calculate the total enthalpy of the stream at this temperature in BTU/h and draw a straight line between this temperature/enthalpy point and the one for the draw-off.

- *Step 5* Using the feed enthalpy curve and starting with the condenser stream 1 overlay the stream enthalpy curves on the feed graph. Select a reasonable approach temperature for each stream against feed and measure in terms of enthalpy the duty for each exchange. A reasonable approach temperature is on the cold side of the exchange. This should not be less than 20 °F for distillates and 40–60 °F for the heavier residue stream.

- *Step 6* For the first trial use the sequence of product exchange as they are drawn from the tower. Then using a ruler and a set square move the product enthalpy curves around to make up other configurations. The worked example shown here is the second trial or sequence developed. Note the feed temperature at the end of each trial configuration. However, the highest feed temperature obtained is not necessarily the best configuration. Selecting the best of the configurations requires more analysis.

- *Step 7* Analysing the configurations requires obtaining some rough costs for heat exchanger surface area, heater duty size, cooler costs (water cooling or air, whichever is appropriate), fuel cost and either cooling water cost or power cost.

- *Step 8* Calculate the rough surface area required for the heat exchangers in the configuration. This is done using the equation

$$A = \frac{Q}{U \cdot \Delta T_{\mathrm{m}}}$$

where

A = surface area (ft^2)
Q = heat duty of exchanger (BTU/h)
U = overall heat transfer coefficient (BTU/h.ft^2.°F)
ΔT_{m} = log mean temperature difference (°F)

Add these areas and multiply by the budget cost figure obtained in step 7.

- *Step 9* Using the budget cost figures obtained in step 7, calculate the other equipment costs associated with the configuration, that is, the cost of the heater and the total cost in terms of BTU/h duty of the product coolers. For this latter item it will be necessary to calculate the duty in BTU/h of cooling required for each product to battery limit temperatures.

- *Step 10* Calculate the fuel required by the heater in each configuration. This is accomplished by dividing the heater duty by an efficiency factor (use 0.7) to obtain the heat required to be supplied by the fuel. Use the calorific value for the fuel (See Figure A1.1 in Appendix 1) to obtain the mass per hour.

- *Step 11* Compare the estimated capital costs and the respective operating cost for each case to arrive at the most economic of the configurations.

A worked example showing the development of one configuration now follows. The evaluation and economic analysis of this configuration compared with others is also given.

Example calculation

A heat balance of the fractionator shown in the diagram above is as follows. The fractionation calculations to give feed enthalpy and vaporization required has already been completed. Also the draw-off temperatures of the product streams have been determined.

	V/L	°API	°F	K	lb/h	Enthalpy BTU/lb	Enthalpy mmBTU/h
IN							
Feed	$V + L$	–	720	–	379 575	–	162.500
OUT							
Stream 1	V	70	229	12	76 650	258	19.776
Stream 2	L	45.5	339	12	34 860	184	6.414
Stream 3	L	36.3	488	11.5	44 226	255	11.278
Stream 4	L	30	558	11.5	26 754	297	7.946
Stream 5	L	13	704	11.5	197 085	384	75.680
Reflux							41.406
Total out					379 575		162.500

The reflux term in the heat balance was obtained by difference. Thus:

$$\text{Heat in with feed} = \text{heat out with products} + \text{heat out with reflux}$$
$$162.500 \text{ mm BTU/h} = 121.094 + \text{reflux}$$
$$\text{Reflux} = 162.500 - 121.094$$
$$= 41.406 \text{ mm BTU/h}$$

The feed enthalpy curve is plotted on Figure 1.3 as BTU/h versus temperature (°F) from the following data:

Total mass of feed = 379 575 lb/h considered as all liquid throughout

Temp (°F)	BTU/lb	mm BTU/h
60	24	9.11
100	42	15.94
200	80	33.40
300	140	53.14
400	192	72.88
500	260	98.69
600	327	124.12

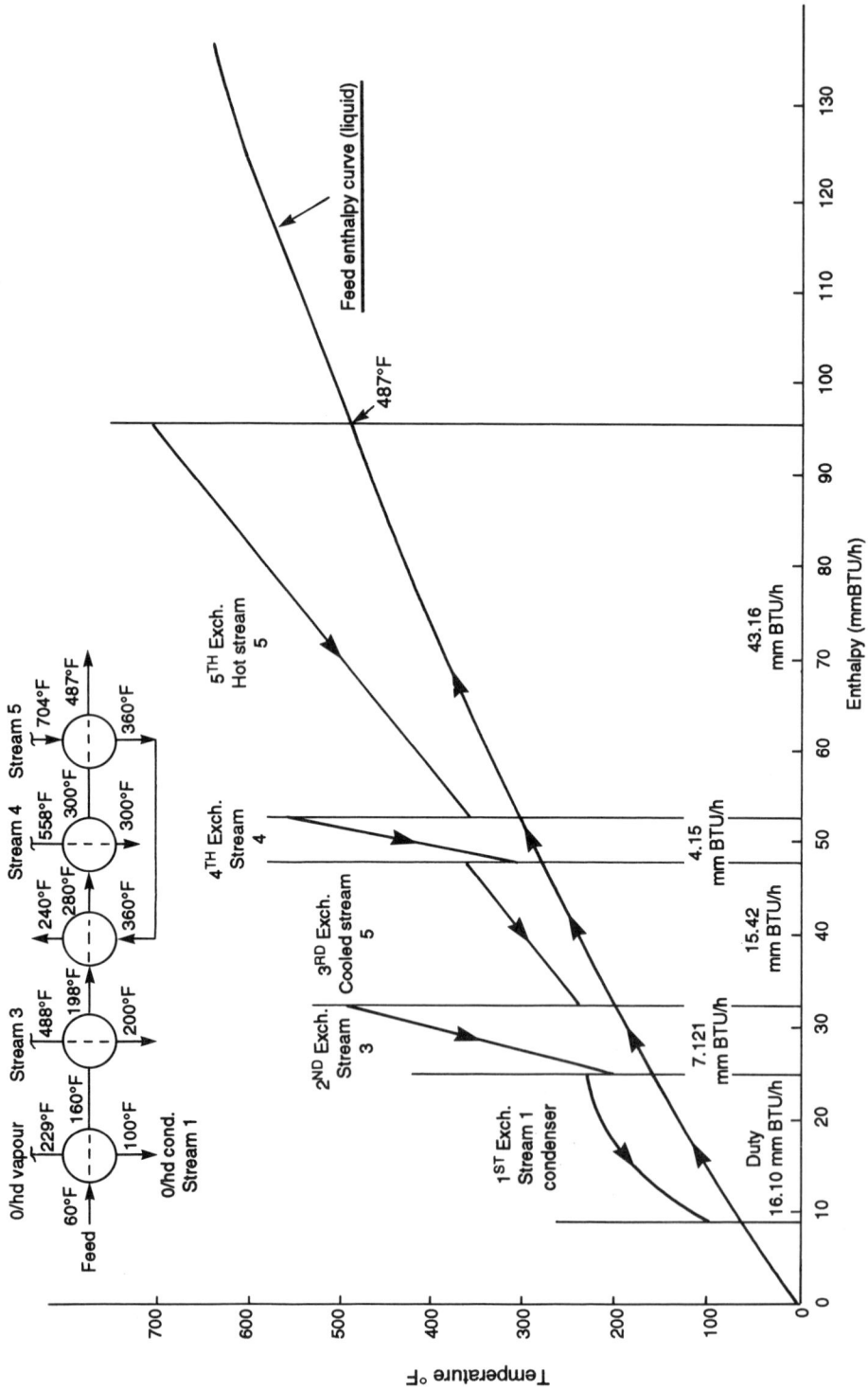

Figure 1.3. Heat exchanger layout sequence — case 2

First exchanger: stream No. 1; overhead condenser
From enthalpy curve of product:

$$\text{Temperature in} = 229\,°F \text{ all vapour}$$

$$\text{Temperature out} = 100\,°F \text{ all liquid}$$

$$\text{Duty of exchanger} = 16.097 \text{ mm BTU/h}$$

$$\text{Temperature in of feed} = 60\,°F$$

$$\text{Temperature out of feed} = 160\,°F$$

Second exchanger: stream No 3

$$\text{Temperature of product stream in} = 488\,°F$$

$$\text{Temperature of product stream out} = 200\,°F$$
$$(40\,°F \text{ approach})$$

$$\text{Enthalpy in} = 11.278 \text{ mm BTU/h}$$

$$\text{Enthalpy out} = 4.157 \text{ mm BTU/h}$$

$$\text{Duty} = 7.121 \text{ mm BTU/h}$$

$$\text{Feed temperature in} = 160\,°F$$

$$\text{Feed temperature out} = 198\,°F$$

Third exchanger: cooled stream No 5
Set inlet temperature of product at 360 °F for this case. This will be changed for other case studies.

$$\text{Temperature in of product stream} = 360\,°F$$

$$\text{Temperature out of product stream} = 240\,°F$$

$$\text{Enthalpy in} = 197\,085 \times 165 = 32.520 \text{ mm BTU/h}$$

$$\text{Enthalpy out} = 197\,085 \times 88 = 17.100 \text{ mm BTU/h}$$

$$\text{Duty} = 15.42 \text{ mm BTU/h}$$

$$\text{Temperature of feed in} = 198\,°F$$

$$\text{Temperature of feed out} = 280\,°F$$

Fourth exchanger: stream No 4

$$\text{Temperature of product stream in} = 558\,°F$$

$$\text{Temperature of product stream out} = 300\,°F$$

$$\text{Enthalpy in} = 26\,754 \times 297 = 7.946 \text{ mm BTU/h}$$

$$\text{Enthalpy out} = 26\,754 \times 142 = 3.800 \text{ mm BTU/h}$$

$$\text{Duty} = 4.146 \text{ mm BTU/h}$$

$$\text{Temperature of feed in} = 280\,°F$$

$$\text{Temperature of feed out} = 300\,°F$$

Fifth exchanger: stream No 5

$$\text{Temperature of product Stream in} = 704 \,°F$$

$$\text{Temperature of product Stream out} = 360 \,°F$$

$$\text{Enthalpy in} = 197\,085 \times 384 = 75.680 \text{ mm BTU/h}$$

$$\text{Enthalpy out} = 197\,085 \times 165 = 32.520 \text{ mm BTU/h}$$

$$\text{Duty} = 43.160 \text{ mm BTU/h}$$

$$\text{Temperature of feed in} = 300 \,°F$$

$$\text{Temperature of feed out} = 487 \,°F$$

Total enthalpy in the feed after fifth exchanger:

$$\text{Enthalpy of feed at } 60 \,°F = \;9.000 \text{ mm BTU/h}$$

$$\text{Duty of exchanger No. 1} = 16.097 \text{ mm BTU/h}$$

$$\text{Duty of exchanger No. 2} = \;7.121 \text{ mm BTU/h}$$

$$\text{Duty of exchanger No. 3} = 15.420 \text{ mm BTU/h}$$

$$\text{Duty of exchanger No. 4} = \;4.146 \text{ mm BTU/h}$$

$$\text{Duty of exchanger No. 5} = 43.160 \text{ mm BTU/h}$$

$$\text{Total enthalpy in feed} = 94.944 \text{ mm BTU/h}$$

Note: Graph gives 95.5 mm BTU/h. Difference is 0.6%, which is acceptable.

$$\text{Total heat in partially vaporized feed to fractionator} = 162.500 \text{ mm BTU/h}$$

$$\text{Then duty required of fired preheater} = 162.500 - 94.944$$

$$= \underline{67.556 \text{ mm BTU/h}}$$

Estimated exchanger heating surface is calculated as follows:
Exchanger No. 1

$$\text{Duty} = 16.097 \text{ mm BTU/h}$$

LMTD:

$$
\begin{array}{ccc}
229 & \rightarrow & 100 \\
160 & \leftarrow & 60 \\
\hline
69 & & 40 \\
\hline
\end{array}
$$

$$\frac{69 - 40}{2.303 \times \log(69/40)}$$

$$= 53.2 \,°F$$

Overall heat transfer coefficient (from Table A1.1 in Appendix 1) (say, 50 BTU/h.ft^2.°F):

$$\text{Area} = \frac{16\,097\,000}{53.2 \times 50}$$
$$= 6052 \text{ ft}^2$$

Exchanger No. 2

$$\text{Duty} = 7.121 \text{ mm BTU/h}$$

LMTD:

$$488 \rightarrow 200$$
$$198 \leftarrow 160$$

$$\overline{}$$

$$290 \qquad 40$$

$$\overline{}$$

$$\frac{290 - 40}{2.303 \times \log(290/40)}$$
$$= 126 \text{ °F}$$

Overall heat transfer coefficient $U = 60$ BTU/h.ft^2.°F

$$\text{Area} = \frac{7\,121\,000}{126 \times 60}$$
$$= 942 \text{ ft}^2$$

Exchanger No. 3

$$\text{Duty} = 15.42 \text{ mm BTU/h}$$

LMTD:

$$360 \rightarrow 240$$
$$280 \leftarrow 198$$

$$\overline{}$$

$$80 \qquad 42$$

$$= \frac{80 - 42}{2.303 \times \log(80/42)}$$
$$= 59 \text{ °F}$$

Overall heat transfer coefficient $U = 35$ BTU/h.ft^2.°F

$$\text{Area} = \frac{15\,420\,000}{59 \times 35}$$
$$= 7467 \text{ ft}^2$$

Exchanger No. 4

$$\text{Duty} = 4.146 \text{ mm BTU/h}$$

LMTD:

$$558 \rightarrow 300$$
$$300 \leftarrow 280$$

$$258 \qquad 20$$

$$= \frac{258 - 20}{2.303 \times \text{Log} \ (258/20)}$$
$$= 93 \,^\circ\text{F}$$

Overall heat transfer coefficient $U = 45 \text{ BTU/h.ft}^2.^\circ\text{F}$

$$\text{Area} = \frac{4\,146\,000}{93 \times 45}$$
$$= 991 \text{ ft}^2$$

Exchanger No. 5

$$\text{Duty} = 43.160 \text{ mm BTU/h}$$

LMTD:

$$704 \rightarrow 360$$
$$487 \leftarrow 300$$

$$217 \qquad 60$$

$$= \frac{217 - 60}{2.303 \times \log \ (217/60)}$$
$$= 122 \,^\circ\text{F}$$

Overall heat transfer coefficient $U = 30 \text{ BTU/h.ft}^2.^\circ\text{F}$

$$\text{Area} = \frac{43\,160\,000}{122 \times 30}$$
$$= 11\,792 \text{ ft}^2$$

Total area of exchanger surface for this case is $27\,244 \text{ ft}^2$. The fuel required for the heater for this case will be

$$\frac{66\,576\,000}{0.7} = 95.109 \text{ mm BTU/h of fuel fired}$$

(assuming 70% efficiency). From Figure A1.1, gross heating value of 15∘API fuel = 18 600 BTU/lb. Then lb/h of fuel required in this case

$$= \frac{95\,109\,000}{18\,600}$$
$$= 5113\,\text{lb/h}.$$

To complete the calculation of this trial it is necessary to calculate the size of the associated final product coolers and the amount of utilities used in each trial configuration. Thus for this trial final cooler sizes are as follows:

- *Stream 1* There will be no final cooler for this overhead product.
- *Stream 2* This product stream is not heat exchanged against feed so it does not enter into the configuration evaluation.
- *Stream 3* It is required to cool this product to a battery limit condition of 90 °F. Air cooling will be used with the air temperature rise of 5 °F. From Figure 1.3, continuing the enthalpy line for product 3 down to 90 °F and measuring this enthalpy difference gives the duty of the cooler as 1.98 mm BTU/h.

LMTD for the cooler is as follows:

$$200\,°\text{F} \rightarrow 90\,°\text{F}$$
$$65\,°\text{F} \leftarrow 60\,°\text{F}$$

$$\overline{135\,°\text{F} \quad\quad 30\,°\text{F}}$$

$$\text{LMTD} = \frac{135 - 30}{2.303.\log 135/30}$$
$$= 70\,°\text{F}$$

A typical overall heat transfer coefficient U from Table A1.1 in Appendix 1 is 100 BTU/h.ft^2.°F.

$$\text{Surface area } A = \frac{Q}{U \times \text{LMTD}}$$
$$A = \frac{1\,980\,000}{100 \times 70}$$
$$= 282\,\text{ft}^2$$

Calculating the fan BHP and the power consumption

$$\text{Quantity of air flow} = \frac{Q}{0.24 \times \Delta T}$$
$$= \frac{1\,980\,000}{0.24 \times 5}$$
$$= 1\,650\,000\,\text{lb/h}$$

Actual cubic feet per minute of the air:

$$\frac{1\,650\,000 \times 378}{60 \times 29} = 358\,448 \text{ ft}^3/\text{min}$$

Pressure drop for the air across the tubes (say, 0.8 in. of water). Assume two fans then

$$\text{bhp/fan} = \frac{\text{acfm/fan} \times \Delta P}{6370 \times 0.7}$$

(see Section 6.4 for a detailed calculation of bhp)

$$= 32.1 \text{ hp}$$

Power consumption per fan $= 0.746 \times 32.1 = 24 \text{ kWh}$

Total for unit $= 48 \text{ kWh}$

- *Stream 4* This product is cooled from 300 °F to battery limit condition of 90 °F. Using the enthalpy curve (Figure 1.3) the duty of the cooler is found to be 4.0 mm BTU/h. Air cooling is again used with an air temperature rise of 8 °F.

$$\text{LMTD} = 99\,°\text{F}$$

$$U = 60 \text{ BTU/h.ft}^2.°\text{F}$$

$$\text{Surface area (bare tube)} = \frac{4\,000\,000}{99 \times 60}$$

$$= 673 \text{ ft}^2$$

$$\text{Air flow} = \frac{4\,000\,000}{0.24 \times 8} \quad (0.24 \text{ BTU/lb is specific heat of air})$$

$$= 2.083 \text{ mm lb/h}$$

Assume ΔP for the air across tubes is again 0.8 in:

$$\text{ACFM of air} = 409\,952 \text{ ft}^3/\text{min}$$

Assume two fans, then BHP/fan

$$= \frac{409\,952 \times 0.8}{2 \times 6370 \times 0.7}$$

$$= 36.8$$

Total power consumption $= 0.746 \times 36.8 \times 2$

$$= 54.9 \text{ kWh}$$

- *Stream 5* This is the cooled stream from exchanger 3 and is further cooled from 240 °F to the battery limit of 180 °F. Air cooling is used with an 8 °F rise in air temperature. Using the enthalpy chart as before, the duty of the cooler is found to be 14.5 mm BTU/h.

$$\text{LMTD} = 144\,°\text{F}$$

$$U = 45 \text{ BTU/h.ft}^2.°\text{F}$$

Surface area on bare tube basis

$$= \frac{14\,500\,000}{144 \times 45}$$

$$= 2238 \text{ ft}^2$$

$$\text{Air flow} = \frac{14\,500\,000}{0.24 \times 8} = 7\,552\,083 \text{ lb/h}$$

$$\text{Acfm} = 1.64 \text{ mm ft}^3/\text{min}$$

Assume four fans and a ΔP of 0.7 in.: then

$$\text{bhp/fan} = 64.4$$

$$\text{Total power consumption} = 64.4 \times 4 \times 0.746$$

$$= 192 \text{ kWh}$$

This completes the details of ancillary equipment influenced by the trial criteria. We must now add cost data to these items to compare this trial with others on a cost basis. Assume the cost data are as follows:

- Shell and tube exchangers = $20 per ft^2 of surface
- Air coolers (including motors, etc.) = $50 per ft^2 of bare tube
- A 50 mm BTU/h heater costs $250\,000 installed. Use a 0.6 exponential for pro-rating.
- Cost of utilities:
  Fuel oil = $0.01 per lb
  Power = $0.042 per kWh

Trial 2 cost data

Capital costs

Shell and tube heat exchangers	
1	6 052 ft^2
2	942 ft^2
3	7 467 ft^2
4	991 ft^2
5	11 792 ft^2
Total	27 244 ft^2

Total cost = 20 × 27244 = $544 880

Air coolers:

Stream 2	282 ft^2
Stream 4	673 ft^2
Stream 5	2238 ft^2
Total	3193 ft^2

Total cost = 50 × 3193 = $159 650

Heater:

$$\text{Duty} = 67.556 \text{ mm BTU/h}$$

$$\text{Cost} = (67.556/50) \times \$250\,000$$

$$= \$300\,000$$

$$\text{Total capital costs} = \$1\,004\,530$$

Operating costs

$$\text{Power} = 48 + 54.9 + 192 = 294.9 \text{ kWh}$$

$$\text{Cost} = 0.042 \times 294.9 = 12.39 \text{ \$/h}$$

Cost/year at 90% service time:

$$12.39 \times 24 \times 365 \times 0.9 = \$97\,682/\text{yr}$$

$$\text{Fuel} = 5113 \text{ lb/h} \times 7884 \times \$0.01/\text{lb}$$

$$= \$403\,109/\text{yr}$$

$$\text{Total operating costs} = \$500\,790/\text{yr}$$

Other configurations were analysed in the same way and these together with the above calculation are summarized below:

Trial number	Heater inlet (°F)	Capital cost ($m)	Δ Capital cost ($m)	Operating cost ($/yr)	Δ Operating cost ($)	Payout (yr)
1.0 Base case	445	0.930	–	655 000	–	–
2.0	452	0.973	0.043	632 000	23 000	1.86
3.0	487	1.005	0.075	500 790	154 210	0.486
4.0	520	1.603	0.673	424 000	231 000	2.913

It can be seen from the table that the difference in capital cost between the lowest and the highest can be paid off in a number of years in each of the cases by the savings incurred through the difference in operating costs. On this basis the Trial No. 3 configuration offers the best solution in terms of payout years. That is, the difference in capital cost between the lowest and Trial No. 3 can be paid off in just 6 months (0.486 yr). This rate of savings would then continue, of course, for the remainder of the economic life of the plant.

1.8   Friction Loss in Piping and Fittings

This section deals with pressure drop in pipelines and fittings caused by resistance to flow by friction. The subject covers the pressure drop for the flow of liquids and gases. This section also includes some guidelines for sizing pipelines and fittings to cater for friction losses. Table 1.1 provides a list of acceptable ranges of pressure drops for various process and utility piping systems. The data are based on 'average' carbon steel lines.

Table 1.1. Acceptable pressure drop ranges

System	Average psi/100 ft	maximum psi/100 ft	Remarks
Pump suction lines	0.25	0.4	High-pressure drop is one of the common causes of poor pump performance
Gravity rundown lines	0.25	0.4	
Pump discharge lines	1.5	2.0	Except high pressure
Pump discharge lines	3.0	4.0	700 psi and higher
Vapour lines	0.2	0.5	Atmos and pressure towers overhead lines
Gas lines	0.2	0.5	Within battery limits
Gas lines (tie-in lines)	–	–	5–10% of available pressure
Compressor suction lines	0.1	0.3	Total pressure drop (inc. fittings) 0.5–1.0
Compressor discharge lines	0.2	0.5	Total pressure drop (inc. fittings) 4–5
High-pressure steam lines	0.5	1.0	Short distances—within battery limits
High-pressure steam lines	0.1	0.4	Long distances
Exhaust steam lines	0.2	0.4	Short distances
Exhaust steam lines	0.05	0.1	Long distances
Water lines	0.1	1.5	Short distances
Water lines	0.25	0.5	Long distances

CALCULATING FRICTION PRESSURE DROP IN GAS LINES

There are several methods available to predict pressure drop in gas lines. Among the more reliable of these is the 'Weymouth' formula[4] method. This is written as follows:

$$Q_s = 433.45\left(\frac{T_s}{P_s}\right) \times d^{2.667} \times \left[\frac{P_1^2 - P_2^2}{LST}\right]^{1/2}$$

where

Q_s = rate of gas flow in ft^3 per 24 hours measured at standard conditions (60 °F, 14.7 psia)
d = internal diameter of pipe (in inches)
P_1 = initial pressure (psia)
P_2 = Terminal pressure (psia)
L = length of line (in miles)
S = specific gravity of gas (air = 1)
T = absolute temperature of flow gas (degrees Rankin)
T_s = standard absolute temperature (60 °F + 460 = 520 °R)
P_s = standard pressure (psia) (14.7 psia)

This equation has been resolved to graphical form and is given in the Figure A1.3 in Appendix 1. This graph shows MCFD (thousand cubic feet per day) plotted against pressure drop per 1000 feet of line for pipe sizes ranging from 1½ in. with 0.2 in. wall to 36 in. Armco, 0.25 in. wall. There are two such graphs. The first ranges from pressure drop 1 psi/1000 ft to 300 psi/1000 ft. The second ranges from

300 psi to 100 000 psi/1000 ft. The use of the chart is self-explanatory with the correction factors fully described on the charts. An example calculation is given below (calculation No. 1).

CALCULATING FRICTION PRESSURE DROP IN LIQUID LINES

In this book the Darcy formula[5] is used as the method to calculate liquid frictional pressure drop in steel lines. This equation is expressed as follows:

$$h_f = f \frac{L}{D} \frac{V^2}{2g}$$

where

h_f = friction loss (ft of liquid)
L = pipe length (feet)
D = inside diameter of pipe (feet)
V = Velocity of fluid (feet per second)
g = 32.2 fps
f = Dimension less friction factor. Based on Reynold's number (Re) and relative roughness of pipe

The relationship of f to Re and the relative roughness of the pipe is given by the Moody Chart[6] (Figure 1.4). This value when applied to the Darcy equation has been reduced to a series of tables for pipe size ranging from 1 in. to 48 in. The tables read of friction loss in feet of liquid per 1000 ft of pipe for flows in GPM or GPH (US dimensions) and at kinematic viscosities (or SSU) ranging from 0.6 cSt to 650 cSt. These are given in the Appendix Table A1.2 and their use is self-explanatory. An example calculation No. 2 shows the use of these tables.

FRICTION LOSS FOR FITTINGS

Friction loss for fittings is measured in terms of equivalent length of pipe. That is the friction loss for, say, a long sweep elbow in a 6-inch line would be equivalent to 10 feet of that line. This then means that when measuring a line length to calculate total pressure drop over a system the equivalent length for fitting must be added on to the measured line length. A chart correlating fittings and the respective equivalent line length is given in the Figure A1.4 in Appendix 1.

Example calculation no. 1

Calculate the pressure drop per 100 feet of 24-inch line (standard) for propane LPG gas at 230 psig and 100 °F flow of gas is 150 000 lb/h. Mol wt of LPG is 46.

Moles/h of gas = 3261

$$MCFD = \frac{3261 \times 378 \times 14.7 \times 560}{244.7 \times 520 \times 1000} \times 24$$

$$= 1914.1$$

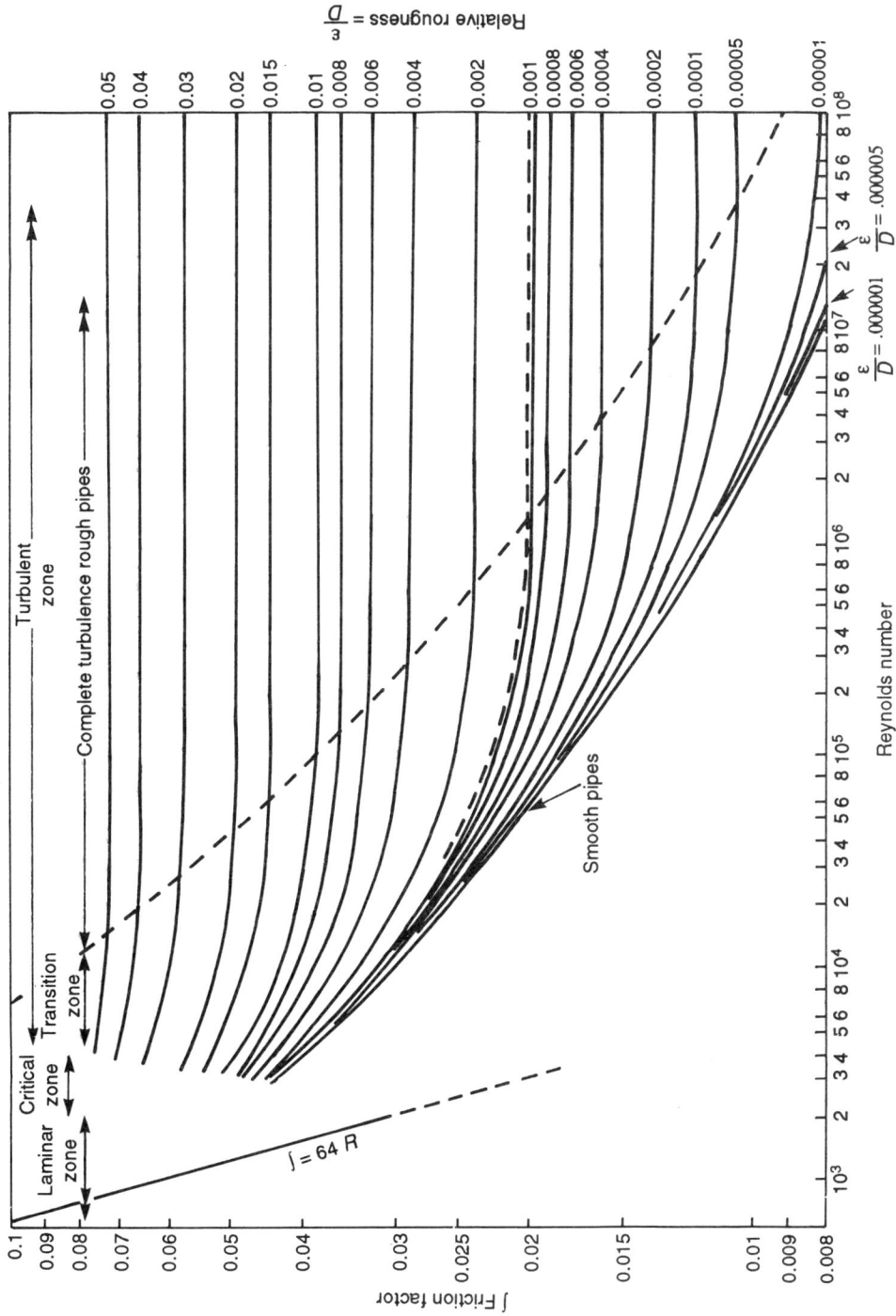

Figure 1.4. The Moody diagram

$$\text{MCFD (chart)} = \frac{P}{14.65}\left(\frac{\text{spgr}}{0.9} \times \frac{560}{550} \times \frac{0.9808}{1}\right)^{1/2} \times 3237$$

$$\text{spgr} = \frac{46}{29} = 1.586$$

$$= \frac{244.7}{14.65}\left(\frac{1.586}{0.9} \times \frac{560}{550} \times 0.9808\right)^{1/2} \times 1914$$

$$= 42\,410 \text{ MCFD (chart)}$$

From Weymouth graph $P_1 - P_2$ per 1000 ft $= 33$ psi psi/100 ft $= 3.3$, which is higher than accepted design.

Example calculation no. 2

Calculate the pressure drop in psi/100 ft of 275 gpm gas oil in a 6-inch schedule 40 suction line to a pump. Viscosity of gas oil is 2.7 CSt at flowing temperature. The SG of the gas oil at flowing temp is 0.825 from the table for 6-inch pipe, frictional pressure drop for a fluid of 2.7 CSt viscosity and at 275 gpm is 6.28 ft/1000 ft

$$6.28 \text{ ft of liquid} = \frac{6.28 \times 62.2 \times 0.825}{144}$$

$$= 2.24 \text{ psi/1000 ft}$$

$$= 0.224 \text{ psi/100 ft}$$

where

$$62.2 = \text{standard density for water at } 60\,°\text{F in lb/ft}^3$$

$$144 = \text{in.}^2/\text{ft}^2$$

which is within acceptable limits for pump suction.

1.9 The Control Valve

Control valves are used throughout the process industry to control operating parameters. These parameters are:

- Flow
- Pressure
- Temperature
- Level

Figure 1.5 shows the conventional control valve which in this case is taken as a double-seated plug type valve. Like most control valves, it is operated pneumatically by an air stream exerting a pressure on a diaphragm which in turn allows the movement of a spring-loaded valve stem. One or two plugs (the diagram shows two) are attached to the bottom of the valve stem which, when closed, fit into valve seats thus providing tight shut-off. The progressive opening and closing of

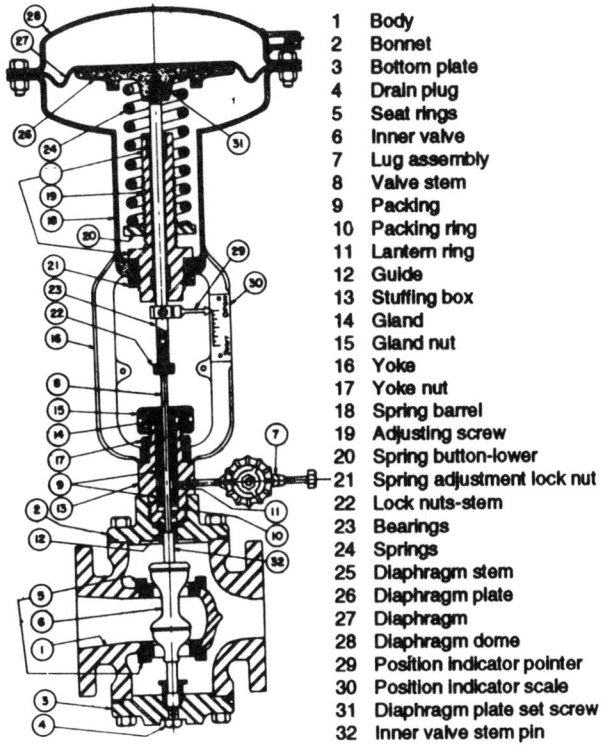

1	Body
2	Bonnet
3	Bottom plate
4	Drain plug
5	Seat rings
6	Inner valve
7	Lug assembly
8	Valve stem
9	Packing
10	Packing ring
11	Lantern ring
12	Guide
13	Stuffing box
14	Gland
15	Gland nut
16	Yoke
17	Yoke nut
18	Spring barrel
19	Adjusting screw
20	Spring button-lower
21	Spring adjustment lock nut
22	Lock nuts-stem
23	Bearings
24	Springs
25	Diaphragm stem
26	Diaphragm plate
27	Diaphragm
28	Diaphragm dome
29	Position indicator pointer
30	Position indicator scale
31	Diaphragm plate set screw
32	Inner valve stem pin

Conventional type

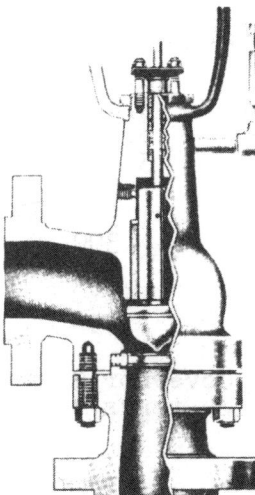

Venturi

Butterfly

Figure 1.5. Control valves

the plugs on the valve seats due to the movement of the stem determines the amount of the controlled fluid flowing across the valve. The pressure of the air to the diaphragm controlling the stem movement is varied by a control parameter, such as a temperature measurement, flow measurement, etc.

There are two types of plug valves used for the conventional control valve function:

- Single-seated valves
- Double-seated valves

Single-seated valves are inherently unbalanced so that the pressure drop across the plug affects the force required to operate the valve. Double-seated valves are inherently balanced valves and are the first choice in most services.

Figure 1.5 also shows two other types of control valves in common service in the industry. These are the venturi and the butterfly. Both types when pneumatically operated (which they usually are) work in the same way as described above for a plug-type valve. The major difference in these two types are in the valve system itself. In the case of the venturi the fluid being controlled is subjected to a 90° angle change in direction within the valve body. The inlet and outlet dimensions are also different, with the inlet having the larger diameter. The valve itself is plug type but seats in the bend of the valve body. Venturi type or angle valves are used in cases where there exists a high-pressure differential between the fluid at the inlet side of the valve and that required at the outlet side.

Butterfly valves operate at very low pressure drops across them. They can operate quite effectively at only inches of water gauge pressure drop, and where the operation of the conventional plug-type valves would be unstable. The action of this valve is by means of a flap in the process line. The movement of this flap from open to shut is made by a valve stem outside the body itself. This stem movement, as in the case of the other pneumatically operated valves, is provided by air and spring loads onto the stem from a diaphragm chamber. The only major disadvantage of this type of valve is that very tight shut-off is difficult to obtain due to the flap type action of the valve.

VALVE CHARACTERISTICS

The conventional control valve predominantly in use can have either an equal percentage or linear characteristics. The difference between these two is shown in Figure 1.6.

With an *equal percentage characteristic*, equal incremental changes in valve stem lift result in equal percentage changes in the flow rate. For example if the lift were to change from 20% to 30% of maximum lift the flow at 30% would be about 50% more than the flow at 20%. Likewise, if the lift increases from 40% to 50% the flow at 50% would be about 50% more than at 40%.

With a *linear* valve having a constant pressure drop across it, equal incremental changes in stem lift result in equal incremental changes in flow rate. For example, if the lift increases from 40% to 50% of maximum, the flow rate changes from

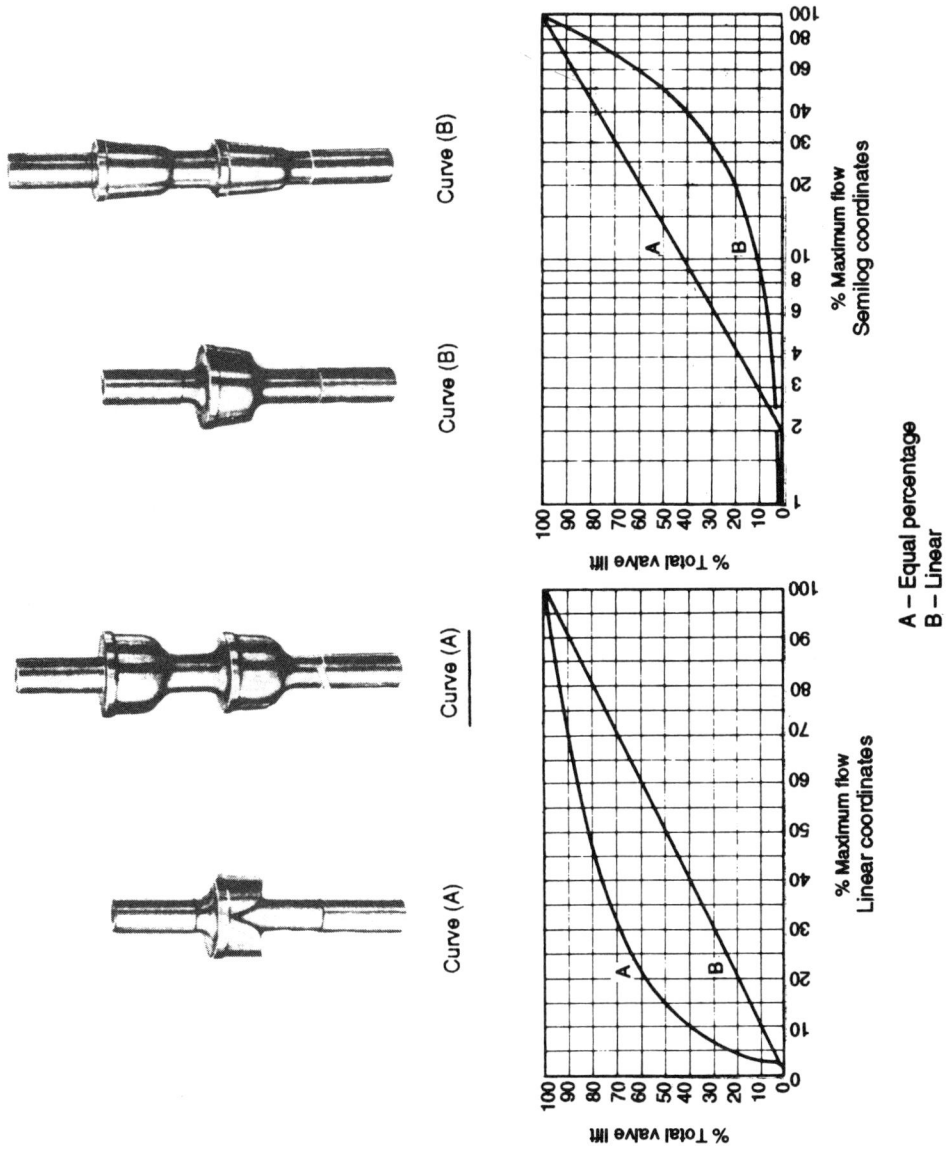

Curve (B)

Curve (B)

Curve (A)

Curve (A)

% Maximum flow
Semilog coordinates

% Maximum flow
Linear coordinates

A – Equal percentage
B – Linear

Figure 1.6. Control valve characteristics

40% to 50% of maximum. Thus equal percentage is the more desirable characteristic for most applications and is the one most widely used.

PRESSURE DROP ACROSS CONTROL VALVES

In sizing or specifying the duty of the control valve the pressure drop across the control valve must be determined for the design or maximum flow rate. In addition, if it is known that a valve must operate at a flow rate considerably lower than the maximum rate, the pressure drop at this lower flow rate must also be calculated. This will be required to establish the range of the valve. As a general rule, the sum of the following pressure drops at maximum flow may be used for this purpose:

(1) 20% of the friction drop in the circuit (excluding the valve). (A circuit generally includes all equipment between the discharge of a pump, compressor or vessel and the next point downstream of which pressure is controlled. In most cases this latter point is a vessel.)
(2) 10% of the static pressure of the vessel into which the circuit discharges up to pressures of 200 psig, 20 psig from 200 psig to 400 psig, and 5% above 400 psig.

The static pressure is included to allow for possible changes in the pressure level in the system (i.e. by changing the set point on the pressure controller on a vessel). The percentage included for static pressure can be omitted in circuits such as recycle and reflux in which any change in pressure level in the receiver will be reflected through the entire circuit. In some circuits the control valve will have to take a much greater pressure drop than calculated from the percentages listed above. This occurs in circuits where the control valve serves to bleed down fluid from a high-pressure source to a low-pressure one. Examples are pressure control valves releasing gas from a tower or streams going out to tankage from vessels operating at high pressure. These are the circumstances where venturi or angle valves are used, as described earlier.

PROCESS FLOW COEFFICIENT (C_V) AND VALVE SIZING

The process flow coefficient C_V is defined as the water flow in GPM through a given restriction for 1 psi pressure drop. C_V can be determined by the following equations:

(a) $C_V = Q_L \sqrt{\dfrac{G_L}{\Delta P}}$ for liquids

(b) $C_V = \dfrac{Q_S}{82} \sqrt{\dfrac{T}{\Delta P \cdot P_2}}$ for steam

(c) $C_V = \dfrac{Q_G}{1360} \sqrt{\dfrac{\mu_2 S T}{\Delta P \cdot P_2}}$ for gases

where

C_V = flow coefficient
Q_L = liquid flow in gpm at conditions
ΔP = pressure drop across valve (psi)
G_L = specific gravity of liquid at conditions
Q_S = steam rate (lb/h)
P_2 = pressure downstream of valve (psia)
Q_G = gas flow in scfh (60 °F, 14.7 psia)
T = temperature of gas in °R (°F + 460)
S = mol weight of gas divided by 29
μ_2 = compressibility factor at downstream conditions

The following are some special considerations that may have to be made in determining process C_V values.

Pressure drop

For compressible fluids the maximum usable pressure drop in equations (b) and (c) is the critical value. As a rule of thumb and for design purposes this value is 50% of the absolute upstream pressure. (The valve can take more than the critical pressure drop, but any pressure drop over the critical takes the form of exist losses.)

Flashing liquids

In the absence of accurate information, it is recommended that for flashing service the valve body be specified as one nominal size larger than the valve port.

Two-phase flow

If two-phase flow exists upstream of the control valve experience has shown that for fluids below their critical point a sufficiently accurate process C_V value can be arrived at by adding the process C_V values for the gas and liquid portions of the stream. The calculation is based on the quantities of gas and liquid at upstream conditions. The valve body is specified to be one nominal size larger than the port to allow for expansion.

Valve rangeability

The rangeability of a control valve is the ratio of the flow coefficient at the maximum flow rate to the flow coefficient at the minimum flow rate ($R = C_{Vmax}/C_{Vmin}$). Valve rangeability is a criterion which is used to judge whether a given valve will be in a controlling position throughout its required range of operation (neither wide open nor fully closed). In practice, the selection of the valve to be installed is the responsibility of the instrument engineer. As the process engineer is usually the person responsible for the correct operation of the process itself he or she must be satisfied that the item selected meets the control criterion

8

required. The process engineer must therefore be satisfied that the valve will control over the range of the process flow.

Control valves are usually limited to a rangeability of 10:1. If R is greater than 10:1 then a dual valve installation should be considered in order to ensure good control at the maximum and minimum flow conditions.

In some applications, particularly on compressor or blower suction, butterfly valves have been specified to be line size without considering that, as a result, the valve may operate almost closed for long periods of time. Under this condition there have been cases of erosion resulting from this. It is recommended therefore that butterfly valves be sized so that they will not operate below 10% open for any appreciable period of time and not arbitrarily be made line size.

Valve flow coefficient (C'_V)

In order to ensure that the valve is in a controlling position at the maximum flow rate, the valve C'_V is the maximum process value determined above, divided by 0.8. The reasons for using this factor are that:

(1) It is not desirable to have the valve fully open at maximum flow since it is not then in a controlling position.
(2) The valves supplied by a single manufacturer often vary as much as 10–20% in C_V.
(3) Allowance must be made for pressure drop, flow rate, etc, values which differ from design.

CONTROL VALVE SIZING

Control valve sizes are determined by the manufacturers from the process data submitted to them. However, there are available some simple equations to give a good estimate of the required valve sizes to meet a process duty. Three of these are given below:

$$S \text{ (inches)} = \left(\frac{\text{valve } C_V}{9}\right)^{1/2}$$

Double-seated control valve sizes may be estimated by:

$$S \text{ (inches)} = \left(\frac{\text{valve } C_V}{12}\right)^{1/2}$$

Butterfly valve sizes may be estimated by:

$$S \text{ (inches)} = \left(\frac{\text{valve } C_V}{20}\right)^{1/2}$$

The constants (9, 12 or 20) in these equations can vary by as much as 25% depending on the valve manufacturer.

A control valve should be no larger than line size. A control valve size that is calculated to be greater than line size should be carefully checked together with the calculation used for determining line size. Ideally a control valve size should be one size smaller than line size.

Once the valve size is estimated and the valve C_V known, then the percentage opening of the valve at minimum flow and maximum flow can be obtained by dividing the respective process C_V values conditions by the valve C_V. This information is normally required to check the percentage opening of a butterfly at minimum flow. It is not normally necessary to calculate it for any other type of valve.

VALVE ACTION ON AIR FAILURE

In the analysis of the design and operation of any process or utility system the question always arises on the action of control valves in the system on instrument air failure. Should the control valve fail open or closed is the judgement decision of the process engineer after evaluating all aspects of safety and damage in each event. For example, control valves on fired heater tube inlets should always fail open to prevent damage to the tubes through low or no flow through them when they are hot. On the other hand, control valves controlling fuel to the heaters should fail closed on air failure to avoid overheating of the heater during the air failure.

The failed action of the valve is established by introducing the motive air to either above the diaphragm for a failed open requirement or below the diaphragm for a failed shut situation. The air failure to the valve above the diaphragm allows the spring to pull up the plugs from the valve seats. Air failure to valves below the diaphragm forces the spring to seat the valves in the closed position.

1.10   Hydraulic Analysis of a System

This section deals with calculating a pressure profile in a process system. Such a calculation is used to size pipelines and to determine the pumping requirements for the system. The following calculation is an example of a typical system process engineer's encounter in a design of a plant or in checking out an operating plant's process flow. The following diagram is used as the basis for this example.

Example calculation

$$\text{Total flow to P-103 A\&B} = 519\,904\,\text{lb/h}$$
$$\text{API gravity} = 20.7 = \text{SG of } 0.930 \text{ at } 60\,°\text{F}$$
$$\text{Stream temperature} = 545\,°\text{F}$$
$$\text{SG at } 545\,°\text{F} = 0.755 = 6.287\,\text{lb/gal}$$

$$\text{Gallons/min at stream temp.} = \frac{519\,904}{6.287 \times 60}$$
$$= 1378\,\text{gpm}$$
$$\text{Viscosity of the oil at } 545\,°\text{F} = 1.2\,\text{cSt}$$

The suction line to P-103 A&B is 10″ schedule 40.

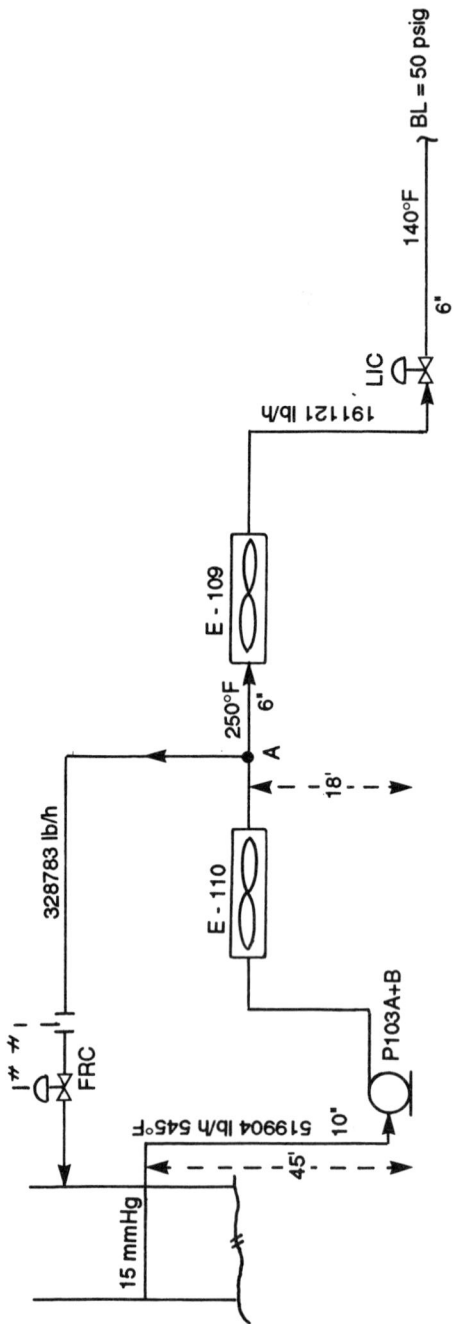

From Table A1.2 in Appendix 1:
Friction loss in feet/1000 ft of pipe is 9.3 feet (equivalent to 0.30 psi/100 ft).

SUCTION LINE EQUIVALENT LENGTH

The following information would normally be obtained from a piping general arrangement drawing (a piping GA). For this calculation this information is fictional:

Number of standard elbows in the line = 8
Number of gate valves (all open) in the line = 3
Total straight length of 10-inch line = 85 feet

From Figure A1.4 in Appendix 1:

The equivalent length for 10-inch elbows is 22 ft per elbow = 22 × 8 = 176 ft
The equivalent length for 10-inch gate valves is 5 ft per valve = 5 × 3 = 15 ft

Total equivalent line length: 85 + 176 + 15 = 276 ft

$$\text{Head loss to pump suction due to friction} = \frac{9.30 \times 276}{1000} = 2.57 \text{ ft}$$

$$\text{In terms of psi this is } \frac{2.57 \times 62.2 \times 0.755}{144} = 0.83 \text{ psi}$$

PUMP SUCTION PRESSURE

Source pressure at vacuum tower draw-off = 15 mmHg
= 0.29 psia

$$\text{Static head} = 45 \text{ feet} = \frac{45 \times 62.2 \times 0.755}{144}$$

= 14.7 psia
Line loss = 0.83 psi
Total pressure at the pump suction flange = 0.29 + 14.7 − 0.83
= 14.21 psia (which is −0.49 psig)

CALCULATING THE PRESSURES AT THE PUMP DISCHARGE

Destination pressure at battery limits = 50 psig.
Temperature of the oil at battery limits = 140 °F
Viscosity of the oil at 140 F = 20 cSt
SG of the oil at 140 F = 0.900; lbs/gal = 7.495
From a material balance or from plant data: flow of oil = 191 121 lb/h

$$\text{Flow rate} = \frac{191\,121}{7.495 \times 60}$$

$$= 425 \text{ gpm}$$

1   Line pressure drop from E-109 to battery limits

Equivalent length of line:

$$\text{Straight line} = 126\,\text{ft}$$
$$\text{Number of elbows} = 18 \text{ equiv. length} = 18 \times 16 = 288\,\text{ft}$$
$$\text{Number of gate valves} = 6 \text{ equiv. length} = 6 \times 3.5 = 21\,\text{ft}$$
$$\text{Number of TEEs} = 1 \text{ equiv. length} = 1 \times 30 = 30\,\text{ft}$$
$$\text{Total equivalent length} = 465\,\text{ft}$$

Line to BL is a 6-inch schedule 40. Then from Table A1.2 in Appendix 1 loss due to friction is 21.4 ft/1000 ft, which is equivalent to 0.83 psi/100 ft. Then line friction loss is

$$\frac{21.4 \times 465}{1000} = 9.95\,\text{ft}$$

$$\text{or } \frac{9.95 \times 62.2 \times 0.900}{144} = 3.86\,\text{psi}$$

2   Control valve pressure drop

There is a battery limit Level Control Valve (LICV) between E-109 and the BL. It is required to calculate the pressure drop required by this valve at design flow. The rule of thumb given in Section 1.9 will be used to determine this. Thus the pressure drop will be estimated as 20% of the circuit frictional pressure drop plus 10% of the static head of the receiving vessel.

The oil discharges into a surge drum of another downstream unit. This drum is pressurized by a blanket of inert gas. The net static head to this drum is 15 feet above valve outlet flange. This is equivalent to 6 psi.

The total line pressure drop for the whole circuit is estimated at three times that calculated above. (This will be checked and may be revised when the analysis of the whole circuit is complete). This pressure drop therefore is $3 \times 3.86 = 11.58$ psi for line losses. In addition to this line loss there are also two air coolers which have pressure drops as follows:

$$\text{E-109} = 6\,\text{psi (from data sheets)}$$

$$\text{E-110} = 8\,\text{psi}$$

Then total system pressure drop is estimated as:

$$11.58 + 14 = 25.58\,\text{psi}$$

Then control valve pressure drop is:

$$(0.1 \times 6.0) + (0.2 \times 25.58) = 5.2\,\text{psi}$$

Check the control valve size based on this pressure drop:

$$C_V = Q_L \times \sqrt{\frac{SG}{\Delta P}}$$

$$C_V = 425 \times \sqrt{\frac{0.9}{5.2}}$$

$$= 176.8$$

The valve is a single-seated type. Therefore valve size is

$$\text{Size} = \sqrt{\frac{C_V}{9}}$$

$$\text{Size (inches)} = \sqrt{\frac{176.8}{9}}$$

$$= 4.4 \text{ in.}$$

This is less than line size and is reasonable. Use 5.2 psi as pressure drop.

3   Calculate pressure at point A which is the reflux stream take-off

Length of line between point A and inlet to E-109 is as follows:

$$\text{Straight line} = 121 \text{ ft}$$
$$4 \text{ elbows} = 4 \times 16 = 64 \text{ ft}$$
$$2 \text{ valves} = 2 \times 3.5 = 7 \text{ ft}$$
$$\text{Total 6-inch schedule 40 line equiv.} = 192 \text{ ft}$$

Temperature of stream at this point is 250 °F.

$$\text{Viscosity at } 250\,°F = 3.5 \text{ cSt}$$

SG at 250 °F is 0.860 which is 7.16 lb/gal

$$\text{Rate of flow into E-109} = \frac{191\,121}{7.16 \times 60}$$

$$= 445 \text{ gpm}$$

From Table A1.2 in Appendix 1 friction loss in 6-inch pipe $= 16.1$ ft/1000 ft. Total line loss:

$$\frac{192 \times 16.1}{1000} = 3.09 \text{ ft or } 1.15 \text{ psi}$$

Pressure at point A therefore is the sum of:

$$\text{Destination pressure at BL} = 50 \text{ psig}$$
$$\text{Pressure drop for E-109} = 6 \text{ psi}$$
$$\text{Loss in line from E-109 to battery limits} = 3.86 \text{ psi}$$
$$\text{Loss in line from point A to E-109} = 1.15 \text{ psi}$$
$$\text{Control valve pressure drop} = 5.2 \text{ psi}$$
$$\text{Flow meter (not shown) say} = 0.2 \text{ psi}$$
$$\textit{Total pressure at point A} = 66.41 \text{ psig}$$

4   Calculating the pressure at the pump discharge flange

Total flow from the pump is 519 904 lb/h
Oil temperature at outlet of E-110 is 250 °F
gpm of flow from E-110 is

$$\frac{519\,904}{7.16 \times 60} = 1210\,\text{gpm}$$

Line size at this point is 8-inch schedule 40.
Head loss in this line is 31.1 ft/1000 ft

Equivalent length of line:

$$\begin{aligned}
\text{Straight line} &= 82\,\text{ft} \\
\text{1 tee} &= 30\,\text{ft} \\
\text{Total equiv. length} &= 112\,\text{ft}
\end{aligned}$$

$$\text{Total head loss in line} = \frac{112 \times 31.1}{1000}$$
$$= 3.5\,\text{ft or } 1.3\,\text{psi}$$

Pressure at E-110 inlet will be:

$$\begin{aligned}
\text{Pressure at point A} &= 66.41\,\text{psig} \\
\text{Head loss in line} &= 1.3\,\text{psi} \\
\text{Pressure drop across 11-E-10} &= 8\,\text{psi} \\
&= 75.7\,\text{psig}
\end{aligned}$$

This pressure is 18 ft above grade as these air coolers are located above the piperack. Allowing 1.5 feet from grade to pump centre line, the static head at pump discharge flange is 16.5 ft. Equivalent length of line from pump to E-110 is:

$$\begin{aligned}
\text{Straight length} &= 155\,\text{ft} \\
\text{12 elbows} = 12 \times 20.5 &= 246\,\text{ft} \\
\text{3 gate valves} = 3 \times 4.8 &= 14.4\,\text{ft} \\
\text{1 non-return valve} = 1 \times 51 &= 51\,\text{ft} \\
\text{Total equivalent length} &= 466.4\,\text{ft}
\end{aligned}$$

Head loss in 8-inch schedule 40 pipe at a flow rate of 1378 gpm.

Pump temperature is 545 °F
Viscosity at pump temperature is 1.2 cSt
SG at pump temperature is 0.755

$$\text{Head loss from Table A1.2} = 28.8\,\text{ft/1000 ft}$$
$$\text{Total line head loss} = \frac{28.8 \times 466.4}{1000}$$
$$= 13.4\,\text{ft or } 4.38\,\text{psi}$$

Then the pump discharge pressure is the sum of:

$$\begin{aligned}
\text{Pressure at E-110 inlet} &= 75.7 \text{ psig} \\
\text{Line pressure drop} &= 4.38 \text{ psi} \\
\text{Static head (16.5 ft)} &= 5.38 \text{ psi} \\
\text{Total discharge pressure} &= 85.46 \text{ psig}
\end{aligned}$$

5 Calculating the pressures in the reflux line from point A

At point A the pressure has been calculated at 66.41 psig. Flow of the reflux stream (from the material balance) is:

$$519\,904 - 191\,121 = 328\,783 \text{ lb/h}$$

Temperature of the stream is 250 °F.

$$\begin{aligned}
\text{Viscosity at 250 F} &= 3.5 \text{ cSt} \\
\text{SG at 250 °F} &= 0.860 \text{ and lb/gal is } 7.16 \\
\text{Rate of flow} &= \frac{328\,783}{7.16 \times 60} \\
&= 765 \text{ gpm}
\end{aligned}$$

Line to tower from point A is a 6-inch schedule 40 and head loss in this line is found from Table A1.2 to be 40.2 ft/1000 ft.

Equivalent line lengths:
To the flow controller inlet flange:

$$\begin{aligned}
\text{Straight line} &= 16 \text{ ft} \\
\text{2 elbows} = 2 \times 16 &= 32 \text{ ft} \\
\text{1 valve} = 1 \times 3.5 &= 7 \text{ ft} \\
\text{Total equiv. length} &= 81.5 \text{ ft}
\end{aligned}$$

From the flow controller to tower:

$$\begin{aligned}
\text{Straight line} &= 71 \text{ ft} \\
\text{6 elbows} = 6 \times 16 &= 96 \text{ ft} \\
\text{1 valve} = 1 \times 3.5 &= 3.5 \text{ ft} \\
\text{Total equiv. length} &= 171 \text{ ft}
\end{aligned}$$

Total line length from point A to tower:

$$81.5 + 171 = 252.5 \text{ ft}$$

Total head loss in line due to friction:

$$\frac{252.5 \times 40.2}{1000} = 10.15 \text{ ft or } 3.77 \text{ psi}$$

The pressure required to deliver 765 gpm of reflux to the tower excluding the pressure drop across the control valve at this rate is the sum of the following:

Destination pressure = 0.29 psia
Static head = 38 ft = 14.11 psi
Distributor (tower internals) = 2.0 psi (from data sheet)
Flow meter pressure drop = 0.5 psi (from data sheet)
Head loss in line = 3.77 psi
Total required = 20.67 psia or 5.97 psig

Then valve pressure drop at design flow of 765 gpm is:

Pressure at point A − required pressure

$$= 66.41 - 5.97 \text{ psig} = 60.44 \text{ psi}$$

6 Checking flow control valve size based on calculated pressure drop

Control valve will be a single-seated plug type.

$$C_V = Q_L \sqrt{\frac{SG}{\Delta P}} \text{ for liquids}$$

$$C_V = 765 \sqrt{\frac{0.86}{60.4}}$$

$$= 91.2$$

$$\text{Size of valve (inches)} = \sqrt{\frac{91.2}{12}}$$

$$= 2.76 \text{ in. (say, 3.0 in., which is satisfactory)}$$

1.11 Engineering Flow Diagrams

Diagrams are used extensively by all disciplines of engineers to convey ideas and data. Process engineers use and in some cases develop three types of flow diagrams to project their work and responsibilities:

- The process flow diagram
- The mechanical flow diagram (sometimes called the P&I diagram)
- The utilities diagram

The process flow diagram is originated by the process engineer who retains sole responsibility for its future development and update. The mechanical flow diagram may also be initiated by the process engineer or it may be developed by others from the process flow diagram. In many companies, however, the process engineer remains responsible for its technical content, development and completeness. The utilities diagram shows the routing, sizing, and specification of all the utility flow lines between units and within units of a process. This diagram is usually superimposed on the plot layout diagram. The piping engineers or those engineers who are responsible for initiating the plot plan drawing usually initiate the utilities drawing and administer it. The process engineer in this case is responsible for sizing the flow lines, establishing flow conditions of temperature and pressure, and ensuring that all lines have been included.

THE PROCESS FLOW DIAGRAM

This diagram is usually the first drawing that will be produced for an engineering or development project. In some cases it may be preceded by a process block flow diagram but it is the process flow diagram that is the basis for

- A process definition
- A budget cost estimate
- An equipment list
- A mechanical flow diagram
- Process equipment data sheets

The flow diagram supports the material and energy balances for the process and establishes the sequence and direction of the process flow. It also shows the control philosophy that will be adopted for the process and the salient temperature/pressure conditions within the process. As a minimum therefore process flow diagrams should contain the following:

(1) *Vessels* The outline of all major vessels, such as towers, drums and tanks are shown. Their equipment item number and their overall dimensions are indicated on the diagram. Where vessels contain special internals such as trays, demisters, packing, etc. these too should be simply indicated on the vessel drawing. For example, the number of trays in a tower may be indicated by showing the top, feed, and bottom trays only but including their respective tray numbers. The main temperature and pressure conditions are also shown on the vessel drawing. For example, on fractionation towers the tower top and bottom operating temperatures will be shown but only the top pressure is normally given.

(2) *Heat exchangers* All heat exchangers are shown as single shells on the process flow diagram. That is, no attempt is made to show the number of passes or the type (i.e. shell and tube, double pipe, etc.) on this diagram. The process flows to and from these items are shown as flowing through the shell side or the tube side. Tube side flow within the heat exchanger diagram is shown as a dotted line. The exchanger item number only is indicated on the flow diagram adjacent to the equipment drawing. Its equipment name is normally not given. The heat duty of the exchanger is also shown on the flow diagram again adjacent to the equipment drawing and below the equipment number. The temperature conditions for the exchanger are shown on the process lines in and out of the equipment. No pressures are normally indicated for this equipment.

(3) *Air coolers* These are shown simply as a narrow rectangular block in a process line with a fan symbol appearing inside the rectangle as a dotted

outline. The item number is shown directly below the rectangle and below that is given the operating duty of the item in mmBTU/h or kcal/h, etc.

(4) *Heaters* Fired heaters are shown as box type with a stack outlet. The specific type of heater or the number of tube passes are not shown on the process flow diagram. Again only the equipment item number and the heater duty is given in the flow diagram adjacent to the item drawing. The heating medium is shown as a single line entering the bottom of the equipment and marked only as 'fuel'. The line would contain a control valve with an instrument control line to the process coil outlet showing the firing control philosophy. The temperatures connected with the heater are shown on the process line in and out of the equipment. Normally no pressure levels are indicated for this item.

(5) *Pumps* These are shown in as simple a manner as possible. Most companies carry their own symbols for equipment. Many, however, show a centrifugal pump as a circle with the suction line proceeding to the centre of the circle and the discharge line leaving tangentially from the top. Other symbols are used for positive displacement pump types. Only these two types are differentiated on the process flow diagram. The various types within these categories are not indicated. The pump item number and an indication as to whether the pump is spared is shown adjacent to the pump drawing on the flow diagram. If the pump is spared then the item number will be followed by 'A + B'. If it is not spared it will be followed by the letter 'A' only. The capacity of the pump as gallons per minute (gpm) or cubic metres per hour will be shown under the item number. This is the normal or operating capacity.

(6) *Compressors* These are shown as either centrifugal or reciprocating machines. The centrifugal type are shown similar to the centrifugal pump and the reciprocating type as two small square boxes connected by a double line. The item number appears below the item and in some cases indicates the number of stages. The capacity in standard or normal units is given below the item number while the temperature/pressure conditions to and from the item is shown on the process lines to and from it.

(7) *Process lines* All major process lines interconnecting equipment, recycle, or bypasses are shown on the process flow diagram. These lines will show direction of flow as black arrows on the lines. Control valves and major block valves are also indicated on these lines. The temperature and pressure conditions of the flowing material are given on the lines at appropriate positions in the drawing, that is, in a location on the diagram which establishes the operating parameters of equipment that handles the material. Details of the material flowing in the process lines are given in the material balance which is also included in the process flow diagram. Reference is made to the respective column in the material balance that gives this material data at the various points in the process flow lines. This reference is in the form of a figure (1 to n) contained in a diamond symbol placed on the process line. The figure refers to the column number in the material balance that gives the details (flow, SG, temperature, etc.) of the material at that location in the process. Utility lines are not shown on the process diagram

but indication of utility streams are made by showing short lengths of lines with the utility symbol marked on them.

(8) *Instruments*   Only major control instrumentation are shown and the instrument symbols are kept as simple as possible. Details of instrument 'hook-ups' are confined to showing the control valve being activated from either a level, flow, pressure or temperature elements by a dotted line to the valve. The measurement instruments that affect control are shown as circles with the type of instrument (e.g. FC for flow controller, TC for temperature controller, etc.) printed in the circle.

(9) *The material balance*   The material balance for the process represented by the process flow diagram is shown either in table form on the bottom of the flow sheet or on an attached but separate table. Preferably it should appear on the flow sheet itself. The table should contain at least the following:
- The stream number—This is the number given to the process stream and referenced on the respective process line in the diagram. This initiates the columns that will make up the table.
- The stream identification—This next line should identify the stream, such as 'Debutanizer bottoms' for each of the columns.
- The items following down the title column consists of the flow rate for each stream as wt per unit time, the stream temperature, the specific gravity at a standard temperature for liquids, the mole wt for gas streams, volume flow at standard temperature (and pressure for gasses), stream pressure for gases.

An example of a process flow diagram is shown in Figure 1.7. Note that companies may use their own symbols for some equipment and instrumentation which differ from those shown in Figure 1.7. They may also elect to increase the scope of the contents of a process flow diagram. The description of the contents given here is, however, considered to be the minimum that will achieve the objective of the flow diagram.

THE MECHANICAL FLOW DIAGRAM

The mechanical flow diagram (MFD) is developed from the process flow diagram. The detail provided by the MFD is sufficient for other engineering disciplines to:

- Initiate a plot layout
- Prepare a line list (piping design)
- Initiate piping arrangement drawings (piping design)
- Prepare a preliminary piping material take-off
- Prepare an instrument register (instrument engineering)
- Initiate instrument 'hook up drawings' (instrument design)
- Prepare electrical 'one-line drawings' (electrical engineering)
- Prepare preliminary instrument and electrical material take-off
- Initiate civil and structural design (civil engineering)
- Develop the project execution plan (planning engineers)
- Prepare a semi-definitive cost estimate (cost estimators).

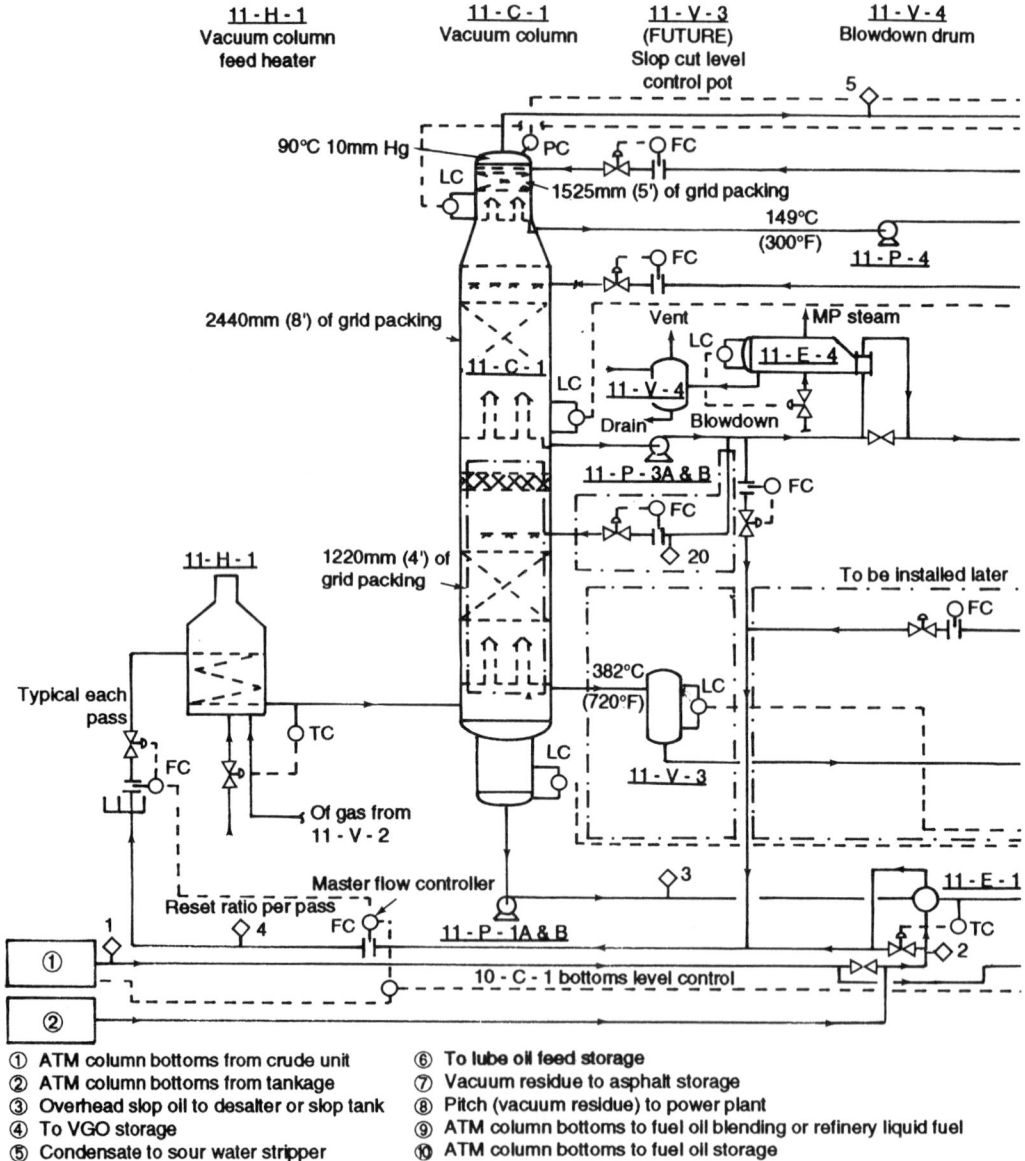

Figure 1.7. A typical process flow diagram

11 - H - 1
Vacuum column
feed heater

11 - C - 1
Vacuum column

11 - V - 3
(FUTURE)
Slop cut level
control pot

11 - V - 4
Blowdown drum

90°C 10mm Hg

PC

FC

5

LC

1525mm (5') of grid packing

149°C
(300°F)

11 - P - 4

FC

2440mm (8') of grid packing

Vent

MP steam

LC

11 - E - 4

11 - C - 1

LC

11 - V - 4

Drain

Blowdown

11 - P - 3A & B

FC

FC

20

To be installed later

11 - H - 1

1220mm (4') of
grid packing

FC

Typical each
pass

382°C
(720°F)

LC

FC

TC

LC

11 - V - 3

11 - E - 1

Of gas from
11 - V - 2

3

TC

Master flow controller

Reset ratio per pass

2

1

4

FC

11 - P - 1A & B

10 - C - 1 bottoms level control

① ATM column bottoms from crude unit
② ATM column bottoms from tankage
③ Overhead slop oil to desalter or slop tank
④ To VGO storage
⑤ Condensate to sour water stripper
⑥ To lube oil feed storage
⑦ Vacuum residue to asphalt storage
⑧ Pitch (vacuum residue) to power plant
⑨ ATM column bottoms to fuel oil blending or refinery liquid fuel
⑩ ATM column bottoms to fuel oil storage

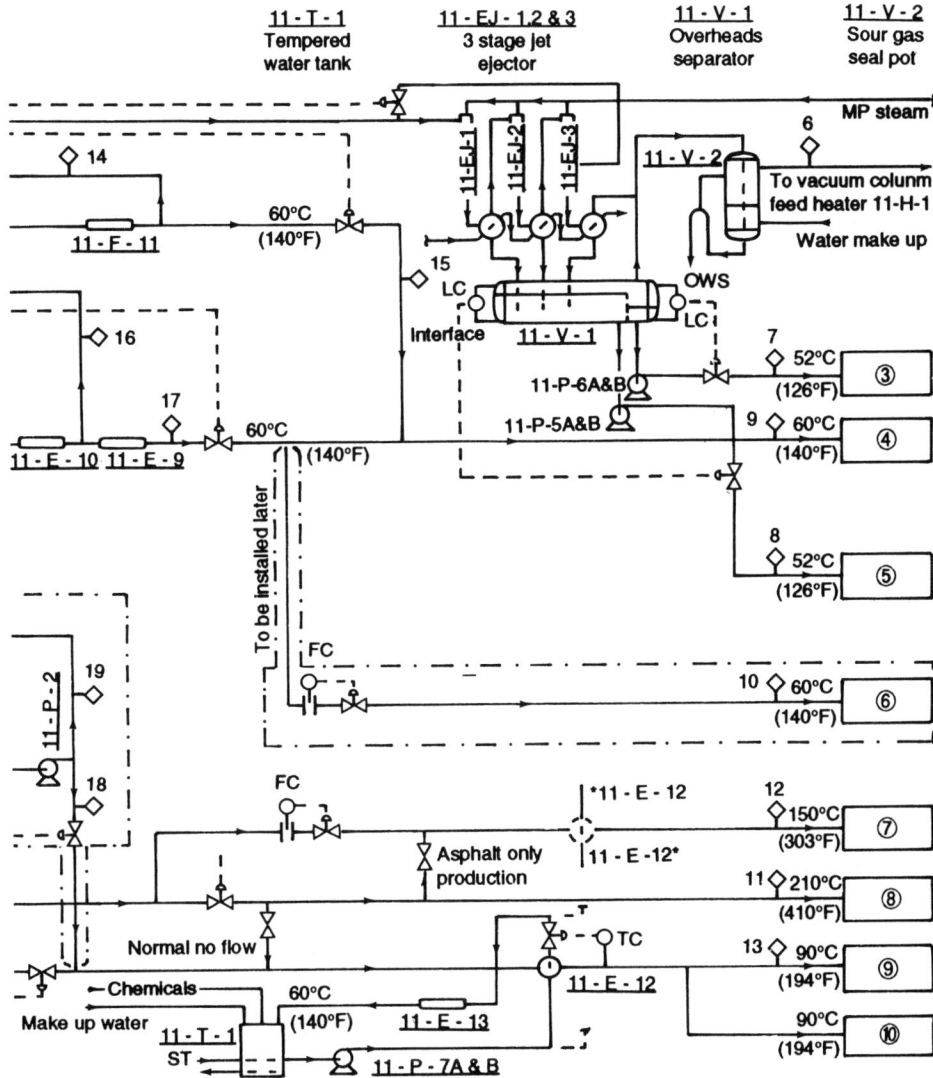

11 - T - 1
Tempered
water tank

11 - EJ - 1,2 & 3
3 stage jet
ejector

11 - V - 1
Overheads
separator

11 - V - 2
Sour gas
seal pot

MP steam

14

11 - F - 11

60°C
(140°F)

11-EJ-1 11-EJ-2 11-EJ-3

11 - V - 2

6

To vacuum colunm
feed heater 11-H-1

Water make up

15

LC

Interface

11 - V - 1

OWS

LC

7 52°C
(126°F) ③

16

17

11 - E - 10   11 - E - 9

60°C

(140°F)

11-P-6A&B

11-P-5A&B

9 60°C
(140°F) ④

8 52°C
(126°F) ⑤

To be installed later

FC

10 60°C
(140°F) ⑥

11 - P - 2

19

18

FC

Asphalt only
production

*11 - E - 12

11 - E -12*

12 150°C
(303°F) ⑦

11 210°C
(410°F) ⑧

Normal no flow

TC

13 90°C
(194°F) ⑨

11 - E - 12

Chemicals

Make up water

11 - T - 1
ST

60°C
(140°F)

11 - E - 13

11 - P - 7A & B

90°C
(194°F) ⑩

Stream up	1	2	3	4	5
Stream identification	ATM column bottoms from crude column	Feed to vacuum column	Vacuum column residue	Feed to vacuum column	Vacuum column overhead vapour
Normal Kg/h	470,269	391,836	245,982	391,836	2971
BPSD based on standard conditions	71,800	60,000	36,000	60,000	
Cut	371°C+	371°C+	80 pen	371°C+	94.1 (Av. mol/wt)
Sg @ 15%	0.959	0.989	1.033	0.989	
API	11.5	11.5	5.5	11.5	
Max operating Kg/h	470,300	457,070 (69,000 BPSD of 385°C+)	329,460 (48,500 BPSD @ 800 pen)	457,070 (385°C+ resid)	3000

Stream up	11	12	13	14	15
Stream identification	Combined slop cut and vac. resid streams	Asphalt	ATM column botoms to fuel blending	Top circulating reflux	Top cut TBP 370 to 427°C (700 to 800°F)
Normal Kg/h	251,565		78,433	174,603	51,279
BPSD based on standard conditions	36,900		11800		8593
Cut			371°C+		
Sg @ 15%			1.033	0.906	0.906
API	5.5	5.5	5.5	24.6	24.6
Max operating Kg/h		11,900	457,000 (385°C+ resid)		51,280

Figure 1.7. (*continued*)

Stream up	6	7	8	9	10
Stream identification	Vacuum. column inerts exit stream	Vacuum column overhead slop	Condensate water from electors	Combined top side cut and heavy V.G.O.	Heavy vavuum gas oil
Normal Kg/ h	600	2371		137,900	
BPSD based on standard conditions		407		22,693	
Cut	29 (Av. mol/wt)	218 (Av. mol/wt)	18		
Sg @ 15%		0.883			
API		28.7		22.1	20.7
Max operating Kg/ h	600	2371	7710	137,901	

Stream up	16	17	18	19	20
Stream identification	H.G.O. circulating reflux	Heavy vacuum gas oil	Slop cut vacuum column product	Vac. column bottoms recycle	Metals wash out
Normal Kg/ h	149.206	86621	5583	0 to 16748	8481
BPSD based on standard conditions		14100	900	2700	
Cut					
Sg @ 15%	0.930	0.930	0.942	0.948	0.930
API	20.7	20.7	18.7	18.7	20.7
Max operating Kg/ h		86621	16,748	16,748	15,964

To meet these objectives the mechanical flow diagram will contain more detail of the process and the equipment included in it than the process flow diagram. This is described briefly as follows:

(1) *Vessels* These are shown in approximate relative size to one another wherever possible. Again only the top and bottom trays need be shown, unless other trays are required to indicate the location of side draw-offs, instruments, sample points or changes in type of tray layout. Trays are numbered sequentially either from top to bottom or bottom to top. Catalyst beds, packing, demisters, and the type of tray (such as single-pass, double-pass, etc.) are shown. The height of packing etc. is shown adjacent to the vessel and the height of all vessels above grade is also indicated on the vessel. The following detail is usually shown on the top of the flow sheet directly above the vessel:
   - Vessel item number (this should also appear in or near the vessel drawing)
   - Title
   - Size (inside diameter, and length tan to tan)
   - Design temperature and pressure.
   - Insulation (i.e. whether the vessel insulated or not)
   - Trim number (line specification for miscellaneous vessel connections).

(2) *Heat exchangers* The arrangement and type of heat exchangers are shown on the MFD. Shell and tube exchangers are still shown as circles with tubeside flows in dotted lines as in the process flow diagram. Here, however, the actual number of shell passes are indicated. Double pipe and reboilers are shown as their specific format, and again the actual number of shell passes. The following data for each exchanger is shown at the top of the flow diagram:
   - The heat exchanger item number (this is also shown adjacent to the exchanger drawing)
   - The title
   - The duty of the exchanger (in BTU/h, kcal/h, etc.)
   - Insulation

(3) *Air coolers* As in the case of the heat exchangers, air coolers are shown in more detail on the MFD than on the process flow diagram. Usually on the MFD the air cooler is given as a narrow rectangle. The fan in this case is drawn below the rectangle for a forced-draught cooler or above the rectangle in the case of an induced air cooler. Only one process line is shown to and from the cooler but the number of passes are shown on these lines. Any temperature control by louvres or variable pitch fans are indicated on the diagram together with the appropriate instrumentation. The following data are given at the top of the MFD above the cooler:
   - The cooler item number (this is also shown adjacent to the drawing)
   - The title
   - Duty of the cooler

(4) *Fired heaters* The outline of the furnace is shown as being a cylindrical or a cabin-type heater. Most companies carry their own symbols to portray this

feature. The actual number of passes are shown on an inlet and outlet header together with the control system for the process flow. Only one pass is shown as entering and leaving the heater and flowing through the item. The internal flow is shown as a dotted line in the fire box. All instrumentation for the heater is given in detail such as temperature points on the heater coil and in the fire box, oxygen analysers in the chimney and the chimney damper control. Snuffing steam and decoking manifolds are usually shown as separate details. The firing control system is illustrated in detail. Usually this follows a standard adopted by the company for all the heaters. All the instrumentation interlocks and safety features are detailed in the MFD for all heaters. The following data are provided for this item at the top of the flow diagram and above each heater:

- The item number
- Title
- Duty of the heater

(5) *Pumps* These are shown in much greater detail on the MFD than on the process flow diagram. The pump drawing itself indicates the type of pump and the type of driver. For example, a centrifugal pump is shown as a narrow vertical parallel lines curved at both ends and resting on a base plate. A motor driver is shown as a small horizontal rectangle curved at both ends also resting on the base plate and connected to the pump by a short line. Process suction lines enter the centre of the pump case and the discharge line leaves from the top of the casing. All pumps are shown in the same detail, that is, both the normal operating pump and its spare are shown. Process (and utility) lines connecting the pumps in a set are shown together with the valving configuration. Isolating block valves, non-return valves, etc. on the pumps are shown together with cooling systems where required. Instrumentation for automatic pumpstart and stop are also detailed together with electrical switches for manual start/stop facilities. Pumps and other rotary equipment are normally drawn on the bottom of the flow diagram. The data supplied on the MFD for pumps are given below the item on the bottom of the diagram and are as follows:

- Pump item number (followed by 'A' for normally operating and 'B' for spare pump)
- Title
- Flow rate (gpm, m^3/h, etc.)
- Differential pressure (psi, kPa, etc.)
- SG of pumped fluid at pumping temperature
- Miscellaneous auxiliary piping (cooling water, flushing oil, seal oil, etc.).

(6) *Piping* The MFD is among the most important documents that are developed in the course of a fully engineered project that can proceed to construction and finally to operation. Its importance is probably highest in the case of the piping engineering and design function of a project. For this discipline the MFD is the basis for all their work. Any piping detail that is required to define the work must appear on this flow diagram if it is to be included at all in the constructed facility. The diagram then becomes a major communication tool for multi discipline interface.

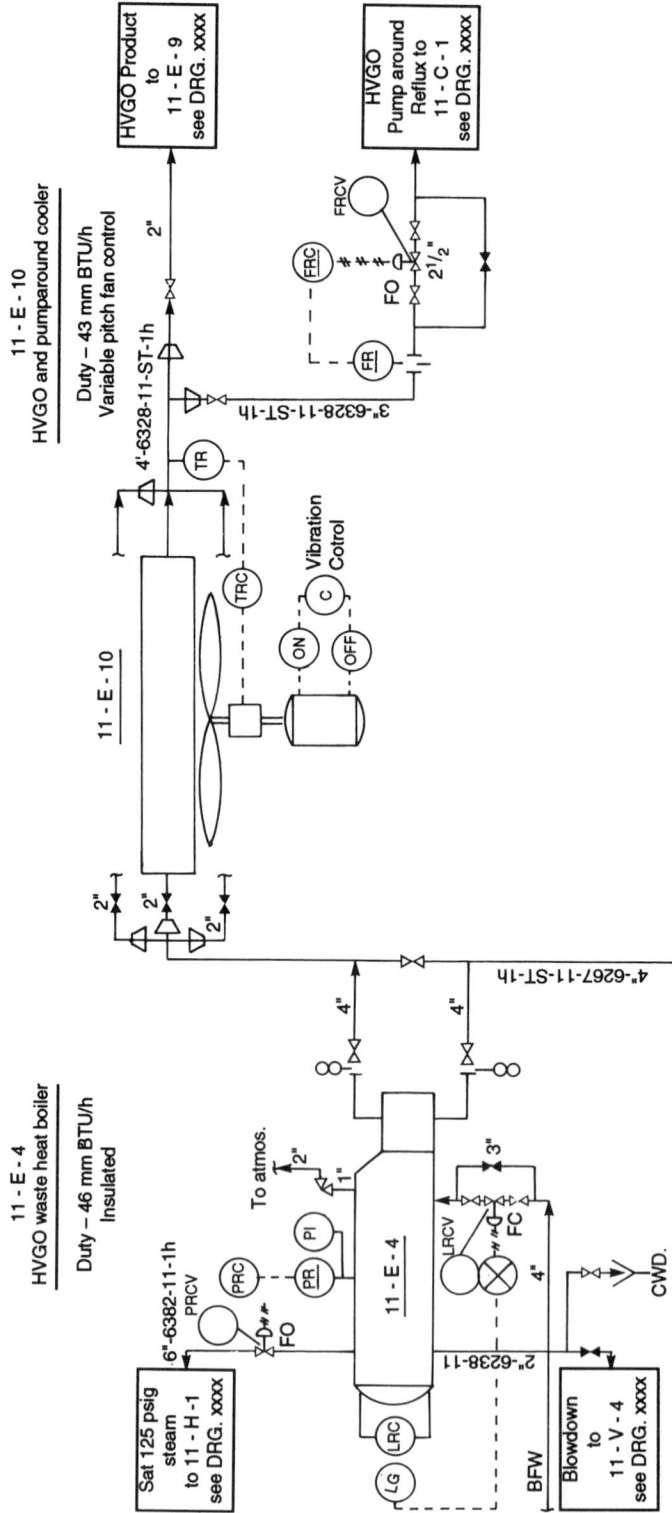

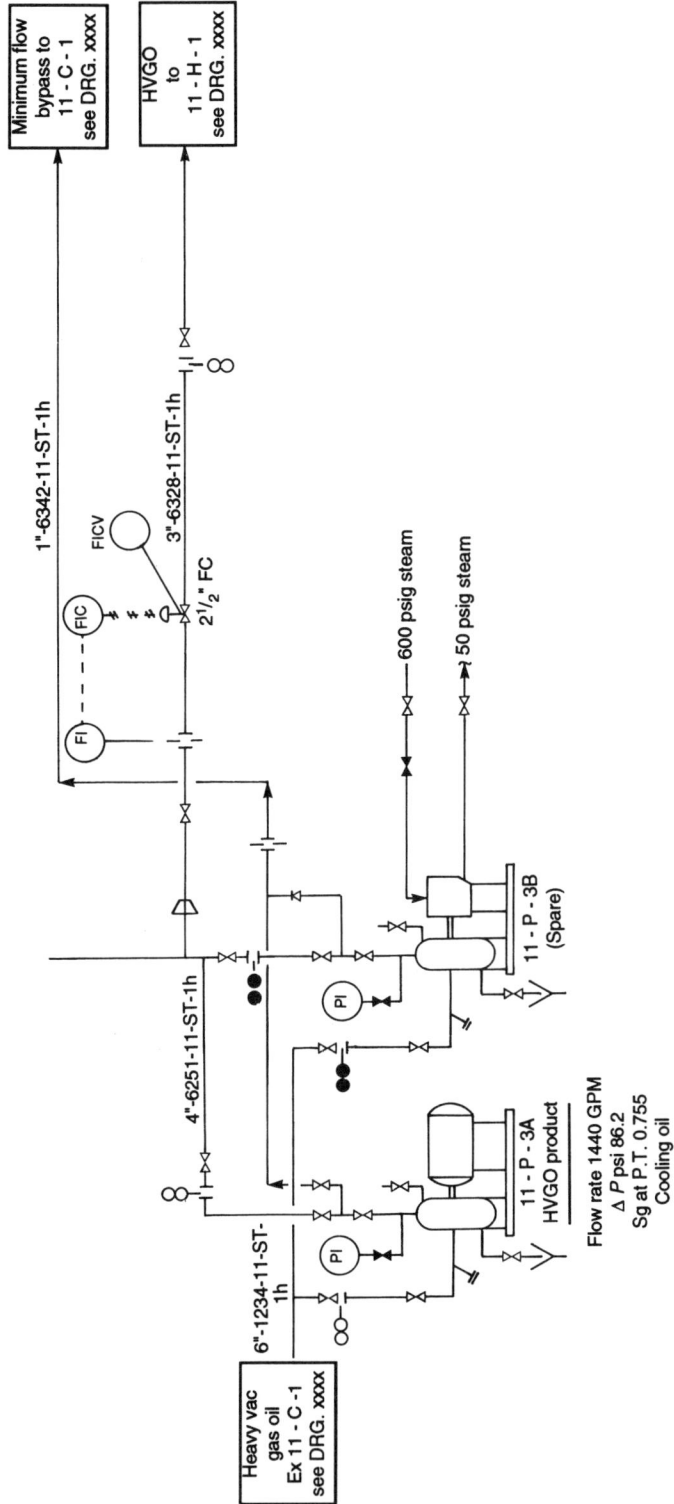

Figure 1.8. Section of a typical mechanical flow diagram

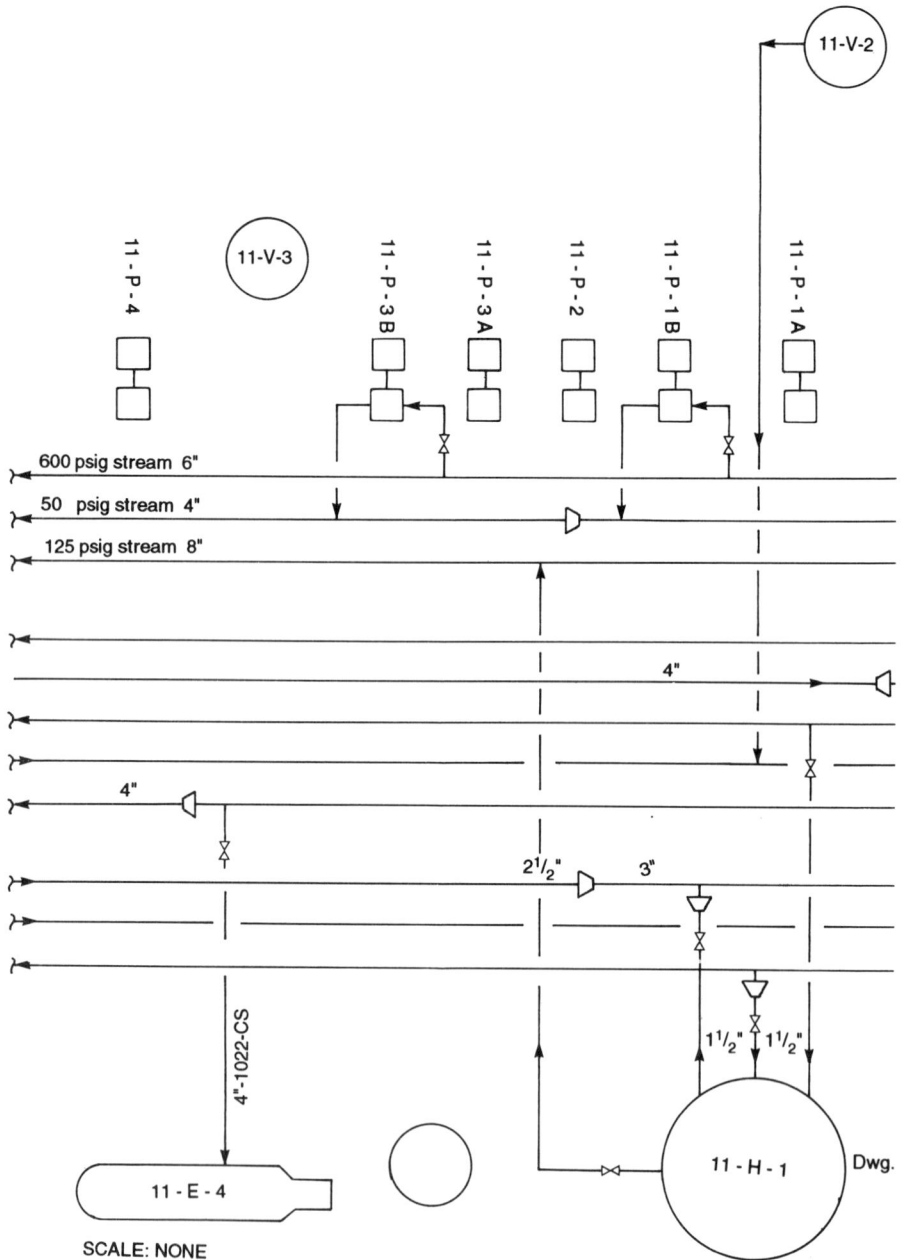

Figure 1.9. Section of a typical utility flow diagram

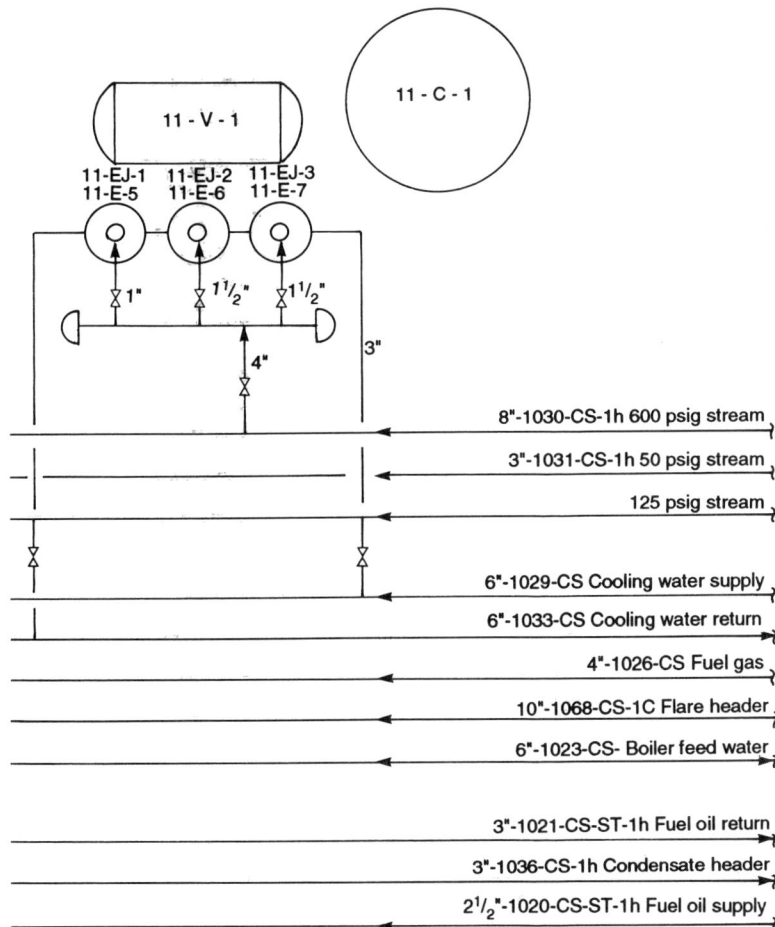

11 - V - 1

11 - C - 1

11-EJ-1   11-EJ-2   11-EJ-3
11-E-5   11-E-6   11-E-7

1"   1½"   1½"

4"   3"

8"-1030-CS-1h 600 psig stream

3"-1031-CS-1h 50 psig stream

125 psig stream

6"-1029-CS Cooling water supply

6"-1033-CS Cooling water return

4"-1026-CS Fuel gas

10"-1068-CS-1C Flare header

6"-1023-CS- Boiler feed water

3"-1021-CS-ST-1h Fuel oil return

3"-1036-CS-1h Condensate header

2½"-1020-CS-ST-1h Fuel oil supply

See detail
11-01-1234-008
Sheet 2 of 2

(6)   In the layout of a MFD, process feedlines should originate at the left-hand
       end of the drawing and the process product lines terminate at the right-hand
       end. Where this is impractical origin and terminus of the lines are located for
       clarity and convenience. The origin and terminus of each process line in any
       case are identified by a box which shows the descriptive title of the line, the
       drawing number, and section number of any reference drawing.
            Where possible, process lines between equipment drawings are located
       either above or below the line of equipment. Every effort in the layout
       should be made to avoid breaking lines around equipment. Piping high-point
       vents or low-point drains are not shown on the MFD unless they have some
       significant process requirement. Any pertinent requirement for process
       reasons on any line must be noted. This includes such requirements as no
       pocketing, sloping, etc. These types of notes are clearly marked on the
       respective lines.
            Utility lines originate and terminate adjacent to the equipment involved.
       Only the length of line necessary for valving, instrumentation, and line
       numbering is shown. The utility line origin and terminus are identified by a
       descriptive title only (e.g. 'LP steam' and 'LP condensate'). Main utility
       headers are not shown—these will appear on the 'utility flow diagram'.
            All line sizes are shown on the lines to which they refer. Where there is a
       change in a line size this is also indicated by a 'swage' up or down. All valve
       sizes are indicated on the MFD even if they are line size. Flow direction on
       all lines (whether process or utility) are clearly shown by directional arrows
       on the lines. Steam or electrical heat tracing are also indicated on the lines
       that require it. A section of a typical MFD is shown in Figure 1.8.
(7)   *Instrumentation*   All instrumentation is shown on the MFD in detail. This
       detail includes:
       ● Size of control valves
       ● Instrument hook-up method
       ● Vessel surge levels and level range
       ● Type of instrument activation (i.e. pneumatic or electronic)
       ● Computer interface if it applies
       ● Instrument identification number
       ● Position of the control valve on air failure (i.e. failed open or closed).
       Instrument symbols are usually to ISA standard with some minor modifica-
       tions to meet the respective company's needs.

THE UTILITY FLOW DIAGRAM

Process input to the utility flow diagram development is minor compared with
that for the PFD and MFD. It involves the sizing of the utility lines and valves
only. The process engineer is responsible for ensuring that all the necessary utility
lines to satisfy the process have been included. The UFD itself is based on the
approved plot layout of the plant and/or plants and is usually prepared by others.
It shows the direction of flow and sequence of equipment geographically just as
they appear on the plot plan. A section of a typical UFD is shown in Figure 1.9.
   Although process input to the UFD as such is minor, there will be considerable

detail necessary to complete the entire utility picture. For example, the UFD will show an instrument air header serving all the units in the process. This header will originate at the instrument air compressor set and driers. This origin will only be indicated on the UFD with reference to a detail drawing for the instrument air compressor set. This detail drawing will be a MFD and the process engineer will have the same input and responsibility for this diagram as for any other MFD.

2 COMMON SYSTEMS IN CHEMICAL PROCESSES

There are systems which are common to all chemical processes. The major ones that apply to process engineering are:

- Control systems
- Safety systems
- Utility systems
- Offsite systems

This chapter describes these systems and provides guidelines for the analysis and design of the more important parts of them.

Process engineers are often required to develop suitable support systems for new plants, expand existing systems or provide a suitable design for costing. It is therefore important to know the fundamental criteria in these systems and their purpose in the overall process operation. This chapter begins with a section on process controls. This system is probably the most important in the design and the future operation of the process itself. Without a properly designed control system the operation of the process is at a great disadvantage even though the equipment and other systems contained in the process have been well designed and specified.

2.1 Control Systems

Control systems in a process are required to maintain the correct conditions of flow, temperature, pressure and levels in process equipment and piping. There are therefore four major types of controls:

- Flow control
- Temperature control
- Pressure control
- Level control

Before looking at these in some detail it is necessary to define some of the more common terms met with in control systems.

DEFINITIONS

- *Surge volumes*   This is the volume of liquid between the normal liquid level (NLL) to the bottom of a vessel.
- *Level control range*   This is the distance between the high liquid level (HLL) and the low liquid level (LLL) in the vessel. When using a level controller the signal to the control valve at HLL will be to open the valve fully. At LLL the signal will be to close the valve fully.
- *Proportional band*   This determines the response time of the controller. Normally a proportional band is adjustable between 5% and 150%. The wider the proportional band, the less sensitive is the control. If a slower response time is required a wider proportional band is used.
- *Control valve response*   The minimum times that should be allowed between HLL and LLL to permit the control valve to respond effectively to changes in level are:

CV size (inches)	Response (seconds)
1	6
2	15
3	25
4	35
6	40
8	45
10	50

The above times allows for air signal lags and for operating at a proportional band of about 50%.
- *Surge volume*   This is the volume retained in a vessel during operation at a set level. It is used for protecting:
  —Equipment from damage caused by flow failures
  —Downstream processes from fluctuating flows which cause poor process performance
  —Downstream processing from fluctuations in feed composition or temperature
  —Equipment from damage due to coolant failure

TYPES OF SURGE VOLUMES

There are basically two types of surge volumes:

(1) *Upstream protection surge*   This is a surge volume provided to protect the upstream equipment and its associated pump from feed failure.
(2) *Downstream flow surge*   This is a surge volume provided to protect downstream equipment from feed failures or fluctuations.

Examples of surge types

(1) *Process feed* Feeds to process units are almost invariably on flow control. Many units also have a feed surge drum, particularly those that are sensitive to flow fluctuations or where complete flow failure can cause equipment damage. This is an example of 'downstream flow surge'. The surge volume of the drum will depend on
  - Source and reliability of source
  - Type of control at source
  - Variations and fluctuations in source rate

(2) *Column feed from an upstream column* This feed stream will usually be controlled by the level in the source column, hence it will be fluctuating. If surge volume is provided only in the source column it must be sufficient to cater for 'upstream protection' and 'downstream protection'. The use of a surge vessel would be recommended for this case.

(3) *Feed to fired heaters or boilers* The failure of flow through the tubes of fired heaters or steam boilers can cause serious damage through overheating of the tubes. Consequently 'Downstream protection' is required in this case. Invariably flows to heaters are on flow control.

Reflux drums

(1) When the drum provides only reflux or reflux and product to storage all that is needed in terms of surge volume is sufficient to provide 'upstream protection'. That is, the surge volume required is only to protect the reflux pump from losing suction in the case of column feed failure. The pump will be required to circulate reflux and cool down the column during an orderly shutdown period.

(2) When the reflux drum supplies reflux and the feed stream to another unit then the drum must provide 'upstream protection' surge and 'downstream protection' surge.

(3) If there is a vapour product from the drum additional volume must be provided in the drum to allow vapour/liquid disengaging. This will be such as to retain the same liquid surge capacity as described above.

(4) If the vapour phase from the reflux drum is routed to the suction of a compressor an even large-volume reflux drum will be required. This is to ensure complete disengaging of the vapour/liquid. Internal baffles or screens are also used in the drum's vapour outlet section to ensure complete phase separation.

QUANTITY OF SURGE VOLUME

The amount of surge volume will vary with the types given above and with the specific case in question. Sometimes this amount is set by company specifications or, in the case of engineering contractors, by the client. Generally, however, the process engineer will be responsible for setting a safe and economic surge volume. In doing this the engineer needs to analyse each case in terms of why the surge volume is being provided then decide how much, based on this answer. Figure 2.1 provides some guidelines to the amount of surge that should be applied.

Some useful equations used in setting and handling surge volumes

For surge volume: vol ft^3 = (gpm) (minutes)/7.48

For vessel size:

$$\text{Dia. } D = \sqrt[3]{(\text{ft}^3/2.35)} \quad \text{at } L/D = 3$$
$$\text{Dia. } D = \sqrt[3]{(\text{ft}^3/1.96)} \quad \text{at } L/D = 2.5$$
$$\text{Dia. } D = \sqrt[3]{(\text{ft}^3/1.57)} \quad \text{at } L/D = 2.0$$

For line size:

$$\text{Dia. } D = \sqrt[3]{(\text{gpm}/25)} \quad \text{at velocity} = 10\,\text{ft/s}$$
$$\text{Dia. } D = \sqrt[3]{(\text{gpm}/17)} \quad \text{at velocity} = 7\,\text{ft/s}$$
$$\text{Dia. } D = \sqrt[3]{(\text{gpm}/12)} \quad \text{at velocity} = 5\,\text{ft/s}$$

For flow rate:

$$\text{ft/s} = \text{gpm}/450$$

For approx. control valve size:
  One size smaller than line size. Thus:

Line (inches)	CV (inches)
4	3
6	4
8	6
10	8

LEVEL CONTROL

Having decided why surge volume is being provided and how much it should be, the process engineer must now decide how best to control the surge volume. There are a number of choices:

(1) *Control the surge liquid outlet on level control* This will give a close level control but a fluctuating outlet flow. The level control valve (LCV) will close completely at LLL (low liquid level). Thus flow through the outlet line will be completely shut off at LLL.

(2) *Control the surge liquid outlet on flow control and provide a low-level alarm* This will give a fluctuating level but will eliminate flow fluctuations. As there is no LCV to restrict flow at LLL then operators must physically reset the flow controller to maintain the surge volume.

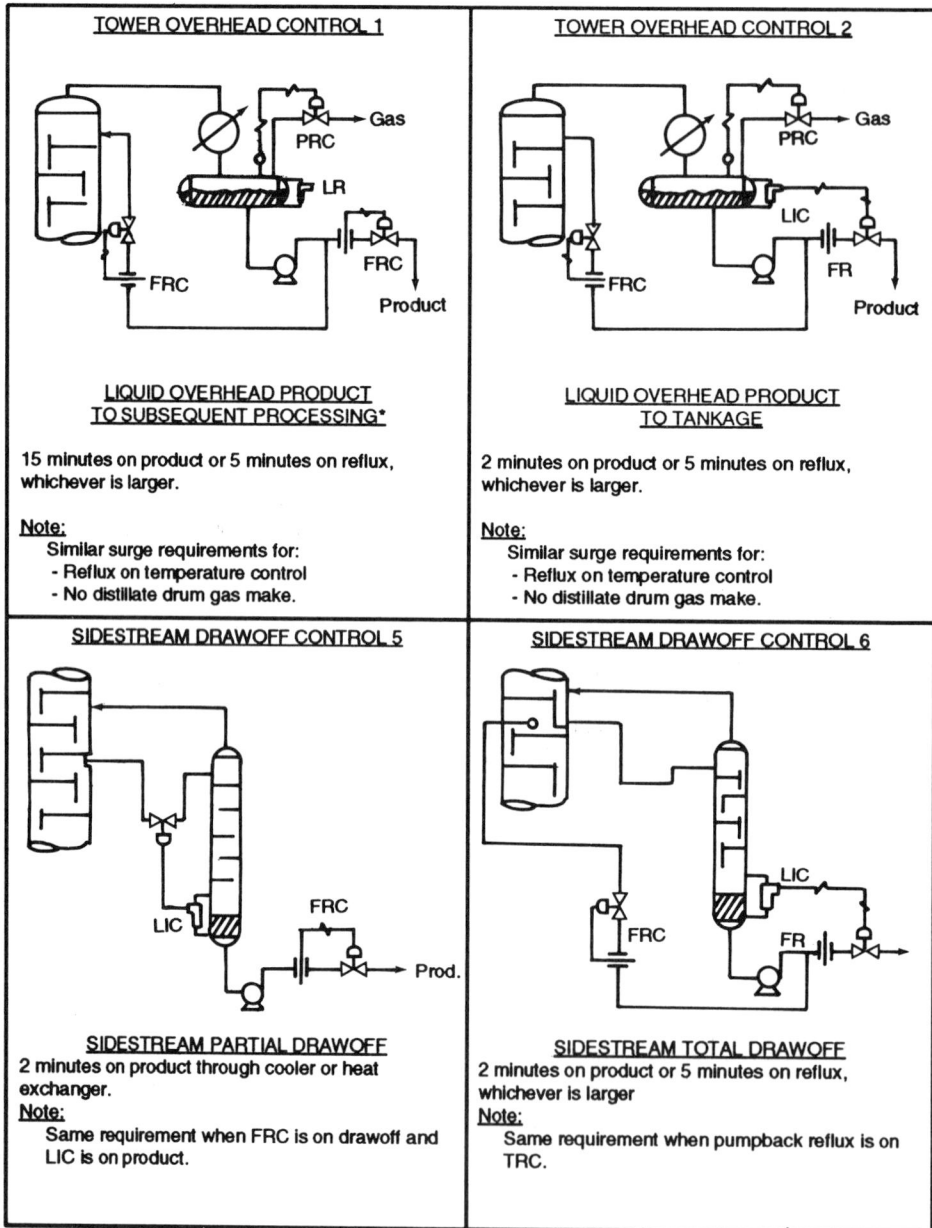

Figure 2.1. Process surge requirements

TOWER OVERHEAD CONTROL 3

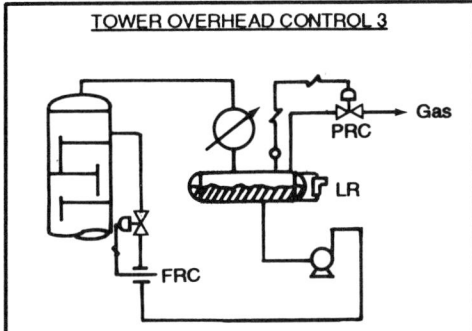

REFLUX CONDENSED - GAS PRODUCT

5 minutes on reflux.

Note:
  Similar surge requirements for:
  - Reflux on temperature control
  - Level control on cooling water.

TOWER OVERHEAD CONTROL 4

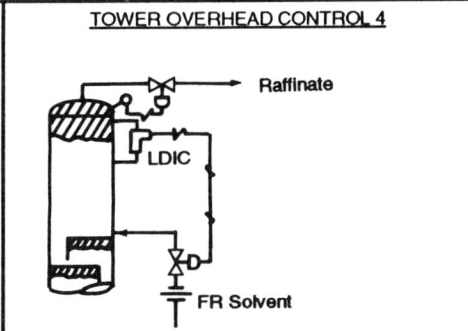

LIQUID - LIQUID EXTRACTION TOWER SOLVENT PHASE CONTINUOUS, RAFFINATE ACCUMILATES IN TOWER TOP

Caustic towers - set by 14" displacer.
DEA towers - 5 minutes on DEA.
Phenol treaters - 10 minutes on phenol.

Note:
  Feed and spent solvent streams are on flow control.

SIDESTREAM DRAWOFF CONTROL 7

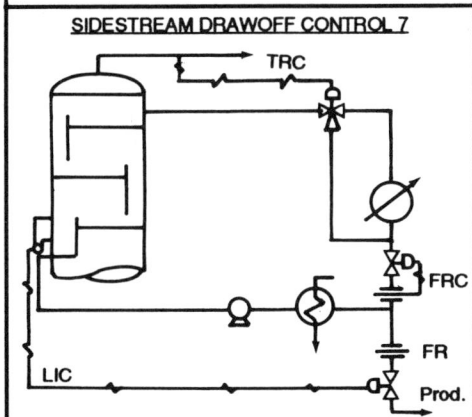

PUMPAROUND PRODUCT CIRCUIT

5 seconds or more on product.
Note:
  When product goes to subsequent processing*
  and when holdup feeding above tower is less
  than 15 minutes then pan must be installed with
  15 minutes hold-up on product or 5 minutes on
  pumparound whichever is larger

* Where constant inflow rate is required.

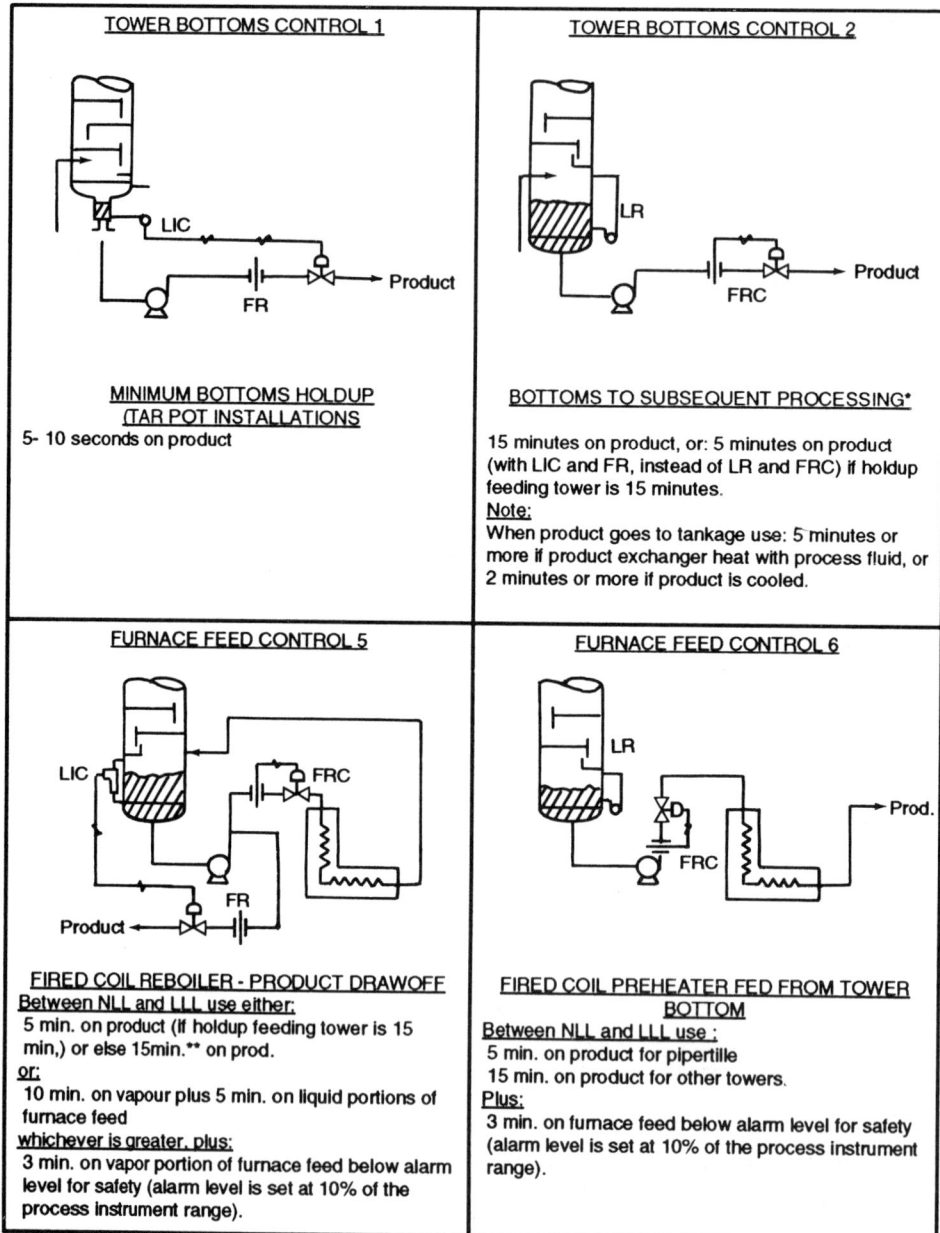

TOWER BOTTOMS CONTROL 1

LIC
FR
Product

MINIMUM BOTTOMS HOLDUP
(TAR POT INSTALLATIONS
5- 10 seconds on product

TOWER BOTTOMS CONTROL 2

LR
FRC
Product

BOTTOMS TO SUBSEQUENT PROCESSING*

15 minutes on product, or: 5 minutes on product
(with LIC and FR, instead of LR and FRC) if holdup
feeding tower is 15 minutes.
Note:
When product goes to tankage use: 5 minutes or
more if product exchanger heat with process fluid, or
2 minutes or more if product is cooled.

FURNACE FEED CONTROL 5

LIC
FRC
FR
Product

FIRED COIL REBOILER - PRODUCT DRAWOFF
Between NLL and LLL use either:
 5 min. on product (if holdup feeding tower is 15
 min,) or else 15min.** on prod.
or:
10 min. on vapour plus 5 min. on liquid portions of
furnace feed
whichever is greater, plus:
3 min. on vapor portion of furnace feed below alarm
level for safety (alarm level is set at 10% of the
process instrument range).

FURNACE FEED CONTROL 6

LR
FRC
Prod.

FIRED COIL PREHEATER FED FROM TOWER
BOTTOM
Between NLL and LLL use :
 5 min. on product for pipertille
15 min. on product for other towers.
Plus:
3 min. on furnace feed below alarm level for safety
(alarm level is set at 10% of the process instrument
range).

Figure 2.1. (*continued*)

TOWER BOTTOMS CONTROL 3

THERMO -SYPHON REBOILER
BOTTOMS TO SUBSEQUENT PROCESSING*
15 minutes on product, or: 5 minutes on product
(with LIC and FR, instead of LR and FRC) if holdup
feeding tower is 15 minutes.
Note:
When product goes to tankage use: 5 minutes or
more if product exchanger heat with process fluid, or
2 minutes or more if product is cooled.

TOWER BOTTOMS CONTROL 4

KETTLE - TYPE REBOILER
BOTTOMS TO TANKAGE OR SUBSEQUENT
PROCESSING
5-10 seconds on product
Note:
Separate surge capacity with 15 minutes holdup is
required if product is fed to subsequent processing*
and when holdup feeding tower is less than 15 minutes.

FURNACE FEED CONTROL 7

FIRED COIL PREHEATER - PRODUCT DRAWOFF
Between NLL and LLL use either:
5 min. on product (if holdupfeeding tower is 15 min.)
or else 15 minutes on product.
or:
5 minutes on furnace feed
whichever is greater, plus:
3 min. on furnace feed below alarm level for safety
(alarm level is set at 10% of the process instrument
range).

* Where constant inflow rate is required.
** When 15 minutes on product is used provide
FRC and LR instead of LIC and FR.
HLL - High Liquid Level
LLL - Low Liquid Level

(3) *Control the surge liquid outlet on flow control reset by a level control*   This will give a fluctuating level but a smooth outlet flow. The LCV reset can still, however, cut off the flow completely on LLL.

(4) *Control the outlet flow by flow control and the system feed by surge volume level*   This will give close level control and also close outlet flow control. The outlet flow also will not be closed off by a low liquid level of the surge. In the case of the feed being that to a fractionating tower level, control on the feed stream could cause tower upset conditions. This would be particularly undesirable on fractionators that operate close to critical conditions such as de-ethanizers.

(5) *Control the surge liquid on level control to an intermediate surge vessel*   The liquid from the surge vessel may then be flow controlled.

LEVEL CONTROL RANGE

If the decision is now to use an LC on the surge outlet (as in (1) or (5) above) the range of the instrument must be decided. The range is the vertical distance between the high liquid level (HLL) and the low liquid level (LLL). Now if the liquid outlet is feeding another unit which requires a smooth flow it is possible to achieve this by using a wide proportional band and a large range. The larger the instrument range and the wider the proportional band, the less sensitive is the level controller. Consequently the flow becomes smoother. However, the larger the range, the more expensive is the controller and, of course, the larger is the tower or vessel in order to accommodate the greater distance between HLL and LLL.

The selection of level control system and the level control range depends therefore on:

(1) The number of outlet streams there are from the surge vessel
(2) Which streams cannot tolerate complete shut off before all the available surge is used
(3) The degree to which the outlet stream requires a smooth flow

PRESSURE CONTROL FOR GASES

Pressure control for gases is similar to level control for liquids in that it is really a material balance control. If gas production rate varies, which is the usual case, a pressure controller will relieve the system of gas by holding a given pressure in a drum or tower. In this case the entire space above the liquid level actually constitutes a surge volume.

To summarize, there are two very general rules to follow in selecting a control system:

• If it is permissible for the product outflow rate to vary, use level control and a relatively small amount of surge capacity.
• If the product goes to a subsequent process where feed rate must be held constant, use flow control and considerably more surge.

Example calculations

The following examples are used to illustrate surge requirements and surge control.

Example 1

The bottom product from a fractionating tower is to feed a fired heater. Fluctuation in the flow of feed to the fractionator is not critical to the tower operation in this case. Select the type of control system and calculate the surge requirement. Tower diameter is 12 ft and the outlet flow is 1000 gpm.

Solution This problem is a 'downstream flow surge'. The bottom stream must be on a flow control to protect the heater coil. Consider the following three possibilities overleaf:

From Figure 2.1 the system requires 15 min surge on output flow. Then surge volume = $(1000 \times 15)/7.48 = 2005 \text{ ft}^3$

$$\text{Required NLL} = 2005/(0.786).144$$
$$= 17.7 \text{ ft (say 17 ft 6 in.)}$$

- *System B* If feed to the tower is lost, the flow controller will be able to use all of the 17 ft 6 in. surge volume. Set the LL alarm at about 5–6 ft. This gives the operator 5 min to take corrective action and reset the flow controller. To make the system quite satisfactory add a level recorder and a level gauge to cover at least the range from NLL to LLL (about 13–14 ft).
- *System D* As the tower is not sensitive to flow fluctuations the feed can be controlled by level. A 14 in. LC range can be considered in this case. However, better tower operation can be achieved with a larger range, say a 32 in. LC range. Don't forget this system is only acceptable if the column can stand flow fluctuations in satisfactory operation. There are not many cases of this in process fractionation.
- *System C* To use a level controller reset sufficient range must be provided for the level controller to make full use of the 15 min surge volume *without* shutting off the feed to the heater. This means a range of 35 ft which is twice the previous 17 ft 6 in. between NLL and LLL. Also a space of at least 17 ft 6 in. must be left above the NLL to prevent flooding the bottom fractionator trays. This is much too expensive both in instrument cost and in additional tower cost.

The solution in this case therefore is to use system B.

Example 2

Consider the control system for a sidestream stripper on a fractionating column. Here feed is taken off a tray in the main tower to be stripped of light ends in the stripper. Fractionation is not normally an issue here and the product in this case is routed to storage. Surge volume in the stripper is required only to protect the

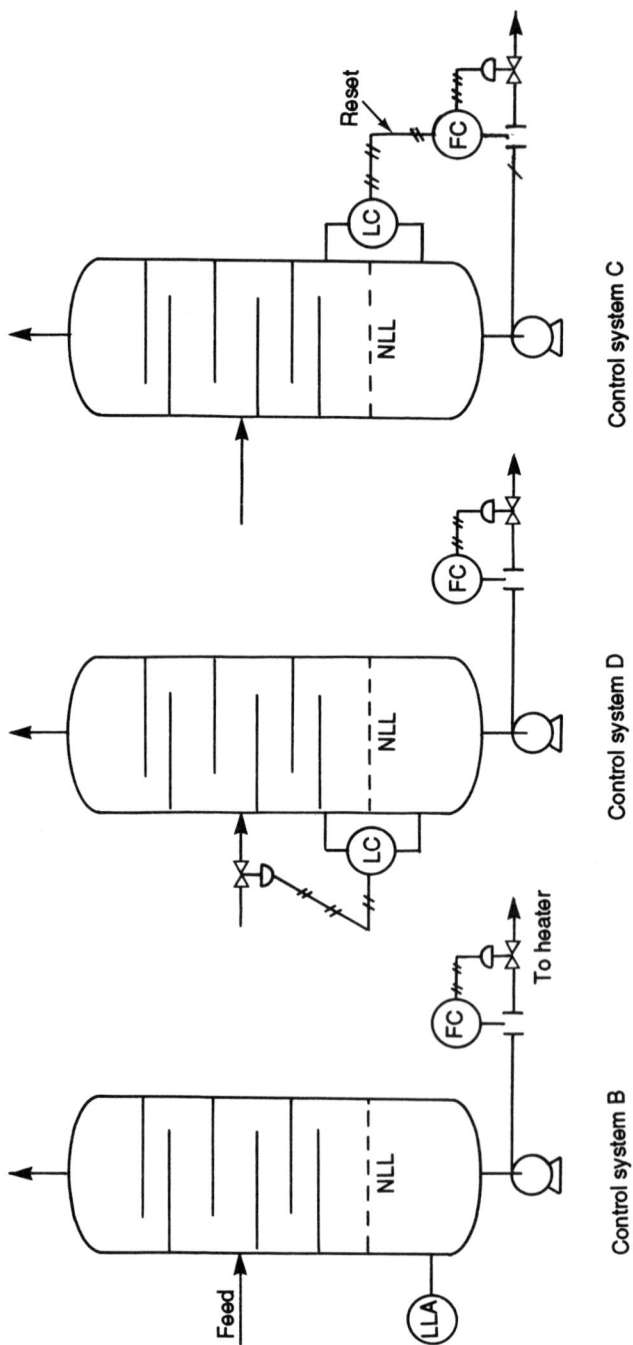

product pump. Two systems are considered and these are given in the diagrams below.

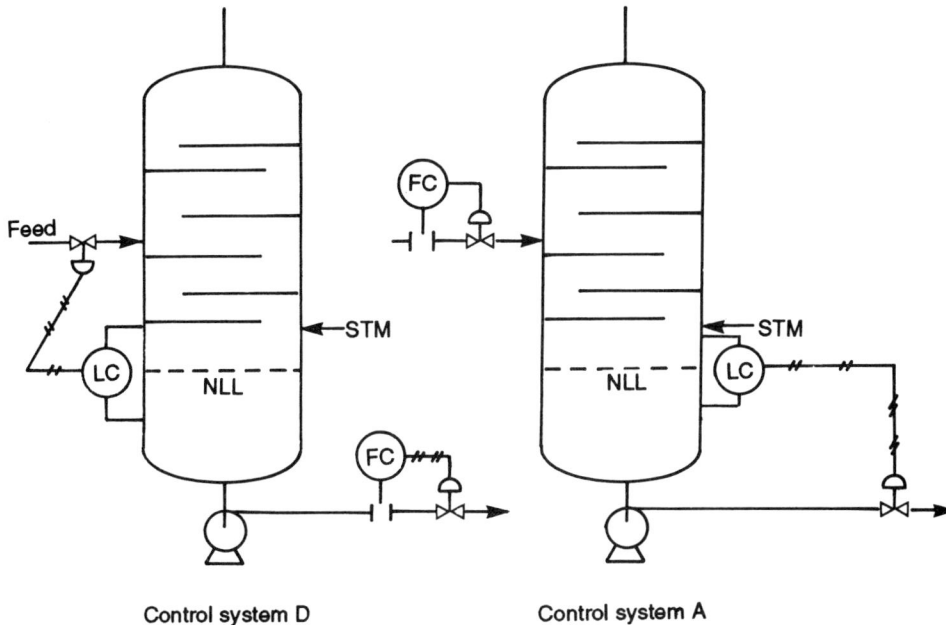

Control system D   Control system A

The stripper tower has a diameter of 36-in. and the flow is 150 gpm. From Figure 2.1 the hold-up time required for the product is 3 min. Then

$$\text{Surge volume} = 3 \times 150/7.48 = 60\,\text{ft}^3$$
$$\text{Required NLL} = 60/0.786 \times 9 = 8\,\text{ft}\,6\,\text{in.}$$

- *System D*   A 14-in. LC range can be used as fractionation performance is not an issue. However, if there is some concern regarding the main tower draw-off this range could be increased to 32-in. without a great deal of extra cost. This system is perfectly satisfactory and allows sufficient latitude for the operator to switch pumps or shut the pump down in case of 'no flow'.
- *System A*   This is not a practical system for this operation. To maintain a 3-min surge it will be required to use a 17 ft LC range to completely use this surge. This is far too costly in instrumentation and tower dimensions.

Example 3

This example examines a fairly widely used system whereby the outflow from a fractionator reflux drum splits. One stream is returned to the tower while the other is fed to another fractionator. The type of surge in this case is 'upstream protection' coupled with a 'downstream flow surge' for smooth flow. There are three systems that can be considered to meet this requirement. These are as follows:

Control system A

Control system B

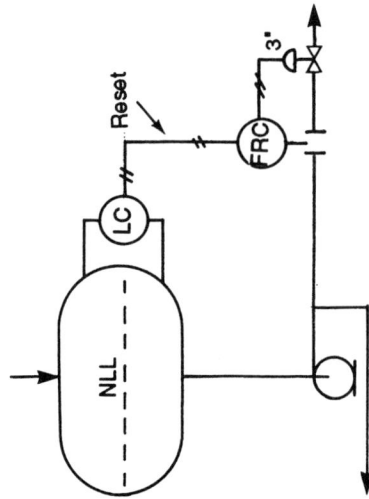

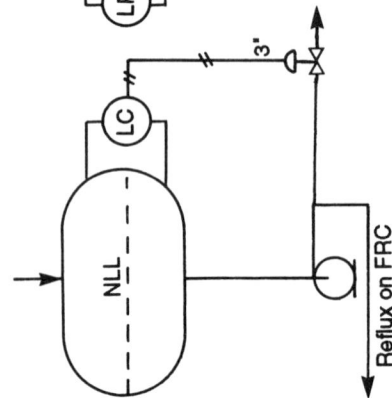

Control system C

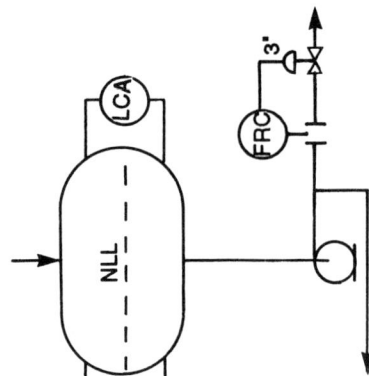

The reflux in all three cases = 500 gpm. and the product = 350 gpm. From Figure 2.1 the surge time will be 5 min on reflux rate = 2500 gpm. The drum dimensions are 7.0 ft i/d × 17 ft 6 in. T-T. volume = 680 ft^3. Let NLL = 3 ft 6 in. = 42 in.

$$\text{Outflow rate} = 850/450 = 1.9 \text{ ft}^3/\text{s}$$

- *System A* This system is satisfactory if

  (1) The next column can tolerate flow fluctuations.
  (2) The LC range covers the entire diameter of the vessel to prevent flow cut-off. Therefore the LC range in this case is 7.0 ft.

  Since the cost of a 7 ft range LC in this case is only that of an instrument and does not involve additional vessel cost it will be acceptable.

- *System B* In this system two flow controllers are used and level control is effected by visual and alarm only. While this system provides smooth flow to both reflux and product streams it does put a burden on the operator in maintaining a good stable operation. This would be particularly disadvantageous during plant start-up and shutdown. Many companies would avoid this system.

- *System C* This is a more expensive version of system A. Here the flow to the next unit is smoothed out by using a flow control reset by a level control. Fluctuation in level therefore is taken up by periodic adjustment to the flow control setting. This system is highly favoured particularly when the next unit is a light hydrocarbon fractionator.

Example 4

This example covers the control system around the bottom of a fractionator which has a fired heater as reboiler. Whichever system is considered, one criterion is firmly fixed, and that is the feed to the reboiler must be on flow control. Consider the following diagram overleaf:

The tower diameter is 6 ft (72 in.) and the flows are:

$$\text{Product} = 570 \text{ gpm}$$
$$\text{Reboiler feed} = 800 \text{ gpm}$$
$$\% \text{ vaporization} = 30$$

From Figure 2.1. surge volume is based as follows:

From HLL to LLL = 5 min (assume tower feed has 15 min hold-up on product)

$$= \frac{5 \times 570}{7.48}$$
$$= 381 \text{ ft}^3$$
$$= \frac{381}{(0.786 \times 36)}$$
$$= 13 \text{ ft } 6 \text{ in.}$$

Instrument range therefore is 13.5 ft. LLA will be set 1.3 ft below LLL. In

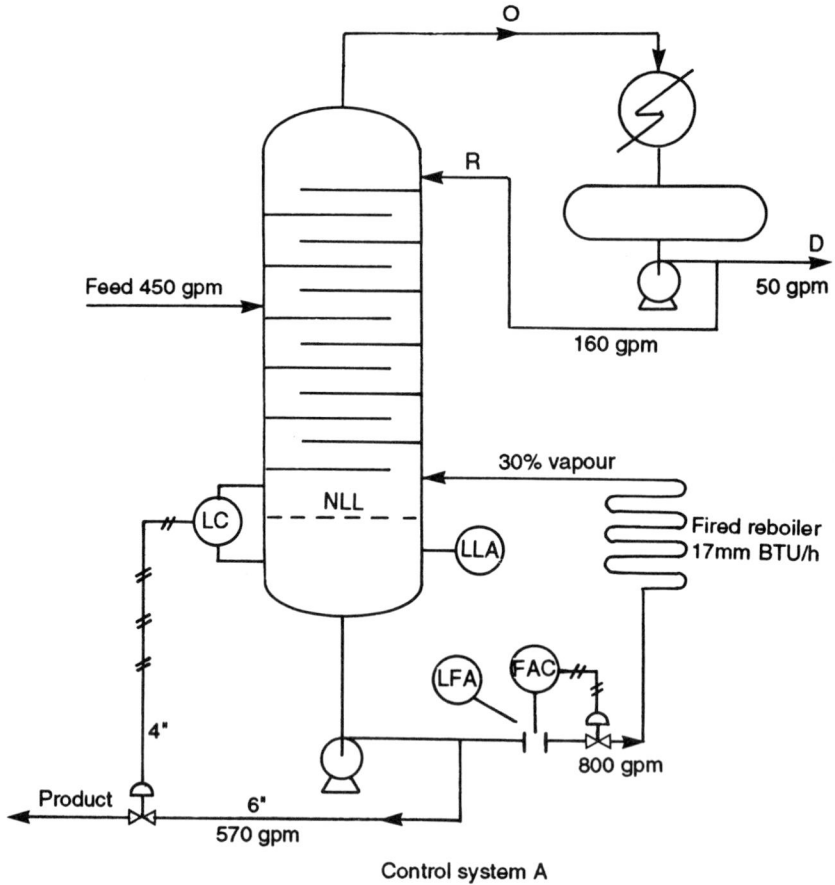

Control system A

addition, a volume equivalent to 3 min of the vaporized volume of the feed will be allowed below LLA. Thus:

$$\text{Volume of feed vaporized} = 0.3 \times 800 = 240 \text{ gpm}$$

$$\text{Volume below LLA} = \frac{240 \times 3}{7.48} = 96 \text{ ft}^3$$

$$= \frac{96}{(0.786 \times 36)} = 3.4 \text{ ft}$$

Then bottom of the tower levels are:

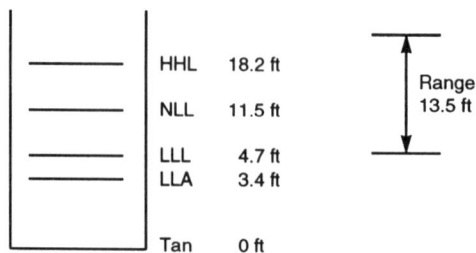

This system is entirely satisfactory particularly if the product is being routed to storage. Should the product be feed to another unit then it could be set on a flow control with a level control reset (control system C) to establish a smooth flow. Many companies relax the requirements for the surge range between HLL and LLL. This does, of course, reduce the cost of tower and the instruments considerably, but the risk of flow failure to the product stream increases.

2.2 Safety Systems

A major activity in the process design of a plant or system is to ensure its safe operation. Much of the effort in this respect is directed to determining the pressure limits of equipment and to protect that equipment from dangerous overpressuring. Pressure relief valves are normally used for this protection service, although under certain conditions bursting disks may be used. This section covers the various types of relief valves, the procedure for calculating the correct orifice size of the valve and valve selection.

DETERMINATION OF RISK

The cost of providing facilities to relieve all possible emergencies simultaneously is prohibitive. Every emergency arises from a specific cause and the simultaneous occurrence of two or more emergencies or contingencies is unlikely. Hence, an emergency which can arise from two or more contingencies (e.g. simultaneous failure of a control valve and cooling water) is not considered when sizing safety equipment. Likewise, simultaneous but separate emergencies are not considered.

Every unit or piece of equipment must be studied individually and every contingency must be evaluated. The safety equipment for an individual unit is sized to handle the largest load resulting from any possible single contingency. If a certain emergency involves more than one unit, then all must be considered as an entity. The equipment judged to be involved in any one emergency is termed 'single risk'. The single risk which results in the largest load on the safety facilities in any system is termed 'largest single risk' and forms the design basis for the equipment. (Note: the emergency which results in the largest single risk on the overall basis may be different from the emergency(ies) which form the basis for individual pieces of equipment.)

Contingencies generally fall into one of two categories—fire (external) or operating failure. Operating failure covers such contingencies as utility failure, mechanical failure, or maloperation.

DEFINITIONS

The terms used and the descriptions given in this section are based on data given in two API publications, API RP520 and 521. References are also made to Part 1 ANSI Proposed Standard, Terminology for Pressure Relief Devices and to ASME PTC 25.2. These publications are the safety standards commonly in current use.

The following definitions of terms used in the design of safety systems help us to understand the design and criteria of safety systems:

- *Accumulation* the pressure increase over the maximum allowable working pressure of the vessel during discharge through the pressure relief valve expressed as a percentage of that pressure or in psi.
- *Atmospheric discharge* the release to the atmosphere of vapours and gases from pressure relief and depressurizing devices.
- *Back pressure* the pressure existing at the outlet of the pressure relief device due to pressure in the discharge system.
- *Balanced safety relief valves* incorporate means for minimizing the effect of back pressure on the performance characteristics—opening pressure, closing pressure, lift and relieving capacity.
- *Blowdown* the difference between the set pressure and the resealing pressure of a pressure relief valve, expressed as a percentage of the set pressure or in psi.
- *Burst pressure* the value of inlet static pressure at which a rupture disk device functions.
- *Conventional safety relief valve* a closed bonnet pressure relief valve that has the bonnet vented to the discharge side of the valve. The performance characteristics—opening pressure, closing pressure, lift and relieving capacity—are directly affected by changes of the back pressure on the valve.
- *Design pressure* the pressure used in the design of a vessel to determine the minimum permissible thickness or physical characteristics of the different parts of a vessel.
- *Flare* a means for safe disposal of waste gases by combustion. With an *elevated flare* the combustion is carried out at the top of a pipe or stack where the burner and igniter are located. A *ground flare* is similarly equipped except that combustion is carried out at or near ground level. A *burn pit* differs from a flare in that it is normally designed to handle both liquids and vapours. Flare systems are described in more detail in section 2.4.
- *Lift* the actual travel of the disk away from closed position when the valve is relieving.
- *Overpressure* the pressure increase over the set pressure of the primary relieving device; it would be termed *accumulation* when the relieving device is set at the maximum allowable working pressure of the vessel. (Note: When the set pressure of the first (primary) pressure relief valve to open is less than the maximum allowable working pressure of the vessel, the overpressure may be greater than 10% of the set pressure of the valve.)
- *Pilot-operated pressure relief valve* a valve that has the major flow device combined with and controlled by a self-actuated auxiliary pressure relief valve. This type of valve does not utilize an external source of energy.
- *Pressure relief valve* a generic term applied to relief valves, safety valves, or safety relief valves.
- *Relieving conditions* pertain to pressure relief device inlet pressure and temperature at a specific overpressure. The relieving pressure is equal to the valve set pressure (or rupture disk burst pressure) plus the overpressure. The

temperature of the flowing fluid at relieving conditions may be higher or lower than the operating temperature.

- *Set pressure* in psig is the inlet pressure at which the pressure relief valve is adjusted to open under service conditions. In a safety or safety relief valve in gas, vapour, or steam service the set pressure is the inlet pressure at which the valve pops under service conditions. In a relief or safety relief valve in liquid service the set pressure is the inlet pressure at which the value starts to discharge under service conditions.
- *Superimposed back pressure* is static pressure existing at the outlet of a pressure relief device at the time the device is required to operate. It is the result of pressure in the discharge system from other sources. This type of back pressure may be constant or variable; it may govern whether a conventional or balanced-type pressure relief valve should be used in specific applications.
- *Vapour depressing system* a protective arrangement of valves and piping intended to provide for rapid reduction of pressure in equipment by release of vapours. Actuation of the system may be automatic or manual.
- *Vent stack* the elevated vertical termination of a disposal system which discharges vapours into the atmosphere without combustion or conversion of the relieved fluid.

TYPES OF PRESSURE RELIEF VALVES

The following is a list of those types of relief valves commonly used in industry. These have been approved according to the ASME VIII, Boiler and Pressure Vessel code:

- *Conventional safety relief valves* (see Figure 2.2) In a conventional safety relief valve the inlet pressure to the valve is directly opposed by a spring closing the valve, the back pressure on the outlet of the valve changes the inlet pressure at which the valve will open.
- *Balanced safety relief valves* (see Figure 2.3) Balanced safety valves are those in which the back pressure has very little or no influence on the set pressure. The most widely used means of balancing a safety relief valve is through the use of a bellows. In the balanced bellows valve, the effective area of the bellows is the same as the nozzle seat area and back pressure is prevented from acting on the top side of the disk. Thus the valve opens at the same inlet pressure even though the back pressure may vary.
- *Pilot-operated safety relief valves* A pilot-operated safety relief valve is a device consisting of two principal parts, a main valve and a pilot. Inlet pressure is directed to the top of the main valve piston, and with more area exposed to pressure on the top of the piston than on the bottom; pressure, not a spring, holds the main valve closed. At the set pressure the pilot opens reducing the pressure on top of the piston and the main valve goes fully open.
- *Resilient seated safety relief valves (see Figure 2.4)* When metal-to-metal seated conventional or bellows type safety relief valves are used where the operating pressure is close to the set pressure, some leakage can be expected

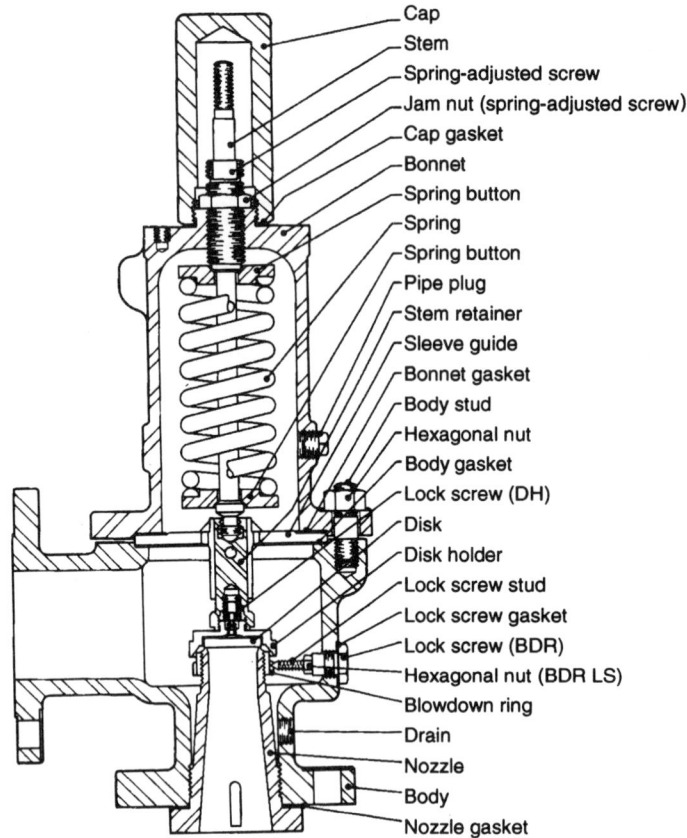

Figure 2.2. A typical conventional safety relief valve

through the seats of the valve (refer to API Standard 527, Commercial Seat Tightness of Safety Relief Valves with Metal-to-Metal Seats). Resilient seated safety relief valves with either an O-ring seat seal or a plastic seat such as Teflon provide seat tightness. Limitations of temperature and chemical compatibility of the resilient material must be considered when using these valves.

- *Rupture disk* A rupture disk consists of a thin metal diaphragm held between flanges. The disk is designed to rupture and relieve pressure within tolerances established by the ASME code.

CAPACITY

The maximum amount of material to be released during the largest single-risk emergency determines the size of the safety relief valves in any given system. Any calculation to determine valve sizing must therefore be preceded by a calculation or some determination of the maximum amount. Among the most common sizing criteria is the event of fire and its effect on the contents of exposed vessels. There

Cap
Stem
Spring-adjusted screw
Jam nut (spring-adjustment screw)
Cap gasket
Bonnet
Spring button
Spring
Spring button
Stem retainer
Vent
Sleeve guide
Lock screw (DH)
Bonnet gasket
Body stud
Hexagonal nut
Body gasket
Bellows
Bellows gasket
Disk
Disk holder
Lock screw stud
Lock screw (BDR)
Hexagonal nut (BDR LS)
Lock screw gasket
Blowdown ring
Drain
Nozzle
Body
Nozzle gasket

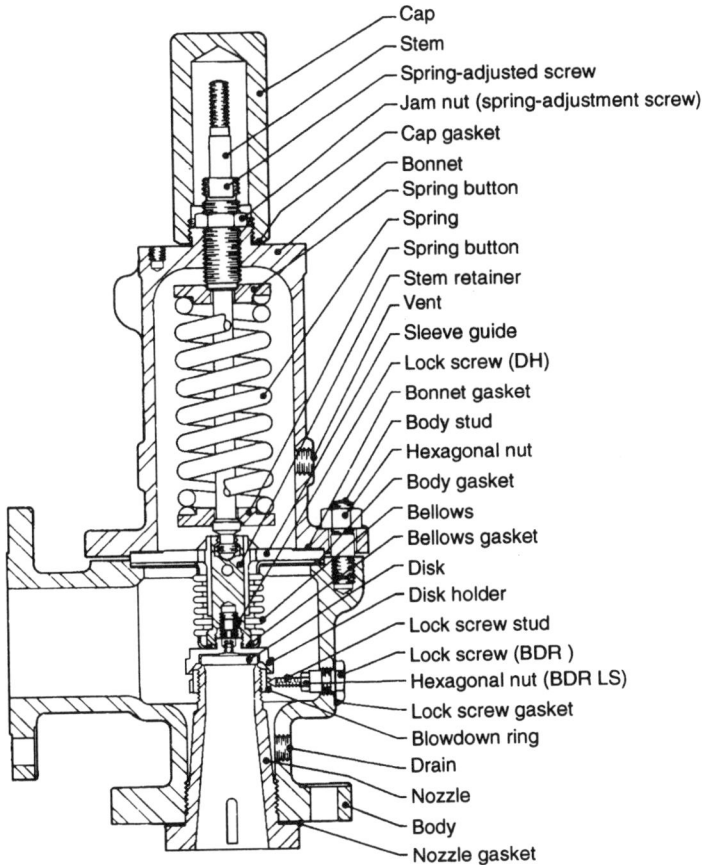

Figure 2.3. A typical balanced bellows safety relief valve

are also other criteria which can determine maximum release that are attributable to operational failure.

Capacity due to fire

The exact method of making this calculation must be established from the appropriate codes which apply, API RP-520, Part I, API Standard 2510, NFPA No. 58 or local codes. Each of the listed codes or standards approach the problem in a slightly different manner.

A majority of the systems that are encountered will contain liquids or liquids in equilibrium with vapour. Fire relief capacity in this situation is calculated on the basis of heat energy from the fire translated in terms of vapour generated in the boiling liquid. API RP-520, Part I, applies to refineries and process plants. It expresses requirements in terms of heat input.

$$Q_a = 21\,000 FA^{0.82}$$

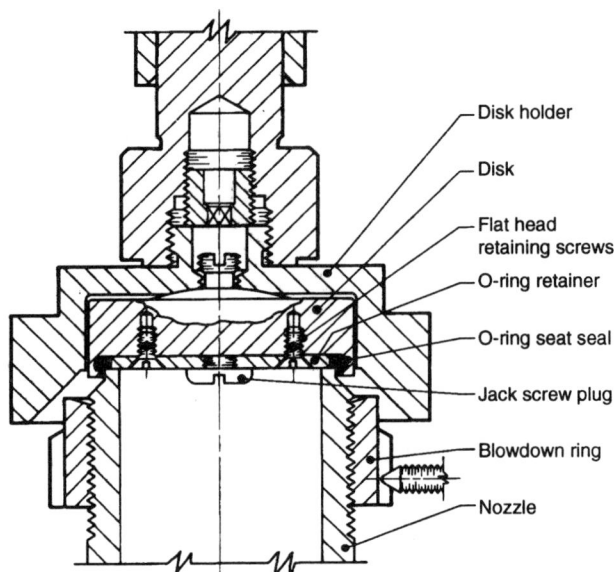

Figure 2.4. An 'O'-ring seal safety relief valve

where

Q = BTU/h heat input
A = area in square feet of wetted surface of the vessel up to 25 ft above grade.
    Wetted surface is calculated at the maximum fill level. Grade is the
    ground level under the vessel
F = fireproof factor due to insulation becoming 1.0 for a bare vessel (see
    Figure 2.5).

The amount of vapour generated with this is then calculated from the latent
heat of the material at the relieving pressure of the valve. For fire relief only this
may be calculated at 120% of maximum allowable working pressure. All other
conditions must be calculated at 110% of maximum allowable working pressure:

$$Q/H_L = W$$

where

Q = BTU/h heat input to the vessel
H_L = BTU/lb latent heat of the material being relieved
W = lb/h of vapour to be relieved by the relief valve

The latent heat of pure and some mixed paraffin hydrocarbons materials may
be estimated using the data given in API RP-520. A more accurate latent heat
evaluation for mixed hydrocarbons will be found by utilizing vapour–liquid
equilibrium K data and making a flash calculation. Mixed hydrocarbons will
fractionate, beginning with the lowest-temperature boiling mixture and progress
to the highest-temperature mixture. Therefore, consideration must be given to the

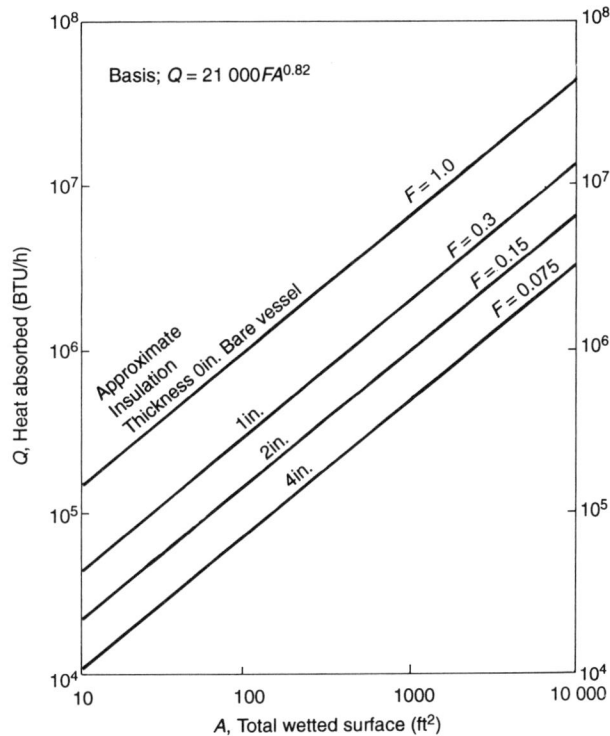

Figure 2.5. Insulation factors

condition which will cause the largest vapour generation requirements from the heat input of a fire.

Latent heat will approach a minimum value near critical conditions. However, the effect does not go to zero. An arbitrary minimum value that may be used is 50 BTU/lb.

SIZING OF REQUIRED ORIFICE AREAS

The safety relief valve manufacturers have standard orifice designation for area and the valve body sizes which contain these orifices (API Standard 526, Flanged Steel Safety Relief Valves for Use in Petroleum Refineries). The standard orifices available, by letter designation and area, are (in square inches):

D	orifice	0.110
E	orifice	0.196
F	orifice	0.307
G	orifice	0.503
H	orifice	0.785
J	orifice	1.287
K	orifice	1.838

L orifice   2.853
M orifice   3.60
N orifice   4.34
P orifice   6.38
Q orifice   11.05
R orifice   16.0
T orifice   26.0

Note, however, that many small safety relief valves are manufactured with orifice areas smaller than 'D', and many pilot-operated types contain orifice areas larger than 'T'.

Sizing for gas or vapour relief, sonic or critical flow

Safety relief valves in gas or vapour service may be sized by use of one of the following equations:

$$A = \frac{W\sqrt{T}\sqrt{Z}}{CKP_1 K_b \sqrt{M}}$$

$$A = \frac{V\sqrt{T}\sqrt{M}\sqrt{Z}}{6.32\ CKP_1 K_b}$$

$$A = \frac{V\sqrt{T}\sqrt{G}\sqrt{Z}}{1.175 CKP_1 K_b}$$

where

W = flow through valve (lb/h)
V = flow through valve (scfm)
C = coefficient determined by the ratio of the specific heats of the gas or vapour at standard conditions. This can be obtained from Table A1.3 in Appendix 1
K = coefficient of discharge, obtainable from the valve manufacturer (usually 0.6–0.7)
A = effective discharge area of the valve (in^2)
P_1 = upstream pressure (psia). This is set pressure plus overpressure plus the atmospheric pressure
K_b = capacity correction factor due to back pressure. This can be obtained from Figure 2.6 for conventional valves or pilot operated valves, and from Figure 2.7 for balanced bellows valves
M = molecular weight of gas or vapour
T = absolute temperature of the inlet vapour in °R (°F +460)
Z = compressibility factor for the deviation of the actual gas from a perfect gas.
G = specific gravity of gas referred to air = 1.00 at 60 °F and 14.7 psia

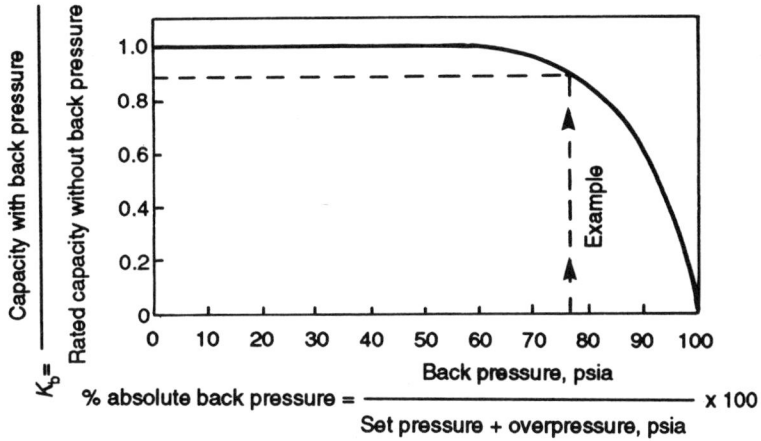

Figure 2.6. Constant back-pressure sizing factor K_b (conventional valves)

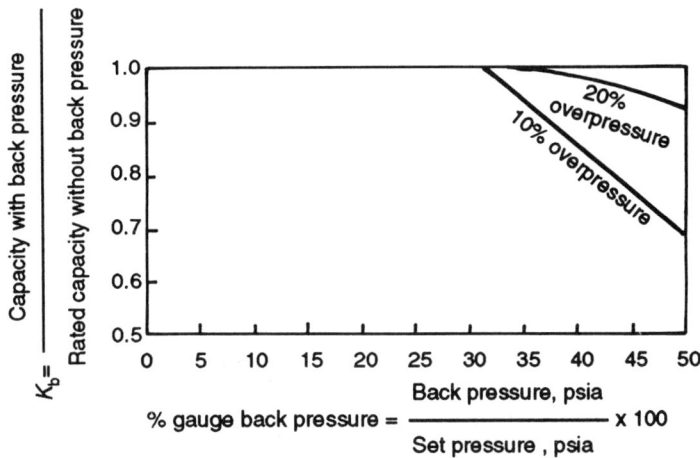

Figure 2.7. Back-pressure sizing factor K_b (balanced bellows type)

Sizing for liquid relief

Conventional and balanced bellows safety relief valves in liquid service may be sized by use of the following equation. Pilot-operated relief valves should be used in liquid service only after determining from the manufacturer that they are suitable for the service. (Note: A coefficient of discharge of 0.62 is normally used for K.)

$$A = \frac{\text{gpm}\sqrt{G}}{38\,K(K_P K_W K_V)(1.25P - P_b)^{1/2}}$$

where

gpm = flow rate at the selected percentage of overpressure, in US gpm
  A = effective discharge area (in^2)
  K_P = capacity correction factor due to overpressure. For 25% overpressure
     K_P = 1.00. The factor for other percentages of overpressure can be
     obtained from Figure 2.8
  K_W = capacity correction factor due to back pressure and is required only when
     balanced bellows valves are used. K_W can be obtained from Figure 2.9
  K_V = capacity correction factor due to viscosity. For most applications,
     viscosity may not be significant, in which case K_V = 1.00. When viscous
     liquid is being relieved see method of determining K_V as described below
  P = set pressure at which relief valve is to begin opening (psig)
  P_b = back-pressure (psig)
  G = specific gravity of the liquid at flowing temperature referred to
     water = 1.00 at 70 °F.

When a relief valve is sized for viscous liquid service, it is suggested that it be
sized first as for non-viscous-type application in order to obtain a preliminary
required discharge area, A. From manufacturers' standard orifice sizes, the next
larger orifice size should be used in determining the Reynold's number, R, from
either of these relationships:

$$R = \frac{\text{gpm}(2\,800\ G)}{\mu\sqrt{A}}$$

$$R = \frac{12\,700\ \text{gpm}}{U\sqrt{A}}$$

where

gpm = flow rate at the flowing temperature, in US gpm
  G = specific gravity of the liquid at the flowing temperature referred to water
     = 1.00 at 70 °F
  μ = absolute viscosity at the flowing temperature (in centipoise)
  A = effective discharge area, in square inches (from manufacturers' standard
     orifice areas)
  U = viscosity at the flowing temperature, in Saybolt Universal seconds

After the value of R is determined, the factor K_V is obtained from Figure 2.10
Factor K_V is applied to correct the 'preliminary required discharge area'. If the
corrected area exceeds the 'chosen standard orifice area', the above calculations
should be repeated using the next larger standard orifice size.

Sizing for mixed-phase relief

When a safety relief valve must relieve both liquid and gas or vapour it may be
sized by the following steps:

(1) Determine the volume of gas or vapour and the volume of liquid that must
     be relieved.

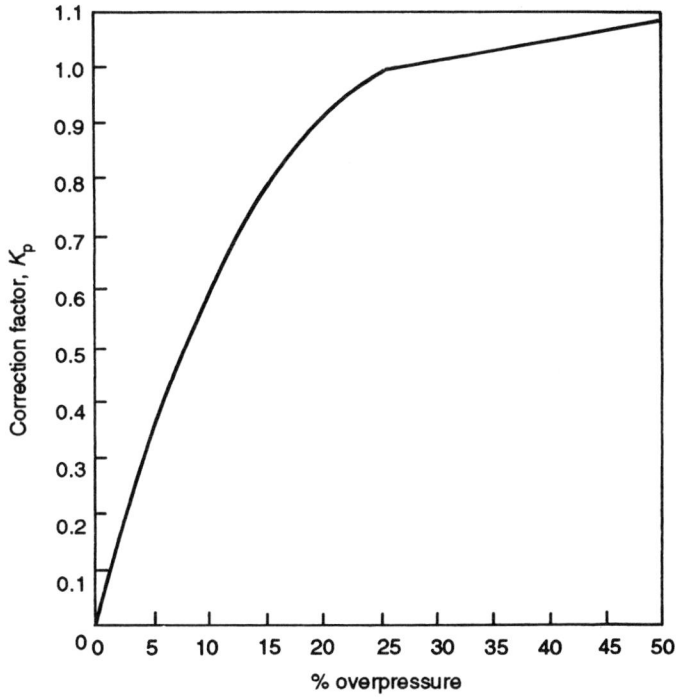

Figure 2.8. Capacity correction factors due to overpressure K_p

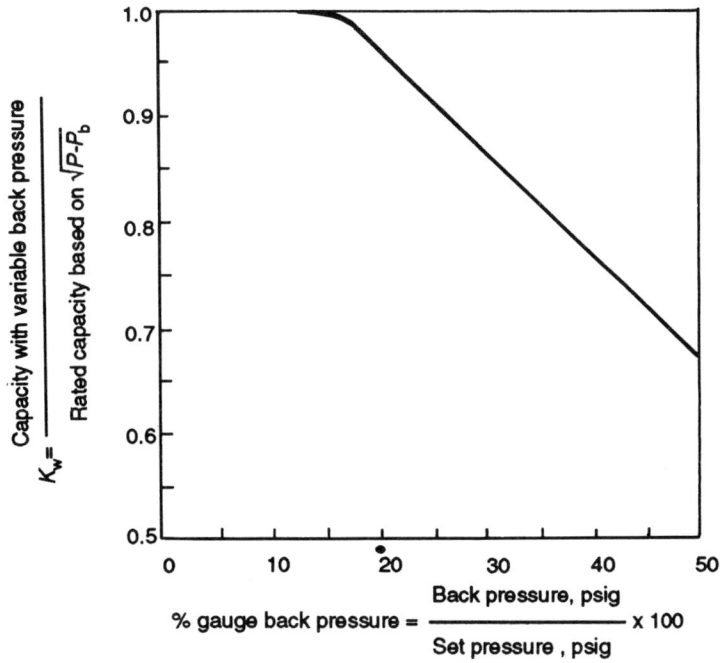

$$\% \text{ gauge back pressure} = \frac{\text{Back pressure, psig}}{\text{Set pressure, psig}} \times 100$$

Figure 2.9. Back pressure sizing factor K_w

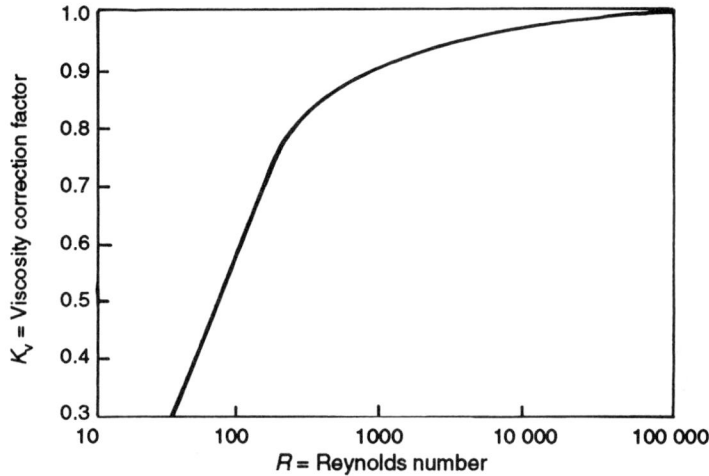

Figure 2.10. Capacity correction factor due to viscosity

(2) Calculate the orifice area required to relieve the gas or vapour as previously outlined.

(3) Calculate the orifice area required to relieve the liquid as previously outlined.

(4) The total required orifice area is the sum of the areas calculated for liquid and vapour.

Sizing for flashing liquids

The conventional method is to determine the percentage flashing from a Mollier diagram or from the enthalpy values. Then consider the liquid portion and vapour portion separately as in mixed-phase flow. A method to calculate the percent flashing is shown in the following equation:

$$\text{Percentage flashing} = \frac{h_f(1) - h_f(2)}{h_{fg}(2)} \times 100$$

where

$h_f(1)$ = enthalpy in BTU/lb of saturated liquid at upstream temperature
$h_r(2)$ = enthalpy in BTU/lb of saturated liquid at downstream pressure
$h_{fg}(2)$ = enthalpy of evaporation in BTU/lb at downstream pressure

Sizing for gas or vapour on low-pressure sub-sonic flow

When the set pressure of a safety relief valve is very low, such as near atmospheric pressure, the K_b values obtainable from Figure 2.6 are not accurate. Safety relief valve orifice areas for this low-pressure range may be calculated:

$$A = \frac{Q_v \sqrt{GT}}{863\, KF\sqrt{(P_1 - P_2)P_2}}$$

where

A = effective discharge area of the valve (in^2)
Q_V = flow through valve (scfm)
G = specific gravity of gas referred to air = 1.00 at 60 °F and 14.7 (psia)
T = absolute temperature of the inlet vapour, °R = °F + 460
K = coefficient of discharge (available from the valve manufacturer)
F = correction faction based on the ratio of specific heats. This can be obtained from Figure 2.11
P_1 = upstream pressure in psia = set pressure plus overpressure plus atmospheric pressure
P_2 = downstream pressure at the valve outlet in psia

Example calculation

A vessel containing naphtha C_5–C_8 range is uninsulated and is not fireproofed. The vessel is vertical and has a skirt 15 ft in length. Dimensions of the vessel are I/D 6 ft 0 in. T-T 20 ft liquid height to HLL = 16 ft. Calculate the valve size for fire condition relief. Set pressure is 120 psig.

Latent heat of naphtha at 200 °F is 136 BTU/lb = H_L

$Q = 21\,000\ FA^{0.82}$
A = wetted area and is calculated as follows:
liquid height above grade = 15 + 16 ft
                          = 31 ft

Therefore wetted surface of vessel need only be taken to 25 ft above grade which is 25 − 15 = 10 ft of vessel height.

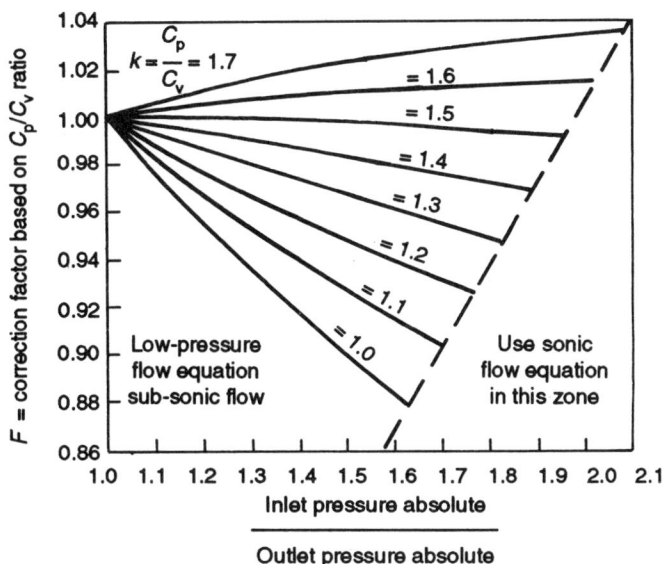

Figure 2.11. Flow correction factor F based on specific heats

$$\text{Wetted surface} = \pi D \times 10 \text{ for walls}$$
$$= 188.5 \text{ ft}^2$$
$$\text{Plus } 28.3 \text{ ft}^2 \text{ for bottom}$$
$$= 216.8 \text{ ft}^2$$
$$Q = 21\,000 \times 1.0 \times (216.8)^{0.82}$$
$$= 1.729 \text{ BTU/h} \times 10^6$$

$$Q/H_L = \frac{1.729 \times 10^6}{136} = 12\,713 \text{ lb/h} = W$$

$$A = \frac{W\sqrt{T}\sqrt{Z}}{CKP_1 K_b \sqrt{M}}$$

where

A = effective discharge area of the value (in.^{2})
W = flow through valve in lb/hr = 12 713
T = absolute temp. of inlet vapour = 460 + 200
$\qquad\qquad\qquad\qquad\qquad = 660\,°\text{R}$
$\qquad$ (bubble point of C_5 to C_8 and, say, 10 psig)
Z = 0.98 (NC$_5$)
C = 356.06 (based on $CA/CV = 1.4$)
K = 0.65 P_i = 134.7
K_b = 0.9M = 100 (use C$_7$)

$$A = \frac{12\,713 \times 25.7 \times 0.99}{356 \times 0.65 \times 134.7 \times 0.9 \times 10}$$

$$= 1.153 \text{ in.}^2 \text{ nearest orifice size} = \text{'J' at } 1.287 \text{ in.}^2$$

The size and rating of this value is read off the Crosby and Farris selection table given in Table A1.4 in Appendix 1. This is a Crosby JO25 2 in. $\times$ 3 in. with 150 lb flanges.

2.3 Utility Systems

All chemical processes require utilities in order to function. The more common utility systems are:

- A steam and condensate system
- A fuel system
- Water systems including cooling water, potable water and boiler feed water
- A compressed air system
- A power supply system

The engineering and design of the first four of these systems is initiated by the process engineer. On operating plants and processes the process engineer undertakes the responsibility for the correct and efficient operation of utility facilities. The duties associated with the power systems are usually left to the electrical engineer or department, although the process engineer does initiate the sizing of the system by developing a 'motor list'.

THE UTILITY BALANCE

As in the case of process systems, definition of the system begins with a balance. In the case of process systems the mass balance of materials in and out is the start point. In utility systems the start point is the utility balance. These balances differ from the process type mass balance insofar as they are not tied to a flow sheet as such. Utility balances are usually developed as block flow sheets. These show in block form the unit the utility is supplying, the quantities involved, the quantities of any utility exported, the condition or designation of the utility stream(s). Figure 2.12 is an example of a steam and condensate balance for a small aromatics production unit.

In this plant there are three levels of steam utilized:

- High-pressure steam at 600 psig
- Medium-pressure steam at 125 psig
- Low-pressure steam at 50 psig

There is also a condensate-recovery system which collects the condensate to be returned to the steam-generation plant. The blocks represent the processing areas within the plant configuration. These are in the sequence in which they occur geographically in the plant. The routing of the utility stream to each and from each area is clearly shown together with the total supply or export of the respective stream. The end user of the utility is shown within the block by its equipment number or by another recognizable designation (such as 'cryogenic' plant). The quantities utilized and exported by individual items of equipment are also shown. The positive numbers are used to denote supply and the negative numbers to denote export or generated. Thus in the case of pump turbines (shown as 'PT' items) the supply is shown as 600 psig steam and the export as either 125 psig or 50 psig steam. The compressor drivers in the example are condensing turbines (shown as 'CT'). Thus the supply is 600 psig steam and the export is condensate routed to the condensate header.

The data in the utility balance were initially developed from equipment horsepower calculations, and in the case of stripping and atomizing steam from the process heat and material balances. Later, as more definitive data such as equipment manufacturers' certified data become available, the utility balance is fine-tuned with these data. The balance itself is used to size the source utility generating units, such as the plant's steam generator in the case of the example and the distribution system (pipelines, valves, and instrumentation of the distributing system).

UTILITY FLOWSHEETS

Utility systems may be divided into two parts: the generation plant and the distribution system. Both are defined by flowsheet drawings. In the case of the utility source or generation unit the flowsheet has the same format as that used in the process plants as the 'mechanical flow diagram' (see Section 1.11). The

Isomerate and O-xylene recovery

Item	HP	MP	LP	Cond.	Loss
PT 302	1670	-1670	0	0	0
E 301	0	0	1700	-1700	0
E 302	0	2600	0	-2600	0
H 301	0	0	65	0	-65
PT 311	2010	0	-2010	0	0
Total	3680	930	-245	-4300	-65

Isomerization unit (licensed)

Item	HP	MP	LP	Cond.	Loss
CT 201	6500	0	0	-6500	0
V 201	0	1260	0	-1260	0
H 201	0	0	65	0	-85
Total	6500	1260	65	-7760	-85

High pressure steam-600 psig

Medium pressure steam-125 psig

Low pressure steam-50 psig

Condensate-5 psig

Benzene-toluene splitter

Item	HP	MP	LP	Cond.	Loss
PT 402	840	0	840	0	0
H 421	0	0	58	0	-58
E 401	0	0	2200	-2200	0
E 402	0	1670	0	-1670	0
Total	840	1670	1418	-3870	-58

De-alkylation unit (licensed)

Item	HP	MP	LP	Cond.	Loss
PT 502	600	-600	0	0	0
PT 506	850	0	-850	0	0
CT 501	2300	0	0	-2300	0
Cryogen	0	0	1560	-1560	0
V 502	0	0	1300	0	-1300
E 503	0	825	0	-825	0
H 501	0	0	160	0	-160
H 502	0	0	172	0	-172
H 503	0	0	165	0	-165
Total	3750	225	2507	-4685	-1797

OVERALL BALANCE

HP steam	21 555	lb\h	(supply)
MP steam	2935	lb\h	(supply)
LP steam	-975	lb\h	(export)
Condensate	-21 400	lb\h	(export)
Loss	-2115	lb\h	

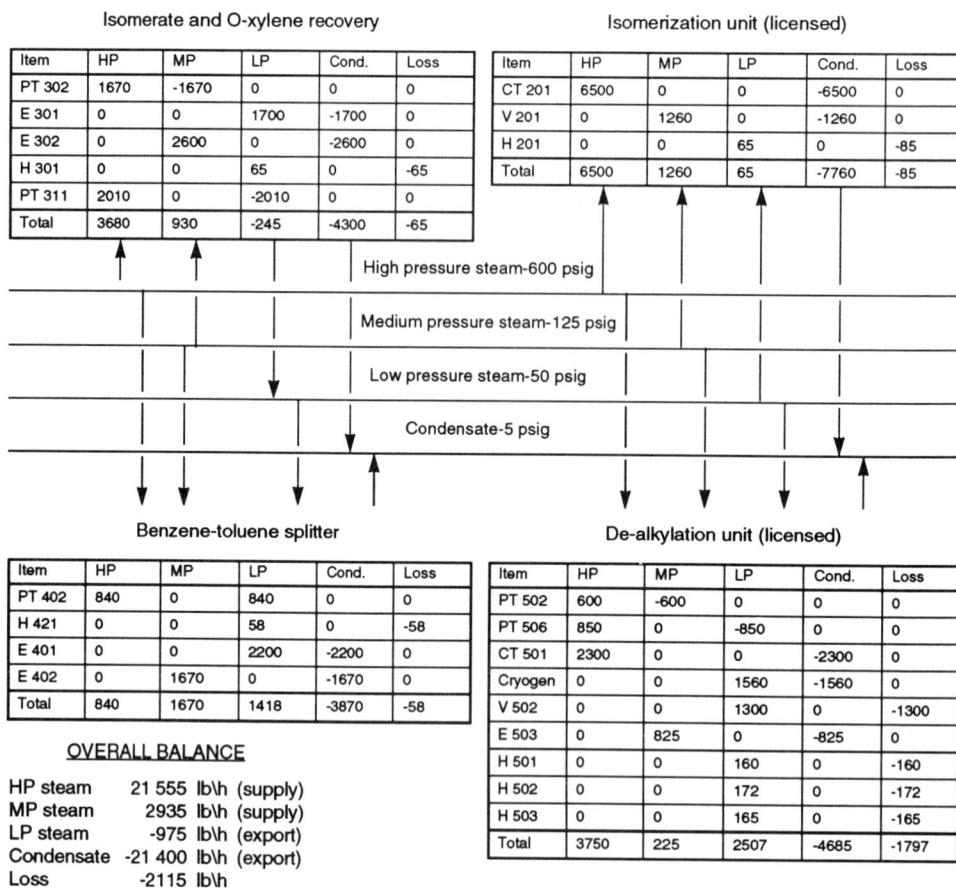

Figure 2.12. A typical steam and condensate balance

distribution flowsheet shows the routing of the utility lines to and from process areas and the equipment in them. Figures 2.13 and 2.14 are examples of these two drawings defining an utility system.

Figure 2.13 is the mechanical flow diagram of a cooling water pond, pumps and treating process. It shows in detail the pump types, cooling tower, the pond and the water-treatment streams. The cooling water supply and return headers are shown at the right of the drawing, and is the limit of the data provided by this drawing. The routing of the supply and return headers are then shown in Figure 2.14. Now, unlike the mechanical flow diagram which is diagrammatically precise, the utility distribution flow diagram is geographically precise. That is, each unit and equipment item are shown exactly in their location in the plant. To achieve this, the plant layout drawing is used, and the respective utility pipelines transposed on the layout drawing. The equipment item is shown only as a block but properly labeled with the item number.

Feed and xylene splitter

Item	HP	MP	LP	Cond.	Loss
PT 101	1250	0	-1250	0	0
PT102	1150	-1150	0	0	0
PT 104	2100	0	-2100	0	0
H 101	0	0	60	0	-60
H 102	0	0	50	0	-50
Total	4500	-1150	-3240	0	-110

21 555 lb\h (supply)

2935 lb\h (supply)

-975 lb\h (export)

-21 400 lb\h (export)

Alkylate stabiliser

Item	HP	MP	LP	Cond.	Loss
PT 601	1500	0	-1500	0	0
PT 602	785	0	0	-785	0
Total	2285	0	-1500	-785	0

All quantities in lb/h

BRIEF DESCRIPTIONS OF TYPICAL UTILITY SYSTEMS

The following paragraphs describe typical utility systems found in the chemical industries. The details of these systems may vary from company to company but their format and general layout will be as described here.

Steam and condensate systems

In most plants steam condensate accumulated in the various processes is collected into a single header and returned to the steam-generating plant. It is stored separately from the treated raw water because condensates may contain some oil contamination. A stream of treated water and condensate are taken from the respective storage tanks and pumped to the deaerator drum. The pumps in this case are usually vertical pumps set in a pit near the storage tanks. The condensate

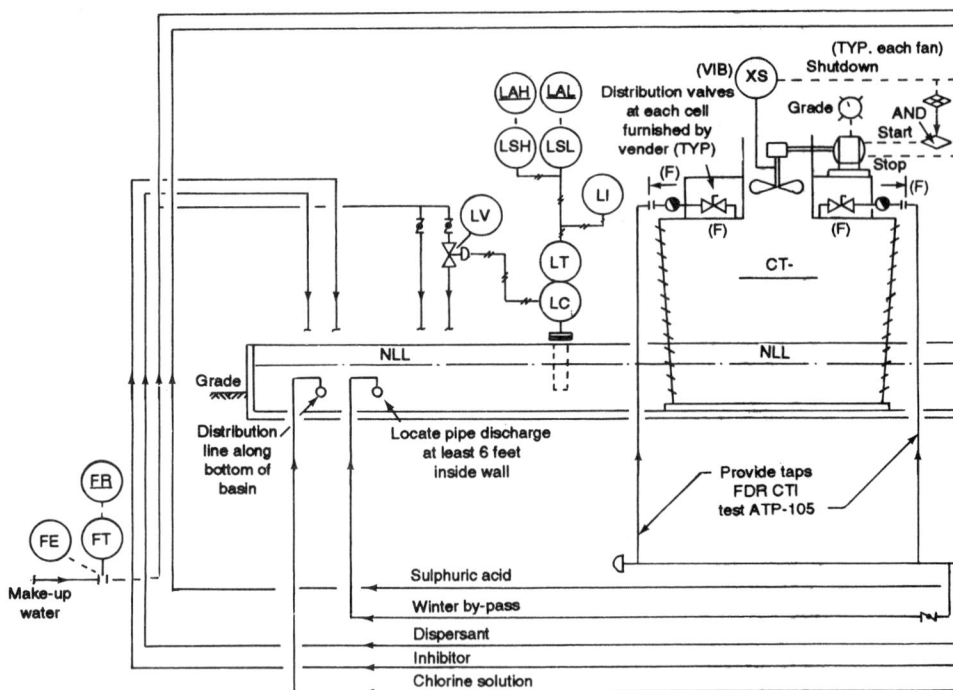

Figure 2.13. Water cooling—mechanical flow diagram

stream passes through a simple filter en route to the deaerator to remove any oil contamination. The combined water and condensate streams enter the top of a packed section of the deaerator drum. Low-pressure steam is introduced immediately below the packing in the drum to flow upwards countercurrent to the liquid streams. Any air entrained in the water is removed by this countercurrent flow of steam to be vented to atmosphere.

The deaerated boiler feed water (BFW) is pumped by the boiler feed water pumps into the steam drum of the steam generator. There will normally be three 60% pumps for this service. Two will be operational and one will be standby. Those pumps normally operating are usually motor driven while the standby pump is very often driven by an automatically start-up diesel engine. These pumps are quite large in capacity, operating at high head and discharge pressure. The main steam lines in most plants are higher pressure (at least 700 psig at the generator coil outlet), so the pump discharge pressure will be much greater than

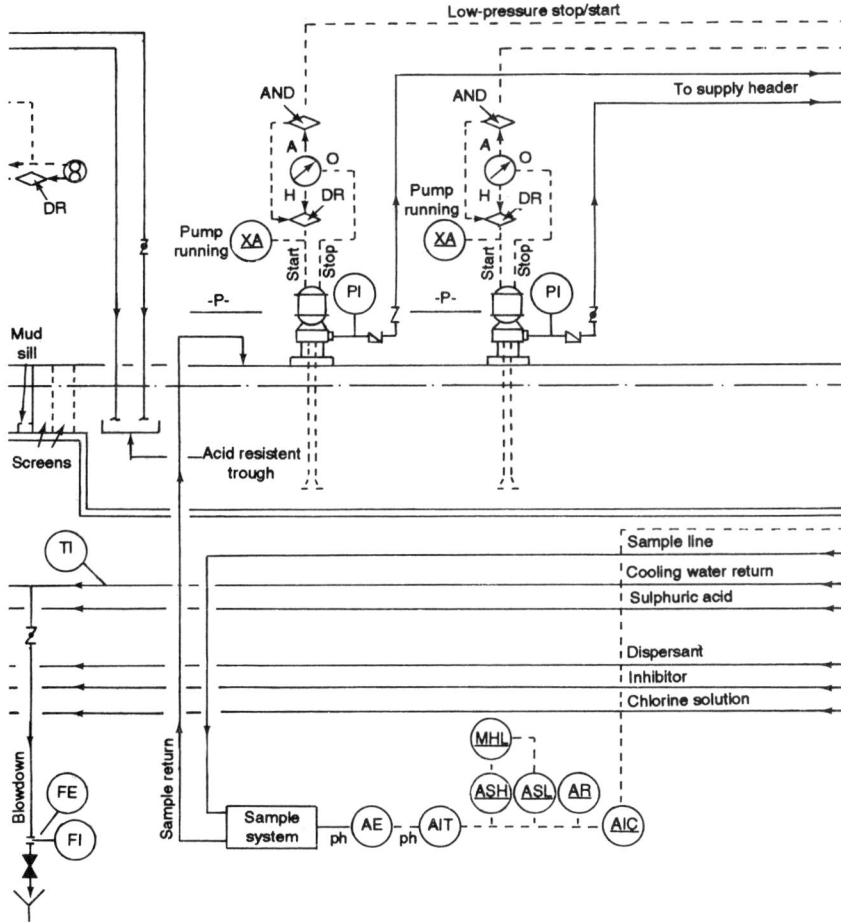

the HP steam outlet. These pumps are the most important in any chemical plant. If they fail, no steam can be generated and the whole complex is in danger of total shutdown or worse. Therefore three separate pumps are used to cater for the normal high head and high capacity and a separate pump driver operating on a completely different power source than electrical power or steam is mandatory to minimize the danger of complete shutdown.

The steam drum is located above the generator's firebox. The liquid in the drum flows through the generator's coils located in the firebox by gravity and thermosyphon. A mixture of steam and water is generated in the coils and flows back to the steam drum. Here the steam and water are separated with the steam leaving the drum to enter the superheater coil. The steam is heated to the plant's steam main temperature in this coil and enters the high-pressure steam header for distribution to the various users. The steam pressure is controlled by a pressure controller on the steam outlet to the header. Steam to the lower-pressure heads is

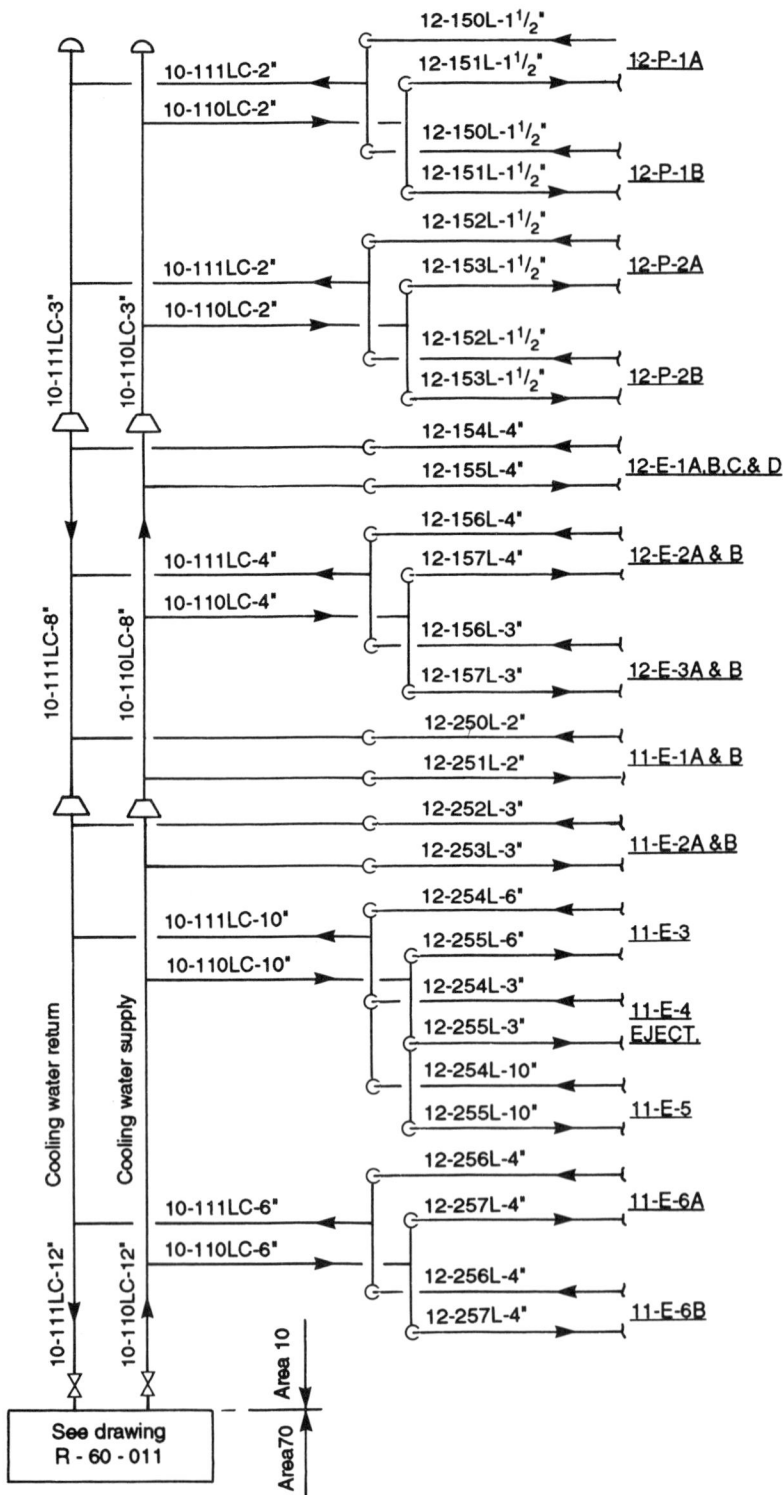

Figure 2.14. Cooling water distribution—utility flow diagram

generated through turbines where possible. Where lower-pressure steam is required and it is not possible to produce it through equipment then let-down stations are located in suitable places in the system. When steam pressure is reduced to the lower-pressure headers the associated increase of temperature above that specified for the lower-pressure steam may need to be reduced. Desuperheaters are used for this purpose.

Desuperheaters consist of a chamber in the steam line into which cold condensate is injected. These items are purchased equipment with specially designed injection nozzles for the condensate. The amount of condensate delivered is controlled by the downstream temperature of the steam. Desuperheaters are located at critical locations of the plant where relatively large quantities of high-pressure steam are reudced to low pressures such as the discharge from turbines.

The condensate return header is usually operated at a positive pressure of between 5 and 10 psig. The condensate is stored at atmospheric pressure, and very often the small amount of steam flashed from the header pressure to the storage pressure is used in the deaerator instead of the low-pressure steam (the deaerator operates at atmospheric pressure). Figure 2.15 is a schematic flow diagram of a typical steam-generation unit.

2 Fuel systems

Most major chemical complexes and certainly most oil refineries have two separate fuel systems:

- A fuel gas system and
- A fuel oil system

The user burners in these plants are of dual-purpose design. They fire either the fuel gas or the fuel oil stream and can be easily switched over from one to the other. The pilot burners, however, must be fuel gas, and the system is designed such that if the pilot burner is extinguished the whole burner system is shut-down. Generally, the design of the burner system in most plants has many safety and shutdown features. After all, in processes that handle flammable material the heater burners are the one feature in the plant design that are a major fire hazard source. Thus the design of the burner operation is such as to shut-down on:

- Low fuel pressure
- High process outlet temperature
- Pilot burner extinguished
- Atomizing steam failure (oil burners)
- Low process feed to heater

Fuel gas system

This is the simplest of the two systems. Waste gas streams from the process plants are gathered and directed to the plant's fuel gas drum. This drum operates at 30 psig pressure and close to ambient temperature or 60 °F. A small steam coil is

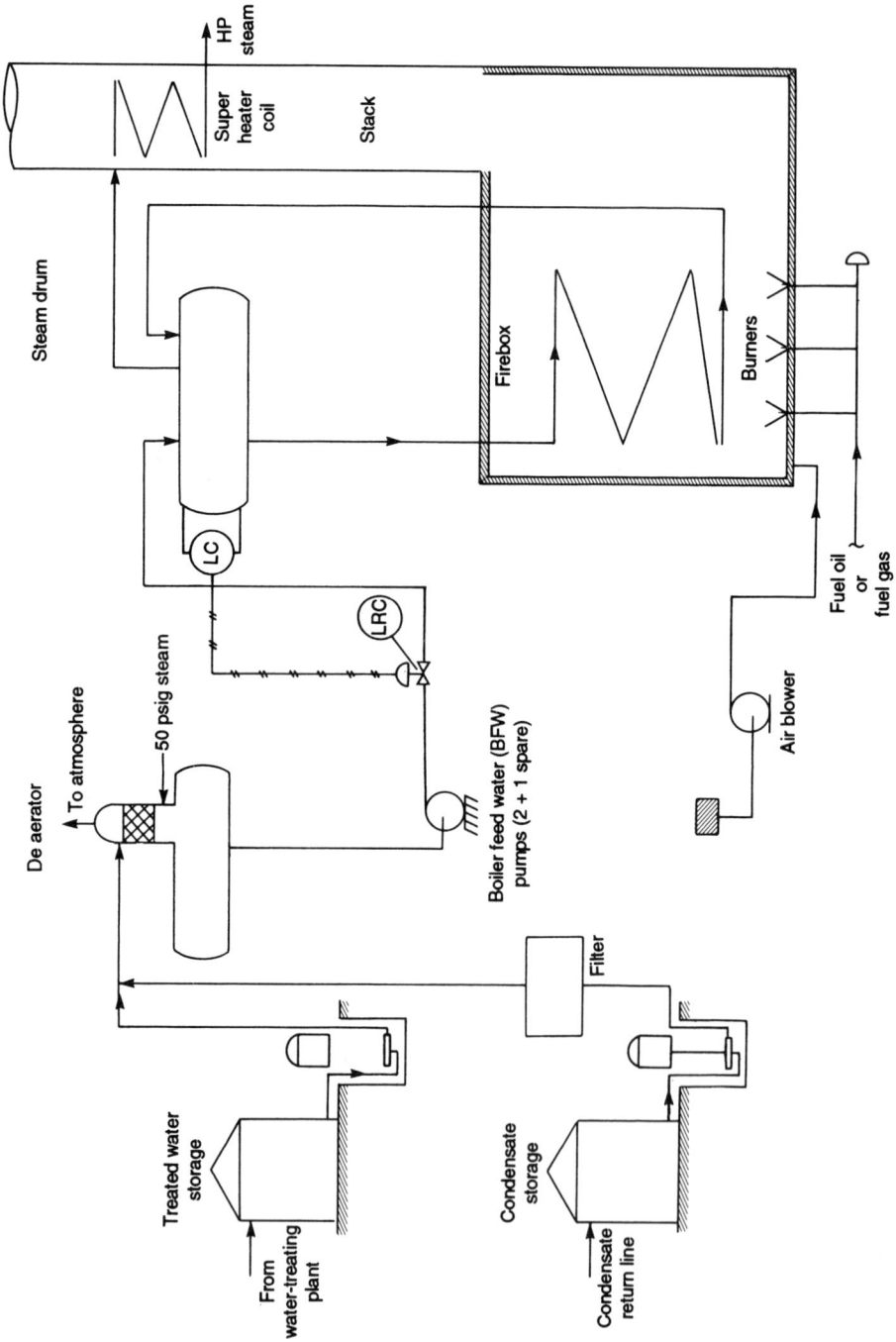

Figure 2.15. Schematic of steam generation plant

usually installed in the drum to gasify any 'below dew point' material that may have condensed out. The drum is held at the set pressure by pressure control valves which allow surplus gas to flow to flare and activate an emergency source of gas on low pressure. This emergency source may be in the form of imported LPG (liquified petroleum gas) or some other plant gas stream that may be diverted to fuel. If LPG is used as a secondary fuel source it is routed to the fuel gas drum via a vaporizer. This item of equipment is a kettle-type reboiler, heating and vaporizing the LPG at the drum pressure. Medium- or low-pressure steam is used as the heating medium for the vaporizer.

The gas burner at the process fired heater operates at or close to atmospheric pressure. The burner draws fuel from a separate header from that used to supply gas to the pilot burner. Many heaters contain an automatic switch-over from gas firing to fuel oil firing on low gas flows or when manually selected. The fuel gas system is 'dead ended', that is, there is no return system to the fuel gas drum, the gas header is pressured up and gas flows to the burners by means of this differential pressure and intermediate control valves. The fuel gas flow to the heater burners is controlled by a temperature control valve activated by a temperature controller on the heater coil process outlet line. The same controller also regulates the oil firing arrangement when the heater is operated on fuel oil.

Fuel oil system

Figure 2.16 is a schematic of a typical fuel oil system. Most plants use petroleum residues as fuel oil. These types of fuel are high in viscosity and very often have a high pour point. (Pour point temperature is the temperature when the oil ceases to flow.) For this reason the fuel oil is stored in insulated and heated cone roofed tanks. Heating may be accomplished by steam coils located in the base of the tank or by an external steam heat exchanger through whih the fuel oil is continually circulated.

Positive displacement pumps (usually rotary type) are used to deliver the fuel oil from the tank, through the distribution system to the heater burners. These pumps are always spared and the spare pump is driven by a steam turbine while the operating pump is motor driven. The fuel oil passes through a duplex filter before entering the suction of the pumps. This filter is included to remove any solid contaminants that may be in the oil such as fine coke particles which would foul the fuel oil burner. The discharge pressure of the pumps is controlled by a slipstream routed back to the storage tanks through a pressure control valve. This valve is activated by a pressure control element on the pump discharge header.

The pumps discharge the fuel oil via the pressure controller to a preheater. This preheater may be a simple double-pipe heat exchanger for relatively small units or regular shell and tube for the larger systems. Double-pipe exchangers are favoured in this service when economical because they are easier to clean and maintain. The fuel oil leaves the preheater to enter the fuel oil distribution system at a temperature high enough to maintain a viscosity low enough for the oil to flow easily and to be easily atomized by steam at the fuel oil 'gun' (or burner). Table A1.5 in Appendix 1 lists the temperature/viscosity of fuel oils and

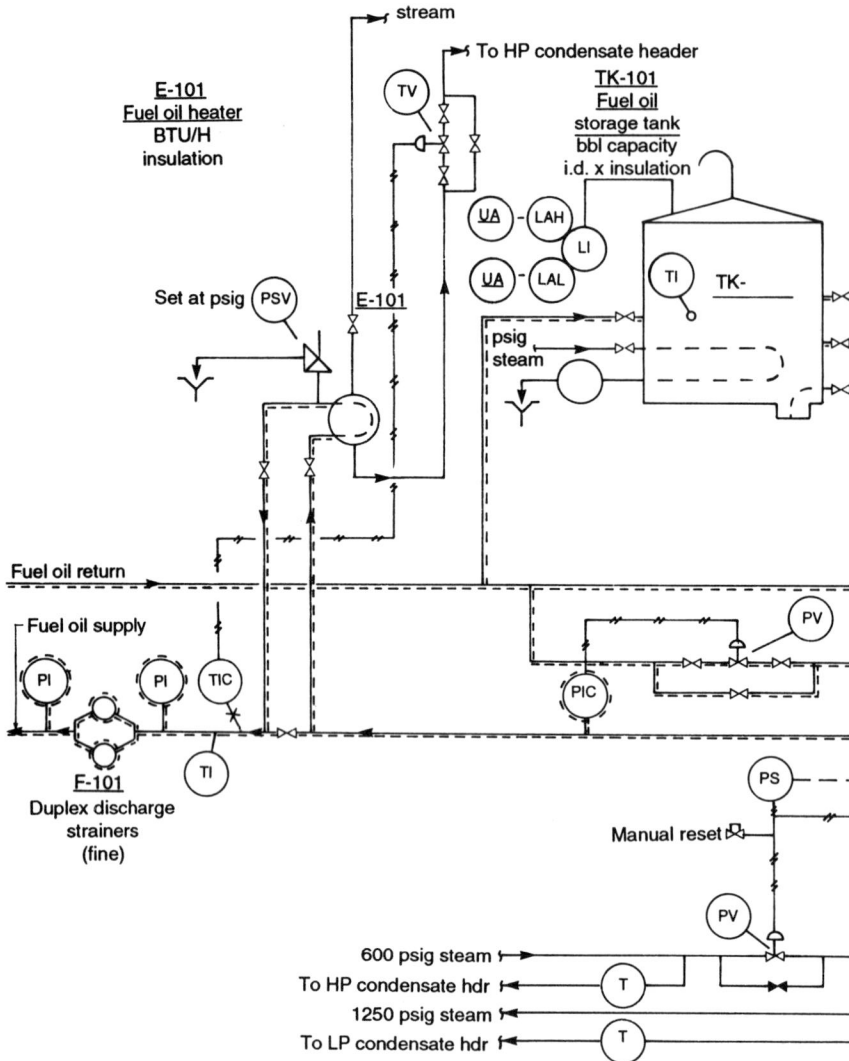

Figure 2.16. Schematic of a fuel oil system

manufacturers of the oil burner will specify the viscosity range suitable for the operation of the particular burner.

All the piping associated with residual fuel systems are heavily insulated and steam traced. The distribution systems of residual fuel oils is usually the recirculating type. That is, the fuel leaves the preheater to circulate to all the user plants in a loop where the quantity used is taken off the stream and the remainder allowed to return to the system. The return header is routed back to the storage tanks. The circulation system handles between one to three times that amount of oil that is actually burnt.

Table 2.1 gives a brief description and properties of the fuel oils normally used

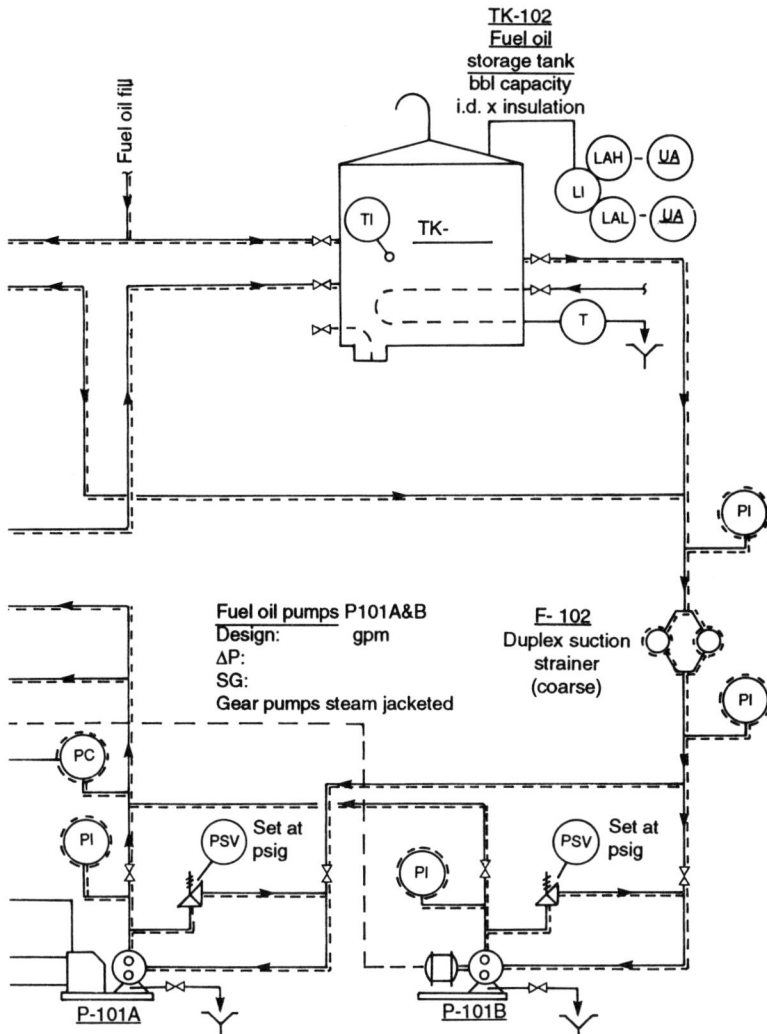

in the chemical and petroleum industry. This table also gives the suggested range of temperatures used in the various systems.

Fuel oils are introduced into the fire box and ignited through a fuel oil burner sometimes called a fuel oil gun. In order to ensure combustion in a manner suitable for a fire box operation the fuel oil needs to be dispersed into small droplets or spray at the burner tip. In heavy residual oils this is always accomplished by a steam or air stream. This atomizing stream is introduced into the gun chamber and comes into contact with the oil stream just before the burner tip. The kinetic energy in the atomizing medium forces the oil into suitable droplets as it leaves the burner. Steam is normally used as the atomizing material

Table 2.1. General classification of fuel oils

Trade No.	Principal use	Gravity °API	lb/gal	BTU/gal	Kinematic viscosity SSU at 100 °F		SSF at 122 °F	
					Max.	Min.	Max.	Min.
1	A distillate oil intended for vaporizing pot-type burners and other uses requiring volatile fuel	35–40	6.879–7.085	135 800–138 800	–	–	–	–
2	A distillate oil for general-purpose domestic heating for use in burners not requiring No. 1. Moderately volatile	26–34	7.128–7.490	139 400–144 300	40	–	–	–
4	An oil for burner installation not equipped with preheating facilities	24–25	7.538–7.587	145 000–145 600	125	45	–	–
5	A residual-type oil for burners equipped with preheating facilities. Sold as Bunker B. Preheat suggested 170–220 °F	18–22	7.686–7.891	146 800–149 400	–	150	40	
6	An oil for use in burners equipped with preheaters permitting a high-viscosity fuel. Bunker C. Preheat suggested 220–260 °F	14–16	7.998–8.108	150 700–152 000	–	–	300	45

because it is usually cheaper, more readily available, and has a more reliable source than air from a compressor.

The steam pressure for atomising should be 15 to 25 psig higher than the fuel oil pressure. The quantity of steam will range from between 1.5 to 5 lb per gallon of oil. Dry steam with a superheat of about 50 °F is preferred for atomizing. In order to control the process heater operation oil burners require a turndown ability. That is, they require to operate satisfactorily over a prescribed range of flow. In keeping with an operating range for oil flow the atomizing medium must also have a similar operating range. Burner pressure is a critical requirement for turndown. Table 2.2 gives the burner pressure requirement for the oil at various turndown ratios. The steam (or air) pressure should be 15 psi or 10% (whichever is the

Table 2.2. Required burner pressures

Turndown ratio	Burner pressure (psig)
3–4/1	60
5/1	75
7–8/1	100
9–10/1	150

greater) higher than oil pressure. The fuel oil supply system should be 100 psi higher than the burner requirement.

The oil burner operation, as with the fuel gas burner, is controlled by the heater's process stream outlet temperature. A temperature control valve activated by the coil outlet temperature increases or decreases the oil flow from the circulating oil stream to the burner. A proportional control valve on the atomizing steam line regulates the flow of steam to the burner in keeping with the oil flow. Figure 2.17 shows a typical burner control system. There are various methods for safety shutdown. The one shown here shuts down the oil flow on steam failure and on loss of the pilot burner. In some systems there is also an automatic change-over to gas firing on low oil pressure.

3 Water systems

The major water systems generated in most chemical plants are:

- Cooling water
- Treated water for BFW

Potable water as raw water is usually drawn from the municipal supply. Where water is required for cleaning or drinking this potable water is used without further processing.

Cooling water

Figure 2.13 is a mechanical flowsheet showing the arrangement around the base of a cooling tower for the collection and supply of cooling water. The cooling water system is a circulating one. That is, there is a cold supply line with an associated warmer return line from all its users. Figure 2.14 shows a section of this distribution system. The water returned to the cooling tower by the return header enters the top of the tower and flows down across the tower internals countercurrent to an air flow either induced or forced by fans passing up through the tower. The water cooled by the air flow is collected in the cooling tower basin. Make-up water (usually potable water) is added to the basin under level control. Vertical cooling water circulating water pumps take suction from the cooling tower basin sump to deliver the water into the plant's distribution header. These pumps are usually high capacity with a moderate differential head. Because of the critical nature of the water supply the pumps are each rated at around 60% of

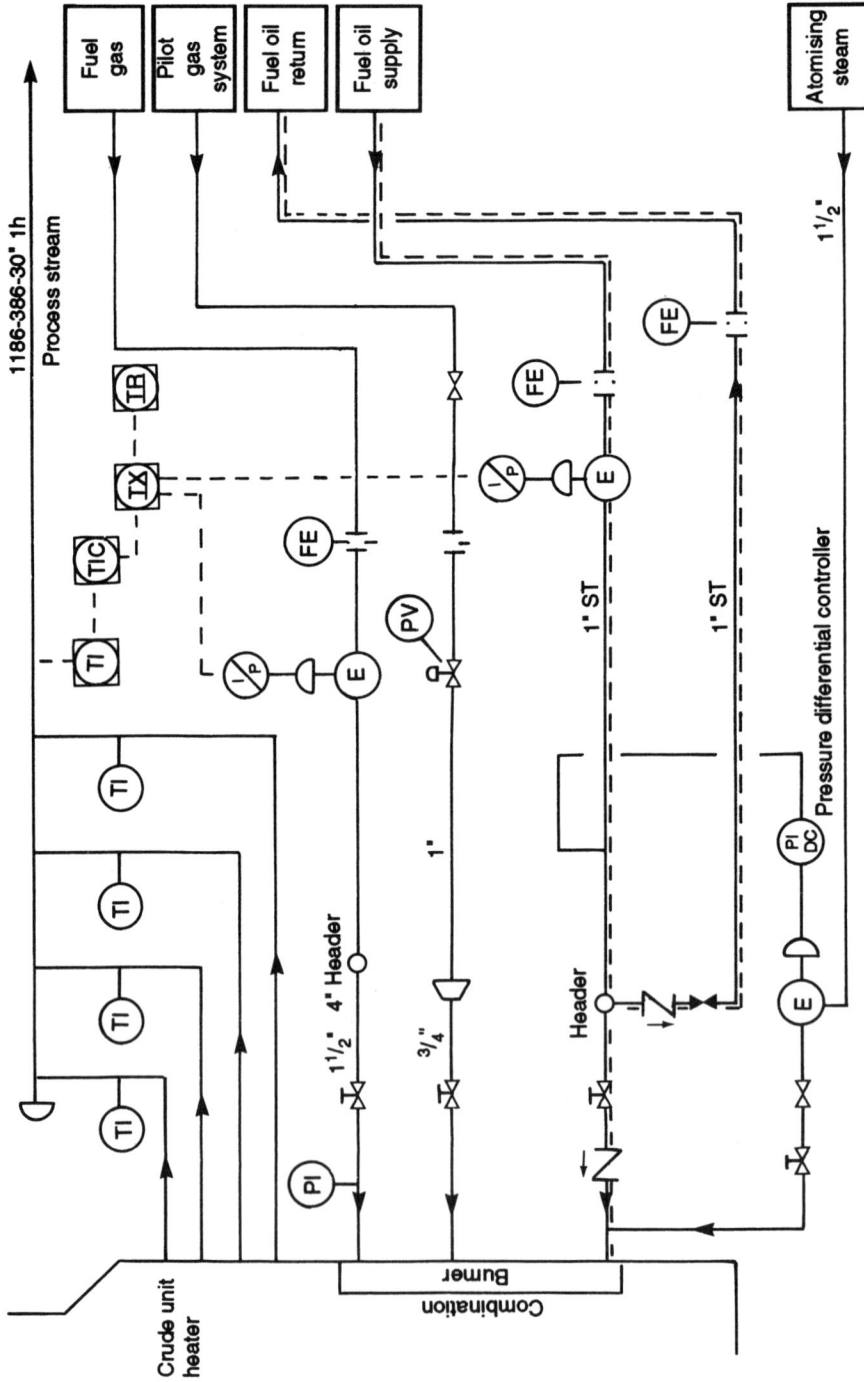

Figure 2.17. A typical burner control system. TI—temperature indicator, TR—temperature recorder, TIC—temperature indicator controller, TX—automatic switch-over control, PI—pressure indicator, E—control valve, PDIC—differential pressure controller

design capacity with two in operation and two on standby. A mixture of motor and steam turbine drivers is quite common.

The supply header pressure is kept at around 30 psig and, very often in large plants covering long distances, booster stations are installed at predetermined locations to maintain the supply header pressure. These booster stations consist of pump pits with again high-capacity vertical pumps, rated smaller, of course, than the main supply pumps. The location of these booster stations is determined by a rigorous hydraulic analysis of the distribution system which also determines the header pipeline sizes. The return flow is collected from each user into the return header and flows back to the cooling tower under the users' outlet pressure.

The water in the cooling tower basin and in the cooling tower itself requires some treatment to prevent the build-up of algae and other undesirable contaminants. A separate small treatment plant is used for mixing the inhibiting chemicals and injecting them into the critical sections of the system.

Cooling tower performance

The following section has been summarized from The Gas Processor's Suppliers Association *Engineering Data Book* (9th edition), and is published here by kind permission of The Gas Processor's Suppliers Association.

The performance characteristics of various types of towers will vary according to height, type of flow, fill configuration, etc. These factors have been taken into consideration in the development of the nomograph given in Figure 2.18. The use of the nomograph is illustrated by the following examples covering typical changes in operating conditions.

Assume a cooling tower is operating at known conditions of 1000 gpm, 105 °F hot water, 85 °F cold water, and 75 °F wet bulb temperature. (Commonly referred to as 105-85-75 or 20° range and 10° approach.)

Example 1 What is the new cooling water temperature (CWT) when the wet bulb (WB) changes from 75 °F to 60 °F with gpm and range remaining constant? Enter the nomograph at 85 °CWT, go horizontally to 75 °WB, then vertically down to 60 °WB. Read new CWT of 76 °F.

Example 2 What is the new CWT when the cooling range is changed from 20° to 30° with GPM and WB held constant (increase of 50% in heat load)? Enter nomograph at 85 °CWT, go horizontally to 75 °WB, vertically to 20° range (R), horizontally to 30 °R, vertically to 75° WB. Read off new CWT of 87.5 °F.

Example 3 What is the new CWT when the water circulation rate is changed from 1000 gpm to 1500 gpm (50% increase in heat load at constant temperature range)? Enter nomograph at 85 °CWT, go horizontally to 75 °WB, vertically to 20 °R, horizontally to performance factor (PF) of 3.1. Obtain new PF by multiplying 3.1 by 1500/1000 = 4.65. Then enter the nomograph at PF 4.65, go horizontally to 20 °R, vertically down to 75 °WB. Read new CWT of 90.5°.

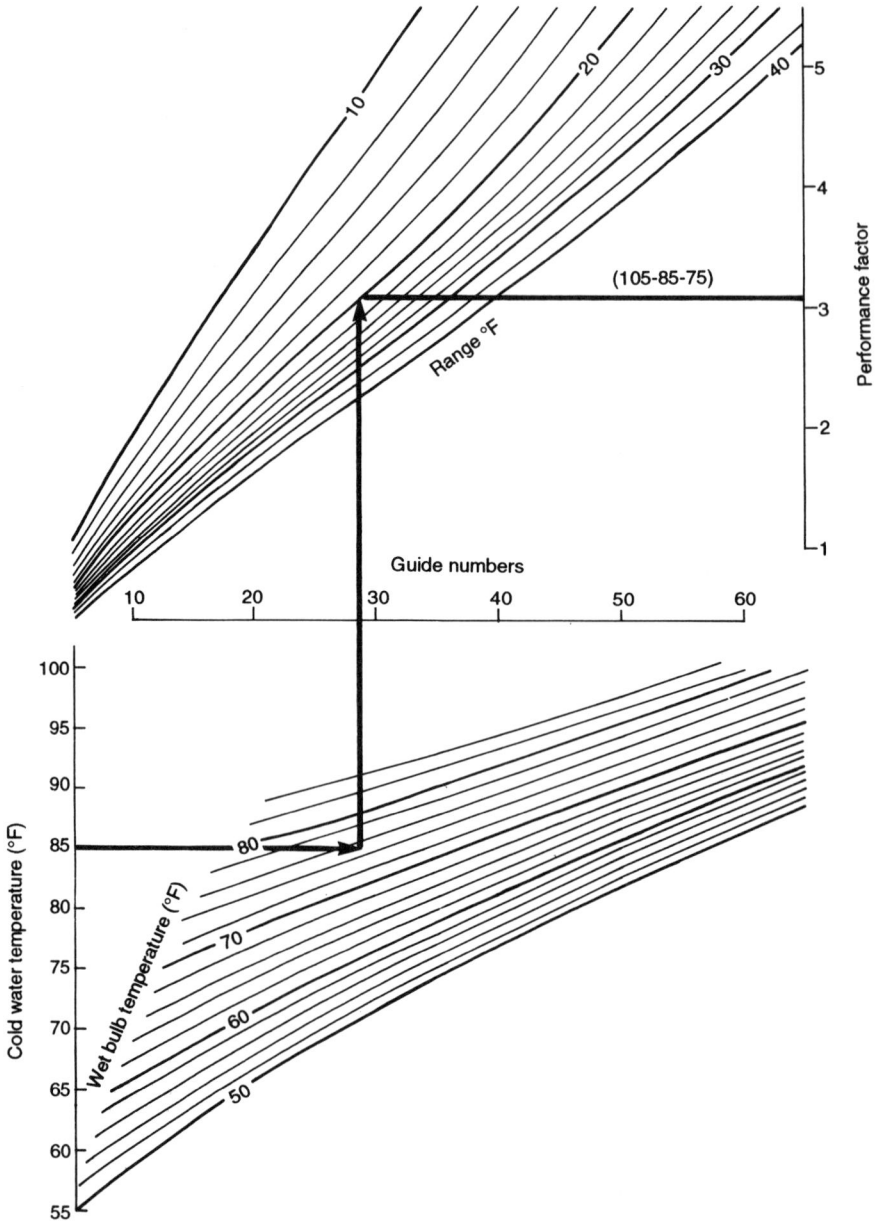

Figure 2.18. Cooling tower performance characteristics (Reproduced by permission of Gas Processors Supplier Association)

Example 4 What is the new CWT when the WB changes from 75° to 60 °R, changes from 20° to 25°, gpm changes from 1000 to 1250 gpm (50% change in heat load.)? Enter the nomograph at 85 °CWT, go horizontally to 75 °WB, vertically to 20 °R, horizontally to read PF as 3.1. Multiply 3.1 by 1250/1000 to give new PF of

3.87. Enter nomograph at PF of 3.87, go horizontally to 25 °R, vertically down to 60 °WB, read 82° as the new CWT.

Example 5 What is new CWT if fan motor hp is changed from 20 hp to 25 hp in Example 4? The air flow rate varies as the cube root of the horsepower, and performance varies almost directly with the ratio of water rate to air rate. Therefore the change in air rate can be applied to the performance factor. Increasing the air flow rate decreases the performance factor. PF correction factor $= (25\,\text{hp}/20\,\text{hp})^{1/3} = 1.077$. Divide PF by the correction factor to obtain new PF. Applying this to Example 4 we have $3.87/1.077 = 3.6$. Enter nomograph at 3.6 PF, go horizontally to 25 °R, vertically down to 60 °WB. Read 81 °F CWT.

Example 6 The correction factor shown in Example 5 could also be used to increase gpm instead of decreasing CWT. In Example 4 we developed a new CWT of 82° when circulating 1250 gpm at 25 °R and 60 °WB. If motor HP is increased from 20 to 25 under these conditions with PF correction factor of 1.077, gpm could be increased from 1250 to $1250 \times 1.077 = 1346$ gpm.

Boiler feed water treating

All water contains impurities no matter from what source the water comes. Table 2.3 gives a listing of the common impurities found in water for industrial use. Normally, in a chemical plant, water with most of these impurities can be used without treating of any kind. However, when generating steam, and particularly high-pressure steam, these impurities become problematic. To operate steam generators effectively and to avoid serious damage to the unit, these impurities have either to be removed or converted into compounds that can be tolerated in the system. Table 2.3 provides a description of the effect of these impurities on steam generators and gives the normal means of treating. Table 2.4 gives the chemical reaction equations to some of these treating processes.

In general, there are three types of soluble impurities naturally present in water and which must be removed or converted in order to make the water suitable for boiler feed purpose:

- *Scale-forming impurities* These are usually salts of calcium, magnesium, silica, manganese, and iron.
- *Compounds that cause foaming* These are usually soluble sodium salts.
- *Dissolved gases* These are usually oxygen and carbon dioxide. The soluble gases must be removed to prevent corrosion.

Solid build-up in the boiler itself is removed or kept at a low level by blowdown. This is the mechanism of draining a prescribed amount of the boiling water from the boiler steam drum at regular intervals. This amount is calculated from the analysis of the solid content of the feedwater and must equal the amount brought into the system by the feedwater. Figure 2.19 gives an example of boiler blowdown.

The American Boiler Manufacturers Association (ABMA) have developed

Table 2.3. Common impurities found in water

Constituent	Chemical formula	Difficulties caused	Means of treatment
Turbidy	None	Imparts unsightly appearance to water. Deposits in water lines, process equipment, boilers, etc. Interferes with most process uses	Coagulation, settling and filtration
Colour	None	May cause foaming in boilers. Hinders precipitation methods such as iron removal, hot phosphate softening. Can stain product in process use	Coagulation and filtration. Chlorination. Adsorption by activated carbon
Hardness	Calcium and magnesium salts expressed as $CaCO_3$	Chief source of scale in heat exchange equipment, boilers, pipe lines, etc. Forms curds with soap interferes with dying, etc.	Softening, Distillation. Internal boiler water treatment. Surface-active agents
Alkalinity	Bicarbonate (HCO_3^-), carbonate $(CO_3^=)$, and hydroxyl (OH^-), expressed as $CaCO_3$	Foaming and carryover of solids with steam. Embrittlement of boiler steel. Bicarbonate and carbonate produce CO_2 in steam, a source of corrosion	Lime and lime-soda softening. Acid treatment. Hydrogen zeolite softening. Demineralization. De-alkalization by anion exchange. Distillation
Free mineral acid	H_2SO_4, HCl, etc. expressed as $CaCO_3$, titrated to methyl orange end-point	Corrosion	Neutralization with alkalies
Carbon dioxide	CO_2	Corrosion in water lines and particularly steam and condensate lines	Aeration. Deaeration. Neutralization with alkalies. Filming and neutralizing amines
pH	Hydrogen Ion concentration defined as $$pH = \log \frac{1}{(H^+)}$$	pH varies according to acidic or alkaline solids in water. Most natural waters have a pH of 6.0–8.0	pH can be increased by alkalies and decreased by acids
Sulphate	$(SO_4)^{--}$	Adds to solids content of water, but, in itself, is not usually significant. Combines with calcium to form calcium sulphate scale	Demineralization. Distillation
Chloride	Cl^-	Adds to solids content and increases corrosive character of water	Demineralization. Distillation
Nitrate	$(NO_3)^-$	Adds to solids content, but is not usually significant industrially. High concentrations cause methemoglobinemia in infants. Useful for control of boiler metal embrittlement	Demineralization. Distillation

Fluoride	F^-	Cause of mottled enamel in teeth. Also used for control of dental decay. Not usually significant industrially	Adsorption with magnesium hydroxide, calcium phosphate, or bone black. Alum coagulation
Silica	SiO_2	Scale in boilers and cooling water systems. Insoluble turbine blade deposits due to silicavaporization	Hot process removal with magnesium salts. Adsorption by highly basic anion exchange resins, in conjunction with demineralization. Distillation
Iron	Fe^{++} (ferrous) Fe^{+++} (ferric)	Discolours water on precipitation. Source of deposits in water lines, boilers, etc. Interferes wih dyeing, tanning, paper manufacturer, etc.	Aeration. Coagulation and filtration. Limesoftening. Cationexchange. Contact filtration. Surface activeagents for ion retention
Manganese	Mn^{++}	Same as iron	Same as iron
Oil	Expressed as oil or chloroform extractable matter	Scale, sludge and foaming in boilers. Impedes heat exchange. Undesirable in most processes	Baffle separators. Strainers. Coagulation and filtration. Diatomaceous earth filtration
Oxygen	O_2	Corrosion of water lines, heat exchange equipment, boilers, return lines, etc.	Deaeration. Sodium sulphite. Corrosion inhibitors
Hydrogen sulphide	H_2S	Cause of 'rotten egg' odour. Corrosion	Aeration. Chlorination. Highly basic anion exchange
Ammonia	NH_3	Corrosion of copper and zinc alloys by formation of complex soluble ion	Cation exchange with hydrogen zeolite. Chlorination. Deaeration
Conductivity	Expressed as micromhos, specific conductance	Conductivity is the result of ionizable solids in solution. High conductivity can increase the corrosive characteristics of a water	Any process which decreases dissolved solids content will decrease conductivity. Examples are demineralization, lime softening
Dissolved solids	None	'Dissolved solids' is measure of total amount of dissolved matter, determined by evaporation. High concentrations of dissolved solids are objectionable because of process interference and as a cause of foaming in boilers	Various softening process, such as lime softening and cation exchange by hydrogen zeolite, will reduce dissolved solids. Demineralization. Distillation
Suspended solids	None	'Suspended solids' is the measure of undissolved matter, determined gravimetrically. Suspended solids plug lines, cause deposits in heat exchange equipment, boilers, etc.	Subsidence. Filtration, usually preceded by coagulation and settling
Total solids	None	'Total solids' is the sum of dissolved and suspended solids, determined gravimetrically	See 'Dissolved solids' and 'Suspended solids'

Table 2.4. Chemical reactions in boiler feed water treatment

$CaSO_4$	$+ Na_2CO_3$	$= \mathbf{CaCO_3}$	$+ Na_2SO_4$
$CaCl_2$	$+ Na_2CO_3$	$= \mathbf{CaCO_3}$	$+ 2NaCl$
$Ca(NO_3)_2$	$+ Na_2CO_3$	$= \mathbf{CaCO_3}$	$+ 2NaNO_3$
$MgSO_4$	$+ Na_2CO_3$	$= \mathbf{MgCO_3}$	$+ 2Na_2SO_4$
$MgCl_2$	$+ Na_2CO_3$	$= \mathbf{MgCO_3}$	$+ 2NaCl$
H_2SO_4	$+ Na_2CO_3$	$= H_2CO_3$	$+ Na_2SO_4$
$Na_2CO_3 + H_2O$	$+$ Boiler heat	$= 2NaOH$	$+ CO_2 - H_2O$
$Ca(HCO_3)_2$	$+$ Boiler heat	$= \mathbf{CaCO_3}$	$+ CO_2 + H_2O$
$Mg(HCO_3)_2$	$+$ Boiler heat	$= \mathbf{MgCO_3}$	$+ CO_2 + H_2O$
$MgSO_4$	$+ 2NaOH$	$= \mathbf{Mg(OH)_2}$	$+ Na_2SO_4$
$MgCl_2$	$+ 2NaOH$	$= \mathbf{Mg(OH)_2}$	$+ 2NaCl$
H_2SO_4	$+ 2NaOH$	$= Na_2SO_4$	$+ 2H_2O$
H_2CO_3	$+ 2NaOH$	$= Na_2CO_3$	$+ 2H_2O$
$Ca(HCO_3)_2$	$+ 2NaOH$	$= \mathbf{CaCO_3}$	$+ Na_2CO_3 + 2H_2O$
$Mg(HCO_3)_2$	$+ 4NaOH$	$= \mathbf{Mg(OH)_2}$	$+ 2Na_2CO_3 + 2H_2O$
$3CaSO_4$	$+ 2Na_3PO_4$	$= \mathbf{Ca_3(PO_4)_2}$	$+ 3Na_2SO_4$
$3MgSO_4$	$+ 2Na_3PO_4$	$= \mathbf{Mg_3(PO_4)_2}$	$+ 3Na_2SO_4$
$3MgCl_2$	$+ 2Na_3PO_4$	$= \mathbf{Mg_3(PO_4)_2}$	$+ 6NaCl$
$CaSO_4$	$+ Na_2SiO_3$	$= \mathbf{CaSiO_3}$	$+ Na_2SO_4$
$MgSO_4$	$+ Na_2SiO_3$	$= \mathbf{MgSiO_3}$	$+ Na_2SO_4$
$MgCl_2$	$+ Na_2SiO_3$	$= \mathbf{MgSiO_3}$	$+ 2NaCl$
H_2CO_3	$+ Ca(OH)_2$	$= \mathbf{CaCO_3}$	$+ 2H_2O$
$Ca(HCO_3)_2$	$+ Ca(OH)_2$	$= \mathbf{2CaCO_3}$	$+ 2H_2O$
$Mg(HCO_3)_2$	$+ 2Ca(OH)_2$	$= \mathbf{Mg(OH)_2}$	$+ 2CaCO_3 + 2H_2O$
$CaSO_4$	$+ Ba(OH)_2$	$= \mathbf{BaSO_4}$	$+ Ca(OH)_2$
$MgSO_4$	$+ Ba(OH)_2$	$= \mathbf{Mg(OH)_2}$	$+ \mathbf{BaSO_4}$
H_2CO_3	$+ Ba(OH)_2$	$= \mathbf{BaCO_3}$	$+ 2H_2O$
$CaSO_4$	$+ BaCO_3$	$= \mathbf{BaSO_4}$	$+ \mathbf{CaCO_3}$
$MgSO_4$	$+ BaCO_3$	$= \mathbf{BaSO_4}$	$+ \mathbf{MgCO_3}$
$MgSO_4$	$+ 2NaCl$	$= MgCl_2$	$+ Na_2SO_4$
$MgCl_2$	$+ 2H_2O$	$= \mathbf{Mg(OH)_2}$	$+ 2HCl$
$FeCl_2$	$+ 2H_2O$	$= \mathbf{Fe(OH)_2}$	$+ 2HCl$
$FeSO_4$	$+ 2H_2O$	$= \mathbf{Fe(OH)_2}$	$+ H_2SO_4$
$CaSO_4$	$+ Na_2Z$	$= CaZ^*$	$+ Na_2SO_4$
$MgSO_4$	$+ Na_2Z$	$= MgZ^*$	$+ Na_2SO_4$
$CaCl_2$	$+ Na_2Z$	$= CaZ^*$	$+ 2NaCl$
$MgCl_2$	$+ Na_2Z$	$= MgZ^*$	$+ 2NaCl$
$Ca(HCO_3)_2$	$+ Na_2Z$	$= CaZ^*$	$+ 2NaHCO_3$
$Mg(HCO_3)_2$	$+ Na_2Z$	$= MgZ^*$	$+ 2NaHCO_3$

*Upon regeneration with a strong salt (NaCl) brine, the zeolite is converted back to Na_2Z. The calcium and magnesium chlorides formed during the 'exchange' regeneration (CaZ or $MgZ + 2NaCl = Na_2Z + CaCl_2$ or $MgCl_2$) are disposed of as waste.

   The formulas in bold type indicate precipitaes or sludges.

limits for the control of various solids in boiler feed water (BFW). These are given in Table 2.5. Other considerations regarding the limits of solids in BFW are:

• *Sludge* This is a direct measurement of feed water hardness (calcium and magnesium salts) since virtually all hardness comes out of solution in a boiler.

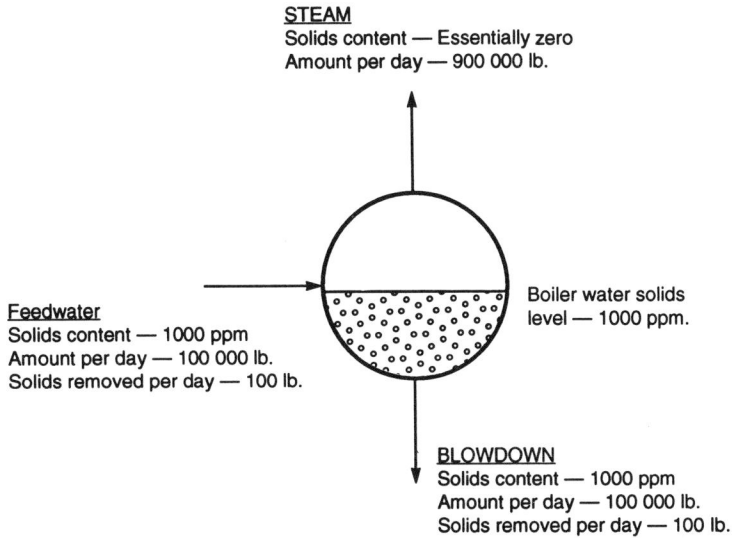

Figure 2.19. How boiler solids build-up is checked by blowdown

Table 2.5. ABMA limits for boiler water constituents

Boiler pressure (psig)	Total solids (ppm)	Alkalinity (ppm)	Suspended solids	Silica
0–300	3500	700	300	125
301–450	3000	600	250	90
451–600	2500	500	150	50
601–750	2000	400	100	35
751–900	1500	300	60	20
901–1000	1250	250	40	8
1001–1500	1000	200	20	2.5
1501–2000	750	150	10	1.0
over 2000	500	100	5	0.5

Note: Silica limits are based on silica in steam of 0.02–0.03 ppm.

- *Total dissolved solids* These consist of sodium salts, soluble silica, and any chemicals added. Total solids do not contribute to scale formation, but excessive amounts can cause foaming.
- *Silica* This may be the blowdown-controlling factor in pre-softened water containing high silica. At elevated pressures high silica content can cause foaming and carry-over.
- *Iron* High concentration of iron in BFW can cause serious deposit problems. Where concentration is particularly high, blowdown may be based on reducing this concentration.

There are two types of boiler feed water treatment in use: external and the internal processes. As the names suggest, the external processes are those that treat the water before it enters the boiler. The internal treatments are those in the form of added chemicals that treat the water inside the boiler. Only the external processes are described here.

The 'hot lime' process

This is a water-softening process which uses a hot lime contact to induce a precipitate of the compounds contributing to the hardness. The sludge formed is allowed to settle out. Very often coagulation chemicals such as alum or iron salts are used to enhance the settling and the removal of the sludge formed. In most plants that use the 'lime' process the reaction by the addition of lime and soda ash is carried out at elevated temperatures. However, the reaction can be allowed to take place at ambient temperatures. The hardness of the water from the 'cold' process will be about 17–35 ppm while that from the 'hot' process will be 8–17 ppm. Clean-up filters containing anthracite are often used to finish the treating process.

The ion exchange processes

As the name implies, this process exchanges undesirable ions contained in the raw water with more desirable ones that produces acceptable BFW. For example, in the softening process, calcium and magnesium ions are exchanged for sodium ions. In dealkylization, the ions contributing to alkalinity (carbonates, bicarbonates, etc.) are removed and replaced with chloride ions. Demineralization in this process replaces all cations with hydrogen ions (H^+) and all anions with hydroxyl ions (OH^-) making pure water ($H^+ + OH^-$).

The ion exchange material needs to be regenerated after a period of operation. The operating period will differ from process to process and will depend to some extent on the amount of impurities in the water and the required purity of the treated water. Regeneration is accomplished in three steps:

- Backwashing
- Regenerating the resin bed with regenerating chemicals
- Rinsing.

Figure 2.20 shows the internals of a typical ion exchange unit. Under operating conditions the raw water is introduced through the top connection and distributor. The water flows through the resin bed where ion exchange takes place. The treated water is removed via the bottom connection. Under regeneration operation, raw water as backwash is introduced through the bottom connection and removed from the top connection. During its passage upward through the resin and support beds it 'fluffs' the beds and removes any waste material that has adhered to them. The backwash water is then sent to the plant's waste-water disposal system. Regenerating exchange chemical is introduced directly above the resin bed through a chemical distributor and allowed to flow downward to be removed at the bottom water outlet. The regenerating cycle is completed with the

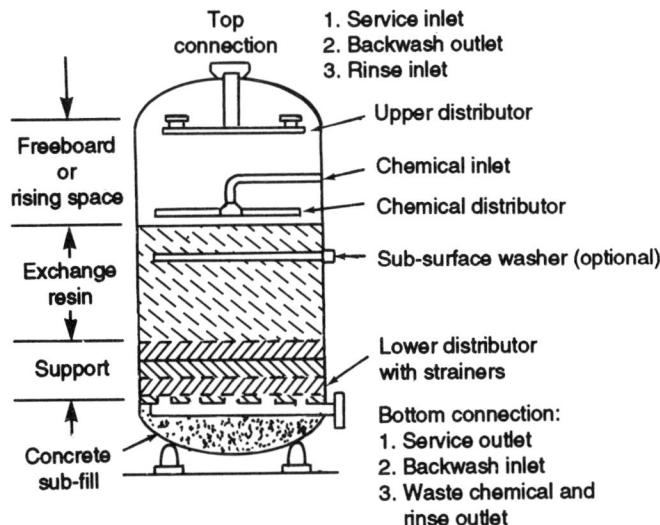

Figure 2.20. A typical ion exchange unit

rinsing of the bed to remove any surplus-regenerating chemicals. This is done by introducing a stream of raw water at the top connection and removing it from the bottom connection. This water is also disposed to waste.

Normally ion exchange units are installed in pairs. When one is operating the other is being regenerated. An automatic switch-over of electronically controlled valves takes the pair of units through the correct cycles at the prescribed time intervals, without disrupting the treating process. Figure 2.21 shows a typical 'hook-up' of an ion exchange unit.

Deaeration

The deaeration process is used in almost all boiler feed water treatments to remove dissolved gases from the water. Normally treated water and returned condensate are routed to a deaerator immediately prior to entering the boiler steam drum. Figure 2.22 is a drawing of a typical deaerator drum layout. The drum consists of a retention vessel surmounted by a degassing tower section. The degassing section contains a packed volume over which the treated water (and condensate) stream is passed. Low-pressure steam, usually let down saturated 50 psig steam or, if available 5 psig steam from condensate flash, is introduced to the bottom of the degassing section. The steam flows upward through the packed section and countercurrent to the water. This action removes the dissolved gases from the water. These gases then leave with the steam from the top of the degassing section to be vented to atmosphere. The gas-free water flows into the main retaining section of the deaerator.

The treated water feed is introduced to the degassing section of the dearator through an atomizing spray/distributor. This reduces the water stream to fine droplets prior to entering the packed section. This enhances the removal of the

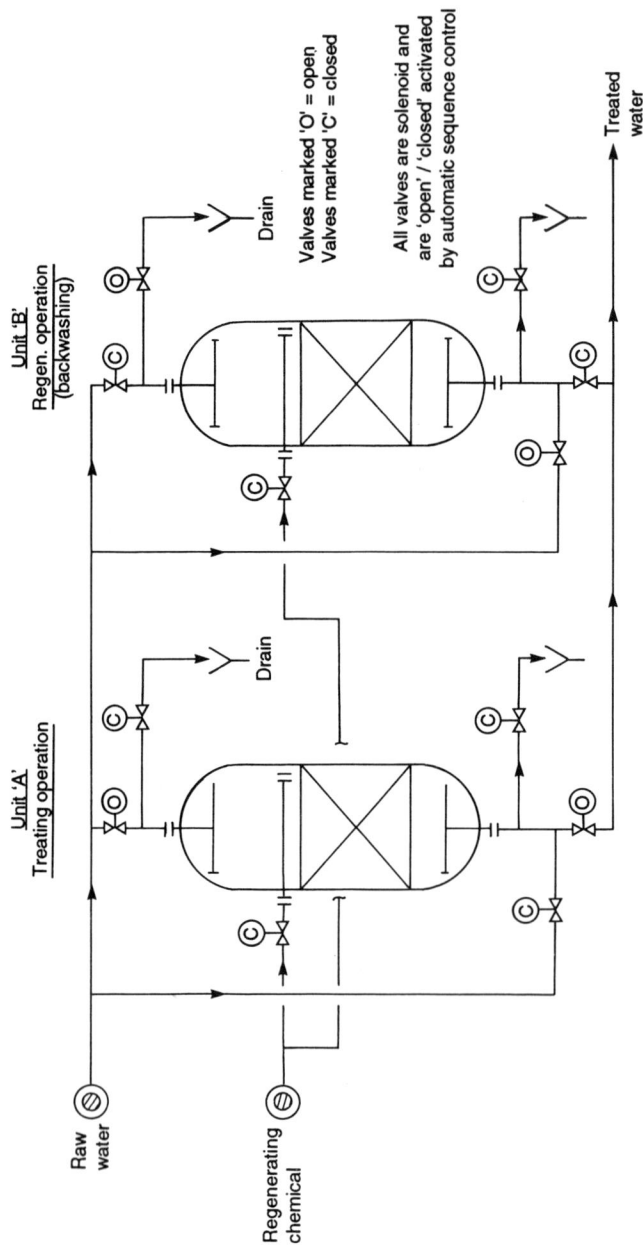

Figure 2.21. A typical ion exchange unit 'hook-up'

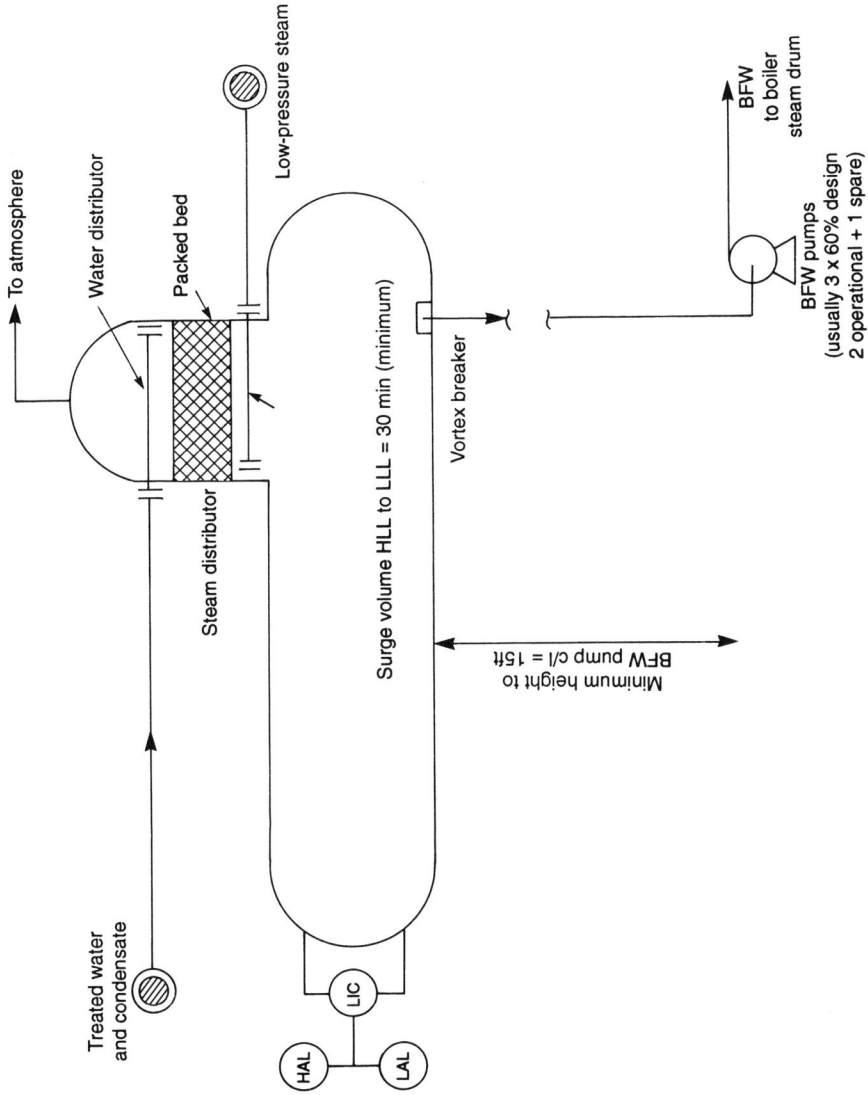

Figure 2.22. A typical deaerator drum

gases in the water. Deaerators operate at about 2–3 psig and at this pressure all the CO_2 contained in the water and most of the oxygen is removed.

The boiler feed water pumps draw suction directly from the dearator. To ensure that there is available sufficient NPSH for the pumps to operate properly, dearator drums are installed on a structure at least 15 ft above the centreline of the pump suction. Most large pumps (as BFW pumps are) usually require a relatively high NPSH when handling hot water.

The retention section of the deaerator should have as a minimum 30 min of surge between HLL and LLL. The water feed to the deaerator is normally on retention section level control. The boiler feed from the BFW pumps will be on flow control wih steam drum level reset. Very often boiler feed flow has a low flow alarm and at very low flow an automatic boiler shutdown device.

INSTRUMENT AIR SYSTEM

All chemical plants require a supply of compressed air to operate the plant and for plant maintenance. There are usually two separate systems:

- Plant air system
- Instrument air system

Plant air is generally supplied by a simple compressor with an aftercooler. Very often when plant air is required only for maintenance this is provided by a mobile compressor connected to a distribution piping system. Air for catalyst regeneration and the like is normally supplied by the regular gas compressor on the unit. Instrument air should always be a separate supply system. Compressed air for instrument operation must be free of oil and dry for the proper function of the instruments it supplies. This is a requirement which is not necessary for most plant air usage. A reliable source of clean, dry instrument air is an essential requirement for plant operation. Failure of this system means a complete shutdown of the plant.

Figure 2.23 shows a typical instrument air supply system. Atmospheric air is introduced into the suction of one of two compressors via an air filter. The compressors are usually reciprocating or screw-type non-lubricating. Centrifugal-type compressors have been used for this service when the demand for instrument air is very high (see Chapter 5 for more details). The air compressors discharge the air at the required pressure (usually above 45 psig) into an air cooler before the air enters one of two driers. One of the compressors is in operation while the other is on standby. The operating compressor is usually motor driven with a discharge pressure operated on/off start-up switch. The standby compressor is turbine (or diesel engine) driven with an automatic start-up on a low/low discharge pressure switch.

The cooled compressed air leaves the cooler to enter the driers. There are two drier vessels each containing a bed of desiccant material. This material is either silica gel (the most common), alumina or, in special cases, zeolite (molecular sieve). One of the two driers is in operation with the compressed air flowing through it to be dried and to enter the instrument air receiver. The desiccant in

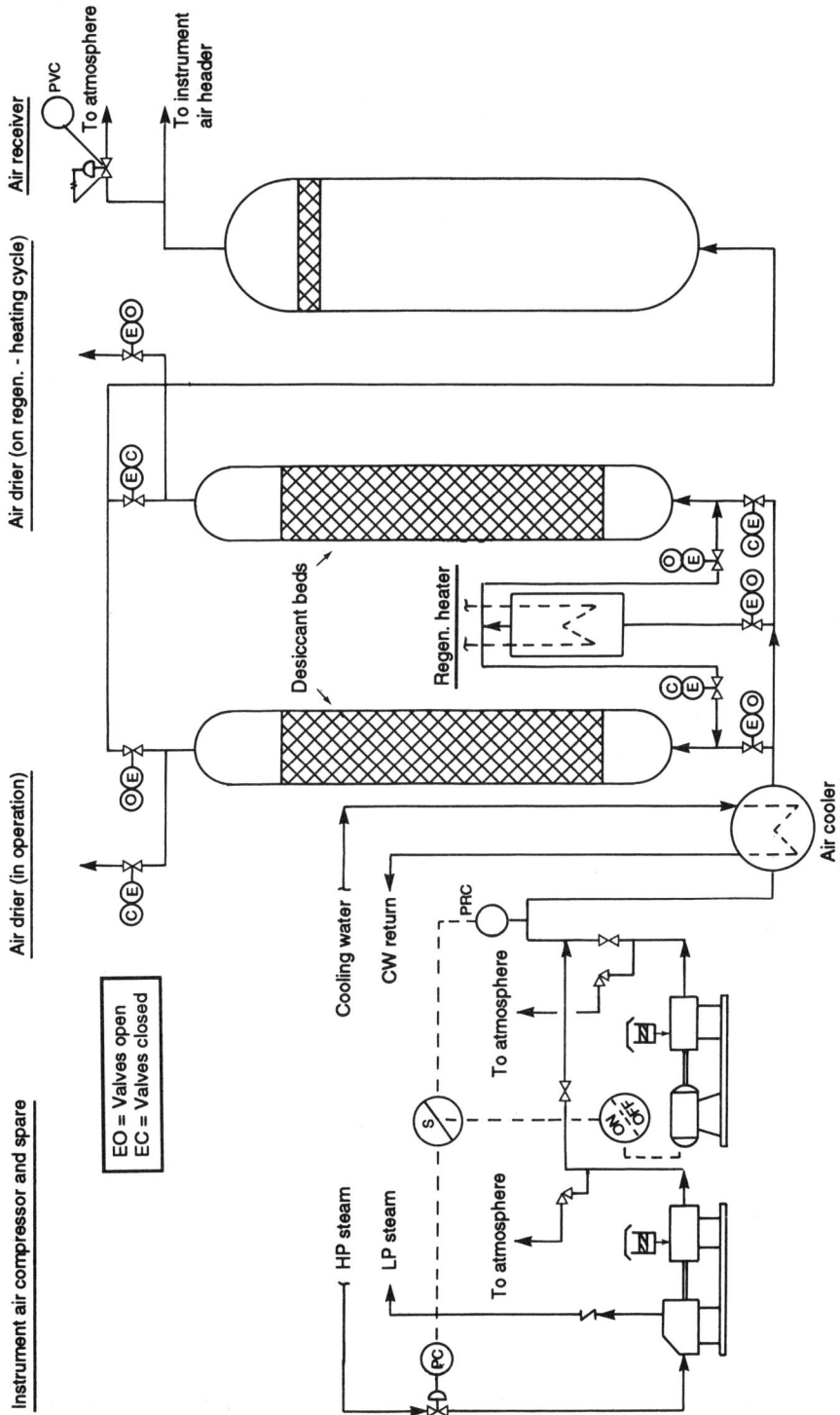

Figure 2.23. Instrument air supply system

the other drier is simultaneously regenerated. Regeneration of the desiccant bed is effected by passing through the bed a stream of heated air and venting the stream to atmosphere. This heated stream removes the water from the desiccant to restore its hygroscopic properties. At the end of this heating cycle cooled air is reintroduced to cool down the bed to its operating temperature. When cool, the unit is then shut in ready to be switched into operation for the first drier to start its regeneration cycle. The various operating and regeneration phases are automatically obtained by a series of selonoid valves operated by a sequence timer switch control. These driers (often including the compressor and receiver items) are packaged units supplied, skid mounted, and ready for operation.

The instrument air receiver vessel is a pressure vessel containing a crinkled wire mesh screen (CWMS) before the outlet nozzle. It is high pressure protected by a pressure control valve venting to atmosphere, and, of course, is also protected by a pressure safety valve (not shown in Figure 2.23). The air leaves the top of this vessel to enter the instrument air distribution system servicing all the plants in the complex.

2.4   Offsite Systems

All chemical complexes require support systems to service the processes contained in the plant. The most common are the utility systems and the offsite systems. The first of these has been discussed in the previous section. This section will therefore deal with some aspects of the offsite systems.

Among the major units found in the offsites of most plants are:

• Storage facilities
• Road and rail loading facilities
• Waste disposal facilities
• Effluent water-treating facilities

In oil refining, offsites may also include product blending, TEL handling facilities, large jetty facilities for tankers and smaller jetty facilities for product barges.

STORAGE FACILITIES

Plant storage facilities can be classified into the following four categories:

• Atmospheric storage
• Pressure storage
• Refrigerated storage
• Heated storage

Cone-roofed tanks

Atmospheric storage are tanks which are open to the atmosphere and are at atmospheric pressure. The simplest of these is the open tank which is generally used for the storage of raw water and fire water. Among the most common is the cone-roofed tank shown in Figure 2.24(a). This tank is used for the storage of non-toxic liquids with fairly low volatility. In its simplest form the roof of the tank will contain a vent, open to atmosphere, which allows the tank to 'breathe' when emptying and filling. A hatch in the roof also provides access for sampling the tank contents. In oil refining this type of tank is used for the storage of gas oils, diesel, light heating oil, and the very light lube oils (e.g. spindle oil).

In keeping with the company's fire protection policy, tanks containing

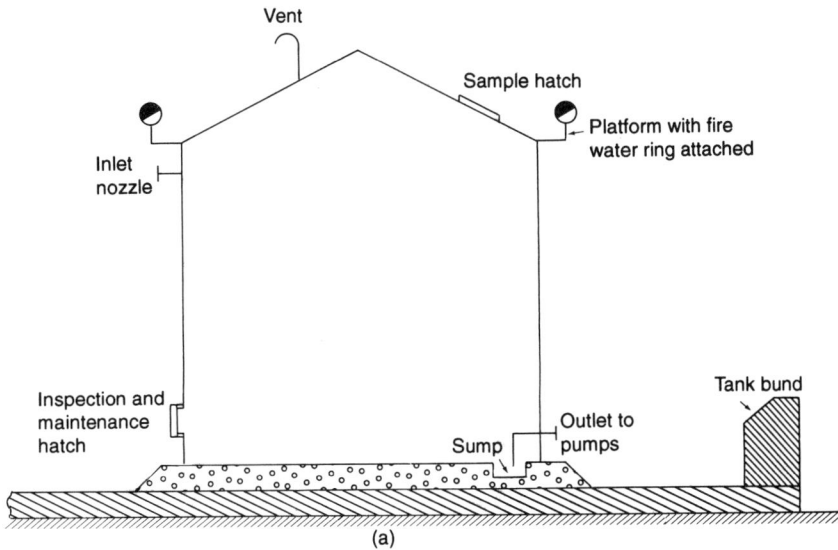

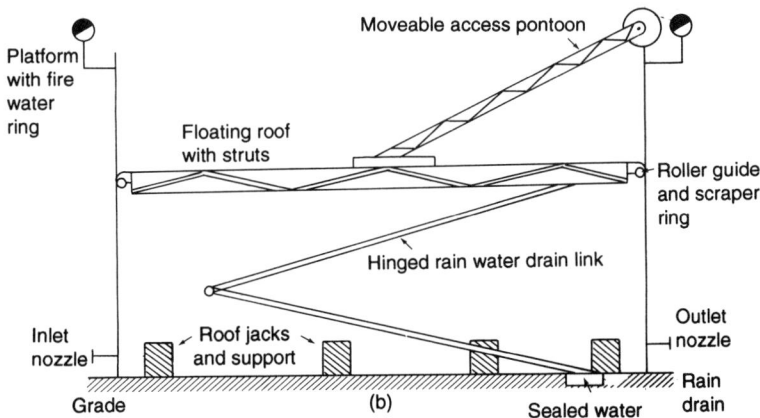

Figure 2.24. Atmospheric storage tanks. (a) A typical cone-roofed tank; (b) a typical floating-roof tank (half-full)

flammable material will be equipped with foam and fire water jets located around
the base of the roof. All storage tanks containing flammable material and material
that could cause environmental damage are contained within a dyked area or
bund. The size of the bunded area is fixed by law and is usually such as to contain
the total contents of one of the tanks included in the area. The number of tanks
per bunded area is also fixed by legislation.

Floating-roofed tanks

Light volatile liquids may also be stored at essentially atmospheric pressure by the
use of 'floating-roof' tanks. A diagram of this type of storage tank is given in
Figure 2.24(b). The roof literally floats on the surface of the liquid contents of the
tank. In this way the air space above the liquid is reduced to almost zero, thereby
minimizing the amount of liquid vaporization that can occur. The roof is specially
designed for this service and contains a top and bottom skin of steel plate, held
together by steel struts. These struts also provide strength and rigidity to the roof
structure. The roof moves up and down the inside of the tank wall as the liquid
level rises when fillings and falls when emptying. The roof movement is enhanced
by guide rollers between the roof edge and the tank wall. A scraper ring around
the edge of the roof top and pressed tightly against the tank wall ensures a seal
between the contents and the atmosphere. It also provides an additional guide to
the roof movement and stability to the roof itself.

When the tank reaches the minimum practical level for the liquid contents the
roof structure comes to rest on a group of pillars at the bottom of the tank. These
provide the roof support when the tank is empty and a space between the roof
and the tank bottom. This space is required to house the liquid inlet and outlet
nozzles for filling and emptying the tank which, of course, must always be below
the roof. The space is also adequate to enable periodic tank cleaning and
maintenance.

A hinged drain line running inside the tank from the roof to a 'below grade'
sealed drain provides the facility for draining the roof of rain water. The hinges in
the drain line allows the line to move up and down with the roof movement. A
pontoon-type access pier from the platform around the perimeter at the top of the
tank provides access to the sample hatch located at the centre of the roof. This
'pontoon' also moves upward and downward with the roof movement.

Liquids stored in this type of tank have relatively high volatilities and vapour
pressures such as gasoline, kerosene, benzene, toluene, etc. In oil refining the
break between the use of cone-roofed and floating-roof tanks is based on the
'flash point' of the material. Flash point is that temperature above which the
material will ignite or 'flash' in the presence of air. Normally this break point is
120 °F.

Pressure storage

Pressure storage tanks are used to prevent or at least minimize the loss of the tank
contents due to vaporization. These types of storage tanks can range in operating

pressures from a few inches of water gauge to 250 psig. There are three major types of pressure storage:

- Low-pressure tanks—These are dome-roofed tanks and operate at a pressure of between 3 in. water gauge and 2.5 psig.
- Medium-pressure tanks—These are hemispheroids which operate at pressures between 2.5 and 5.0 psig, and spheroidal tanks which operate at pressures up to 15 psig.
- High-pressure tanks—These are either horizontal 'bullets' with ellipsoidal or hemispherical heads or spherical tanks (spheres). The working pressures for these types of tanks range from 30 psig to 250 psig.

Although it is possible to store material in tanks with pressure in excess of 250 psig, normally when such storage is required refrigerated storage is usually a better alternate.

Refrigerated storage

Many chemical complexes store liquified gas in their tank farms. The most common that are stored in fairly large bulk are ethane, ethylene, SNG (synthetic natural gas), and LNG (liquified natural gas). Three types of tanks are used for this service: spheres, bullets, and flat-bottom specially designed tanks. All three of these types can be constructed with a single wall or a double-wall configuration. The choice is usually one of economics, and is based on a very thorough economic evaluation, encompassing insulation effectiveness, and the design of the refrigerating cycle. Careful selection of the materials of construction for refrigerated tankage is essential in order to prevent risk of brittle fracture. Consideration should be given to materials such as aluminium, copper, nickel, and stainless steel for this service. For very low temperatures 9% nickel steel is frequently used because of its low cost.

Single-wall tanks can be used effectively to temperatures as low as −50 °F. Below this temperature, double-wall construction with efficient insulating material should be considered. For small volumes, high-pressure refrigerated or non-refrigerated horizontal 'bullets' may prove to be the best choice. By high pressure in this case is meant 20–250 psig. Spheres are used for storing large quantities of light material. These vessels operate at pressures up to 30 psig and may be refrigerated with single- or double-wall construction.

When input to the tank is small in proportion to the tank capacity a flat-bottomed tank properly insulated and refrigerated is usually found to be the most economic design. These tanks operate at between 0.5 psig and 2.0 psig. One distinctive feature of this type of vessel is the use of suspended insulation deck. It consists of a lap-welded metal deck suspended from the tank roof framing with mineral wool spread uniformly over the deck. Simple open pipe vents are installed through the deck to prevent differential pressures across it. Superheated vapours remain stratified in the upper space and colder, nearly saturated, vapours are under the deck.

For storage at very low temperatures double-walled flat-bottomed tank construction may be considered. In the economic evaluation for this type of construction some consideration must be given to the greater effectiveness of the insulation and to its longer life. Double-walled spherical tanks are usually the most economical for storing capacities below 35 000 barrels.

Double-wall flat-bottomed tanks are vertical cylinders with end enclosures designed for pressures near atmospheric. Roofs are either ellipsoidal or spherical segments. The inner container rests on some form of load-bearing insulation which carries the weight of the contents. The most common insulation material used to fill the annular space is expanded perlyte. For these flat-bottomed tanks, foundations present an additional problem. The container is a heat sink, and if no provision is made to supply heat a large quantity of soil below the tank would eventually reach temperatures below that of freezing water. Moisture in the subsoil would then freeze and heaving would occur. To prevent this happening, a heat source is provided just below the outer tank bottom to maintain the temperature above 32 °F. Electric heating cables or circulating hot water (or LP steam) are an effective means for this heating requirement. Alternatively, the tank may be elevated and supported just above grade to prevent this freezing.

Refrigeration cycles

Refrigeration is usually used to reliquefy vapours that have built up in refrigerated storage vessels. Reliquefication is normally affected by circulating a stream of the liquid material from the storage tank through an external refrigeration system to reduce its temperature. With sufficient circulation and with appropriate reduction in its temperature the contents of the tank is reduced to an extent that the total inventory is kept substantially as liquid.

There are many types of refrigeration systems or cycles available from manufacturers specializing in developing, designing, and installing such systems. Only the most common or simple systems are described briefly here and these are as follows:

- *Direct compression* (see Figure 2.25) This is the simplest of all the refrigeration systems. It may be used wherever ambient air or water is available colder than the critical temperature of the stored liquid. In most areas this system can be used for the refrigeration of propane and butane LPG, and ammonia.
- *Simple cycle* (see Figure 2.26) This cycle makes use of external refrigerant circulation. This system is the basis of most refrigeration design. In this cycle, as with most others, the liquid to be cooled is heat exchanged against a cold refrigerant such as ammonia or Freon in a kettle-type heat exchanger. The storage liquid flows tubeside in the kettle and vaporizes the refrigerant on the shell or kettle side. The control of the system is a function of the liquid level in the refrigerant liquid receiver. This cycle is used on a close-coupled system where the condenser can be positioned above the vaporizer, and where only one product condenser is needed.
- *Refrigeration with cascade system* (see Figure 2.27) Cascade refrigeration is used when the temperature difference between evaporator and condenser

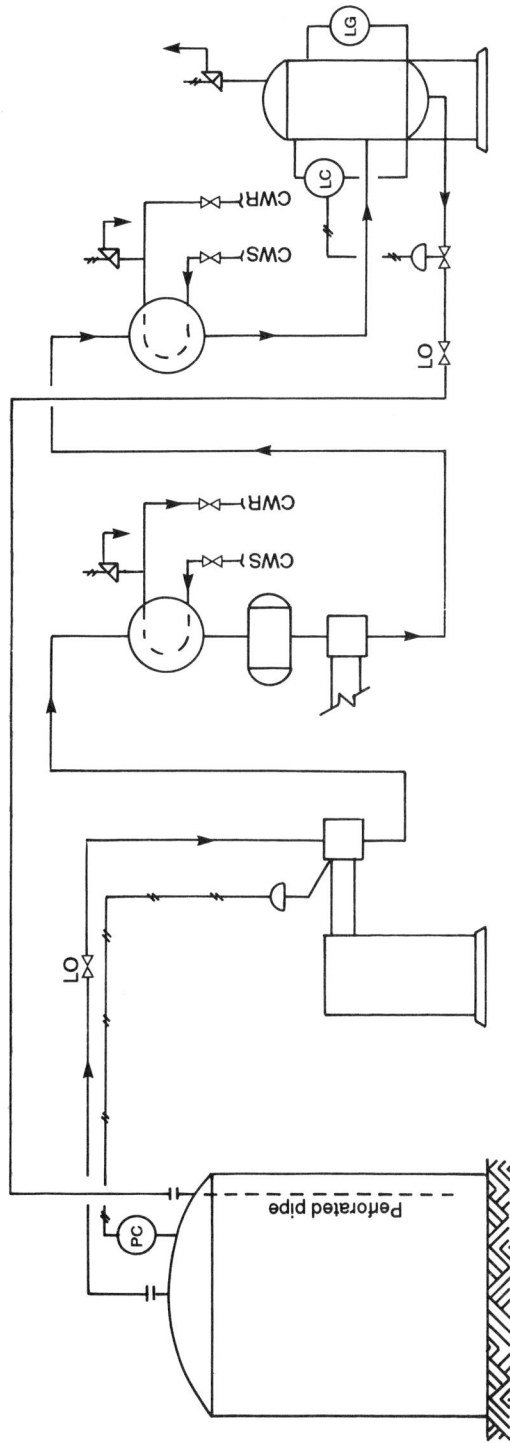

Figure 2.25. Direct compression refrigeration

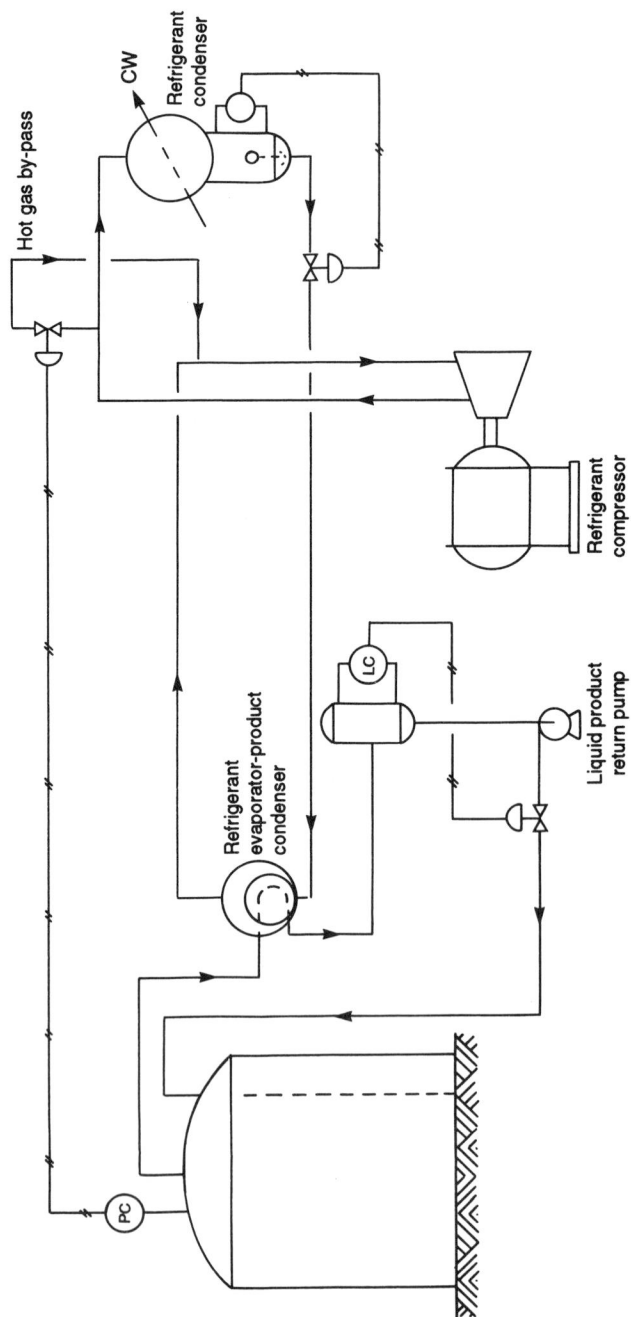

Figure 2.26. A simple cycle refrigeration system

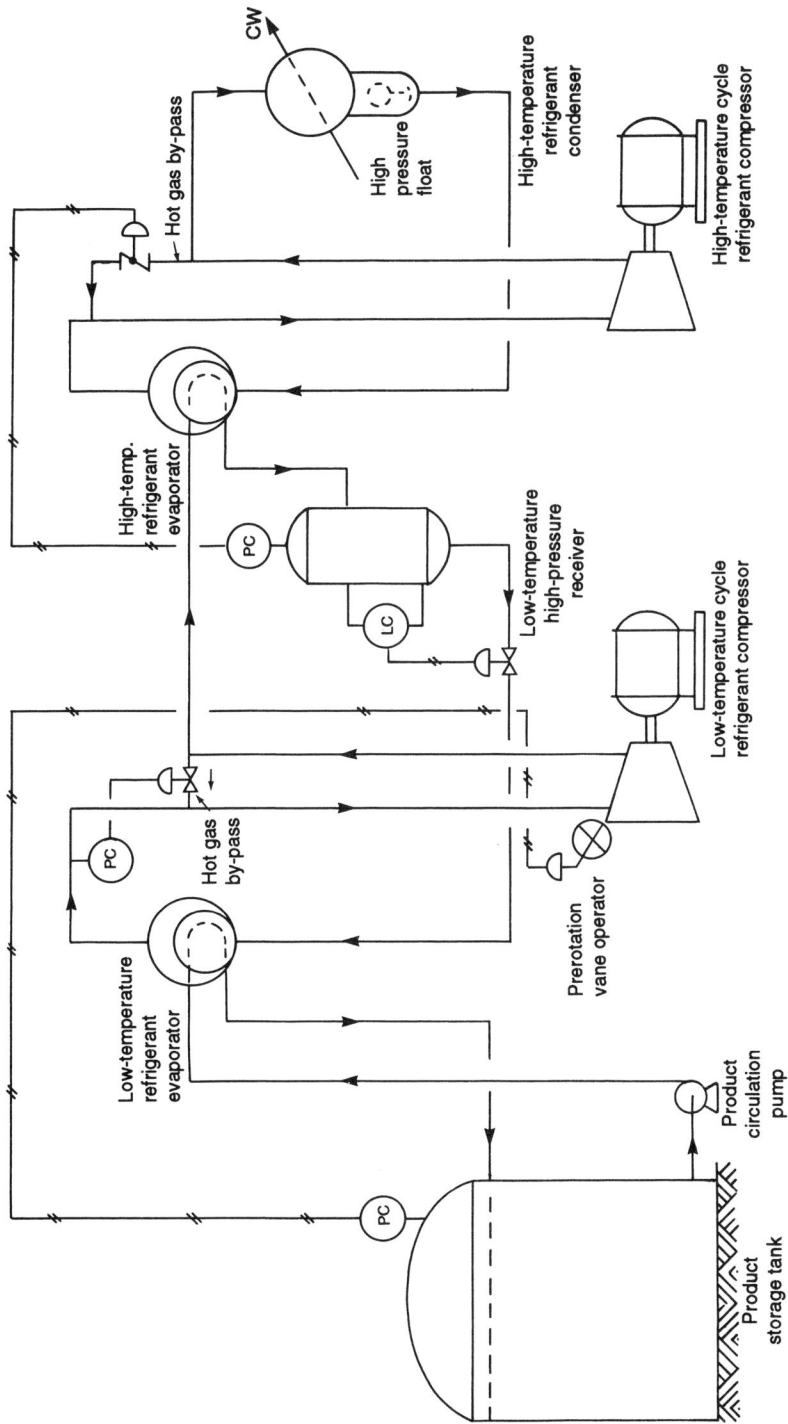

Figure 2.27. Refrigeration with cascade streams

exceeds the practical range of one refrigerant. Two or more systems are then integrated so that the evaporator in the high-temperature cycle condenses the refrigerant in the next lower-temperature cycle. These systems are used with the process evaporator at temperatures of $-80\,°F$ to $-160\,°F$.

Heated storage tanks

Heated storage tanks are more common in the chemical and petroleum industry than most others. They are used to store material whose flowing properties are such as to restrict flow at normal ambient temperatures. In the chemical industry this applies to most heavy olefins and aromatics such as para-xylene and heavier. In the petroleum industry products heavier than diesel oil, such as heavy gas oils, lube oil and fuel oil, are stored in heated tanks.

Generally, tanks are heated by immersed heating coils or bayonet-type immersed heaters. Steam is normally used as the heating medium because of its availability in most chemical or petroleum complexes. Very often where immersed heating is used the tank is agitated usually by side-located propeller agitators for large tanks and a single centrally located paddle-type agitator for the small tanks. Where external circulating heating is used for tanks, the contents are mixed by means of jet mixing. Here the hot return stream is introduced into the tank via a specially designed jet nozzle as shown in Figure 2.28. External tank heating is used when there is a possibility of a hazardous situation occurring if an immersed heater leaks.

Calculating heat loss and heater size for a tank

Heat loss and the heater surface area to compensate for the heat loss may be calculated using the following technique:

- *Step 1* Establish the bulk temperature for the tank contents. Fiz the ambient air temperature and the wind velocity normal for the area in which the tank is to be sited.
- *Step 2* Calculate the inside film resistance to heat transfer between the tank contents and the tank wall. The following simplified equation may be used for this:

$$h_c = 8.5(\Delta t/\mu)^{0.25}$$

where

h_c = inside film resistance to wall in $BTU/h.ft^2.°F$
Δt = temperature difference between the tank and the wall temperature (°F)
μ = the viscosity of the tank contents at the bulk temperature (cP)

The heat loss calculation is iterative with assumed temperatures being made for the tank wall.

- *Step 3* Using the assumed wall temperature made in step 2 calculate the heat loss to atmosphere by radiation[7] using Figure 2.29. Then calculate the heat loss from the tank wall to the atmosphere[8] using Figure 2.30. Note the

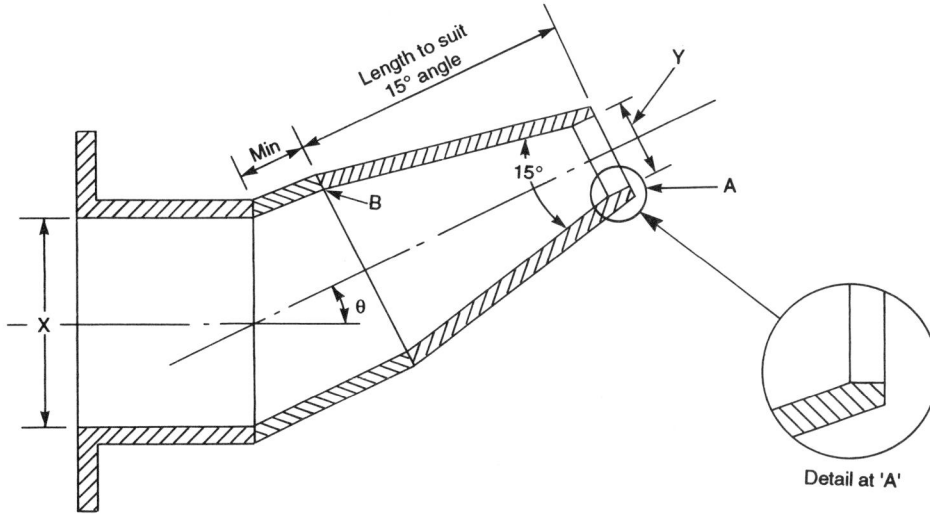

Figure 2.28. A typical jet nozzle for tank mixing. (1) Direct nozzle diametrically across tank. (2) Smooth off rough edges at A, B, etc. (3) Nozzle assembly to fit through standard tank inlet and to be removable from outside of tank

temperature difference in this case is that between the assumed wall temperature and the ambient air temperature. Correct these figures by multiplying the radiation loss by the emissivity factor given in Figure 2.29. Then correct the heat loss by convection figure by the factors as described in step 4 below.

● *Step 4* The value of h_{co} read from Figure 2.30 is corrected for wind velocity and for shape (vertical or horizontal) by multiplying with the following:

Shape factors:

$$\text{Vertical plates} - 1.3$$

$$\text{Horizontal plates} - 2.0 \text{ (facing up)}$$

$$1.2 \text{ (facing down)}$$

Correction for wind velocity:

$$\text{Use } F_W = F_1 + F_2$$

where

$$F_W = \text{wind correction factor}$$
$$F_1 = \text{Wind factor at 200 °F}$$

calculated from:

$$F_1 = (\text{mph}/1.47)^{0.61}$$
$$F_2 = \text{read from Figure 2.31}$$

Then the corrected h_{co} is:

$$h_{co} \times \text{shape correction} \times F_W.$$

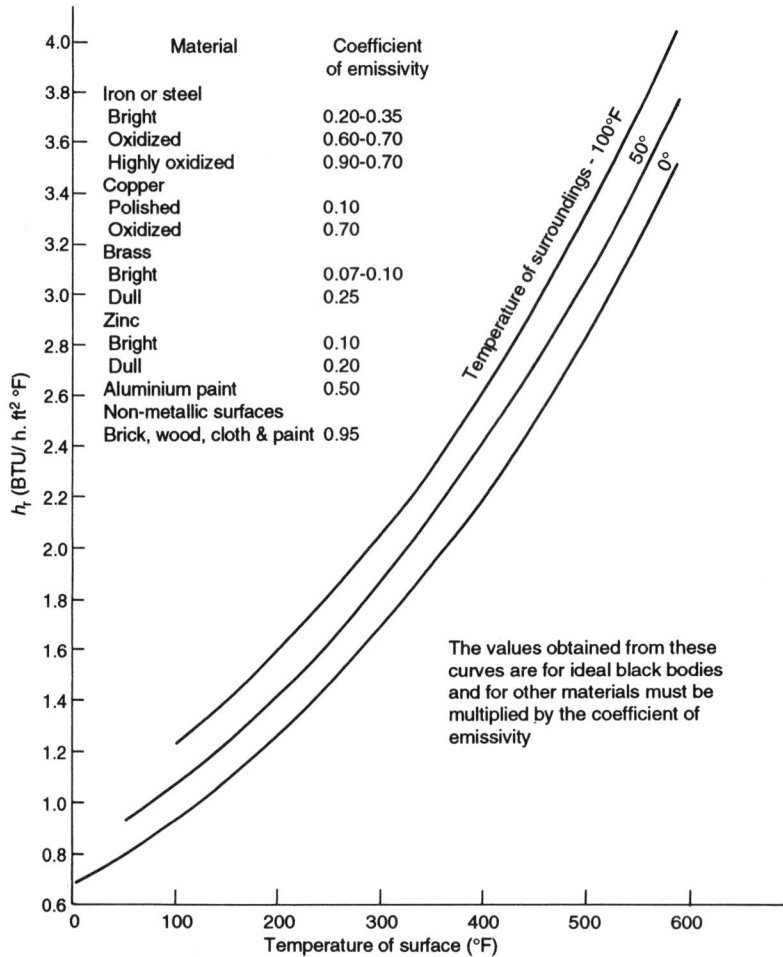

Figure 2.29. Heat loss by radiation. The values obtained from these curves are for ideal black bodies, and for other materials must be multiplied by the coefficient of emissivity (Reproduced by permission of Kreiger Publishing Company, Malabar, Florida, 1950, Maxwell, Data Book on Hydrocarbons)

- *Step 5* The resistance of heat transferred from the bulk of the contents to the wall must equal the heat transferred from the wall to the atmosphere. Thus:

Heat transferred from the bulk to the wall
$$= a$$
$$= h_c \text{ from step 2} \times \Delta t \text{ in BTU/h.ft}^2$$

where Δt in this case is (bulk temp. $-$ assumed wall temp.)

Heat transferred from the wall to the atmosphere
$$= b$$
$$= (h_{co} + h_r) \times \Delta t \text{ in BTU/h.ft}^2$$

where Δt in this case is (assumed wall temp. $-$ air temp.)

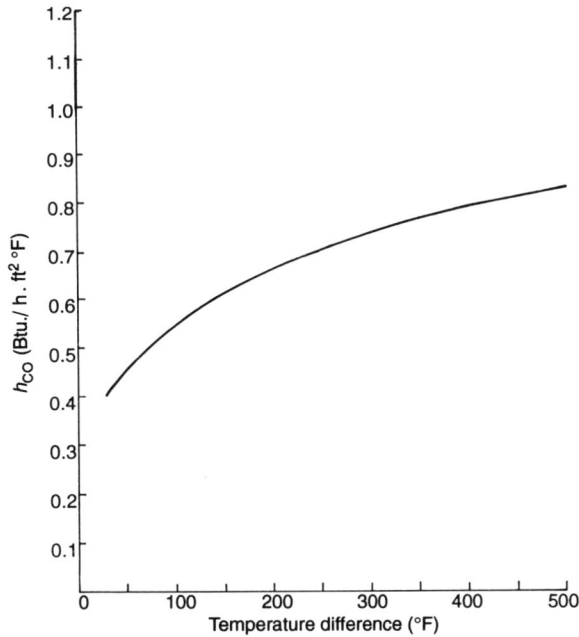

Figure 2.30. Heat loss to atmosphere by natural convection (Reproduced by permission of Kreiger Publishing Company, Malaba, Florida, 1950, Maxwell, Data Book on Hydrocarbons)

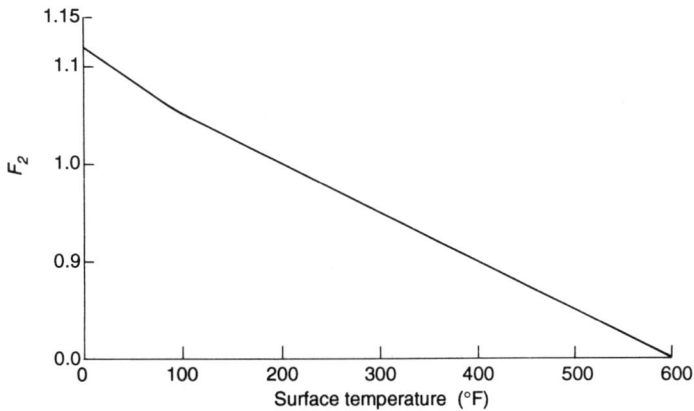

Figure 2.31. Plot of F_2 versus surface temperature

- *Step 6* Plot the difference between the two transfer rates against the assumed wall temperature. This difference $(a - b)$ will be negative or positive but the wall temperature that is correct will be the one in which the difference plotted $= 0$. Make a last check calculation using this value for the wall temperature.
- *Step 7* The total heat loss from the wall of the tank is the value of a or b calculated in step 6 times the surface area of the tank wall. Thus:

$$Q_{wall} = h_c \times \Delta t \times (\pi D_{tank} \times \text{tank height}) \text{ (BTU/h)}$$

- *Step 8* Calculate the heat loss from the roof in the same manner as that for the wall described in steps 2 to 7. Note the correction for shape factor in this case will be for horizontal plates facing upward and the surface area will that for the roof.
- *Step 9* Calculate the heat loss through the floor of the tank by assuming the ground temperature as 50 °F and using

$$h_f = 1.5 \, \text{BTU/h.ft}^2.°\text{F}$$

- *Step 10* Total heat loss then is

$$\text{Total heat loss from tank} = Q_{wall} + Q_{roof} + Q_{floor}$$

- *Step 11* Establish the heating medium to be used. Usually this is medium-pressure steam. Calculate the resistance to heat transfer of the heating medium to the outside of the heating coil or tubes. If steam is used then take the condensing steam value for h as 0.001 BTU/h.ft^2.°F. Take value of steam fouling as 0.0005 and tube metal resistance as 0.0005 also. The outside fouling factor is selected from the following:

Light hydrocarbon = 0.0013
Medium hydrocarbon = 0.002
Heavy hydrocarbon such as fuel oils = 0.005

The resistance of the steam to the tube outside $= \dfrac{1}{h + R}$

where $R = r_{\text{steam fouling}} + r_{\text{tube metal}} + r_{\text{outside fouling}}$.
- *Step 12* Assume a coil outside temperature. Then using the same type of iterative calculation as for heat loss, calculate for a as the heat from the steam to the coil outside surface in BTU/h.ft^2. That is,

$$a = h \times \Delta t_i$$

Calculate for b as the heat from the coil outside surface to the bulk of the tank contents. Use Figure 2.32 to obtain h_0 and again b is $h_0 \times \Delta t_0$ where Δt_0 is the temperature between the tube outside and that of the bulk tank contents. Make further assumptions for coil outside temperature until $a = b$.
- *Step 13* Use a or b from step 12 which is the rate of heat transferred from the heating medium in BTU/h.ft^2 and divide this into the total heat loss calculated in step 10. The answer is the surface area of the immersed heater required for maintaining tank content's bulk temperature.

Example calculation

Problem

It is required to calculate the surface area for a heating coil which will maintain the bulk temperature of fuel oil in a cone-roofed tank at a temperature of 150 °F. The ambient air temperature is an average 65 °F and the wind velocity averaged over the year is 30 mph. The fuel oil data are as follows:

Viscosity (μ) = 36 cP at 150 °F
SG at 150 °F = 0.900

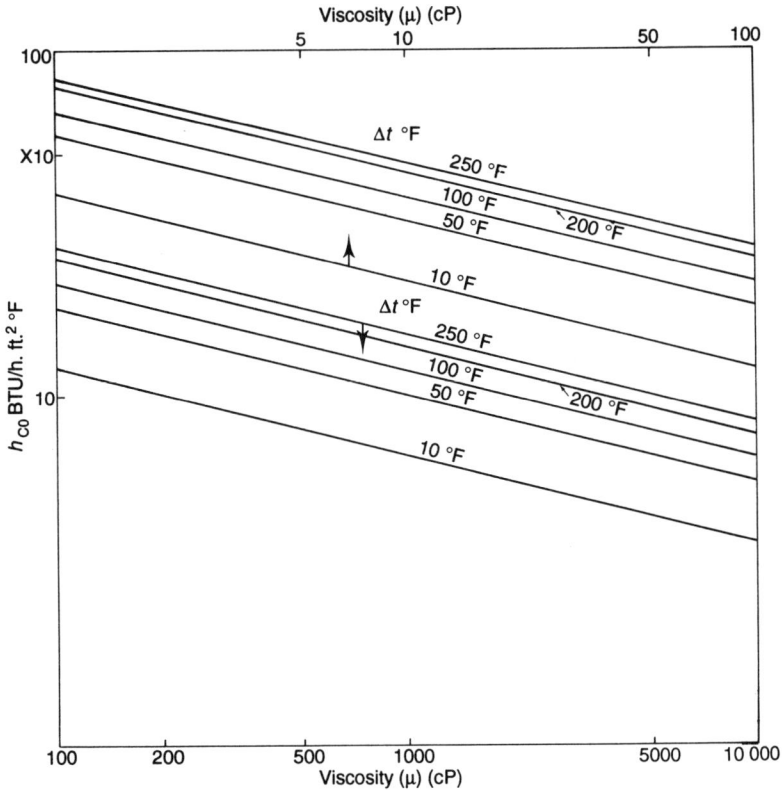

Figure 2.32. Convection heat transfer coefficient (for sizing heating coils)

The tank is to be heated with 125 psig saturated steam. The tank dimensions are 60 ft diameter × 180 ft high. It is not insulated but is painted with non-metallic colour paint.

Solution

1   Calculating the heat loss from the wall

1st trial.   Assume wall temperature is 120 °F

$$h_i = 8.5 \, (\Delta t_i / \mu)^{0.25}$$

where

$$\Delta t_i = 150 - 120 = 30 \, °F$$

$$h_i = 8.5 \times 0.955 = 8.12 \, \text{BTU/h.ft}^2.°F$$

$$'a' = 8.12 \times 30 = 243.6 \, \text{BTU/h.ft}^2$$

Δt_0 is temperature difference between assumed wall temperature and the ambient air $= 120 - 65 = 55 \, °F$

$$h_{c0} = 0.495 \times 1.3 \, (\text{from Figure 2.30})$$

Wind correction factor F_W is as follows:

$$F_1 = (\text{mph}/1.47)^{0.61}$$

$$= 6.29$$

$$F_2 = 1.04 \text{ (from Figure 2.31)}$$

$$F_W = 6.29 + 1.04$$

$$= 7.33$$

$$h_{c0} \text{ (corrected)} = 0.495 \times 1.3 \times 7.33$$

$$= 4.20 \text{ BTU/h.ft}^2.°F$$

Heat loss from wall due to radiation h_{r0} is found from Figure 2.29 = 1.18 BTU/h.ft^2.°F
Corrected for emissivity

$$h_{r0} = 1.18 \times 0.95 = 1.123 \text{ BTU/h.ft}^2.°F$$

$$b = (h_{c0} + h_{r0}) \times \Delta t_0$$

$$= (4.21 + 1.12) \times 55$$

$$= 293 \text{ BTU/h.ft}^2$$

$$a - b = 244 - 293$$

$$= -49 \text{ BTU/h.ft}^2$$

2nd trial. Assume wall temperature is 110 °F

$$h_{i0} = 8.72 \text{ BTU/h.ft}^2.°F$$

$$a = 349 \text{ BTU/h.ft}^2$$

$$h_{c0} \text{ (corrected)} = 4.0 \text{ BTU/h.ft}^2.°F$$

$$h_{r0} \text{ (corrected)} = 1.06 \text{ BTU/h.ft}^2.°F$$

$$b = (4.0 + 1.06) \times 35$$

$$= 177.1 \text{ BTU/h.ft}^2$$

$$a - b = +172 \text{ BTU/h.ft}^2$$

3rd trial. Assume wall temperature is 115 °F

$$h_{i0} = 8.44 \text{ BTU/h.ft}^2.°F$$

$$a = 8.44 \times 35 = 295.4 \text{ BTU/h.ft}^2.°F$$

$$h_{c0} \text{ (corrected)} = 4.09 \text{ BTU/h.ft}^2.°F$$

$$h_{r0} \text{ (corrected)} = 1.11 \text{ BTU/h.ft}^2.°F$$

$$b = (4.09 + 1.11) \times 50 = 260 \text{ BTU/h.ft}^2$$

$$a - b = +35$$

The results of the above trials are plotted linearly below:

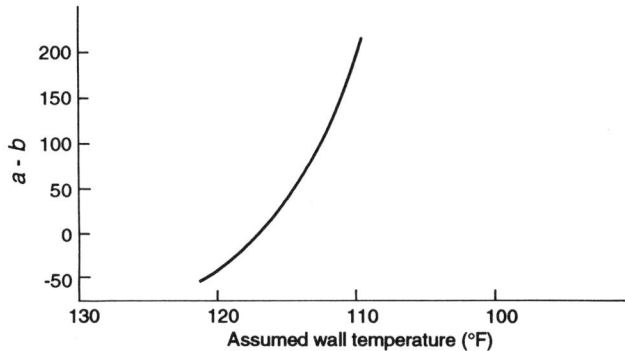

Final trial.   At wall temperature of 117 °F

$$h_{i0} = 8.31 \text{ BTU/h.ft}^2.°F$$

$$a = 8.31 \times (150 - 117) = 274 \text{ BTU/h.ft}^2$$

$$h_{c0} \text{ (corrected)} = 4.18 \text{ BTU/h.ft}^2.°F$$

$$h_{r0} \text{ (corrected)} = 1.12 \text{ BTU/h.ft}^2.°F$$

$$b = (4.18 + 1.12) \times (117 - 65)°F$$

$$= 275.7 \text{ BTU/h.ft}^2$$

a and b are close enough call total heat loss 275.7 BTU/h.ft^2

Surface area of wall = circumference × height

$$= \pi D \times 180 \text{ ft}$$

$$= 33\,929 \text{ ft}^2$$

Total heat loss through wall

$$= 275.7 \times 33929$$

$$= \underline{9.35 \text{ mm BTU/h}}$$

2   Calculating heat loss through roof

Trial 1. . Assume roof temperature is 116 °F.

$$h_i = 8.38 \text{ BTU/h.ft}^2°F$$

$$a = 8.38 \times (150 - 116)°F$$

$$= 284.9 \text{ BTU/h.ft}^2$$

$$h_{c0} \text{ (corrected)} = (0.470 \times 2.0) \times 1.04 \times 6.29$$

(Note: the number read from Figure 2.30 is multiplied by 2.0 in this case as the roof is an upward-facing plate.)

$$= 6.35 \ \text{BTU/h.ft}^2.°\text{F}$$

$$h_{r0} \ (\text{corrected}) = 1.165 \times 0.95$$

$$= 1.11 \ \text{BTU/h.ft}^2.°\text{F}$$

$$b = (6.35 + 1.11) \times (116 - 65)°\text{F}$$

$$= 380 \ \text{BTU/h.ft}^2$$

$$a - b = -95 \ \text{BTU/h.ft}^2$$

Trial 2.   Assume a wall temperature of 110 °F

$a - b$ in this case $= +23$ which is within acceptable limits
The heat loss is taken as an average of a and b

$$= 338 \ \text{BTU/h.ft}^2$$

Total heat loss from the roof

$$= \text{area of roof} \times 338$$

$$= 2827 \ \text{ft}^2 \times 338$$

$$= \underline{0.956 \ \text{mm BTU/h}}$$

3   Calculating the heat loss through the floor

Assume the ground temperature is 50 °F and the heat transfer coefficient is 1.5 BTU/h.ft^2.°F.

$$\text{Then heat loss} = 1.5 \times (150 - 50) \times 2827 \ \text{ft}^2$$

$$= \underline{0.424 \ \text{mm BTU/h}}$$

4   Total heat loss from the tank

$$\text{Heat loss from the walls} = \ 9.350 \ \text{mm BTU/h}$$

$$\text{Heat loss from the roof} = \ 0.956 \ \text{mm BTU/h}$$

$$\text{Heat loss from the floor} = \ 0.424 \ \text{mm BTU/h}$$

$$\overline{\text{Total heat loss} = 10.730 \ \text{mm BTU/h}}$$

5   Calculating the tank heater coil surface area required

The heating medium is saturated 125 psig steam.

Temperature of the steam = 354 °F

Steam side calculations:

Approx. resistance of steam $h_s = 0.001$ h.ft^2.°F/BTU

Fouling factor on steam side $r_1 = 0.005$ h.ft^2.°F/BTU

Tube metal resistance $r_2 = 0.0005$ h.ft^2.°F/BTU

Outside fouling factor $r_3 = 0.005$ h.ft^2.°F/BTU

Heat transfer coefficient for the steam side

$$= \frac{1}{0.001 + 0.0005 + 0.0005 + 0.005}$$
$$= 143 \text{ BTU/h.ft}^2.°F$$

Oil-side heat transfer coefficient is obtained from Figure 2.32

1st trial.   Assume a tube wall temperature of 310 °F.

$$\text{For steam side } a = 143 \times (354 - 310)$$
$$= 6292 \text{ BTU/h.ft}^2$$
$$\text{For oil side } h_0 = 31 \text{ BTU/h.ft}^2.°F \text{ (Figure 2.32)}$$
$$b = 31 \times (310 - 150)$$
$$= 4960 \text{ BTU/h.ft}^2$$
$$a - b = +1332$$

2nd trial. Assume a tube wall temperature of 320 °F.

a in this case was calculated to be 4862

b was calculated to be 5355

$a - b = -493$

Plotted on a linear curve the tube wall temperature to give $a = b$ was 317 °F.

Final trial.   At a tube wall temperature of 317 °F

$$\text{Steam side} = 143 \times (354 - 317)$$
$$a = 5291 \text{ BTU/h.ft}^2$$
$$\text{Oil side } h_0 = 31.2 \text{ (from Figure 2.32)}$$
$$b = 31.2 \times (317 - 150)$$
$$= 5210 \text{ BTU/h.ft}^2$$
$$a - b = +81, \text{ which is acceptable.}$$
$$\text{Make rate of heat transfer} = \frac{5291 + 5210}{2}$$
$$= 5251 \text{ BTU/h.ft}^2$$

$$\text{Then surface area of coil required} = \frac{10.730 \text{ mm BTU/h}}{5251 \text{ BTU/h.ft}^2}$$
$$= 2043 \text{ ft}^2$$

ROAD AND RAIL LOADING FACILITIES

The extent of product shipping facilities required in a chemical or petroleum complex depends on the size of the complex, the local market, the number of different products to be shipped and the market to be supplied. Normally the shipping facilities installed in most plants is sufficient to cater for normal product handling and the flexibility required for seasonal demands. The capacity of these facilities will almost invariably exceed the plant's total production.

The most common method of shipping product is by road or rail in suitably designed tanker cars. In the case of large complexes located on coastal or riverside sites shipping by barge or ships carry the bulk of the plant products. This section, however, will deal only with dispatch by road and rail.

Loading rates

Loading rates for road and rail tankers vary from as low as 150 gpm to as high as 1000 gpm, but most terminals load at rates between 300 and 500 gpm. Road tankers have capacities from 1300 gal to 6500 gal and one tractor can haul two 6500 gal tanks. The number of loading arms required for each product to be loaded varies with:

- Truck size
- Number of loading hours per day
- Number of loading days per week
- Time for positioning, hook-up, and depositioning of the truck.

Figure 2.33 gives the number of arms or spouts required for loading a 3500-gal truck under various conditions. The conditions shown in Figure 2.33 are for filling at a rate of 300 gpm (bottom curve) and 500 gpm (top curve). The loading time is taken as the filling time per tank truck plus 10 min. Thus loading time is:

$$\frac{\text{Tank truck capacity (gal)}}{\text{gpm}} + 10 \text{ min}$$

Tank car capacity is taken as 3500 gal. Thus for the lower curve loading time is 22 min per car, and for the upper curve 17 min per car. Assuming that a single product is loaded over 4 h in an 8 h day 5 days a week, then number of trucks required per barrel/day is

$$\frac{\text{B/D} \times 42 \times 5 \text{ days} \times 8 \text{ h/day}}{3500 \text{ gal/truck} \times 20 \text{ h loading per week}} = 0.024 \text{ trucks per barrel/day.}$$

Then for 1000 B/D number of trucks per working day =

$$0.024 \times 1000 = 24 \text{ trucks/day of 4 h filling}$$

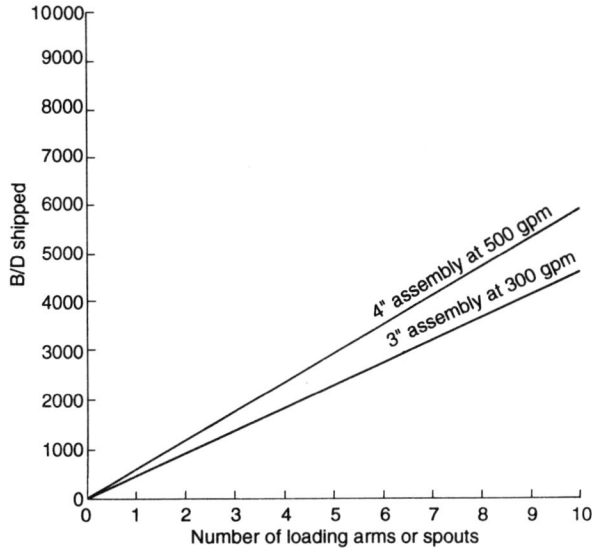

Figure 2.33. Number of loading arms for quantities shipped

Time to load trucks at 300 gpm filling rate =

$$24 \times 22 \text{ min} = 528 \text{ min}$$

To complete loading in 4 h the number of arms required =

$$\frac{528}{240} = 2.2$$

Continuing with this calculation for several more shipping capacities and for filling rates of 300 gpm and 500 gpm produces the data given in Figure 2.33.

Loading equipment

Figure 2.34 is a schematic drawing of a typical road or rail loading facility. The loading pumps which are located close to the product storage tanks take suction from these tanks. The loading pumps are high-capacity, low-head type with flat head/capacity characteristic. They operate at between 35 to 45 psi differential head. These pumps discharge through an air eliminator drum into the loading header. Several loading arm assemblies are connected to the loading header. Each of these assemblies include a remote-operated block valve followed by a desurger (optional), then a strainer located before the loading meter. The product flows from the meter into a swivel-jointed loading arm and nozzle. Tank trucks and rail cars are loaded through their top hatches into which the nozzles of the loading arm fit.

Air eliminators are used to disengage air and other vapours which would interfere with the accuracy of the meters. Disengaging of the vapours is accomplished at about 3 psig. Should there not be sufficient static head at the

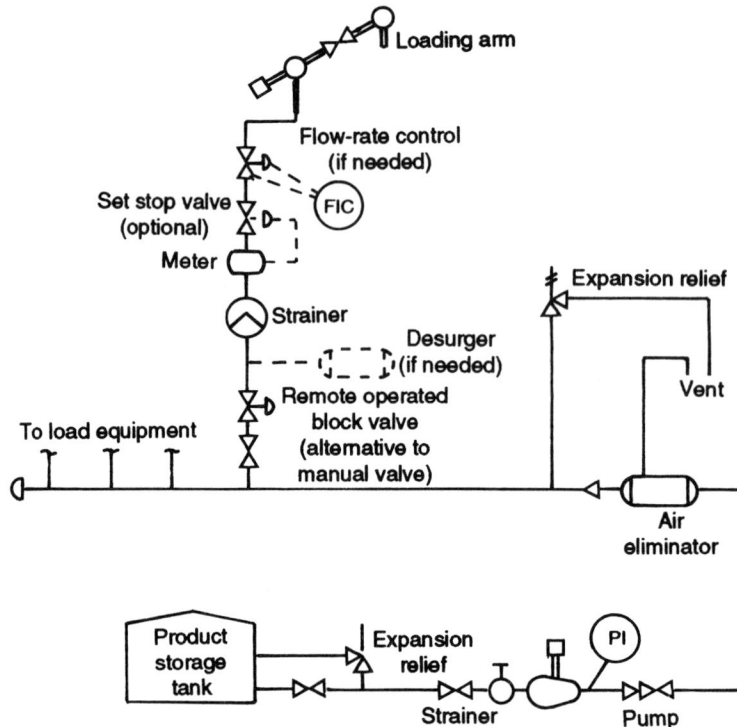

Figure 2.34. Schematic diagram of a loading facility

disengaging vessel a back-pressure valve must be provided to obtain this pressure. The meters are positive displacement type and desurgers are installed to decrease hydraulic shock resulting from quick shut-off.

Loading facilities arrangement

Figures 2.35 and 2.36 show the arrangement of loading facilities for truck and railcar respectively. The dimensions in the figures are applicable to one world area and may not be so to other localities. The equipment and its arrangement shown in the diagrams, however, are standard for most of these facilities.

In truck loading the meters and strainers are located at the loading station. The connection of the loading arm is made by an operator standing on the car itself. In the case of the truck loading facilities the loading arm is operated from an adjacent platform. As in the case of the truck loading the meter and strainer together with the on/off valve is located on the loading site.

WASTE DISPOSAL FACILITIES

All process plants including oil refineries produce large quantities of toxic and/or flammable material during periods of plant upset or emergencies. A properly

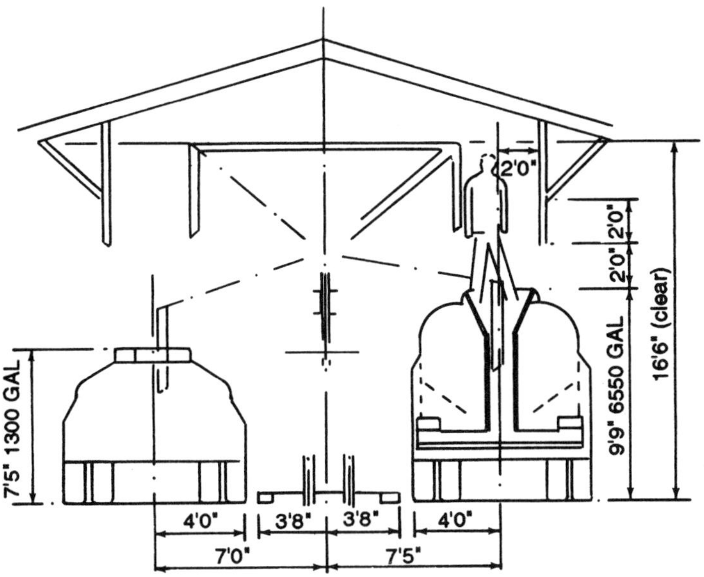

Figure 2.35. Tank truck loading facility arrangement

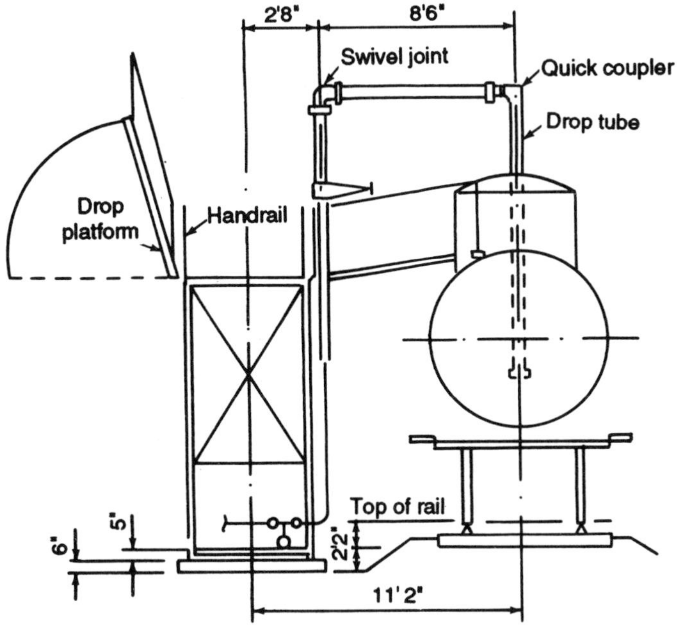

Figure 2.36. Railcar loading facility arrangement

designed flare and slop handling system is therefore essential to plant operation. This section describes and discusses typical disposal systems currently in use in the process industry where the chemical and/or hydrocarbon is immiscible with water. Where the chemical is miscible in water special separation systems must be used.

Figure 2.37 shows a completely integrated waste disposal system for the light end section of an oil refinery. Chemical plants may have similar systems with more segregation than that required for an oil refinery. The system shown here consists of three separate collection systems being integrated to a flare and a slops rerun system. A fourth system is for the disposal of the oily water drainage with a connection to the flare and a separate connection for any oil skimmings. This latter connection would be to route the skimmings to the refinery slop tanks. In

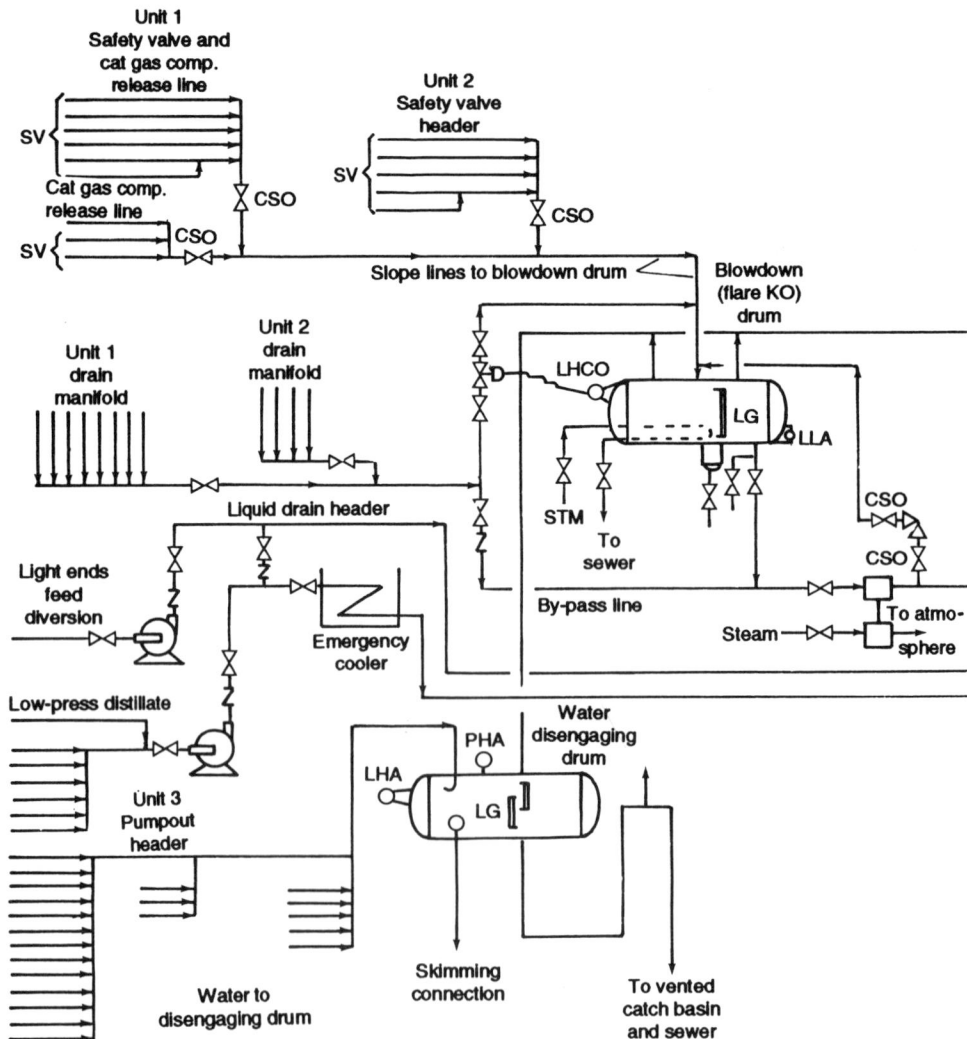

Figure 2.37. An integrated waste disposal system

the three integrated systems, the first collects all the vapour effluent streams from the relief headers. The contents of this stream will be material (normally vapour) at ambient conditions. It would be the collection of the vapours from the relief valve and the plant vapour venting on plant shut down or upset conditions. The second of the three systems is the liquid hydrocarbon drainage. The material in this system is liquid under normal ambient conditions and is collected from drain headers used to evacuate vessels during shutdown or upset conditions. Both the first and the second collection systems are routed to the flare knockout drum. The second (liquid system) may also be routed to the light ends slop storage drum. The liquid phase from the flare knockout drum is also routed to the slops storage drum. The third system is the light ends feed diversion. This allows the light ends

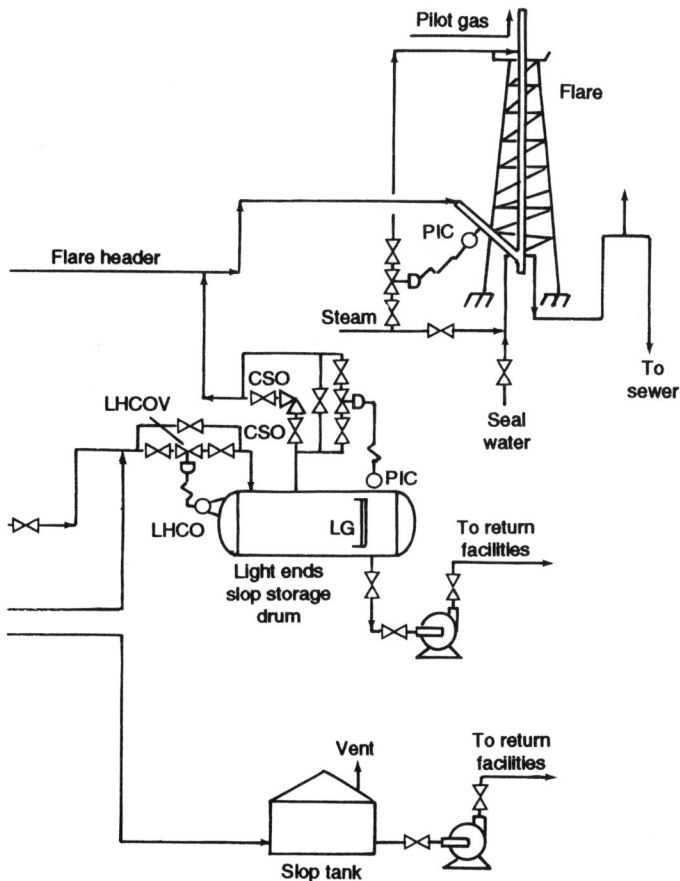

unit to be bypassed temporarily by sending the feed to the slop drum for rerunning later.

Blowdown and slop

This system generally consists of the following drums:

- Non-condensible blowdown drum
- Condensible blowdown drum
- Water disengaging drum

A typical non-condensible blowdown drum is shown in Figure 2.38. These types of drums are provided for handling material normally in the vapour state and high-volatility liquids. These drums receive and disengage liquid from safety valve headers, and drain headers. They are often referred to as flare knockout drums as the disengaged vapour is routed directly to a flare. The drum is basically a surge drum and therefore should be sized as one using the following criteria:

(1) *The liquid hold-up* The sizing of this vessel follows the same design procedure as that given in Section 3.4 of this book using the following liquid surge criteria:
  - Normal liquid surge is based on the daily liquid drawoff to drain per operating day of 24 h. This includes spillage, sample point drainings, etc.
  - The surge capacity between the HLSO (high level shut off) for normal drainage and the HLSO for feed diversion should be such as to contain the total feed to a unit routed to this drum for a period sufficient to shut-down the unit producing the feed stream. Should there be more than one unit routed to the drum then this surge capacity should be for the largest of the feed streams.
  - The capacity between the highest HLSO and the HLA (high level alarm) should be sized to handle the largest liquid volume that can be discharged in 30 min by the relief valves constituting any single risk.
  - The drum must be sized for a vapour velocity above the HLA at a maximum of 100% of the critical figure calculated by:

$$V = 0.157\sqrt{(\rho_\mathrm{l} - \rho_\mathrm{v}/\rho_\mathrm{v})}$$

  where

  V = critical velocity (ft/s)
  ρ_l = liquid density (lb/ft^3)
  ρ_v = vapour density (lb/ft^3) at drum conditions of pressure and temperature

  (see also Section 3.4)

(2) *Drum pressure* The maximum operating pressure for this drum will be about 0.5 psig or that of the water disengaging drum tied to the same flare header.

(3) *Condensable blowdown drum and system* This is used for collection and

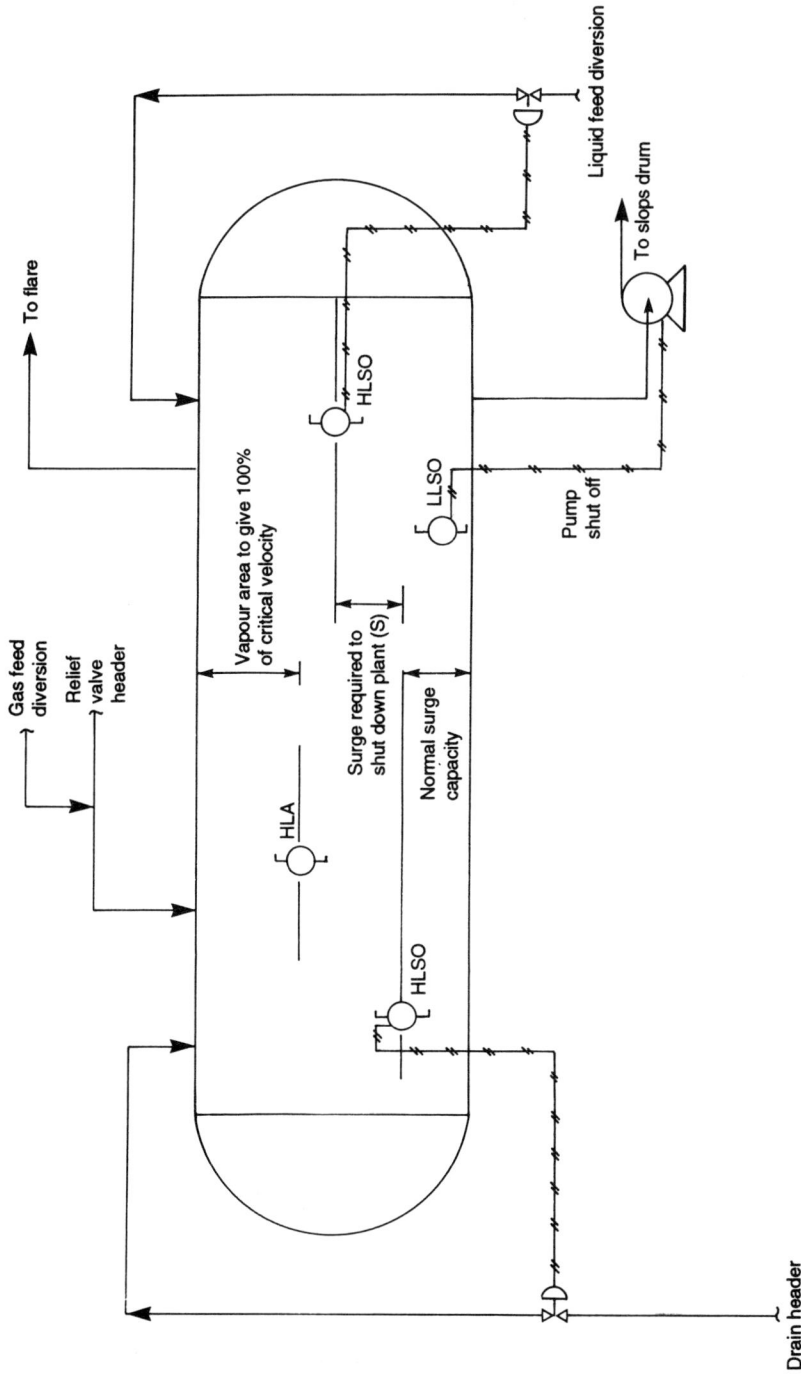

Figure 2.38. A typical flare knockout drum or non-condensible blowdown drum. Drain header HLSO = high liquid shut-off. Liquid feed diversion HLSO = high liquid shut-off. HLA = High liquid level alarm. LLSO = low liquid level pump switch-off

containment of heavier chemicals with low volatility. For example, in petroleum refining this would account for the middle and waxy distillates (kerosene, gas oils, etc.) and in aromatics production this material would include the xylenes and heavier. Figure 2.39 shows a typical blowdown drum and quench.

The material entering this system is generally above ambient temperature. Very often hot streams directly from operating units find their way into this system. To handle these materials the condensable blowdown drum is designed as a direct contact quench drum. The blowdown material leaves the unit in a drain collection system to enter the bottom section of the drum. Cooling water is introduced at the top of the drum and passes over a baffled tray section to contact the hot blowdown stream at the drum base. Any hot vapours rising from the blowdown stream are condensed in the baffle section of the drum and carried down to the

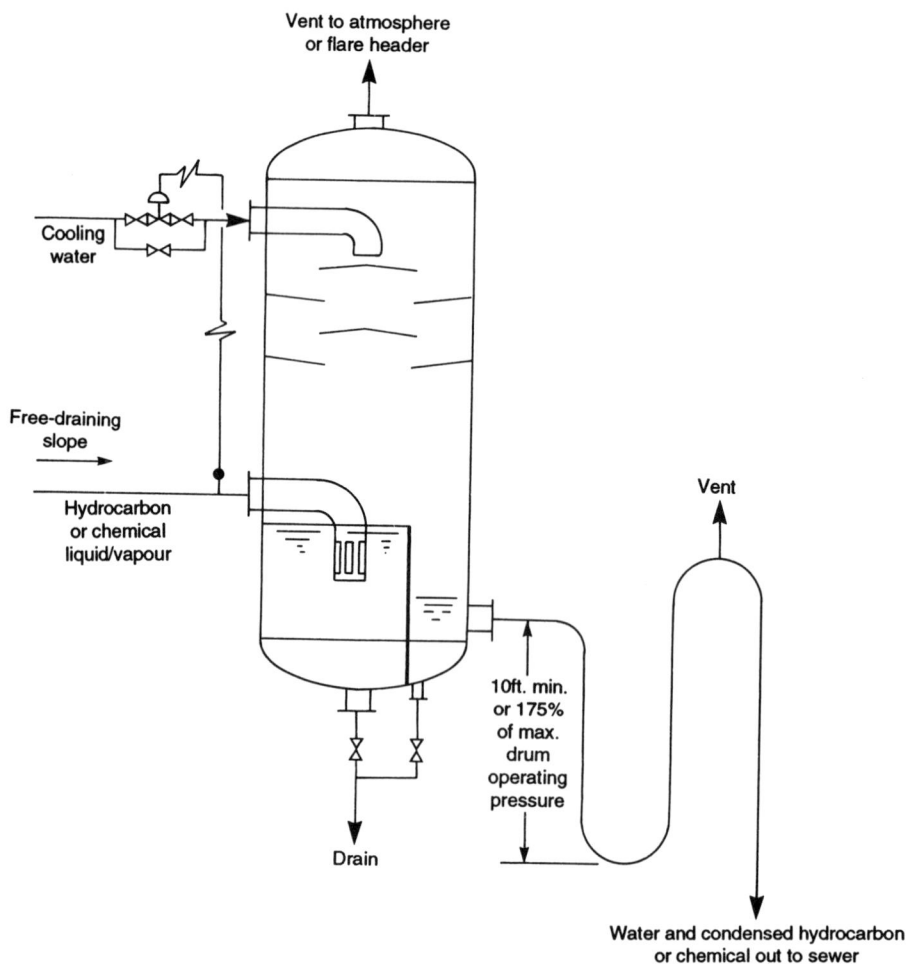

Figure 2.39. A typical condensible blowdown system

OFFSITE SYSTEMS   155

bottom of the drum. Uncondensed material leaving the top of the vessel is routed
to the flare. The aqueous mixture containing the condensed blowdown leaves the
bottom of the drum through a seal system to enter the chemical or oily water
sewer for separation and treatment.

The following criteria are used to size this vessel:

- The vapour load on the drum is based on the safety valve(s) constituting the
  largest single risk.
- The maximum operating pressure for the drum is usually 1–2 psig.
- The stack may vent to atmosphere rather than the flare if desired. However, if
  vented to atmosphere the stack should vent at least 10 ft above the highest
  adjacent structure. In any case, the vent should not release to atmosphere
  below 50 ft above grade. Snuffing steam should also be provided.
- The cooled effluent leaving the drum should be at 150 °F or colder. The cold
  water supply should be controlled either by effluent temperature or inlet
  blowdown stream flow. There should, however, be a bypass flow of water
  entering the drum at all times.
- The drain system from the unit(s) to the drum should be free-draining into the
  drum. The drum therefore should be located at a minimum height to grade.
  Where very waxy materials are likely to be handled, steam tracing of lines and
  a steam coil in the drum should be considered.

The normal design procedures for sizing the drum described in Section 3.4 is used.

The water separation drum arrangement is shown in Figure 2.40. The purpose

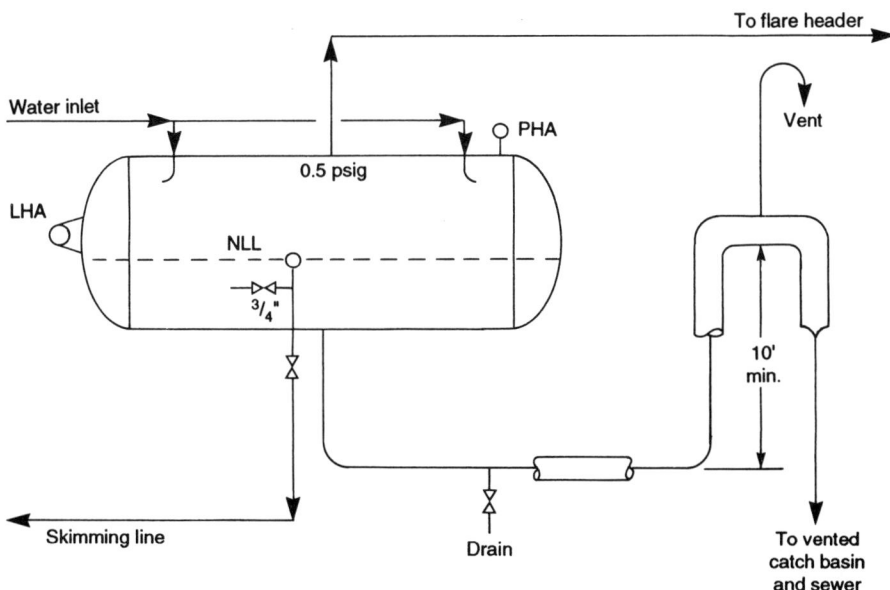

Figure 2.40. A water-disengaging drum

of this drum is to remove any volatile and combustible material from certain water effluent streams before they enter the sewer system. Thus:

- All water from distillate drums which have been in direct contact with flammable material such as light hydrocarbons are sent to the disengaging drum before disposal to the sewer. The exception is where a sour water stripper is included in the plant. Then these streams are sent to this unit.
- Cooling water drainage from coolers and condensers which may have been contaminated with flammable high-volatility material are set to the disengaging drum, so too are steam condensate streams which fall in the same category.

The drum is located at a minimum height above grade. It operates at about 0.5 psig and vents into the flare system. The pressure and the liquid level in the drum are maintained by the free draining of the effluent through a suitable seal.

Design criteria used for the sizing of this drum are as follows:

- The vapour load on the drum will be the result of high volatile material flashing to equilibrium conditions at the drum pressure. This design load is based on the largest amount of vapour arising from a single contingency. For exchangers this contingency will be due to a fractured tube. For liquid from a distillate drum this will be the result of a failed open control valve on the water outlet.
- The liquid seal must be such as to eliminate air from the sewer system and to allow free drainage from the drum.
- An oil or chemical skimming valve is located at the water normal liquid level. This allows for the draw-off of the oil phase from time to time. A high interface level alarm is often included.

The flare

Vapours collected in a closed safety system are disposed of by burning at a safe location. The facilities used for this burning are called flares. The most common of these flares used in industry today are:

- The elevated flare
- The multijet ground flare

The elevated flare is used where some degree of smoke abatement is required. The flare itself operates from the top of a stack usually in excess of 150 ft high. Steam is injected into the gas stream to be burnt to complete combustion and thereby reduce the smoke emission.

The multijet ground flare is selected where luminosity is a problem, for example at locations near housing sites. In this type of flare the vapours are burned within the flare stack thus considerably reducing the luminosity. Steam is again used in this type of flare to reduce smoke emission.

Figure 2.41 shows a typical arrangement of an elevated flare and Figures 2.42

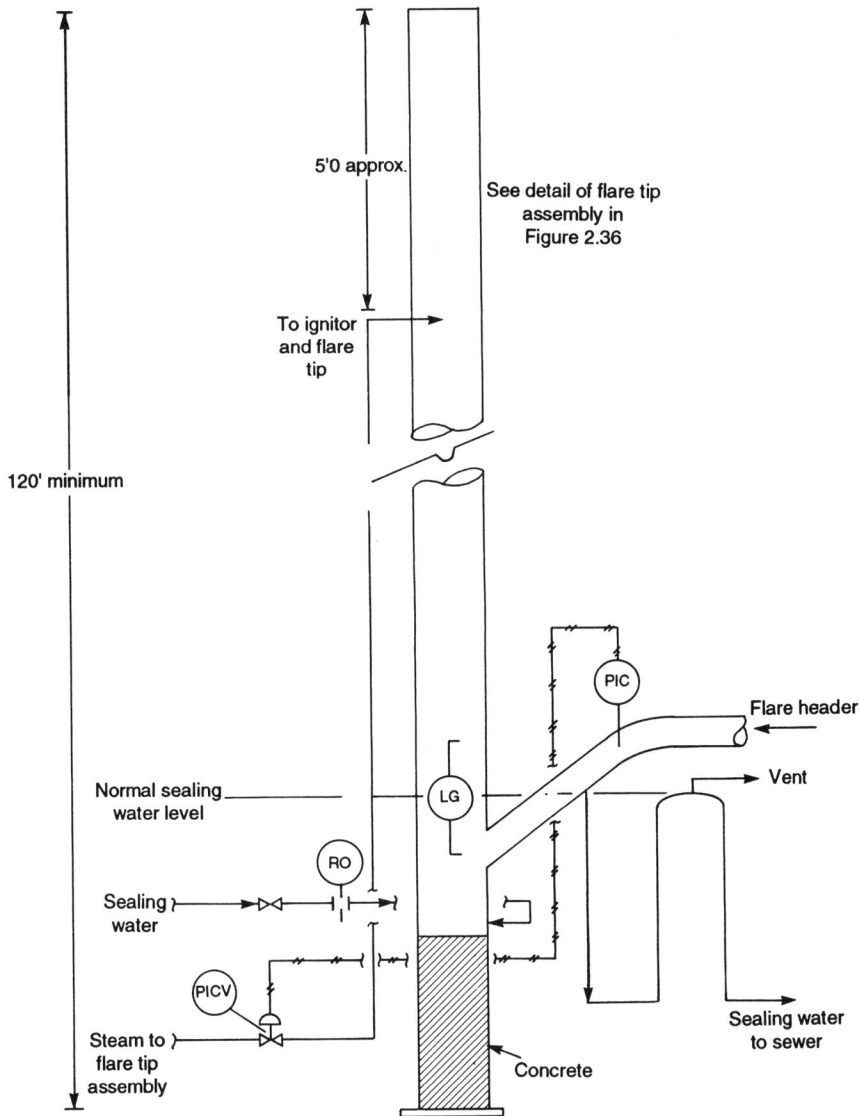

Figure 2.41. A typical elevated flare

and 2.43 that for a multijet ground flare. Table 2.6 gives a comparison of the characteristics of the two flare types and that of a burning pit facility (not normally now used in modern industry).

The elevated flare

This type of flare is the normal choice in the larger process industries such as petroleum refining. It consists of a flare stack over 120 feet in height and which contains an ignitor system, a pilot flame and the flare pipe itself. The flare header

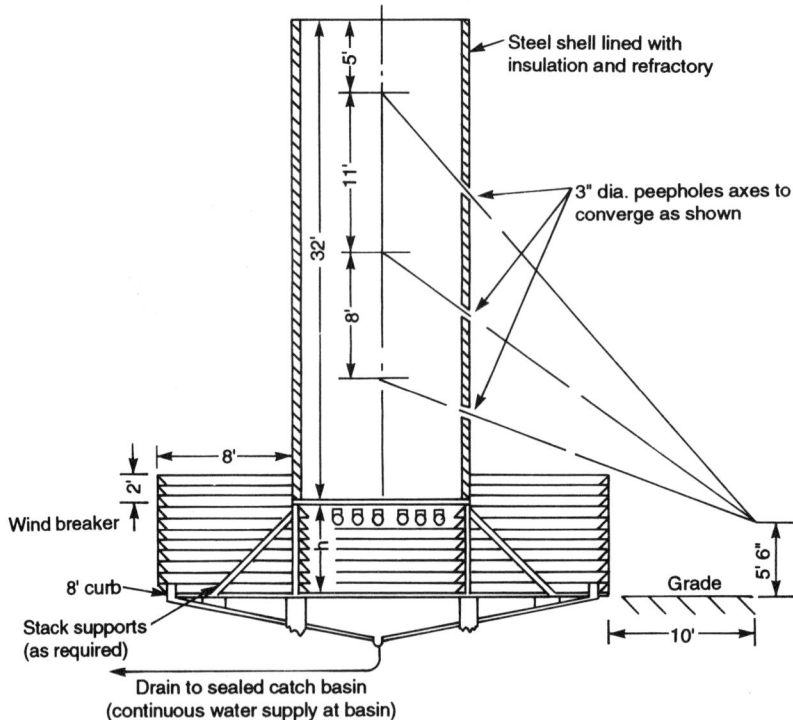

Figure 2.42. A multi-jet ground flare: stack details. Dimension h should equal 6ft or $0.3 \times$ stack i/d whichever is greater

enters the stack through a water seal at the base of the stack immediately above an anchor of concrete plinth. The water seal maintains a back pressure of around 0.5 psig on the flare header. The waste gas to be flared moves up the stack to exit at the top. At the stack top there is an assembly of ignitor and pilot gas which ensures the safe burning of the waste flare gases.

This assembly is shown in Figure 2.44. It consists of three tubes all external to the stack itself and each supplied with the plant fuel gas. The first and largest of these tubes is the ignitor. Here the fuel gas supply is mixed with air (plant or instrument air supply) before passing upwards through a venturi tube to an igniter chamber. A spark is induced in the ignitor chamber by an electric current of 15 Ap. The chamber and the venturi tube are located near grade and a sight glass on the ignitor chamber enables the operator to check on the ignitor's operation. The flame front from the ignitor travels up the ignitor tube to contact the waste gases that are to be flared as they exit the top of the flare stack. The same flame front ignites the 'on and off' pilot burner which is the centre tube of the three and initially ignites the permanent pilot burner at the stack top. The outlet of these three tubes are located at the stack top such that the prevailing wind ensures that the flame from them is blown across the stack exit.

Steam is often injected into the stack at some point near the top to complete combustion and eliminate or at least reduce smoke emission. The amount of

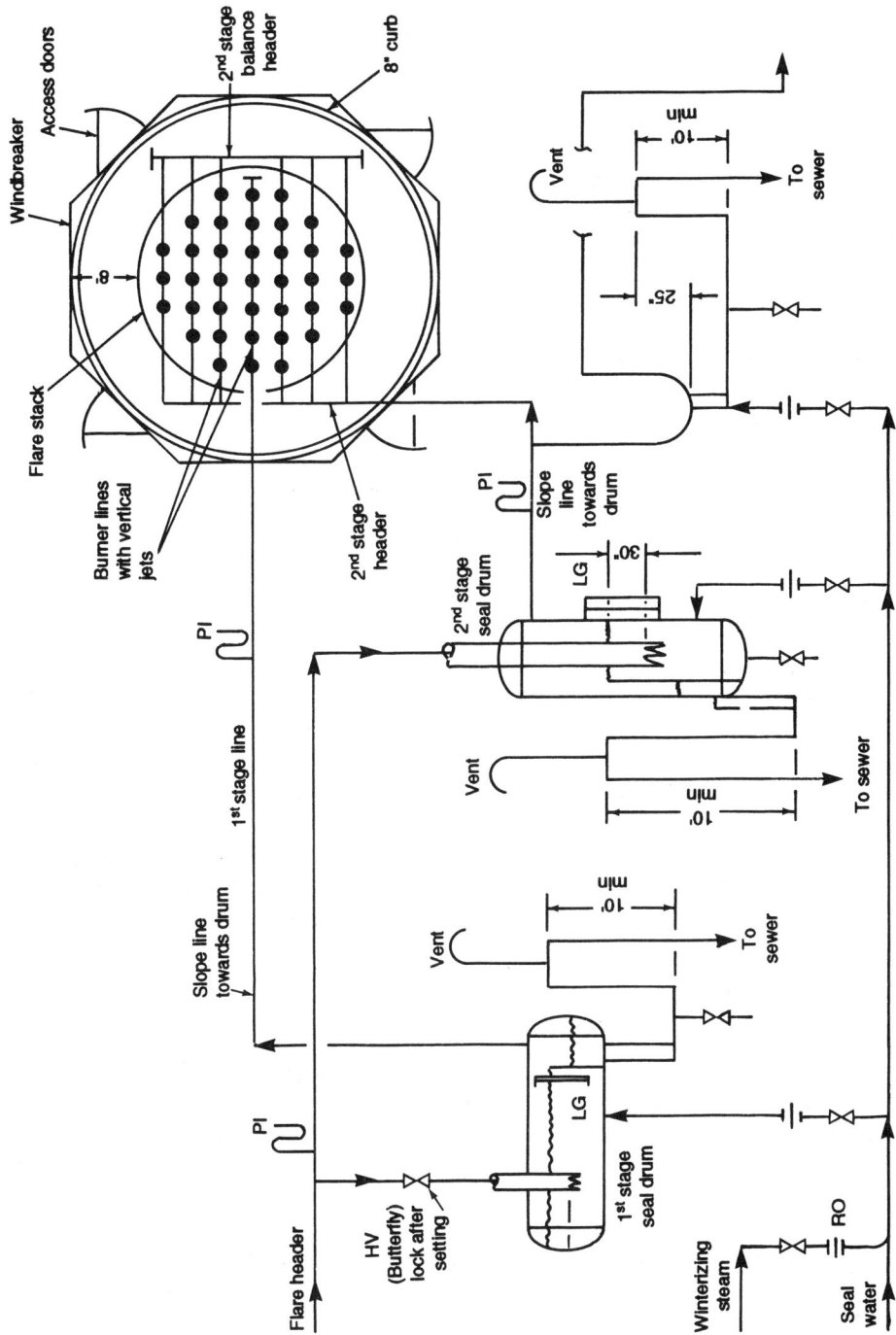

Figure 2.43. A multi-jet ground flare: plan and seal details

Table 2.6. A comparison of different types of flares

Flare type	Smoke	Luminosity	Noise	Steam required	Approx. acreage required	Investment	Operating cost
Burning pit	Very much	Much	None	None	18	Very low	Nil
Elevated							
(Low pressure)	Much	Much	None	None	2	Low	Nil
(High pressure)	Much	Much	Much	None	2	Low	Nil
Elevated, with steam	Little	Some	Some	Moderate to much	2	Low to high	Low to high
Multijet	Little	None	None	None or little	4	High	Nil or little

steam normally used for this purpose depends on the character or composition of the waste gases. Aromatics and olefins when burnt produce a smokey flame. Steam injection allows the free carbon which makes up the smoke to convert into CO and CO_2 which, of course, are invisible gases. An estimate of the amount of steam required for smoke abatement is given in Figure 2.45.

A clear space around an elevated flare is required to allow for the effect of heat radiation from the flare to the ground. Flares which have a heat release of 300 million to 1 billion BTU/h should be located at least 200 ft from the plant property line, or any pond, separator, tankage, or any equipment that could be ignited by a falling spark. The stack also must have a spacing of at least 500 ft from any structure or plant whose elevation is within 125 ft of the flare tip.

The elevated flare stack is designed to maintain a gas velocity of between 100 and 160 ft/s during a major blowdown to flare. This rate is based on the maximum single emergency plus any steam added to improve the burning characteristics. Above a velocity of 160 ft/s noise becomes a problem and the maintenance of ignition also is dubious unless multiple ignition tubes are used. Some proprietary flare tip designs, however, do claim ability to handle satisfactorily velocities up to 400 ft/s.

The multijet ground flare

The multijet flare provides a completely noiseless, non-luminous flaring at a reasonable cost. At normal loads the flare is also essentially smokeless and is particularly useful where continuous flaring is required. Figures 2.42 and 2.43 show the elevation of the flare stack and the plan arrangement of a two stage multijet flare respectively.

The two-stage arrangement shown here shows the flare header being directed to one of two seal drums or to both. The first stage seal drum operates at a back pressure of 20 in. of water at the first stage burner at its design capacity. The second stage burners are activated when the pressure in the flare header reaches 30 in. of water gauge. Very often, particularly in large process complexes, the ground flare is designed to operate in conjunction with an elevated flare. The

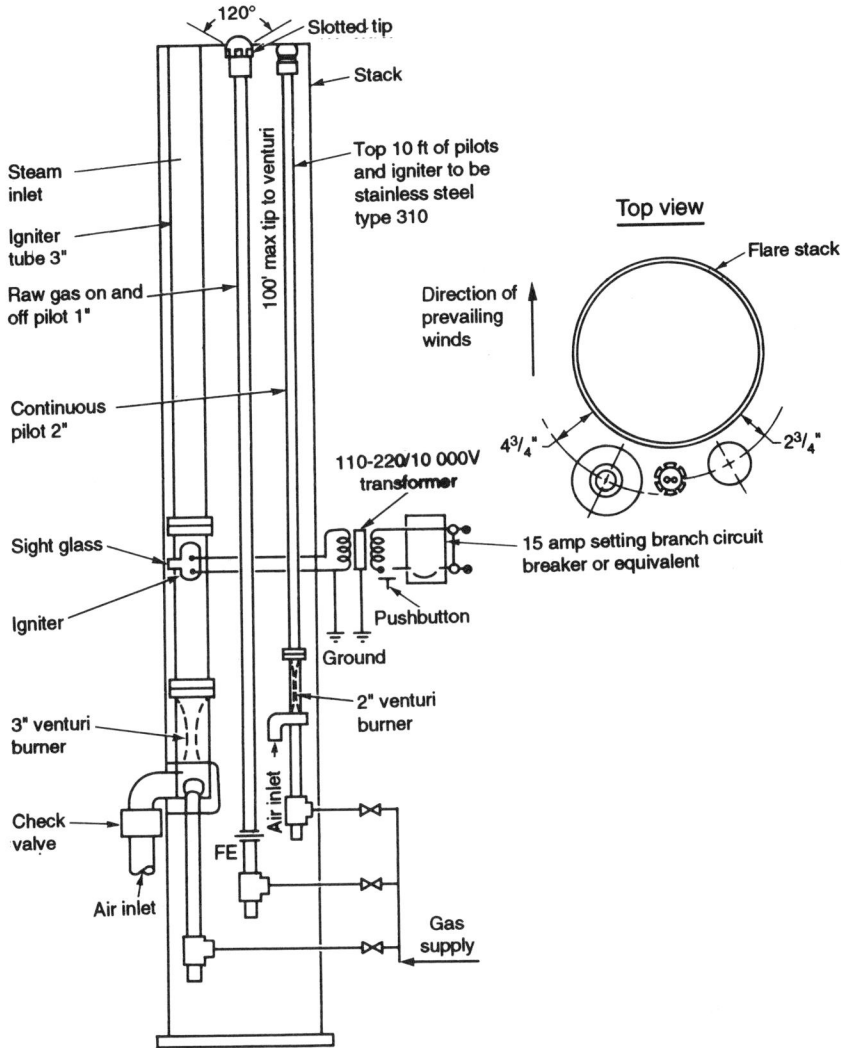

Figure 2.44. A typical flare tip assembly

multijet flare takes a gas stream up to, say, 80% of its rated capacity. Additional flow is then diverted to an elevated flare system. Thus if there is need for the continuous flaring of a reasonably small quantity the ground flare caters for it. In an emergency or surge the elevated flare comes into operation automatically to take the additional load.

The burners of a multijet flare are jet nozzles approximately 15 in. in length of a 1 in. diameter stainless steel pipe. They discharge vertically from the horizontal burner lines which run across the bottom of the stack. The number of jets is based on gas velocity and is expressed by

$$N = 16.4V$$

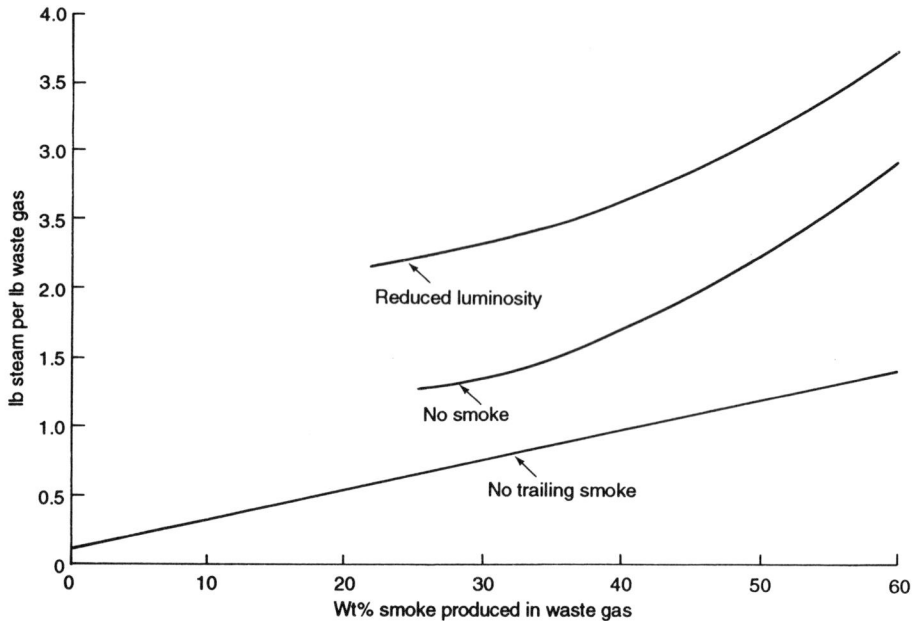

Figure 2.45. Approximate amount of steam for smoke abatement

where

N = number of jets
V = flare design capacity in mm scf/day

The jets are placed on a square or rectangular pitch of 18–24 in. A first estimate of the required pitch may be obtained from the expression

$$P = \frac{100D^2}{NC}$$

where

P = pitch (ins.)
D = stack ID (ft)
N = number of jets.
C = distance between burner centre lines (in.)

No jet should be placed closer than 12 in. to the inside of the stack.

The inside diameter of the stack is based on the rate of heat release at design capacity. It is calculated using the following equation:

$$D = 0.826Q^{0.5}$$

where

D = stack inside diameter (ft)
Q = heat release at max. design (mm BTU/h).

The stack height for diameters up to 25 ft is 32 ft and the steel shell of the stack is lined with 4 in. of refractory material. A wind breaker completes the construction of the stack. This is necessary to prevent high wind gusts from extinguishing the flames.

Flame holders are installed above the burners to prevent the flames 'riding' up to the top of the stack. These are simply solid rods of 1 in. refractory material supported horizontally above each burner line. The position of these flame holders relative to the bottom of the stack is critical to the proper operation of the burners. The stack itself is elevated to allow air for combustion to enter. The minimum space between grade and the bottom of the stack is set at 6 ft or $0.3D$ whichever is the larger.

For any flare a continuous pilot burner is recommended. The proper operation of this pilot is important with respect to multijet type flares because of the danger of unburnt flammable material escaping outside the flare at ground level. A gas pilot is provided at each end of the primary burner to minimize this risk.

EFFLUENT WATER TREATING FACILITIES

This section deals with the treating of waste water accumulated in a chemical process complex before it leaves the complex. Over the years, requirements for safeguarding the environment has demanded close control on the quality of effluents discharged from chemical plants. This includes effluents that contain contaminants that can affect the quality of the atmosphere and those that can be injurious to plant and other life in river waters and the surrounding seas. Effluent management in the chemical and oil industries has therefore acquired a position of importance and responsibility to meet these environmental control demands.

Water effluents that are discharged from the process and other units are collected for treating and removal or conversion of the injurious contaminants. In most oil refineries imported water in the form of ship's ballast water is also collected onshore for treatment before discharging back to the sea. Figure 2.46 is a schematic of the water effluent treating system for a major European oil refinery. Normally, the water effluent treating facilities for a complex would be located at the lowest geographical point in the plant. In this way, very little pumping is required to move the waste water to and from the treating plants. The schematic in Figure 2.46 is for a refinery that was sited below sea level so more than the usual number of pumps are used.

The contaminant that is to be removed in the system shown in Figure 2.46 is, of course, oil. Five separate systems are used in this refinery's treatment plant. The first is that for handling ballast water from sea-going tankers. The second is the handling of clean water. This is included because the system bypasses all the treating processes except the last 'guard' process which, in this case, are the retention ponds. The third system is also for handling non-oily water but water that would be high in certain chemicals. This system also discharges into retention or stormwater ponds. The water is held in these ponds to ensure that there is no contamination. If there is, then the water would be returned into one or the other treating processes for removal of the contaminants.

The last two systems shown are for the handling of contaminated water from

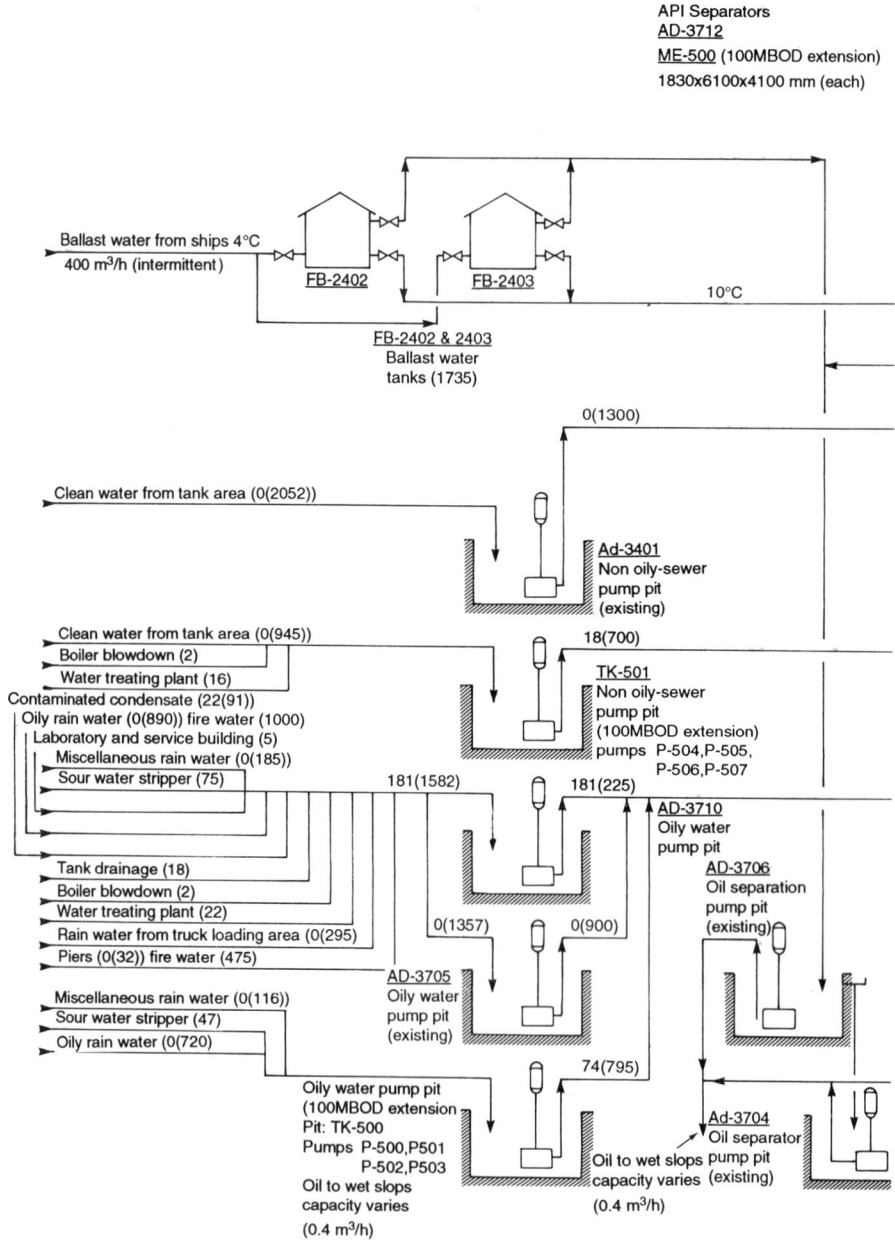

API Separators
AD-3712
ME-500 (100MBOD extension)
1830x6100x4100 mm (each)

Ballast water from ships 4°C
400 m³/h (intermittent)

FB-2402   FB-2403

10°C

FB-2402 & 2403
Ballast water
tanks (1735)

0(1300)

Clean water from tank area (0(2052))

Ad-3401
Non oily-sewer
pump pit
(existing)

Clean water from tank area (0(945))
Boiler blowdown (2)
Water treating plant (16)
Contaminated condensate (22(91))
Oily rain water (0(890)) fire water (1000)
Laboratory and service building (5)
Miscellaneous rain water (0(185))
Sour water stripper (75)

18(700)

TK-501
Non oily-sewer
pump pit
(100MBOD extension)
pumps P-504,P-505,
P-506,P-507

181(1582)   181(225)

AD-3710
Oily water
pump pit

AD-3706
Oil separation
pump pit
(existing)

Tank drainage (18)
Boiler blowdown (2)
Water treating plant (22)
Rain water from truck loading area (0(295))
Piers (0(32)) fire water (475)

0(1357)   0(900)

Miscellaneous rain water (0(116))
Sour water stripper (47)
Oily rain water (0(720))

AD-3705
Oily water
pump pit
(existing)

74(795)

Oily water pump pit
(100MBOD extension
Pit: TK-500
Pumps P-500,P501
P-502,P503
Oil to wet slops
capacity varies
(0.4 m³/h)

Oil to wet slops
capacity varies
(0.4 m³/h)

Ad-3704
Oil separator
pump pit
(existing)

Figure 2.46. A European refinery water treating system

AD-3711
Storm water
holding pond
11300 m³
TK-504
Storm water holding
pond extension
(100MBOD extension)
6200 m³

Retention ponds
AD-3702 and AD-3402 existing
TK-506 A &B (100MBOD extension)
,980x10000x30000 mm each
AD-3701
Parallel plate
interceptor (existing)
7 bays x 50 m³/h each
ME-500
Parallel plate
interceptor extension
(100MBOD extension)
4 bays x 50 m³/h

ME-503
Air flotation
unit
(100MBOD extension)
two section unit
450 m³/h capacity
(existing unit
PA--3703 to be
scrapped

AD-3708
Settling pond
1490 m³

TK-505
Settling pond extension
(100MBOD extension)
980 m³

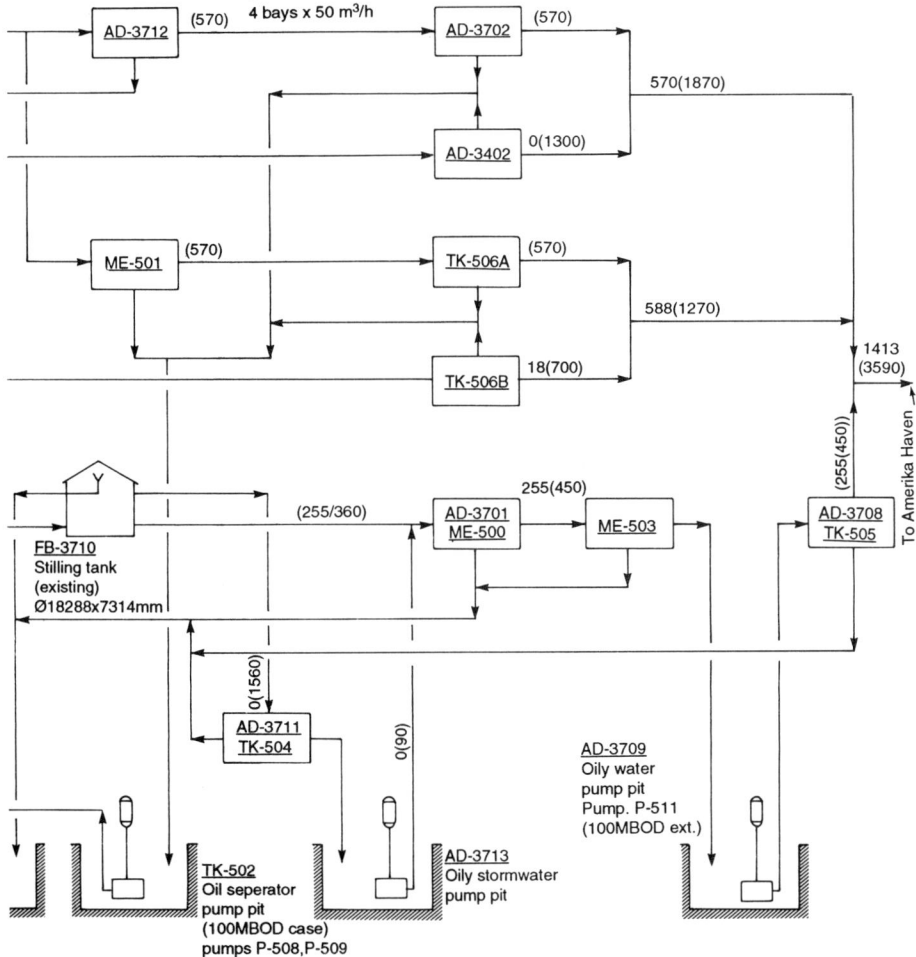

FB-3710
Stilling tank
(existing)
Ø18288x7314mm

AD-3711
TK-504

AD-3709
Oily water
pump pit
Pump. P-511
(100MBOD ext.)

TK-502
Oil seperator
pump pit
(100MBOD case)
pumps P-508,P-509

AD-3713
Oily stormwater
pump pit

refineries' paved areas, various tank and process plant drainage, etc. These oily water systems and the ballast water stream are treated for oil removal. In the case of the ballast water the water drained from the bottom of the holding tanks is routed through an API separator. This is a specially designed pond which reduces the forward velocity of the water stream to allow the separation of oil from the water by settling or gravity.

The water/oil separation for the other refinery streams takes place in a series of settling pods. Final clean-up in this case is accomplished by the use of parallel plate interceptors and an air flotation process. The principle of the parallel plate interceptor is to force the water stream to change direction several times in rapid sequence thus 'knocking out' any oil entrained in the stream. The air flotation unit causes the contaminated water stream to be agitated to force the lighter oil phase to the surface where it can be removed by skimming or by baffled overflow.

Other treating processes

Most chemical plants and indeed a few oil refining plants require more complex methods for clarifying their effluent water to meet environmental requirements for its disposal. The four more common methods are as follows:

- In-line clarification using coagulation, flocculation and filtration
- Plain filtration
- Sedimentation only
- Chemically aided sedimentation using coagulation, flocculation, and settling

Clarification is a process that removes suspended (usually organic) matter that gives the stream colour and turbidity. The removal of this matter, especially in a colloidal form, requires the addition of chemicals to cause coagulation and flocculation to promote settling and separation of suspended solids. Coagulants and coagulant aids added to the influent stream chemically react with impurities to form precipitates. These, together with particles of enmeshed turbidity, are flocculated into large masses that are then readily separated from the bulk liquid.

There are essentially three steps in the chemically aided clarification process:

- Mixing of the additives
- Flocculation
- Settling

Coagulation encompasses the process of mixing the first formation of agglomerates that form the floc. This is carried out in a series of separate compartments with the settling basin occupying the largest volume. Coagulation is the singular most important step in the clarifying process. Because it involves the build-up of colloidal-type particles the chemicals and the process rate is specific to the material that is to be clarified. There are companies that specialize in the design, construction and operation of this type of effluent treating. These companies use their experience in handling the complex electrochemical kinetics associated with flocculation and coagulation principles.

3   EQUIPMENT—VESSELS

This is the first chapter on what is probably the most important role of a process engineer—knowing the equipment that is to be used. Among the most numerous and certainly the largest in size of the equipment are the process vessels. These are divided into the following categories:

- Columns or towers
- Knock-out drums and separators
- Accumulators and surge vessels
- Reactors

Storage tanks operate at atmospheric pressure and therefore are not pressure vessels. These are dealt with in Chapter 2 (Section 2.4).

3.1 Types of Trayed Towers

Columns normally constitute the major cost in any chemical process configuration. Consequently a process engineer needs to exercise utmost care in handling this item of equipment. This extends to the actual design of the vessel or evaluating a design offered by others.

Normally, columns are used in a process for fractionation, extraction or absorption as unit operations. Columns contain internals which may be trays, or packing. Both types of column will also contain suitable inlet dispersion nozzles, outlet nozzles, instrument nozzles and access facilities (such as manholes or handholes). This section deals with trayed towers.

TRAY TYPES

There are three types of trays in common use today:

- Bubble cap

- Sieve
- Valve

Bubble cap tray

This type of tray was in wide use until the mid to late 1950s and then they were displaced by the cheaper sieve and valve tray. The bubble cap tray consists of a series of risers on the tray which are capped by a serrated metal dome. Figure 3.1 shows two types of these caps. One is used in normal fractionation service while the other is designed for vacuum service.

Vapour rises up through the risers into the bubble cap. It is then forced down through the serrated edge or, in some cases, slots at the bottom of the cap. A liquid level is maintained on the tray to be above the slots or serrations of the cap. The vapour therefore is forced out in fine bubbles into this liquid phase thereby mixing with the liquid. Mass and heat transfer between vapour and liquid is enhanced by this mixing action to effect the fractionation mechanism.

- *Capacity*   Moderately high with high efficiency.
- *Efficiency*   Very efficient over a wide capacity range.
- *Entrainment*   Much higher than perforated-type trays due to the 'jet' action that accompanies the bubbling.
- *Flexibility*   Has the highest flexibility for both vapour and liquid rates. Liquid heads are maintained by weirs.
- *Application*   May be used for all services except for those conditions where coking or polymer formation occur. Note: Because of the relatively high liquid level required by this type of tray it incurs a higher pressure drop than most other types of trays. This is a critical factor in tray selection for vacuum units.
- *Tray spacing*   Usually 18–24 in. For vacuum service this should be about 30–36 in.

Sieve tray

This is the simplest of the various types of trays. It consists of holes suitably arranged and punched out of a metal plate. The vapour from the tray below rises through the holes to mix with the liquid flowing across the tray. Fairly uniform mixing of the liquid and vapour occurs and allows for the heat/mass transfer of the fractionation mechanism. The liquid flows across a weir at one end of the tray through a downcomer to the tray below. Sieve trays are usually used with downcomers but they can be used without liquid downcomers.

- *Capacity*   As high as or higher than bubble cap trays at design vapour/liquid rates. Performance drops off rapidly at rates below 60% of design.
- *Efficiency*   High efficiency at design rates to about 120% of design. The efficiency falls off rapidly at around 50–60% of design. This is due to 'weeping' which is the liquid leaking from the tray through the sieve holes.
- *Entrainment*   Only about one third of that for bubble cap trays.
- *Flexibility*   Not suitable for trays operating at variable loads.

Dimension	6"	6" raised	4"	6" vacuum
A	6"	6"	4"	6"
B	$5^3/_4$"	$5^3/_4$"	$3^3/_4$"	$5^3/_4$"
C	4"	4"	$2^1/_2$"	4"
D	$2^3/_4$"	$2^3/_4$"	$2^1/_8$"	$2^3/_4$"
E	$^5/_8$"	$^3/_4$"	$^1/_2$"	$^3/_4$"
F	$3^3/_8$"	$3^1/_2$"	$2^5/_8$"	$3^1/_2$"
G	$2^3/_4$"	$2^1/_2$"	$2^1/_4$"	$2^1/_2$"
H	$1^3/_8$"	$1^3/_8$"	1"	$^3/_4$"
J	$^5/_8$"	$^5/_8$"	$^3/_8$"	1"
K	$^5/_{32}$"	$^5/_{32}$"	$^5/_{32}$"	$^5/_{32}$"
L	$^5/_{32}$"	$^5/_{32}$"	$^1/_8$"	$^5/_{32}$"
Number of slots	28	28	20	28
Total slot area/cap (ft²)	0.105	0.105	0.047	0.056
Peripheral area under cap (ft²)	0.082	0.098	0.044	0.098
Total effective slot area (ft²)	0.187	0.203	0.091	0.154
Chimney area (ft²)	0.087	0.087	0.034	0.087

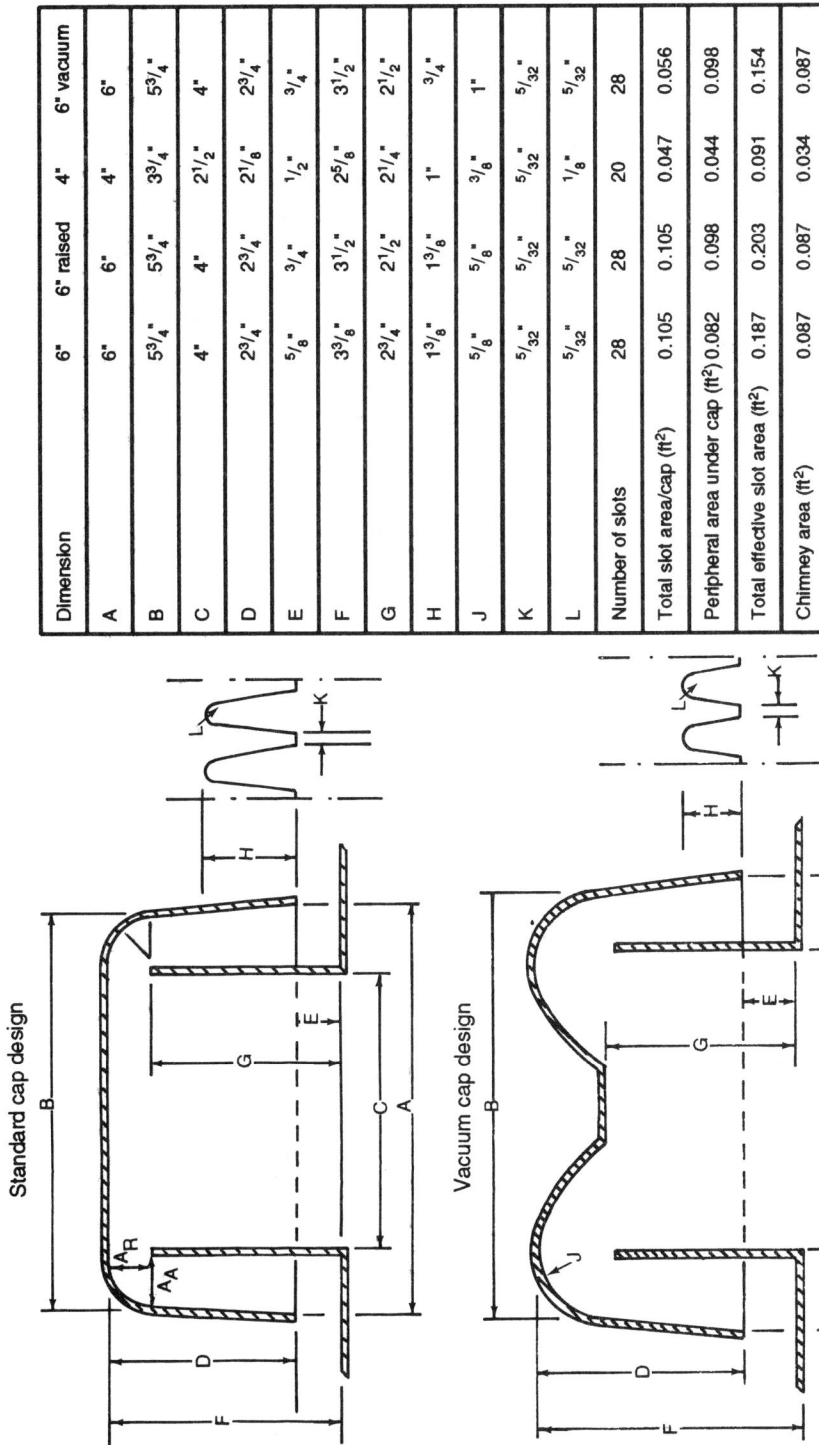

Figure 3.1. Bubble cap design

- *Application* In most mass transfer operations where high capacities in vapour and liquid rates are required. Handles suspended solids and other fouling media well.
- *Tray spacing* Requires less tray spacing than bubble cap. Usually spacing is rarely less than 15 in. although some services can operate at 10 in. and 12 in. In vacuum serivce a spacing of 20–30 in. is acceptable.

Valve tray

These trays have downcomers for the liquid traffic and holes with floating caps that handle the vapour traffic. The holes may be round or rectangular and the caps over the holes are moveable within the limits of the length of the 'legs' which fit into the holes. Figures 3.2 and 3.3 show the type of valves and valve trays offered by Glitsch as their 'ballast' trays.

Valve trays are by far the most common type used in the chemical industry today. The tray has good efficiency and a much better flexibility in terms of turndown than its rivals in the sieve or bubble cap tray. Its only disadvantage over the sieve tray is that it is slightly more expensive and cannot handle excessive fouling as well as the sieve tray. The remainder of this section will now be dedicated to the sizing and analysis of the valve tray tower.

3.2 Trayed Tower Sizing

The height of a trayed tower is determined by the number of trays it is to contain, the liquid surge level at the bottom of the tower, and the tray spacing. The number of trays is a function of the thermodynamic mechanism for the fractionation or absorption duty required to be performed. This is described in Chapter 1 (Sections 1.5 and 1.6). The other criterion for determining the height is described in Chapter 2 (Section 2.1, surge volume).

The diameter of the tower is based on allowable vapour and liquid flow in the tower and the type of tray. This section now deals with determining the tower diameter using valve trays.

THE 'QUICKIE' METHOD

This method is good enough for a reasonable estimate of a tower diameter which can be used for a budget-type cost estimate or initial plant layout studies. The steps used for this calculation are as follows:

- *Step 1* Establish the liquid and vapour flows for the critical trays in the section of the tower that will give the maximum values. These are obtained by heat balances as shown in Chapter 1, Section 1.2. The critical trays are usually:
  - The top tray
  - A sidestream draw-off tray
  - An intermediate reflux draw-off tray
  - The bottom tray

- *Step 2* Calculate the actual ft^3/h at tray conditions of the vapour. Then using the total mass per hour of the vapour calculate its vapour density in lb/ft^3.
- *Step 3* From the heat balance determine the density of the liquid on the tray at tray temperature in lb/ft^3.
- *Step 4* Select a tray spacing. Start with a 24-in. space. Read from Figure 3.4 a value for K on the flood line. Using the equation[9]

$$G_f = K\sqrt{((\rho_v \times (\rho_l - \rho_v))}$$

where

G_f = mass vapour velocity in $lb/h.ft^2$ at flood
K = the constant read from the flood curve in Figure 3.4
ρ_v = density of vapour at tray conditions in lb/ft^3
ρ_l = density of liquid at tray conditions in lb/ft^3

- *Step 5* Multiply G_f by 0.82 to give mass velocity at 82% of flood which is the normal recommended design figure. Divide the actual vapour rate in lb/h by the vapour mass velocity to give the area of the tray. Calculate tray diameter from this area.

Example calculation

Calculate the diameter of the tower to handle the liquid and vapour loads as follows:

Vapour to tray	Liquid from tray
lb/h = 47 700	gph at 60 = 119.7
moles/h = 929.7	Hot gph = 153.0
acfs = 7.83	Hot cfs = 0.339
lb/ft³ ρ_v = 1.69	lb/h = 33273
Temp. (°F) = 167	lb/ft³ ρ_L = 27.3
Pressure (psia) = 220	Temp. (°F) = 162

Tray spacing is set at 24 in. and the trays are valve type. From Figure 3.4

$$K = 1110 \text{ at flood}$$
$$\rho_V = 1.69 \, lb/ft^3 \text{ at tray conditions}$$

To calculate ρ of vapour at tray conditions use:

$$\rho = \frac{wt/h \times 520 \times press.psia}{378 \times 14.7 \times moles/h \times Temp. °R}$$

where

Press. is tray pressure.
Temp. °R is tray temperature in °F + 460

$$\rho_L = 27.3 \, lb/ft^3$$

A-1, A-4

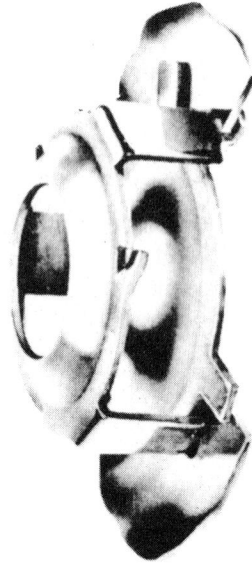

A-2X, A-5X

V-1, V-4

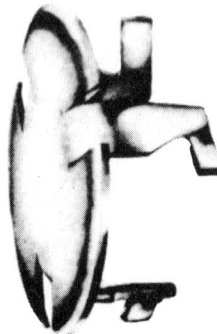

V-1X, V-4X

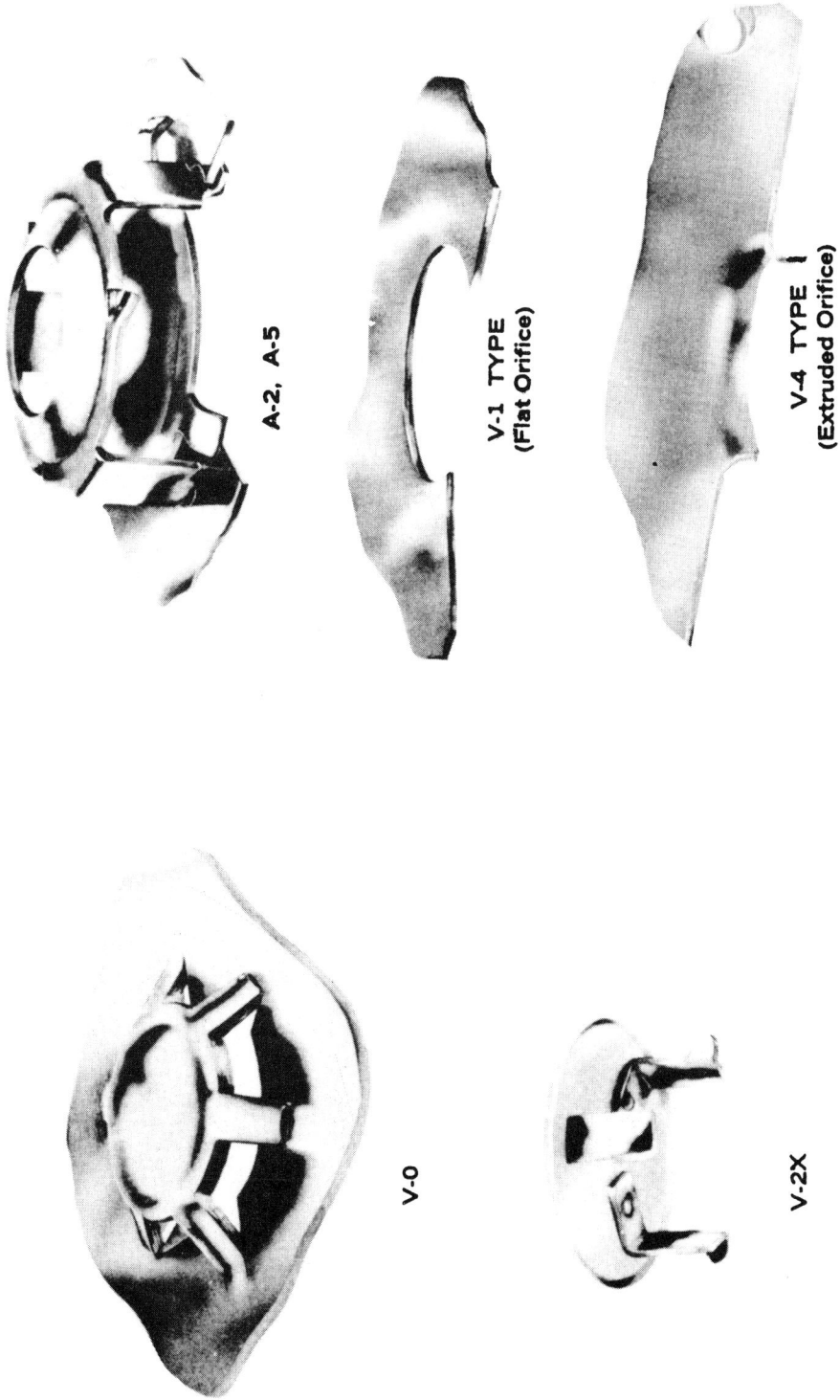

A-2, A-5

V-1 TYPE
(Flat Orifice)

V-4 TYPE
(Extruded Orifice)

V-0

V-2X

Figure 3.2. Valve unit types—Glitsch ballast (Reproduced by permission of Glitsch Inc)

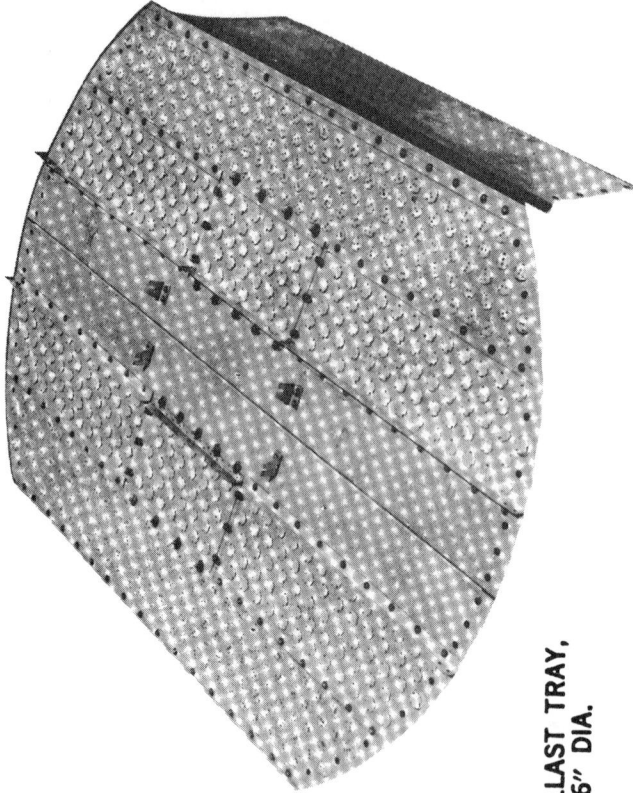

V-1 BALLAST TRAY,
9'-6" DIA.

V-1 BALLAST TRAY
(with Recessed Inlet Sump)
10'-0" DIA.

Figure 3.3. Valve trays—Glitsch ballast (Reproduced by permission of Glitsch Inc)

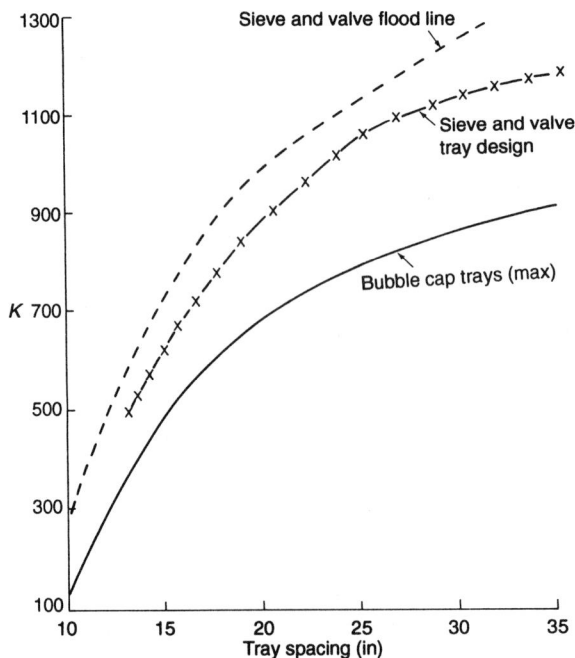

Figure 3.4. Tray spacing versus K factor

Then

$$G_f = 1110\sqrt{((1.69 \times (27.3 - 1.69))}$$

$$= 7302 \text{ lb/h.ft}^2$$

$$\text{Area of tray at 82\% of flood} = \frac{47\,700}{0.82 \times 7302}$$

$$= 7.97 \text{ ft}^2$$

$$\text{Diameter} = \sqrt{(7.97/0.786)}$$

$$= \underline{3.18 \text{ ft (say, 39 in.)}}$$

The tray dimensions and configuration for design purposes is subject to a much more rigorous examination. This is normally undertaken by the tray manufacturer from data supplied by the process engineer. However, the process engineer needs to be able to check the manufacturer's offer before committing to purchase. The following calculation procedure offers a rigorous calculation for this purpose which establishes tray size and geometry. This calculation is based on a method developed by Glitsch Inc., a major manufacturer of valve and other types of trays and packing.

THE RIGOROUS METHOD

A rigorous method used in the design of valve trays is described by the following calculation steps:

- *Step 1* Establish the liquid and vapour flows as described earlier for the 'quickie' method.
- *Step 2* Calculate the downcomer design velocity V_{dc} using the following equations:

  (a) $V_{dc} = 250 \times$ system factor.

  (b) $V_{dc} = 41 \times \sqrt{(\rho_L - \rho_v)} \times$ system factor

  (c) $V_{dc} = 7.5 \times \sqrt{TS} \times \sqrt{(\rho_L - \rho_v)} \times$ system factor
      Where TS = tray spacing

  or by Figure 3.5.

  Use the lowest value for the design velocity in gpm/ft^2. Downcomer system factors are given in Table 3.1.
- *Step 3* Calculate the vapour capacity factor (CAF) using Figure 3.6.

$$CAF = CAF_0 \times \text{system factor}$$

System factors used for this equation are given in Table 3.2.
- *Step 4* Calculate the vapour load using the equation

$$V_1 = \text{cfs}\sqrt{\rho_v/(\rho_1 - \rho_v)}$$

where cfs = actual vapour flow in ft^3/s.
- *Step 5* Establish tower diameter using Figure 3.7. Tray spacing is usually 18, 24 or 30 in. for normal towers operating at above atmospheric pressures. Large vacuum towers may have tray spacing 30–36 in. Note that this diameter may be increased if other criteria of tray design are not met.

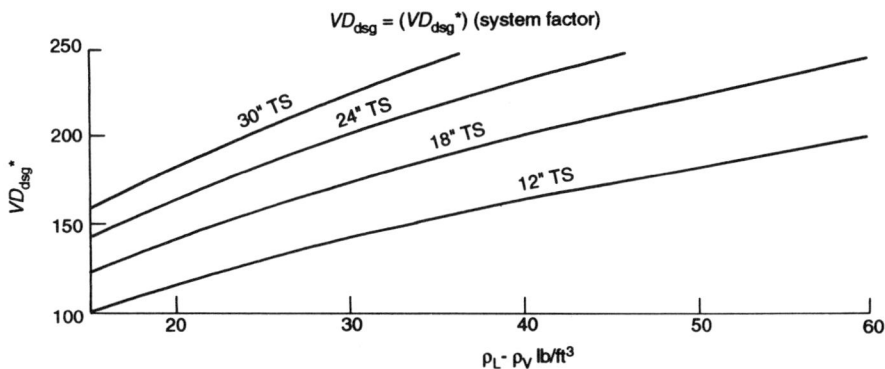

Figure 3.5. Downcomer design velocity (Reproduced by permission of Glitsch Inc)

Table 3.1. Downcomer system factors (Reproduced by permission of Glitsch Inc)

Service	System factor
Non-foaming, regular systems	1.00
Fluorine systems	0.90
Moderate foaming (oil absorbers, amine, etc.)	0.85
Heavy foaming (amine, glycol absorbers)	0.73
Severe foaming (MEK units)	0.60
Foam-stable systems (caustic regenerators)	0.30

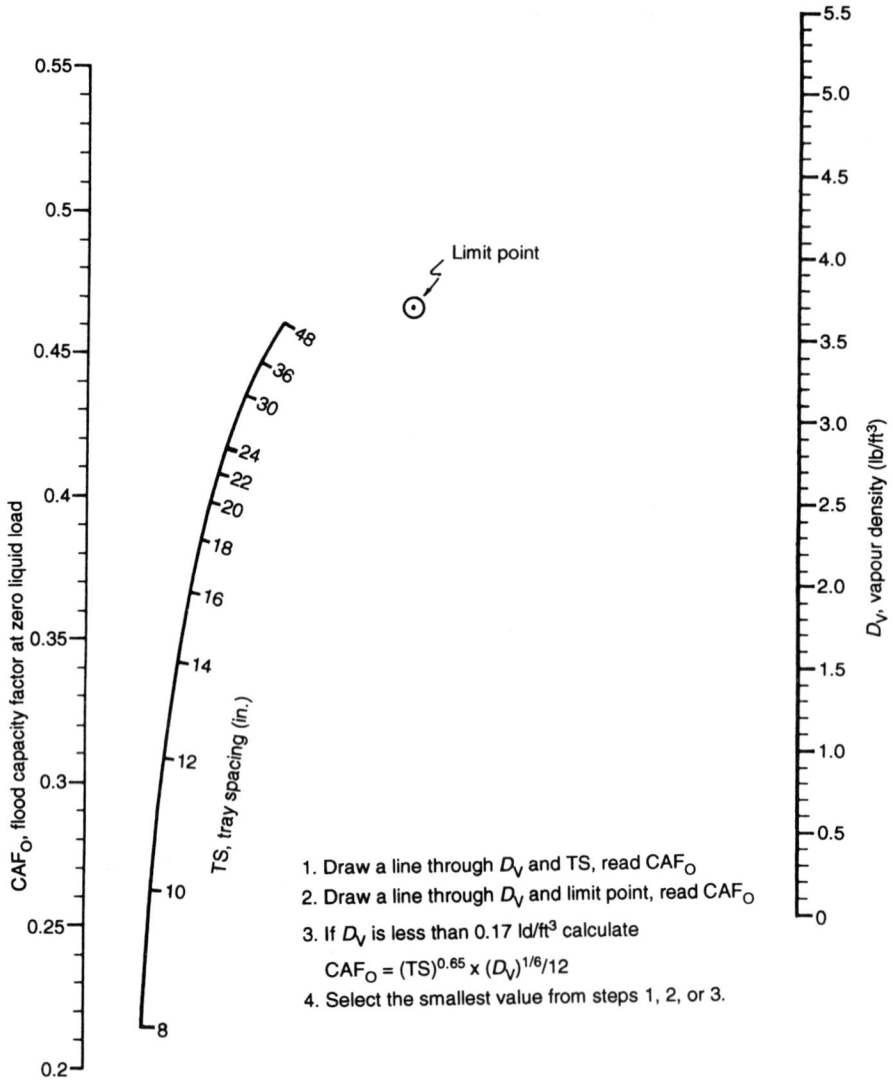

Figure 3.6. Flood capacities of valve trays (ballast) (Reproduced by permission of Glitsch Inc)

Table 3.2. Vapour system factors (Reproduced by permission of Glitsch Inc)

Service	System factor
Non-foaming, regular systems	1.00
Fluorine systems	0.90
Moderate foaming	0.85
Heavy foaming	0.73
Severe foaming	0.60
Foam-stable systems	0.3–0.60

Contained within figure:

1. Draw a line through D_V and TS, read CAF_O
2. Draw a line through D_V and limit point, read CAF_O
3. If D_V is less than 0.17 ld/ft³ calculate
   $CAF_O = (TS)^{0.65} \times (D_V)^{1/6}/12$
4. Select the smallest value from steps 1, 2, or 3.

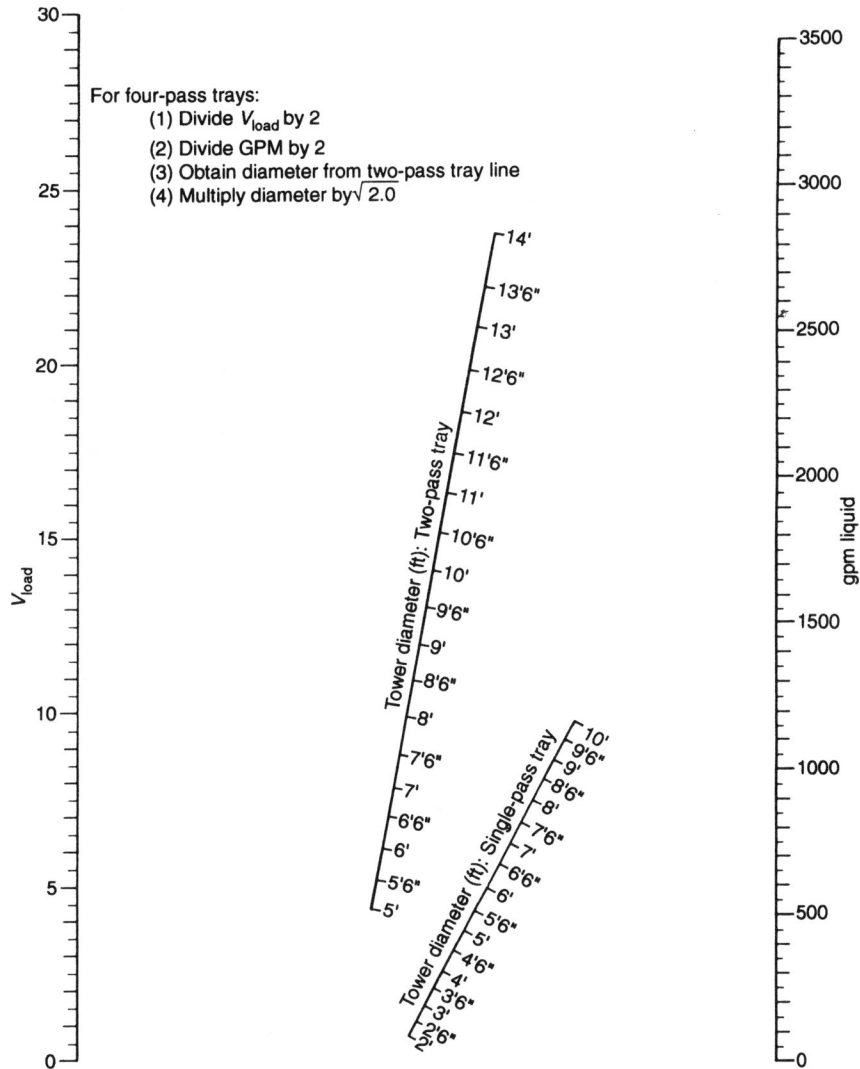

Figure 3.7. Tray diameter versus vapour loads (ballast trays) (Reproduced by permission of Glitsch Inc)

- *Step 6* Calculate the approximate flow path length (FPL) based on tower diameter from step 5 using the equation

$$FPL = 9 \times DT/NP$$

where

    FPL = flow path length (in.)

      DT = tower diameter from step 5 (ft)

      NP = number of passes. For small towers with moderate liquid flows this will be one. For larger towers this will depend on liquid velocities in downcomer. The highest number of passes is usually four.

- *Step 7* Calculate the minimum active area (AA_m) using the expression

$$AA_m = \frac{V_1 + (L \times FPL/13000)}{CAF \times FF}$$

where

AA_m = minimum active area (ft^2)
V_1 = vapour load in CFS
L = liquid flow in actual gpm
FPL = flow path length (in.)
CAF = capacity factor from step 3
FF = flood factor (usually 80–82%)

- *Step 8* Calculate minimum downcomer area (AD_m) using the equation

$$AD_m = \frac{L}{V_{dc} \times 0.8}$$

where

AD_m = minimum downcomer area (ft^2)
L = actual liquid flow (gpm)
V_{dc} = design downcomer velocity from step 2

Note: The downcomer liquid velocity using the calculated minimum downcomer area should be around 0.3–0.4 ft/s.

- *Step 9* Calculate the minimum tower cross-sectional area using the following equations:

$$AT_m = AA_m + 2AD_m$$

or

$$AT_m = \frac{V_1}{0.78 \times CAF \times 0.8}$$

where

AT_m = minimum tower cross-sectional area (ft^2).

- *Step 10* Calculate actual downcomer area using the following equation:

$$AD_c = \frac{AT \times AD_m}{AT_m}$$

where

AD_c = actual downcomer area (ft^2)
AT = tower area (ft^2) from the diameter calculated in step 5

- *Step 11* Determine downcomer width (H_i) from Table A1.6 in Appendix 1 for the side downcomers. For multipass trays use the following equation with the width factors given in Table 3.3:

$$H_i = WF \times \frac{AD}{DT}$$

Table 3.3. Allocation of downcomer area and width factors

| Passes | Fraction of total downcomer area | | | |
	AD_1	AD_3	AD_5	AD_7
2	0.5 ea*	1.00	–	–
3	0.34 ea*	–	0.66	–
4	0.25 ea*	0.50	0.50 ea	–
5	0.20	–	0.40	0.40

| | Width factors (WF) | | |
Passes	H_4	H_5	H_7
2	12.0	–	–
3	–	8.63	–
4	6.0	6.78 ea*	–
5	–	5.66	5.5

* each pass (Reproduced by permission of Glitsch Inc)

where

H_i = width of individual downcomers (in.)
WF = width factor from Table 3.3
AD = total downcomer area (ft^2)
DT = actual tower diameter (ft)

See Figure 3.8 for allocation of downcomers in multi-pass trays.

● *Step 12* Calculate the actual FPL from the equation

$$FPL = \frac{12 \times DT - (2H_1 + H_3 + 2H_5 + 2H_7)}{NP}$$

where

H_1 to H_7 are individual downcomer widths (in.) (see Figure 3.8)
NP = number of passes

● *Step 13* Calculate actual active area (AA) using values for H calculated in step 12 and Table A1.6 in Appendix 1 to establish inlet areas of multi-pass tray downcomers.

$$AA = AT - (2AD_1 + AD_3 + 2AD_5 + 2AD_7)$$

where

AA = actual active area (ft^2)
AT = actual tower area (ft^2)

AD_1 to AD_7 are individual downcomer areas (ft^2) for multi-pass trays corresponding to H_1 to H_7.

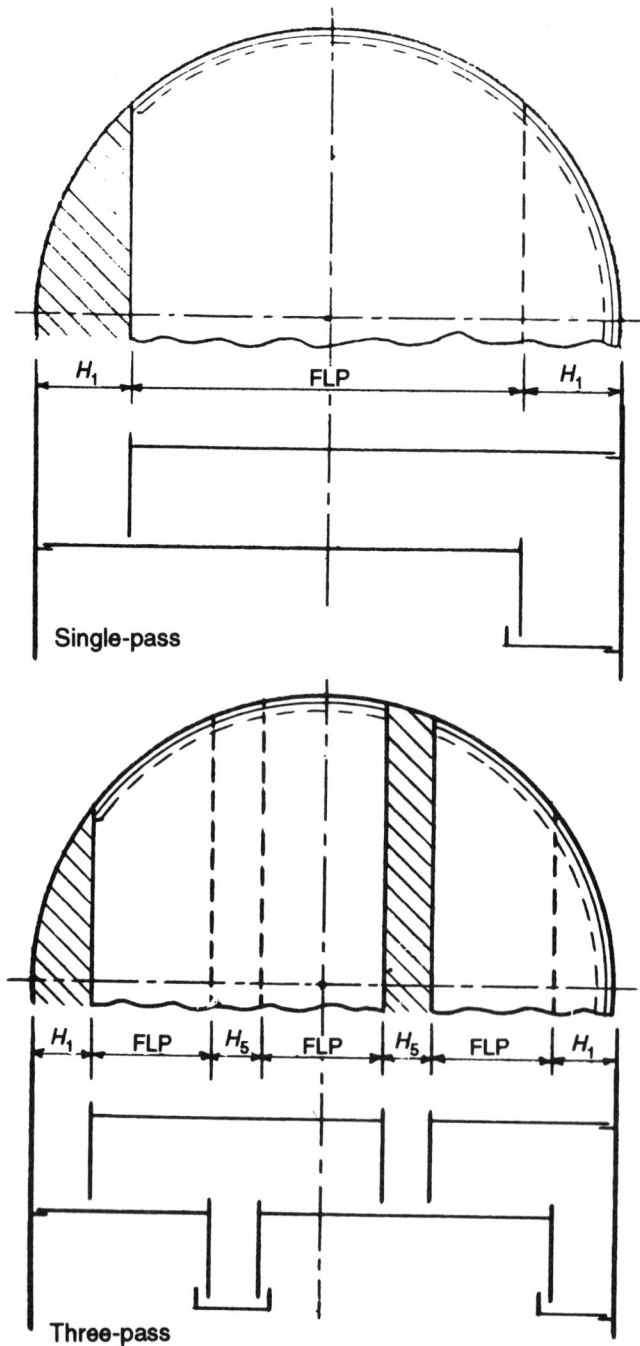

Figure 3.8. Types of tray (Reproduced by permission of Glitsch Inc)

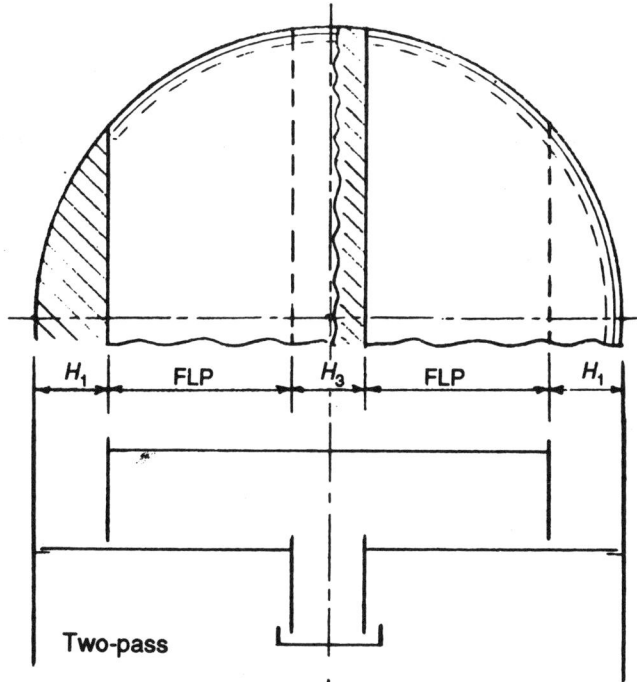

Two-pass

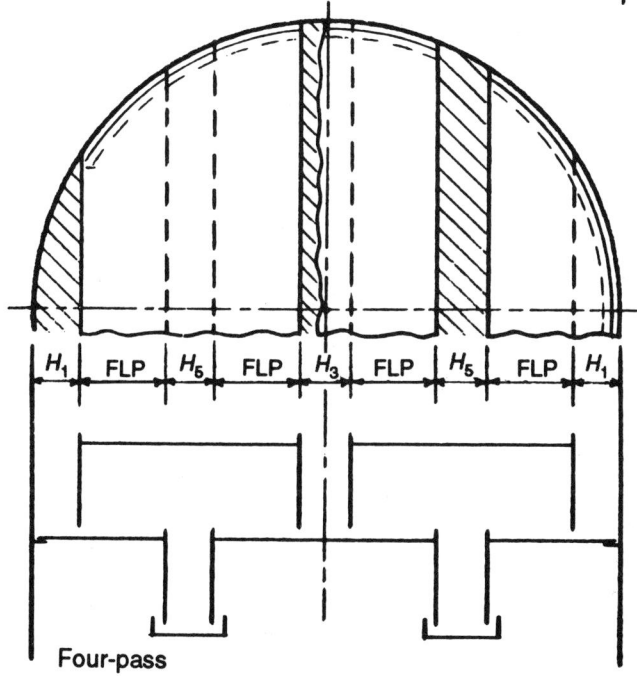

Four-pass

- *Step 14* From the data now developed calculate the actual percentage of flood or flood factor (FF). The following expression is used for this:

$$\% \text{ flood} = \frac{V_1 + (L \times \text{FPL}/13\,000)}{\text{AA} \times \text{CAF}} \times 100$$

- *Step 15* Calculate vapour hole velocity V_h. Assume 12 to 14 units (holes) per square foot of AA. Then

$$V_h = \frac{\text{CFS} \times 78.5}{\text{NU}}$$

where

V_h = hole velocity (ft/s)
CFS = actual ft^3/s of the vapour
NU = total number of units

- *Step 16* Calculate dry tray pressure drops from

Valves partly open: $\Delta P_D = 1.35 t_m \rho_m / \rho_l + K_1(V_h)(\rho_v/\rho_l)$

where

ΔP_D = dry tray valve pressure drop in inches of liquid
t_m = valve thickness in inches (see Table 3.4)
ρ_m = valve metal density in lb/ft^3 (see Table 3.4).
K_1 = pressure drop coefficient (see Table 3.4)

Valves fully open: $\Delta P_D = K_2(V_h)^2 \rho_v/\rho_l$

Table 3.4. Pressure drop coefficients (Glitsch ballast type trays) (Reproduced by permission of Glitsch Inc)

Type of unit	K_1	K_2 for deck thicknesses of (in.):			
		0.074	0.104	0.134	0.25
V-1	0.2	1.05	0.92	0.82	0.58
V-4	0.1	0.68	0.68	0.68	–

Gauge	Thicknesses t_m (in.)	Densities of valve material	
		Metal	Density (lb/ft^3)
20	0.37	CS	490
18	0.50	SS	500
16	0.60	Ni	553
14	0.074	Monel	550
12	0.104	Titanium	283
10	0.134	Hastelloy	560
		Aluminium	168
		Copper	560
		Lead	708

where

K_2 = pressure drop coefficient (see Table 3.4)

- *Step 17* Calculate total tray pressure drop:

$$\Delta P = \Delta P_D + 0.4(L/L_{wi})^{0.67} + 0.4 H_w$$

where

ΔP = total tray pressure drop in inches of liquid
L_{wi} = weir length in inches (from Table A1.6)
H_w = weir height in inches (usually 1–2)

- *Step 18* Calculate height of liquid in downcomer. First calculate the head loss under the downcomer H_{UD}, where $H_{UD} = 0.65\,(V_{UD})^2$. V_{UD} is calculated from the liquid velocity in cfs or gpm/450 divided by the area under the downcomer. Use weir length times weir height for the area. This velocity should be around 0.3 to 0.6 ft/s for most normal towers. Then:

$$H_{dc} = H_w + 0.4(L/L_{wi})^{0.67} + (\Delta P + H_{UD})(\rho_l/(\rho_l - \rho_v))$$

where H_{DC} = height of liquid in downcomer in inches. For normal design this should not exceed 50% of tray spacing.

Example calculation

In this example the same liquid and vapour flows and data will be used as used in the 'quickie calculation'. The objective of this calculation will be to determine the tower diameter, tray pressure drop and configuration and the percentage flood for the design flowrates given.

1 Calculating the downcomer design velocity V_{dc}

System factor in this case is 1.

$$V_{dc} = 250 \times 1.0 = 250$$

or

$$V_{dc} = 41 \times \sqrt{(\rho_l - \rho_v)} \times 1.0$$
$$= 41 \times 5.06 \times 1.0 = 207$$

or

$$V_{dc} = 7.5 \times \sqrt{TS} \times \sqrt{(\rho_l - \rho_v)} \times 1.0$$
$$= 7.5 \times 4.9 \times 5.06 = 186 \; (TS = \text{tray spacing} = 24\,\text{in.})$$

or

$$V_{dc} = 188 \text{ from Figure 3.5.}$$
$$\text{Use } V_{dc} = 186\,\text{gpm/ft}^2$$

2   *Vapour capacity factor (CAF)*

System factor in this case is 1. From Figure 3.6 CAF = 0.43.

3   *Actual vapour load V_1*

$$V_1 = \text{CFS}\sqrt{\rho_v/(\rho_l - \rho_v)}$$
$$= 7.83\sqrt{1.69/(27.3 - 1.69)}$$
$$= 2.01$$

4   *Approximate tower diameter (AT)*

Using Figure 3.7

$$V_1 = 2.01$$
$$\text{TS} = 24 \text{ in.}$$
$$L = 153 \text{ gpm}$$
$$\text{Tower i.d.} = 3.25\text{-ft} = 39 \text{ in.}$$
$$\text{Area} = 8.30 \text{ ft}^2$$

5   *Calculate approx. flow path length (FPL)*

$$\text{FPL} = \frac{9 \times 39}{\text{No. of passes} \times 12} = 29.25 \text{ in.}$$

6   *Calculate minimum active area (AA_m)*

$$AA_m = \frac{V_1 + (L \times \text{FPL}/13\,000)}{\text{CAF} \times \text{FF}}$$
$$= \frac{2.01 + (153 \times 29.25/13\,000)}{0.43 \times 0.8} \text{ (using 80\% flood)}$$
$$= 3.01 \text{ ft}^2$$

7   *Calculating minimum downcomer area (AD_m)*

$$AD_m = \frac{L}{V_{dc} \times 0.8}$$
$$= \frac{153}{186 \times 0.8} = 1.028 \text{ ft}^2$$

8   Calculating the minimum tower cross-sectional area

Either

$$AA_m + 2AD_m = 3.01 + 2.056$$
$$= 5.066 \text{ ft}^2$$

or

$$\frac{V_1}{0.78 \times CAF \times 0.8} = 7.49 \text{ ft}^2$$

Use the larger, which is 7.49 ft^2. Min. diameter therefore is 3.09 ft (say, 37 in.)

9   Calculating actual downcomer area (AD_c)

$$AD_c = \frac{AT \times AD_m}{AT_m} = \frac{8.3 \times 1.028}{7.49}$$
$$= 1.14 \text{ ft}^2$$

Downcomer width $H = AD/AT = 1.14/8.30 = 0.137$
From Table A1.6 $H/D = 0.197$ then $H = 0.197 \times 3.25 = 0.633$ ft $= 7.6$ in.

10   Recalculating flow path length (FPL)

$$FPL = 12 \times D_t - (2H)$$
$$= 12 \times 3.25 - (2 \times 7.6)$$
$$= 23.8 \text{ in.}$$

11   Recalculating active area based on actual downcomer area

$$AA = AT - (2 AD)$$
$$= 8.3 - 2.28$$
$$= 6.02 \text{ ft}^2 \text{ which is greater than min. allowed}$$

12   Checking percentage of flood

$$\% \text{ flood} = \frac{V_1 + (L \times FPL/13\,000)}{AA \times CAF} \times 100$$
$$= \frac{2.01 + (153 \times 23.8/13\,000)}{6.02 \times 0.43} \times 100$$
$$= 88.0\% \text{ which is a little high for design but is acceptable}$$

13   Checking downcomer velocity

$$\text{CFS of liquid} = 0.34 \text{ cfs}$$

$$\text{Area of downcomer} = 1.14 \text{ ft}^2$$

$$\text{Velocity of liquid in downcomer} = \frac{0.34}{1.14}$$

$$= 0.3 \text{ ft/s}$$

14   Calculating pressure drops and downcomer liquid height

Dry tray pressure drop:

  Partially open valves.

$$\Delta P_D = 1.35 T_m \rho_m / \rho_l + K_1 (V_h)^2 (\rho_v / \rho_l)$$

$$V_h = \frac{7.83 \times 78.5}{72} \text{ (assumes 12 units per square foot of AA)}$$

$$= 8.5 \text{ ft/s}$$

$$\Delta P_D = 1.35 \times 0.74 \text{ in.} \times (490/27.3) + (0.2 \times 72.25 \times 0.062)$$

$$= 2.69 \text{ in. liquid}$$

  Fully open valves:

$$\Delta P_D = K_2 (V_h)^2 . (\rho_v / \rho_l)$$

$$= 0.92 \times 72.25 \times 0.062$$

$$= 4.12 \text{ in. liquid. This will be used}$$

  Total tray pressure drop:

$$\Delta P = \Delta P_D + 0.4 \, (L/L_{wi})^{0.67} + (0.4 \times H_w)$$

$$H_w \text{ (weir height) is fixed at 2 in.}$$

$$L_{wi} \text{ (downcomer length) is calculated from Table A1.6 as 30.9 in.}$$

$$\Delta p = 4.12 + 0.4(153/30.9)0.67 + 0.8$$

$$= 6.09 \text{ in of liquid.}$$

  Height of liquid in downcomer.

$$H_{dc} = H_w + 0.4(L/L_{wi})^{0.67} + (\Delta P + H_{UD})(\rho_l / \rho_l - \rho_v)$$

$$H_{dc} = 2 + 0.4 \times 1.16 + (6.09 + 0.405)(27.3/25.61)$$

$$= 10.08 \text{ in. liquid. This is 42.0\% of tray spacing, which is acceptable}$$

(H_{UD} was calculated using a downcomer outlet area of $L_{wi} \times 2$ in. giving a velocity of 0.339 cfs divided by 0.429 ft^2, which is 0.79 ft/s. H_{UD} is then $0.65(0.79)^2 = 0.405$.)

15 Calculating the actual number of valves for tray layout

With truss lines parallel to liquid flow:

$$\text{Rows} = \left[\frac{\text{FPL} - 8.5}{0.5 \times \text{base}} + 1 \right][\text{NP}]$$

where base = spacing of units (usually 3.0, 3.5, 4.0, 4.5, or 6.0 in.).

$$\text{Units/row} = \frac{\text{WFP}}{5.75 \times \text{NP}} - (0.8 \times \text{number of beams}) + 1$$

With truss lines perpendicular to liquid flow:

$$\text{Rows} = \left[\frac{\text{FPL} - (1.75 \times \text{No. of trusses} - 6.0)}{2.5} \right][\text{NP}]$$

$$\text{Units/row} = \frac{\text{WFP}}{\text{Base} \times \text{NP}} - (2 \times \text{No. of major beams} + 1)$$

where

$$\text{WFP} = \text{width of flow path (in.)}$$

$$= \text{AA} \times 144/\text{FPL}$$

Using a base pitch of 3.5 in. the number of rows on the trays with trusses parallel to flow were calculated to be 9.7. Units per row were then calculated to be 8.73. This gives a total number of valves over the active area as 84.7. Thus number of valves per square foot of AA is 14.

The assumption of 12 in the calculation (item 14) gives a more stringent design. Therefore the assumption is acceptable.

Calculation summary

Tower diameter = 3.25 ft or 39 in.
Downcomer area = 1.14 ft^2 (single-pass)
Active area = 6.02 ft^2
Percentage of flood = 88
Tray spacing = 24 in.
Downcomer backup = 42.0% of tray spacing
Number of valves = 85
Number of rows = 10
Valve pitch = 3.5 in.

3.3 Packed Towers and Packed Tower Sizing

Although trayed towers are generally the first choice for fractionation and absorption applications, there is a number of instances where packed towers are preferable. For example, on small diameter towers (below 3 ft diameter) packed towers are generally cheaper and more practical for maintenance, fabrication, and installation. At the other end of the spectrum, packing in the form of grids and large stacked packed beds have superseded trays in vacuum distillation towers whose diameter range up to 30 ft in some cases. This is because packing offers a much lower pressure drop than trays.

The packing in the tower itself may be stacked in beds on a random basis or in a defined structure. For towers up to 10–15 ft the packing is usually dumped or random packed. Above this tower size and depending on its application the packing may be installed on a defined stacked or structured manner. For practical reasons and to avoid crushing the packing at the bottom of the bed the packing is installed in beds. As a rule of thumb, packed beds should be around 15 ft in height. About 20 ft should be a maximum for most packed sections.

Properties of good packing are as follows:

- It should have high surface area per unit volume.
- The shape of the packing should be such as to give a high percentage of area in active contact with the liquid and the gas or in the two liquid phases in the case of extractors.
- The packing should have favourable liquid distribution qualities.
- It should have low weight but high unit strength.
- The packing should have low pressure drop but high coefficients of mass transfer.

Some data on the various common packing available commercially are given in Tables 3.5–3.7. Figure 3.9 shows a sectional layout of a typical packed tower. Note that this tower has bed supports designed for gas distribution and includes intermediate weir liquid distributors between some of the beds.

Other salient points concerning packed towers are as follows:

(1) Reflux ratios, flow quantities, and number of theoretical trays or transfer units are calculated in the same manner as for trayed columns.
(2) Internal liquid distributors are required in packed towers to ensure good distribution of the liquid over the beds throughout the tower.
(3) The packed beds are supported by grids. These are specially designed to ensure good flows of the liquid and the gas phases.
(4) Every care must be taken in the design of the packed tower that the packing is always properly 'wetted' by the liquid phase. Packing manufacturers usually quote a minimum wetting rate for their packing. This is usually around 2.0–2.5 gpm of liquid per ft^2 of tower cross-section. Most companies prefer this minimum to be around 3.0–3.5 gpm/ft^2.

SIZING A PACKED TOWER

The height of the tower is determined by the methods used to calculate the number of theoretical trays required to perform a specific separation. These have been discussed earlier in Chapter 1. A figure equivalent to the height of a

Table 3.5. Physical properties of some common packings

Packing type	Size (in.)	Wall thickness (in.)	OD and length	Approx. no./per ft³	Approx. wt/per ft³	Approx. surface area (ft²/ft³)	% void volume
Raschig rings (ceramic)	$\frac{1}{4}$	$\frac{1}{32}$	$\frac{1}{4}$	88 000	46	240	73
	$\frac{5}{16}$	$\frac{1}{16}$	$\frac{5}{16}$	40 000	56	145	64
	$\frac{3}{8}$	$\frac{1}{16}$	$\frac{3}{8}$	24 000	52	155	68
	$\frac{1}{2}$	$\frac{1}{32}$	$\frac{1}{2}$	10 500	54	111	63
	$\frac{1}{2}$	$\frac{1}{16}$	$\frac{1}{2}$	10 600	48	114	74
	$\frac{5}{8}$	$\frac{3}{32}$	$\frac{5}{8}$	5 600	48	100	68
	$\frac{3}{4}$	$\frac{3}{32}$	$\frac{3}{4}$	3 140	44	80	73
	1	$\frac{1}{8}$	1	1 350	40	58	73
	$1\frac{1}{4}$	$\frac{3}{16}$	$1\frac{1}{4}$	680	43	45	74
	$1\frac{1}{2}$	$\frac{1}{4}$	$1\frac{1}{2}$	375	46	35	68
	$1\frac{1}{2}$	$\frac{3}{16}$	$1\frac{1}{2}$	385	42	38	71
	2	$\frac{1}{4}$	2	162	38	28	74
	2	$\frac{3}{16}$	2	164	35	29	78
	3	$\frac{3}{8}$	3	48	40	19	74
Raschig rings (metal) (1)	$\frac{1}{4}$	$\frac{1}{32}$	$\frac{1}{4}$	88 000	150	236	69
	$\frac{5}{16}$	$\frac{1}{32}$	$\frac{5}{16}$	45 000	120	190	75
	$\frac{5}{16}$	$\frac{1}{16}$	$\frac{5}{16}$	43 000	198	176	60
	$\frac{1}{2}$	$\frac{1}{32}$	$\frac{1}{2}$	11 800	77	128	84
	$\frac{1}{2}$	$\frac{1}{16}$	$\frac{1}{2}$	11 000	132	118	73
	$\frac{19}{32}$	$\frac{1}{32}$	$\frac{19}{32}$	7 300	66	112	86
	$\frac{19}{32}$	$\frac{1}{16}$	$\frac{19}{32}$	7 000	120	106	75
	$\frac{3}{4}$	$\frac{1}{32}$	$\frac{3}{4}$	3 410	55	84	88
	$\frac{3}{4}$	$\frac{1}{16}$	$\frac{3}{4}$	3 190	100	72	78
	1	$\frac{1}{32}$	1	1 440	40	63	92
	1	$\frac{1}{16}$	1	1 345	73	57	85
	$1\frac{1}{4}$	$\frac{1}{16}$	$1\frac{1}{2}$	725	62	49	87
	$1\frac{1}{2}$	$\frac{1}{16}$	$1\frac{1}{2}$	420	50	41	90
	2	$\frac{1}{16}$	2	180	38	31	92
	3	$\frac{1}{16}$	3	53	25	20	95
Raschig rings (carbon)	$\frac{1}{4}$	$\frac{1}{16}$	$\frac{1}{4}$	85 000	46	212	55
	$\frac{1}{2}$	$\frac{1}{16}$	$\frac{1}{2}$	10 600	27	114	74
	$\frac{3}{4}$	$\frac{1}{8}$	$\frac{3}{4}$	3 140	34	75	67
	1	$\frac{1}{8}$	1	1 325	27	57	74
	$1\frac{1}{4}$	$\frac{3}{16}$	$1\frac{1}{4}$	678	31	45	69
	$1\frac{1}{2}$	$\frac{1}{4}$	$1\frac{1}{2}$	392	34	37	67
	2	$\frac{1}{4}$	2	166	27	28	74
	3	$\frac{5}{16}$	3	49	33	19	78

continued overleaf

Table 3.5. (*continued*)

Packing type	Size (in.)	Wall thickness (in.)	OD and length	Approx. no./per ft³	Approx. wt/per ft³	Approx. surface area (ft²/ft³)	% void volume
Berl saddles	$\frac{1}{4}$	–	–	113 000	56	274	60
(ceramic)	$\frac{1}{2}$	–	–	16 200	54	142	63
	$\frac{3}{4}$	–	–	5 000	48	82	66
	1	–	–	2 200	45	76	69
	$1\frac{1}{2}$	–	–	580	38	44	75
	2	–	–	250	40	32	72
Intalox saddles	$\frac{1}{4}$	–	–	117 500	54	300	75
(ceramic)	$\frac{1}{2}$	–	–	20 700	47	190	78
	$\frac{3}{4}$	–	–	6 500	44	102	77
	1	–	–	2 385	42	78	77
	$1\frac{1}{2}$	–	–	709	37	60	81
	2	–	–	265	38	36	79
Pall rings	$\frac{5}{8}$	26 gauge	$\frac{5}{8}$	6 640	37	110	90.2
(car. steel)	1	24 gauge	1	1 440	30	63	93.8
	$1\frac{1}{2}$	22 gauge	$1\frac{1}{2}$	377	27	39	95.3
	2	20 gauge	2	180	25	31	96.4
Pall rings	$\frac{5}{8}$	–	$\frac{5}{8}$	6 630	8	110	88
(plastic)	1	–	1	1 440	5.25	63	90
	$1\frac{1}{2}$	–	$1\frac{1}{2}$	377	4.75	39	90.5
	2	–	2	180	4.50	31	91
Pall rings	2	$\frac{1}{4}$	2	164	38	29	74
(ceramic)	3	$\frac{3}{8}$	3	49	40	20	74

(1) Weights shown are for carbon steel. Raschig rings are also available in aluminum and alloy steels.
(2) Pall rings are also available in stainless, copper, aluminum, and monel, or nickel.

Table 3.6. Coefficients for use in the HETP equation

Packing type	Packing size (in.)	K_1	K_2	K_3
Raschig rings	0.375	2.10	−0.37	1.24
	0.500	8.53	−0.24	1.24
	1.000	0.57	−0.10	1.24
	2.000	0.42	0.00	1.24
Saddles	0.500	5.62	−0.45	1.11
	1.000	0.76	−0.14	1.11
	2.000	0.56	−0.02	1.11

Table 3.7. Recommended packing sizes (in.)

Packing type	Tower diameter (ft)				
	1.0	2.0	3.0	4.0	5.0 or above
Raschig rings	0.50	0.75	1.00	1.50	2.00
Berl saddles	0.75	1.50	2.00	2.00	2.00
Intalox saddles	0.75	1.50	2.00	2.00	2.00
Pall rings	1.00	1.50	2.00	2.00	2.00

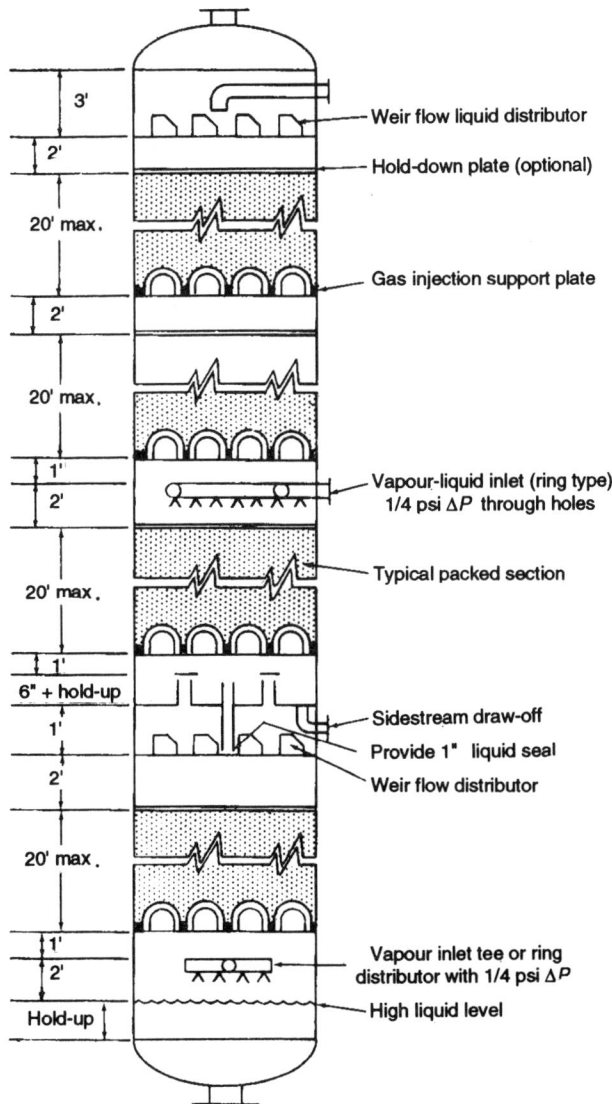

Figure 3.9. A typical packed tower

theoretical tray is then calculated to determine the height of packing required. This is used as the basis to determine the overall height of the tower by adding in the space required for distributors, support trays, etc. The diameter of the tower is calculated using a method which allows for good mass and heat transfer while minimizing entrainment. The same principle of tower flooding is applicable to packed towers as for trayed towers. A calculation procedure for determining a packed tower diameter and the height of packed beds now follows:

- *Step 1*   From examination of the flows of vapour and liquid in the tower determine the critical section of the tower where the loads are greatest. Usually this is at the bottom of an absorption unit and either the top or bottom of a fractionator.
- *Step 2*   Determine the conditions of temperature and pressure at the critical tower section. This is usually accomplished by bubble and dew point calculations as described in Chapter 1. That is, bubble point of the bottoms liquid (in either a fractionator or an absorber) determines the bottom of the tower conditions and dew point calculation of the overhead vapour determines the tower top conditions.
- *Step 3*   Establish the liquid and vapour stream compositions at the critical tray conditions. (See Chapter 1 for determining vapour/liquid streams in absorption and fractionation towers.) Calculate the properties of these streams such as densities, mass/unit time, moles/unit time, viscosity, etc. at the conditions of the critical tower section. Next, select a packing type and size. Use Table 3.7 for this.
- *Step 4*   Commence the tower sizing by calculating the tower diameter. First calculate a value for $(L/G)\sqrt{(\rho_v/\rho_l)}$ where:

  L = mass liquid load (lb/s/ft^2)
  G = mass vapour load (lb/s/ft^2)
  ρ_v = density of vapour in lb/ft^3 at tower conditions
  ρ_l = density of liquid in lb/ft^3 at tower conditions

  Then using Figure 3.10 read off a value for the equation[10]

  $$\frac{L}{G}\left[\frac{\rho_v}{\rho_l}\right]^{1/2} = \frac{V^2}{g} \cdot \frac{S}{F^3} \cdot \frac{\rho_v}{\rho_l} \cdot \mu^{0.2}$$

  where

  V = the vapour velocity at flood (ft/s)
  g = 32.2 ft/s/s
  S = surface area of packing in ft^2/ft^3 of packing. (see Table 3.5)
  F = fraction of void (see Table 3.5)
  μ = viscosity of liquid cP

- *Step 5*   Solve the equation from step 4 to give a value for V. This is the superficial velocity of the vapour at flood. Designing for 80% of flood multiply V by 0.8.
- *Step 6*   Divide the total ft^3/s of the vapour flowing in the tower by 0.8V to

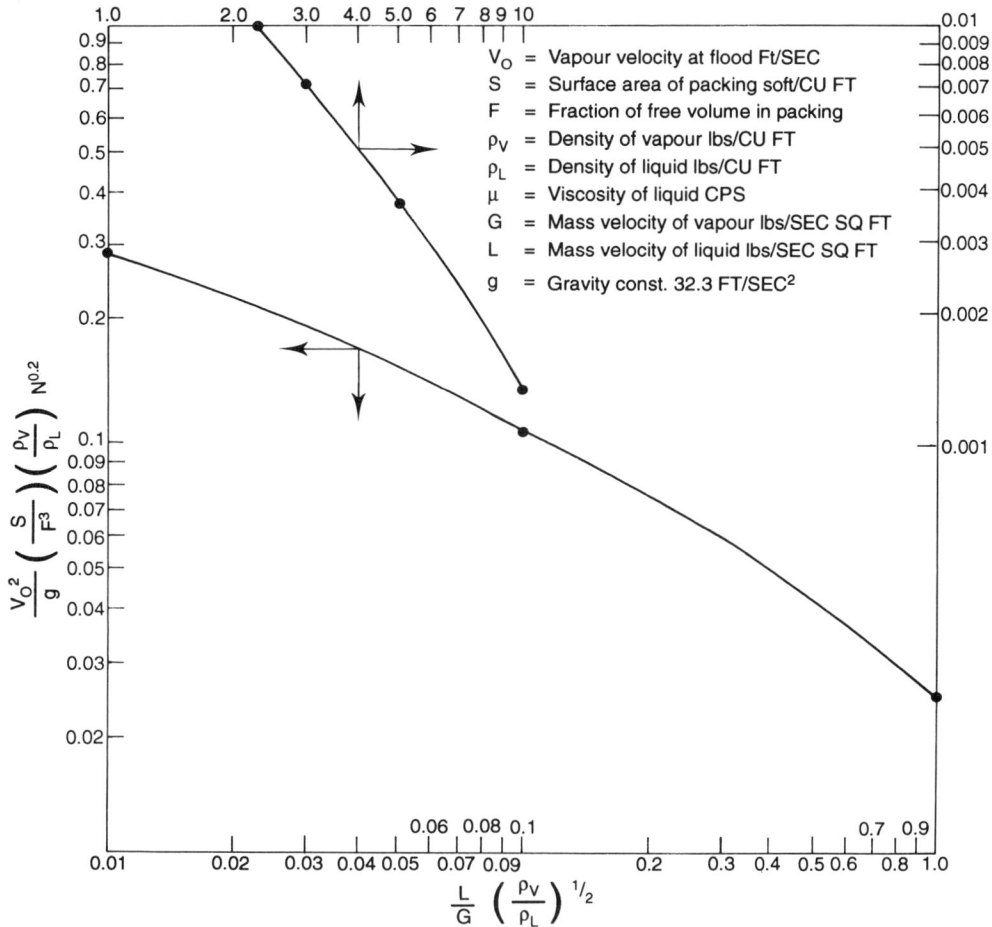

Figure 3.10. Packed tower flooding criteria (Reproduced by permission of Kreiger Publishing Company, Malabar, Florida, 1950, Maxwell Data Book on Hydrocarbons)

give the tower cross-sectional area in ft². Calculate the tower diameter from this area.

- *Step 7* The next part of the calculation is to determine the height of the tower. The number of theoretical trays have been determined by either the fractionation or absorption calculation described in Chapter 1. It is now required to establish either the actual number of trays for a trayed tower or the height of packing in the case of a packed tower. This calculation deals with the second of these. The next step sets out to establish the HETP, which is the height equivalent to a theoretical tray.

- *Step 8* The HETP is calculated from the following equation:

$$ \text{HETP} = K_1 \cdot GH\,K^2 \cdot DK_3 \cdot \frac{62.4 \times \alpha' \times \mu_l}{\rho_l} $$

where

K_1-K_3 = factors from Table 3.6
$\quad D$ = tower diameter (in.)
$\quad \alpha'$ = relative volatility of the more volatile component in the liquid
$\qquad$ phase (see step 9)
$\quad GH$ = mass velocity of the vapour in lb/h.ft^2

- *Step 9* To determine the relative volatility α', select a key light component that is in the lean liquid and the wet gas. The relative volatility is the equilibrium constant of the lightest significant component in the rich liquid divided by the equilibrium constant of the light key component. Solve for a value of HETP and multiply this by the number of theoretical stages to give the total packed height.
- *Step 10* Determine the number of beds to accommodate the packed height. Allow space between the beds for vapour–liquid redistribution and hold-up plates. Use Figure 3.9 as a guide for this. The tower height will be the sum of beds, internal distributors packing support trays, liquid hold-up and vapour disengaging space.

Example calculation

In this example the number of theoretical trays for an absorption unit has been fixed as four. The compositions of the 'wet' gas and the lean liquid have been given and used to determine the composition and quantities of the rich liquid and the lean gas. The quantities to be used in the following calculation are as follows:

Rich liquid leaving the bottom of the absorber = 452.66 moles/h
Wet gas entering the bottom of the tower = 1018.35 moles/h

Their respective composition and conditions are as given or calculated in Section 1.6 of Chapter 1 and are as follows:

Wet gas

	Mole frac.	Mole wt	Weight
H_2	0.467	2.0	0.93
C_1	0.190	16.0	3.40
C_2	0.059	30.0	1.77
H_2S	0.242	34.0	8.24
C_3	0.604	44.0	1.32
iC_4	0.006	58.0	0.35
nC_4	0.006	58.0	0.35
Total	1.000	16.0	16.00

Temperature = 95 °F $\quad$ Pressure = 175 psia

$$Ft^3 = \frac{378 \times 14.7 \times 555}{175 \times 520} = 33.89$$

$$\rho_v = 16/33.89 = 0.473 \text{ lb/ft}^2$$

Rich liquid

	Mole frac.	Mole wt	Wt fact.	lb/gal	Vol. fact.
H_2	0.004	2.0	0.002	—	—
C_1	0.013	16.0	0.208	2.5	0.083
C_2	0.016	30.0	0.480	2.97	0.162
H_2S	0.092	34.0	3.128	6.56	0.477
C_3	0.028	44.0	1.232	4.23	0.291
iC_4	0.011	58.0	0.638	4.68	0.136
nC_4	0.013	58.0	0.754	4.86	0.155
C_9	0.823	128.0	105.344	6.02	17.499
Total	1.000		111.786	5.95	18.803

$$\text{Temperature} = 95\,^\circ\text{F}$$

$$\text{SG at 60} = 0.715$$

$$\text{SG at 95} = 0.696$$

$$\rho_l \text{ at } 95^\circ\text{F} = 43.3\,\text{lb/ft}^3$$

$$\frac{L}{G}\left[\frac{\rho_v}{\rho_l}\right]^{1/2} = \frac{V^2}{g}\cdot\frac{S}{F^3}\cdot\frac{\rho_v}{\rho_l}\cdot\mu^{0.2}$$

where

V = the vapour velocity at flood (ft/s)
g = 32.2 ft/s/s
S = surface area of packing in ft^2/ft^3 of packing (see Table 3.5) = 36 ft^2/ft^3
F = fraction of void (see Table 3.5) = 0.79
μ = viscosity of liquid in cP = 0.56 cP

$$\frac{L}{G}\left[\frac{\rho_v}{\rho_l}\right]^{1/2} = \frac{50\,670}{16\,325} \times \left[\frac{0.473}{43.3}\right]^{1/2} = 0.324$$

From Figure 3.10 = 0.055
Then:

$$\frac{V^2 \times S \times \rho_v \times \mu}{g \times F^3 \times \rho_l} = 0.055$$

$$V^2 = \frac{0.055}{0.022} = 2.5$$

$$v = 1.58\,\text{ft/s}$$

$$\text{at 80\% of flood } V = 1.58 \times 0.8$$

$$= 1.26\,\text{ft/s}$$

$$\text{Total vapour flow} = 16\,325\ \text{lb/h}$$

$$= \frac{16\,325}{0.473} = 34514\ \text{ft}^3/\text{h}$$

$$= 9.59\ \text{ft}^3/\text{s}$$

$$\text{Cross-sectional area} = \frac{9.59}{1.26} = 7.6\ \text{ft}^2$$

$$\text{Tower diameter} = 3.1\ \text{ft (say, 3.25 ft or 39 in.)}$$

To calculate HETP

$$\text{HETP} = K_1{}^{K_2} \cdot GH^{K_3} \cdot D \cdot \frac{62.4 \times \alpha' \times \mu_\text{l}}{\rho_\text{l}}$$

where

$K_1 = 0.56$
$K_2 = -0.02$
$K_3 = 1.11$
α' = relative volatility (neglect H_2 and C_1 in liquid composition as non-condensibles). The key component is C_3. Then α' is KC_2/KC_3. which is $4.6/1.0 = 4.6$.
$GH = 16\,325/7.6 = 2148\ \text{lb/h} \cdot \text{ft}^2$
D = Tower diameter = 39 in.

$$\text{HETP} = \frac{0.56 \times 39^{1.11} \times 62.4 \times 4.6 \times 0.56}{0.02}$$

$$(2148) \times 43.3$$

$$= 8.7\ \text{ft}$$

In the example given in Section 1.6 the number of theoretical trays was fixed at four for this separation. Then using four theoretical trays the total height of packing = $4 \times 8.7\ \text{ft} = 34.8\ \text{ft}$. Two packed beds each, say, 18 in. would satisfy the required duty. Using Figure 3.9 as a guide, the tower height is developed as follows:

- *Bottom tan to HLL (hold-up)*   Liquid is fed to a stripping column. Therefore let the hold-up time be 3 min to NLL. Then NLL = 6.9 ft (say, 7.0 ft) and HLL = 10-ft.
- *HLL to vapour inlet distributor*   This will be set at 2.0 ft.
- *Distributor to bottom bed packing support*   This will be set at 1.0 ft.
- *Bottom packed bed support to top of packed bed*   Packed height, which is 18 ft.
- *Top of bottom bed to bottom of top bed*   Set this at 3.0-ft to allow for a liquid weir-type distributor.
- *Top bed support tray to top of top bed*   Packed height, which is 18 ft.
- *Top of top bed to top tan*   Make this 6-ft to allow for liquid distribution tray and liquid inlet pipe.

$$\underline{\text{Total height tan to tan} = 58\ \text{ft}}$$

3.4 Drums and Drum Design

Drums may be horizontal or vertical vessels. Generally, they do not contain complex internals such as fractionating trays or packing as in the case of towers. They are used, however, for removing material from a bulk material stream and often use simple baffle plates or wire mesh to maximize efficiency in achieving this. Drums are used in a process mainly for:

- Removing liquid droplets from a gas stream (knock-out pot) or separating vapour and liquid streams
- Separating a light from a heavy liquid stream (separators)
- Providing a suitable liquid hold-up time within a process (surge drums)
- Reducing pulsation in the case of reciprocating compressors.

Drums are also used as small intermediate storage vessels in a process.

VAPOUR DISENGAGING DRUMS

One of the most common examples of the use of a drum for the disengaging of vapour from a liquid stream is the steam drum of a boiler or a waste heat steam generator. Here the water is circulated through a heater where it is raised to its boiling point temperature and then routed to a disengaging drum. Steam is flashed off in this drum to be separated from the liquid by its superficial velocity across the area above the water level in the drum. The steam is then routed to a superheater and thus to the steam main. The performance of the steam superheater depends on receiving fairly 'dry' saturated steam, that is, steam containing little or no water droplets. The separation mechanism of the steam drum is therefore critical.

The design of a vapour disengaging drum depends on the velocity of the vapour and the area of disengagement. This is expressed by the equation

$$V_c = 0.157 \sqrt{\frac{\rho_l - \rho_v}{\rho_v}}$$

where

V_c = critical velocity of vapour (ft/s)
ρ_l = density of liquid phase (lb/ft^3)
ρ_v = density of vapour phase (lb/ft^3)

The area used for calculating the linear velocity of the vapour is:

- The vertical cross-sectional area above the high liquid level in a horizontal drum
- The horizontal area of the drum in the case of vertical drums

The allowable vapour velocity may exceed the critical, and normally design velocities will vary between 80% and 170% of critical. Severe entrainment occurs however above 250% of critical. Table 3.8 gives the recommended design velocities for the various services. The minimum vapour space above the liquid

Table 3.8. Some typical drum applications

Service	Liquid surge and distillate drums	Settling drums	Compressor suction KO drums Centrif.	Recip.	Fuel gas KO drums	Steam drums	Water disengaging drums
Allowable vapour velocity Without CWMS, %V_c	170	–	80	80	170	–	170
With 1 CWMS, %V_c	–	–	150	120	–	100	–
With 2 CWMS, %V_c	–	–	–	–	–	150	–
Liquid hold-up set by	(1) Water settling requirements (below low oil level) (2) Minimum instrument dimensions (3) Hold-up requirement for controlling process. For normal circumstances the hold-up between high and low liquid level should be 2 min on product to storage, 15 min on product feeding a subsequent tower or 5 min on reflux—	(1) Settling requirements of individual phases (2) Minimum instrument dimensions (3) Hold-up requirement for controlling process rate (4) Inventory requirements for start-up, shutdown make-up, etc.	10 min liquid spill from largest single producing unit preceding compressor* For interstage KO drums, 10 min based on maximum interstage condensate production rate should be provided between high liquid level and a point one pipe diameter below the inlet nozzle. When taking suction from absorbers, 5 min based on total lean oil circulation rate*		Should be at least equivalent to the volume of a 20 ft slug of condensate in the adjacent fuel gas main* Following an absorber, 5 min. on total lean oil circulation.*	One third the heater and steam piping volume or 2 min on feed water rate, whichever is greater. If harm can occur due to loss of water level provide greater hold-up, depending on process. In recent designs as much as 15 min hold-up on feed water has been provided	Hold-up below low liquid level based on 50 in. per min settling rate for hydrocarbon vapors from water. Minimum height at low liquid level = $1\frac{1}{2}$ ft

whichever is the greatest.
(4) Inventory requirements for start-up shut-down, make-up, etc.

For refrigeration systems, 5 min based on normal refrigerant flow rate to largest cooling unit in the system.*

Normal drum position	Horizontal	Horizontal	Vertical	Vertical	Vertical	Horizontal
Type of nozzle						
Inlet	90° bend	90° bend	Tee distributor	Flush	Tee distributor	90° Bend
Outlet (vapour)	Flush	–	Flush	Flush	Flush	Flush
Outlet (liquid)	Flush or straight extension	Flush	Flush	Flush	Flush	Flush

* Measured between the bottom tangent line and a point one pipe diameter below the inlet nozzle.

level in a horizontal drum should not be less than 20% of drum diameter or 12 in. whichever is greater.

Crinkled wire mesh screens (CWMS) are effective entrainment separators and are often used in separator drums for that purpose. When installed they improve the separation efficiency so vapour velocities much above critical can be tolerated. They are also a safeguard in processes where even moderate liquid entrainment cannot be tolerated.

CWMS are now readily available as packages which include support plates and installation fixtures. Normally for drums larger than 3 ft in diameter the 6-in. thick open mesh type screen is used.

LIQUID SEPARATION DRUMS

The design of a drum to perform this duty is based on one of the following laws of settling:

Stokes law

$$V = 8.3 \times 10^5 \times \frac{d^2 \Delta S}{\mu}$$

when the *Re* number is < 2.0.

Intermediate law

$$V = 1.04 \times 10^4 \times \frac{d^{1.14} \Delta S^{0.71}}{S_c^{0.29} \times \mu^{0.43}}$$

when the *Re* number is 2 − 500.

Newton's law

$$V = 2.05 \times 10^3 \times \left[\frac{d.\Delta S}{S_c}\right]^{1/2}$$

when the *Re* number is > 500. Here

$$Re \text{ number} = \frac{10.7 \times d.v.S_c}{\mu}$$

V = settling rate in in. per minute
d = droplet diameter (in.)
S = droplet specific gravity
S_c = continuous phase specific gravity
ΔS = specific gravity differential between the two phases
μ = viscosity of the continuous phase (cP)

The following may be used as a guide to estimating droplet size:

Lighter phase	Heavy phase	Minimum droplet size
0.850 SG and lighter	Water	0.008 in.
Heavier than 0.850	water	0.005 in.

The hold-up time required for settling is the vertical distance in the drum allocated to settling divided by the settling rate. Some typical applications of drums for this service are given in Table 3.8.

Settling baffles, are often used to reduce the hold-up time and the height of the liquid level.

SURGE DRUMS

This type of drum, the calculation of hold-up time and surge control have been described fully in Section 2.1.

PULSATION DRUMS OR POTS

This type of drum is described in some detail in Section 5.8.

Example calculation

It is required to provide the dimensions and process data for the design of a reflux drum receiving the hydrocarbon distillate, water and uncondensed hydrocarbon vapour from a distillation column. Details of flow and drum conditions are as follows:

Vapour:	12 000 lb/h, 40 mol wt, 300 mole/h
Distillate product:	76 650 lb/h, SG at 100 °F 0.682
Reflux liquid:	61 318 lb/h, SG at 100 °F, 0.682
Water:	1738 l
Temperature of drum:	100 °F
Pressure of drum:	30 psia

The drum is to be a horizontal vessel located on a structure 45 ft above grade. The liquid product is to feed another fractionating unit and therefore requires a hold-up time of 15 min between LLL and HLL. The vapour leaving the drum is to be routed to fuel gas via a compressor, therefore complete disengaging of liquid droplets is required. Complete separation of water from the oil is needed. However, as the water is routed to a desalter separator from the drum separation of oil from the water is not critical.

In all probability the surge volume required by the product will be the determining feature of this design. Setting the liquid levels in the drum will depend on the settling out of the water from the hydrocarbon phase. The design will be checked for satisfactory vapour disengaging.

1   Calculating the surge volume for the distillate product

$$\text{Hold-up time} = 15 \text{ min}$$

$$\text{Product rate} = \frac{76\,650\,\text{lb/h}}{0.682 \times 62.2} = 1807\,\text{ft}^3/\text{h}$$

$$\text{Hold-up volume} = \frac{1807 \times 15}{60} = 452\,\text{ft}^3$$

Then volume of liquid between HLL and LLL is 452 ft^3. Let this be 60% of total drum volume. Then drum volume 452/0.60 = 753 ft^3.

Using a length to diameter ratio (L/D) of 3, diameter and length are calculated as follows:

$$753\,\text{ft}^3 = \frac{\pi.D^2}{4} \times 3D$$

$$D = 3\sqrt{\frac{753 \times 4}{3\pi}}$$

$$= 6.8 \text{ ft (make it 7.0 ft)}$$

$$L = 3 \times 7.0\,\text{ft} = 21.0\,\text{ft}$$

2   Calculating water settling rate

Using the 'intermediate law' then:

$$V = 1.04 \times 10^4 \times \frac{d^{1.14}\Delta S^{0.71}}{S_c^{0.29} \times \mu^{0.43}}$$

$$V = \text{settling rate (in./min)}$$

$$d = \text{droplet size (in.)} = 0.008 \text{ in.}$$

$$S_c = \text{SG of continuous phase} = 0.682$$

$$S_w = \text{SG of water} = 0.993$$

$$\Delta S = 0.311$$

$$V = 1.04 \times 10^4 \times \frac{0.004 \times 0.92}{0.895 \times 0.78}$$

$$= 54.8 \text{ in./min}$$

Check *Re* number

$$Re = \frac{10.7 \times 0.008 \times 54.8 \times 0.682}{0.56}$$

$$= 5.7, \text{ so use of the 'intermediate law' is correct}$$

3   Setting the distance between bottom Tan and LLL (see also Figure 3.12)

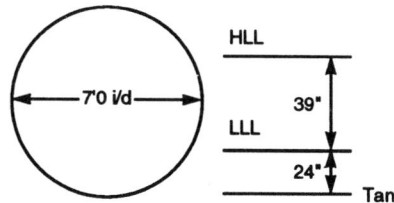

Sufficient distance or surge should be allowed below LLL to provide a LLL alarm at a point about 10% below LLL and bottom Tan. The remaining surge should be sufficient to provide the operator with some time to take emergency action (such as shutting down pumps).

Let LLL be 2 ft above bottom tan. Then surge volume in this section is as follows:

$$R = 2/7 = 0.286 \text{ from Table A1.6 } A = 0.230 \text{ area of section}$$

$$= 0.237 \times 38.48 = 9.1 \text{ ft}^2$$

$$\text{Volume} = 21 \times 8.85 = 191 \text{ ft}^3$$

$$\text{Total flow rate} = \text{product} + \text{reflux} + \text{water}$$

$$= 3531.4 \text{ ft}^3/\text{h} = 58.86 \text{ ft}^3/\text{min}$$

$$\text{Minutes of hold-up below LLL} = 3.25$$

By the same calculation hold-up after alarm = 2.9 min, which is satisfactory.

4   Checking settling time for the water

At the LLL a distance of 2 ft from Tan
Residence time for liquid below LLL

$$= 3.25 \text{ min}$$

Minimum settling time required:

$$\frac{\text{Vert. distance to bottom of drum}}{\text{Settling rate}}$$

$$\frac{24 \text{ in.}}{54.8 \text{ in./min}}$$

$$= 0.44 \text{ min}$$

which is adequate.

5 Calculating height of HLL above LLL

$$\text{Total volume to HLL} = 191 + 452 = 643 \text{ ft}^3$$

$$\text{Area above HLL} = \frac{808 \text{ ft}^3 - 643 \text{ ft}^3}{21 \text{ ft}}$$

$$= 7.58 \text{ ft}^2$$

Using Table A1.6

$$\frac{A_D}{A_s} = \frac{7.58}{38.48} = 0.197 \quad R = 0.251$$

$$r = 0.251 \times 7.0 = 1.76 \text{ ft}$$

$$\text{Height of HLL above LLL} = 7 - (1.76 + 2.0) = 3.25 \text{ ft}$$

6 Checking the vapour disengaging space

$$V_c = 0.157 \sqrt{\frac{\rho_1 - \rho_v}{\rho_v}}$$

where

V_c = critical velocity of vapour in ft/s
ρ_1 = density of liquid phase in lb/ft^3 = 42.42
ρ_v = density of vapour phase in lb/ft^3 = 0.216
V_c = 2.28 ft/s

Actual velocity of vapour is as follows:

$$\text{Cross-sectional area of vapour space above HLL} = 7.58 \text{ ft}^2$$

$$\text{Vapour linear velocity} = \frac{59\,840 \text{ ft}^3/\text{h}}{7.58 \times 3600}$$

$$= 2.21 \text{ ft/s which is 97\% of}$$
$$\text{critical}$$

The drum design meets all necessary criteria and will be used.

3.5 Specifying Pressure Vessels

A process engineer's responsibility extends to defining the basic design requirements for all vessels. These data include:

- The overall vessel dimensions
- The type of material to be used in its fabrication
- The design and operating conditions of temperature and pressure
- The need for insulation for process reasons
- Corrosion allowance and the need for stress relieving to meet process conditions

- Process data for internals such as trays, packing, etc.
- Skirt height above grade
- Nozzle sizes, ratings and location (not orientation)

Typical process data sheets used for specifying columns and horizontal drums are given in Figures 3.11 and 3.12 (with their attachments) respectively. These data sheets have been completed to reflect the examples calculated in Section 3.2 for trayed towers and Section 3.4 for drums. The following paragraphs describe and discusses the contents of these data sheets.

DATA SHEET FOR COLUMNS (FIGURE 3.11)

This particular vessel is a light ends fractionator and has a single tray diameter (i.e. it is not swaged). The tower contains 36 valve trays on a 24-in. tray spacing and has a calculated diameter of 39 in. for the trayed section. (For this calculation see Section 3.2.) This diameter will be specified as 41 in. i/d. This can be met by a standard 42-in. schedule 'X' pipe and this will reduce the cost of the vessel. The overall dimension for the tower is completed by setting the height of the tower from Tan to Tan. In the example here this has been done as follows:

$$\text{Height of trayed section} = (\text{No. of trays} - 1.0) \times \text{tray spacing}$$

The fractionation calculations (see Chapter 1 for this type of calculation) has determined 36 actual trays for this tower. Thus the trayed height is $(36 - 1) \times 24$ in. $= 70$ ft. Add another 3 ft to accommodate the feed inlet distributor on tray 20. Then total trayed height is 73 ft.

The bottom of the tower must accommodate the liquid surge requirement. As the tower diameter is relatively small a swaged section of 4 ft diameter will be considered below the bottom tray for surge. The liquid product goes to storage, therefore the surge requirement need not be more than 3 min on product (see Section 2.1).

From the unit's material balance the bottom product is as follows:

$$\text{Weight per hour} = 117\,513\,\text{lb}$$
$$\text{Temperature} = 440\,°F$$
$$\text{Density at } 440\,°F = 40\,\text{lb/ft}^3$$
$$\text{Then hot ft}^3/\text{min of product} = 48.96$$

The product goes to storage, therefore only 2–3 min surge is required. This will be set at 3 min surge to NLL.

$$\text{Total surge to NLL} = 48.96 \times 3 = 146.9\,\text{ft}^3$$
$$\text{Cross-sectional area of surge section} = \pi/4 \times 4^2$$
$$= 12.6\,\text{ft}^2$$
$$\text{Height of NLL above Tan} = \frac{146.9}{12.6}$$
$$= 11.7\,\text{ft (make it 12 ft)}$$

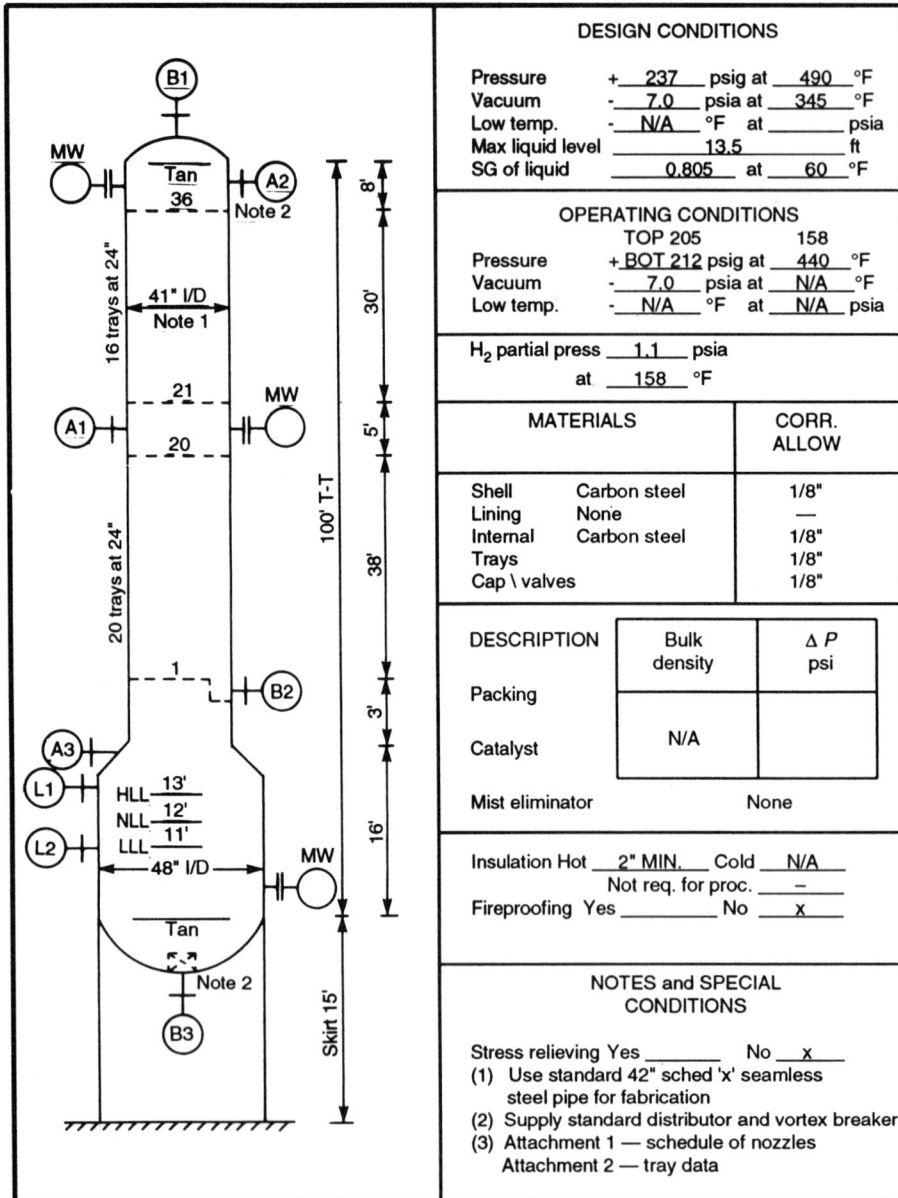

Figure 3.11. A typical process data sheet for columns

Nozzle schedule (attachment 1)

Ref.	Description	Size (in.)	RTG (see below)
A1	Feed inlet nozzle	6	150 RF
A2	Reflux inlet nozzle	4	150 RF
A3	Inlet from reboiler	6	300 RF
B1	Overhead vapour outlet	8	150 RF
B2	Outlet to reboiler	4	300 RF
B3	Bottom product outlet	3	300 RF
L1 L2	Instrument nozzles	$\frac{3}{4}$	300 RF
MW	Manways	24	150 RF

Table 3.9 gives flange ratings for carbon steel material

Tray data sheet (attachment 2)

Vessel no.	C 401	
Vessel name.	Reformate stabilizer	
Description of material	Unstabilized light aromatics	
Section	Top trays 21–36	Bottom trays 1–20
Total trays in section	16	20
Maximum ΔP per tray (psi)	0.25	0.25
Conditions on tray no. (1 = bottom)	Top tray	Bottom tray
	36	1
Vapour:		
Temp. (°F)	167	440
Pressure (psig)	205	212
Density (lb/ft³)	1.69	2.0
Rate (lb/h)	47 700	71 021
ACFS	7.83	9.81
Liquid		
Temp. (°F)	162	430
Viscosity (cP)	0.3	0.85
Density (lb/ft³)	27.3	38.2
Mol wt	57	100
Rate (lb/h)	33 273	104 950
Rate (ft³/min)	20.34	45.79
Tower diameter (ft)	3'5" (41")	
Number of passes	One	
Type of tray	Valve	
Tray spacing	24"	

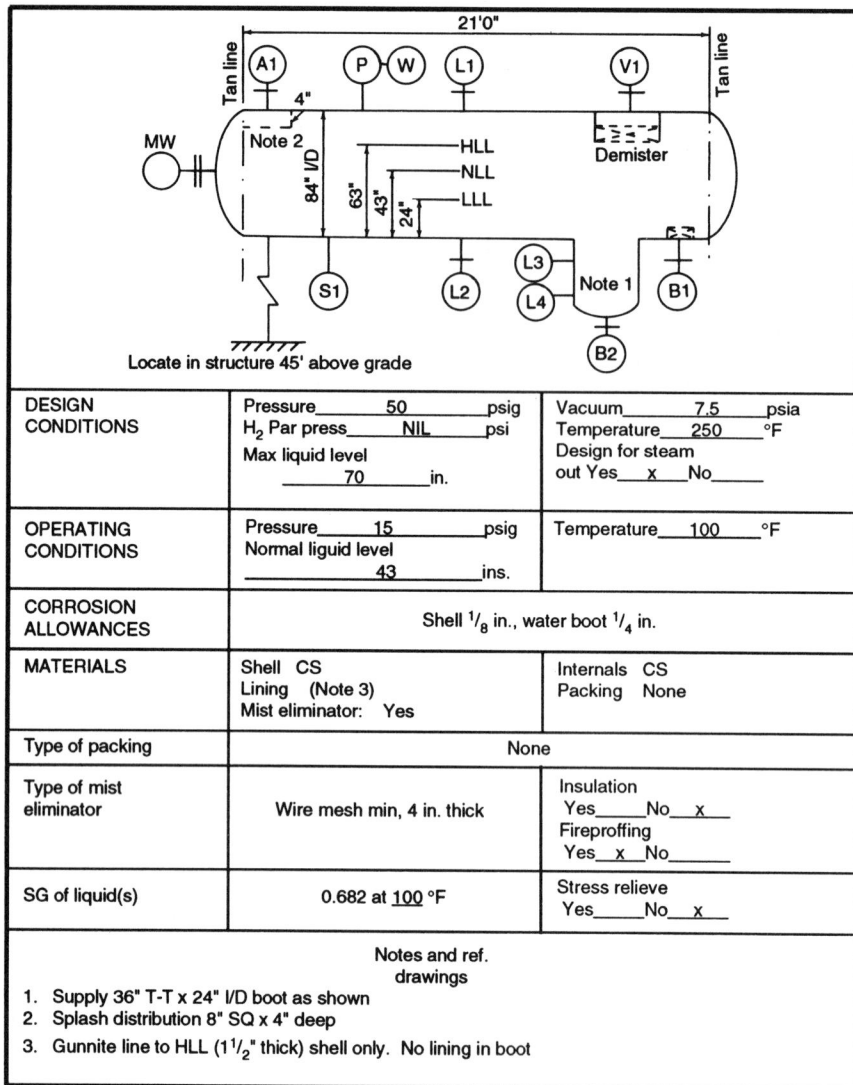

DESIGN CONDITIONS	Pressure_____50_____psig H_2 Par press____NIL____psi Max liquid level _____70_____in.	Vacuum_____7.5_____psia Temperature___250___°F Design for steam out Yes___x___No_____
OPERATING CONDITIONS	Pressure_____15_____psig Normal liquid level _____43_____ins.	Temperature____100____°F
CORROSION ALLOWANCES	Shell $\frac{1}{8}$ in., water boot $\frac{1}{4}$ in.	
MATERIALS	Shell CS Lining (Note 3) Mist eliminator: Yes	Internals CS Packing None
Type of packing	None	
Type of mist eliminator	Wire mesh min, 4 in. thick	Insulation Yes_____No__x___ Fireproffing Yes__x__No_____
SG of liquid(s)	0.682 at <u>100</u> °F	Stress relieve Yes_____No__x__

Notes and ref. drawings

1. Supply 36" T-T x 24" I/D boot as shown
2. Splash distribution 8" SQ x 4" deep
3. Gunnite line to HLL ($1\frac{1}{2}$" thick) shell only. No lining in boot

Nozzle schedule (attachment 1)

Ref.	Description	Size (in.)	RTG
A1	Feed inlet (V + L)	8	150 R.F.
V1	Vapour outlet	8	150 RF
B1	Hydrocarbon liquid outlet	6	150 RF
B2	Water outlet	$2\frac{1}{2}$	150 RF
L1 L2	Level instrument	$\frac{3}{4}$	150 RF
L3 L4	Level instrument	$\frac{1}{2}$	Screwed fitting
MW	Manway	24	150 RF
S1	Steam out	$\frac{1}{2}$	Screwed fitting
PW	PI and RV connection	$\frac{1}{2}$	Screwed fitting

Figure 3.12. A typical process data sheet for horizontal vessels

Level range will be 24 in. Then HLL will be $12 + 1 = 13$ ft above NLL. Let the reboiler inlet nozzle be 3 ft above HLL = 16 ft above Tan. Allow another 3 ft between reboiler inlet nozzle to bottom tray. This provides adequate space for vapour separation from the high liquid level.

At the top of the column space must be provided between the top tray and the top vapour outlet nozzle to accommodate the reflux return distributor and vapour disengaging from the top tray. Let this be 8 ft from top tray to the tower top Tan line. Then

Total tower height:

$$\begin{aligned}
\text{From bottom Tan to bottom tray} &= \quad 19.0\,\text{ft} \\
\text{The trayed section} &= \quad 73.0\,\text{ft} \\
\text{From top tray to Top Tan} &= \quad\;\; 8.0\,\text{ft} \\
\text{Total} &= 100.0\,\text{ft}
\end{aligned}$$

These overall dimensions are now inserted on the vessel diagram as shown in Figure 3.11. An attached sheet will be included to give the nozzle description, size and flange rating referring to those shown in the small circles on the sketch. A schedule of flange ratings for carbon steel is given in Table 3.9.

The only other dimension that will be shown on the sketch is that of the skirt. Now the vessel is installed onsite supported by a metal skirt fixed to the concrete foundation of the vessel. The height of this skirt is fixed by the process engineer. If the product from the bottom of the tower is to be pumped (as is usual), the skirt height must be such as to accommodate the suction conditions for the pump. The most important of these conditions is the head required to meet the pump's NPSH (net positive suction head). (See Chapter 4 for details). Usually a skirt

Table 3.9. Schedule of flange ratings for carbon steel

Flange ratings (psig) Design temp. (°F)	150	300	400	600	900
		Max. design pressure (psig)			
100	275	720	960	1440	2160
150	255	710	945	1420	2130
200	240	700	930	1400	2100
250	225	690	920	1380	2070
300	210	680	910	1365	2050
350	195	675	900	1350	2025
400	180	665	890	1330	2000
450	165	650	870	1305	1955
500	150	625	835	1250	1875
550	140	590	790	1180	1775
600	130	555	740	1110	1660
650	120	515	690	1030	1550
700	100	470	635	940	1410
750	100	425	575	850	1275
800	92	365	490	730	1100
850	82	300	400	600	900
900	70	225	295	445	670
950	55	155	205	310	465
1000	40	85	115	170	255

height of 15 ft meets most NPSH requirements. The second consideration dictating skirt height is that head required be a thermosyphon reboiler. If the vessel is new and being designed the skirt height of 15 ft remains adequate with properly designed piping to and from the reboiler.

Design conditions

This particular tower operates at 212 psig and 440 °F in the bottom and 205 psig and 158 °F at the top. It is therefore classed as a pressure vessel and will be fabricated to meet a pressure vessel code. The most common of these codes is the ASME code, either section 1 or section 8. Most vessels in the chemical industry are fabricated to ASME, section 8. When fabricated, inspected, and approved it will be stamped to certify that its construction conforms to this pressure vessel code.

Process engineers generally are not required to become familiar with pressure vessel codes as such. They must, however, provide such data that vessel engineers and fabricators can calculate vessel details in accordance with the code. Among these data are the design conditions of temperature and pressure for the vessel.

Design pressure

The design pressure is based on the maximum operating pressure at which the relief valve will open plus a suitable safety increment. The following table provides a guide to this increment:

Maximum operating pressure (psig)	Design pressure (psig)
Full or partial vacuum	50
0–5	50
6–35	50
36–100	Operating + 15
101–250	Operating + 25
251–500	Operating + 10%
501–1000	Operating + 50
over 1000	Operating + 5%

In cases where vessels relieve to a flare header it may be necessary to add a little more to the differential between operating and design pressures to accommodate for the flareback pressure.

Design temperatures

The following table may be used as a guide to the max and min design temperatures. Very often companies will have their own standards for these design criteria. The table given here may be used if there are no company standards.

	Maximum operating temperature (°F)	*Design temperature (°F)*
Columns	Ambient–200	250
	201–450	Operating + 50
	Over 450	Divide into zones add 50 to each operating zone
Vessels	Up to 225	250
	226–600	Operating + 25
	Over 600	Operating + 50

	Minimum operating temperatures (°F)	*Minimum design temperatures (°F)*
	15–Ambient	Operating − 25
	14 to −10	−20
	−10 to −80	Operating − 10
	Below − 80	Operating

A vacuum condition can exist in a tower during normal steam out if the tower is accidentally shut in and the steam valve closed. Normally a design vacuum pressure of 7 psia is specified at the steam saturated temperature to cover this contingency.

Low temperature

This applies to towers in cryogenic services (such as demethanizers and LNG plants). There may be a situation in a non-cryogenic service where rapid depressurizing causes sub-zero temperatures. If this is a situation that can exist for several hours and occurs frequently this condition should be entered. Otherwise make an appropriate remark in 'Notes and Special Conditions'.

Max. liquid level

This is the liquid level under operating conditions that will:

Either activate the high liquid level alarm

Or shut down the feed pump.

whichever the system is applicable to protect the plant operation. Usually this is a quoted as the HLL and 1–2% of surge.

SG of liquid

Quote this as the SG of the liquid on which the surge volume was based. This SG is usually quoted at 60 °F.

Operating conditions

In most fractionation towers there will be two distinctly different conditions of temperature and pressure—those for the tower top and those for the bottom of the tower. Both these conditions must be quoted in this case. The same situation may not necessarily arise in an absorption column.

Operating temperatures and pressures

Quote the calculated data as they will appear also on the process flow diagram. Show the tower top pressure and temperature first followed by the bottom set of conditions. If the tower has been sized on data for more than one design case, show the highest numbers calculated for top and bottom. Also make a note in the 'Notes and special conditions' section of the cases the data were based on.

Other operating data

Vacuum conditions in this case only apply if the tower operates normally at sub-atmospheric pressure. In this case quote the lowest pressure the tower will be operated on together with the normal operating temperature(s). Note in many vacuum fractionators there will be a spectrum of these conditions along the tower. These should be quoted for critical locations in the tower. Such locations would be feed inlet (flash zone), side stream and pumparound draw-offs, tower top. Low temperatures and the associated pressure apply only to cryogenic plants in this case.

Hydrogen partial pressure

This item is important to the metallurgist who will select the grade of metal to be used in the fabrication of the vessel. Generally the hydrogen partial pressure that will be quoted will be the one that exists at the tower top under normal operating conditions. For example, the dew point calculation used in the sizing of the tower given in Figure 3.11 was based on the following tower top vapour composition.

	Mole fraction
H_2	0.005
C_1	0.021
C_2	0.117
C_3	0.378
iC_4	0.207
nC_4	0.268
iC_5	0.004
Total	1.000

The tower top pressure is 220 psia and the temperature is 158 °F.

$$\text{Hydrogen partial pressure} = \frac{\text{Moles } H_2}{\text{Total moles}} \times \frac{\text{Total}}{\text{pressure}}$$

$$= 0.005 \times 220$$

$$= 1.1 \text{ psia}$$

Materials and corrosion allowance

The process engineer will state the type of materials required to meet the process condition. For example, where carbon steel only is to be used the process engineer indicates 'CS'. The engineer is not normally required to state the grade of steel to be used as this is the responsibility of the vessel specialist or the metallurgist. However, if the process engineer has a special knowledge of the material to be used and its specifics it should be noted it on the process data sheet.

The same applies to the corrosion allowance. Normally $\frac{1}{8}$ in. is used for this allowance. However, there may be some mild corrosive condition existing for which the process engineer feels justified to use a higher number.

Description of internals

This is self-explanatory as it refers to packed columns. In the case of trayed towers a separate data sheet giving sufficient data for tray rating and sizing is attached to the process data sheet front page. This is shown in the attachments to Figure 3.11.

Other common internals, such as distributors, vortex breakers, etc. are not normally shown on this data sheet. These are normally standard to a particular design office and will be added to the engineering drawing developed from this process data sheet later.

Insulation and fireproofing

The insulation requirement for heat conservation is specified by the process engineer as required. An approximate thickness is shown. This will be checked later by the vessel specialist. In the case of fireproofing the process engineer indicates its need or not. Remember, the process engineer's relief valve sizing based on a fire condition takes into consideration the inclusion or not of fireproofing.

Notes and special conditions

This item is a 'catch-all' and is used to make note of whatever other information the process engineer may wish to add to the data sheet to ensure the equipment item will meet the process requirement. The question of stress relieving of the vessel is an item which is most important to the proper fabrication of the vessel and to its cost. The process engineer usually has knowledge whether this is needed or not to handle the process material at the conditions specified and must therefore indicate this in this section of the data sheet. Other entries in this item should be a list of the attachments to the data sheet.

DATA SHEET FOR HORIZONTAL VESSELS (FIGURE 3.12)

Most of the process data used to define the requirements for a horizontal drum are the same as those applied to a column, and these have already been discussed for Figure 3.11. The data included in the example given in Figure 3.12 have been calculated in Section 3.4. In the data sheet, however, a 'boot' measuring 2 ft i/d × 3 ft high has been added to the outlet end of the vessel to accumulate the water phase for better control of its level and to allow the disengaging of the hydrocarbon from the water.

4   PUMPS

4.1   Types of Pumps

Pumps in the process industry are divided into two general classifications:

- Variable head capacity
- Positive displacement

The variable head capacity types include centrifugal and turbine pumps while the positive displacement types cover reciprocating and rotary pumps.

CENTRIFUGAL PUMPS

Centrifugal pumps comprise a very wide class of pumps in which pumping of liquids or generation of pressure is effected by a rotary motion of one or several impellers. The impeller or impellers force the liquid into a rotary motion by an impelling action, and the pump casing directs the liquid to the impeller at low pressure and leads it away under a higher pressure. There are no valves in centrifugal type pumps: flow is uniform and devoid of pulsations. Since this type of pump operates by converting velocity head to static head, a pump impeller operating at a fixed speed will develop the same theoretical head in feet of fluid flowing, regardless of the density of the fluid. A wide range of heads can be handled. The maximum head (in ft of fluid) that a centrifugal pump can develop is determined primarily by the pump speed (rpm), impeller diameter and number of impellers in series. Refinements in impeller design and the impeller blade angle primarily effect the slope and shape of the head-capacity curve and have a minor effect on the developed head. Multistage pumps are available which will develop very high heads; up to 5000 ft and up to 1200 gpm. This versatility in handling high-pressure head makes the centrifugal pump the most commonly used type in the process industry.

TURBINE PUMPS

Turbine pumps are a type of centrifugal pumps designed to recover power in systems of high flow and high differential pressure. These pumps transmit some of the kinetic energy in the fluid into brake horsepower. The actual energy recovery is about 50% of the hydraulic horsepower available. This type of pump is expensive and is therefore not as widely used as the centrifugal pump.

ROTARY PUMPS

Rotary pumps are positive displacement pumps. Unlike the centrifugal type pump, these types do not throw the pumping fluid against the casing but push the fluid forward in a positive manner similar to the action of a piston. These pumps, however, do produce a fairly smooth discharge flow unlike that associated with a reciprocating pump. The types of rotary pumps commonly used in a process plant are:

- *The gear pump* This pump consists of two or more gears enclosed in a closely fitted casing. The arrangement is such that when the gear teeth are rotated they are enmeshed on one side of the casing. This allows the fluid to enter the void between gear and casing. The fluid is then carried around to the discharge side by the gear teeth, which then push the fluid into the discharge outlet as the teeth mesh again.
- *Screw pumps* These have from one to three suitably threaded screwed rotors of various designs in a fixed casing. As the rotors turn, liquid fills the space between the screw threads and is displaced axially as the threads mesh.
- *Lobular pumps* The lobular pump consists of two or more rotors cut with two, three, or more lobes on each rotor. The rotors are synchronized for positive rotation by external gears. The action of these pumps is similar to gear pumps but the flow is usually more pulsating than that from the gear pump.
- *Vane pumps* There are two types of vane pumps: those that have swinging vanes and those that have sliding vanes. The swinging vane type consists of a series of hinged vanes which swing out as the rotor turns. This action traps the pumped fluid and forces it into the pump discharge. The sliding vane pump employs vanes that are held against the casing by the centrifugal force of the pumped fluid as the rotor turns. Liquid trapped between two vanes is carried around the casing from the inlet and forced out of the discharge.

RECIPROCATING PUMPS

These are positive displacement pumps and use a piston within a fixed cylinder to pump a constant volume of fluid for each stroke of the piston. The discharge from reciprocating pumps is pulsating. Reciprocating pumps fall into two general categories: the simplex type and the duplex type. In the case of the simplex pump there is only one cylinder which draws in the fluid to be pumped on the back stroke and discharges it on the forward stroke. External valves open and close to enable the pumping action to proceed in the manner described. The duplex pump has a similar pumping action to the simplex pump. In this case, however, there are two parallel cylinders which operate on alternate stroke to one another. That is, when the first cylinder is on the suction stroke the second is on the discharge stroke.

Reciprocating pumps may have direct-acting drives or may be driven through a crankcase and gear box. In the case of the direct-acting drive the pump piston is connected to a steam drive piston by a common piston rod. The pump piston therefore is actuated by the steam piston directly. Reciprocating pumps driven by electric motor, turbines, etc. are connected to the prime mover through a gearbox and crankcase.

OTHER POSITIVE DISPLACEMENT PUMPS

There are other positive displacement pumps commonly used in the process industry for special services. Some of these are:

- *Metering or proportioning pumps*   These are small reciprocating plunger type pumps with an adjustable stroke. They are used to inject fixed amount of fluids into a larger stream or vessel.
- *Diaphragm pumps*   These pumps are used for handling thick pulps, sludge, acid or alkaline solution, and fluids containing gritty solid suspensions. They are particularly suited to these kinds of service because the working parts are associated with moving the diaphragm back and forth to cause the pumping action. The working parts therefore do not come into contact with this type of fluid, which would be harmful to them.

CHARACTERISTIC CURVES

Pump action and the performance of a pump are defined in terms of their *characteristic curves*. These curves correlate the capacity of the pump in unit volume per unit time versus discharge or differential pressures. Typical curves are shown in Figures 4.1–4.3.

Figure 4.1 is a characteristic curve for a reciprocating simplex pump which is direct driven. Included also is this reciprocating pump on a power drive.

Figure 4.2 gives a typical curve for a rotary pump. Here the capacity of the pump is plotted against discharge pressure for two levels of pump speed. The curves also show the plot of brake horsepower versus discharge pressure for the two pump speed levels.

Figure 4.3 is a typical characteristic curve for a centrifugal pump. This curve usually shows four pump relationships in four plots. These are:

- A plot of capacity versus differential head. The differential head is the difference in pressure between the suction and discharge.
- The pump efficiency as a percentage versus capacity.

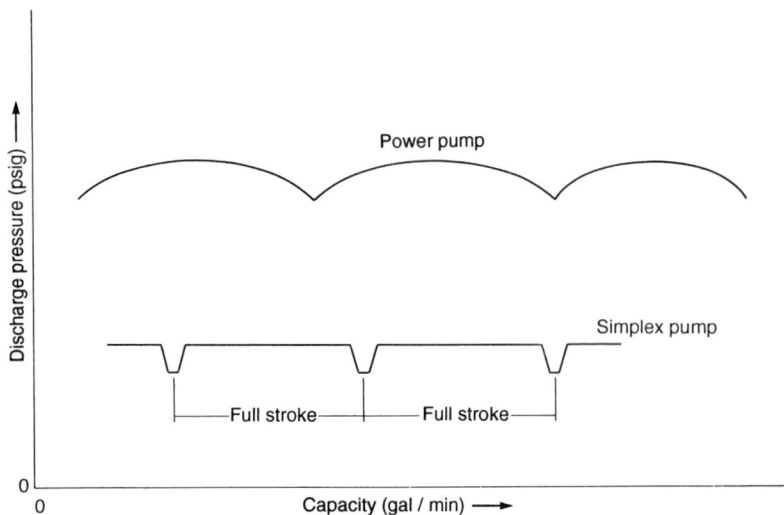

Figure 4.1. Characteristic curves for a reciprocating pump

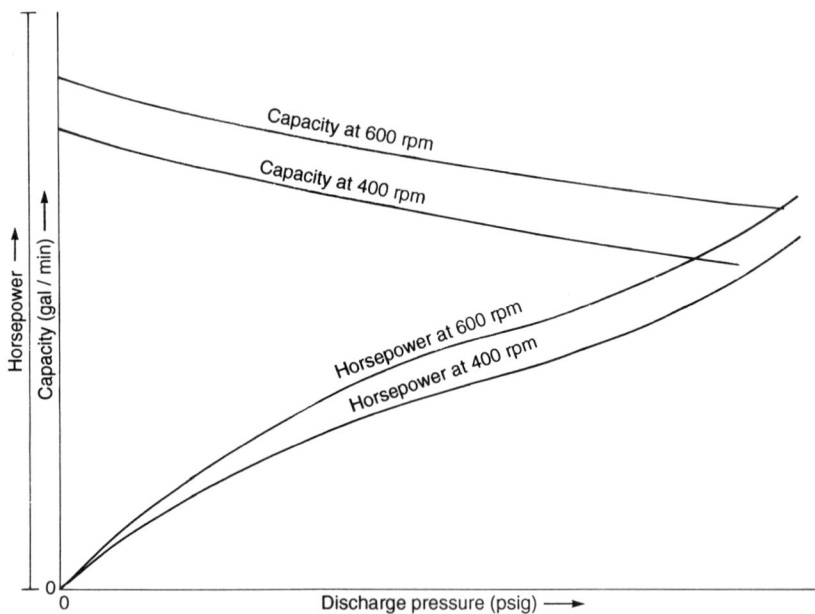

Figure 4.2. Characteristic curves for a rotary pump

- The brake horsepower of the pump versus capacity.
- The net positive head (NPSH) required by the pump versus capacity. The required NPSH for the pump is a characteristic determined by the manufacturer.

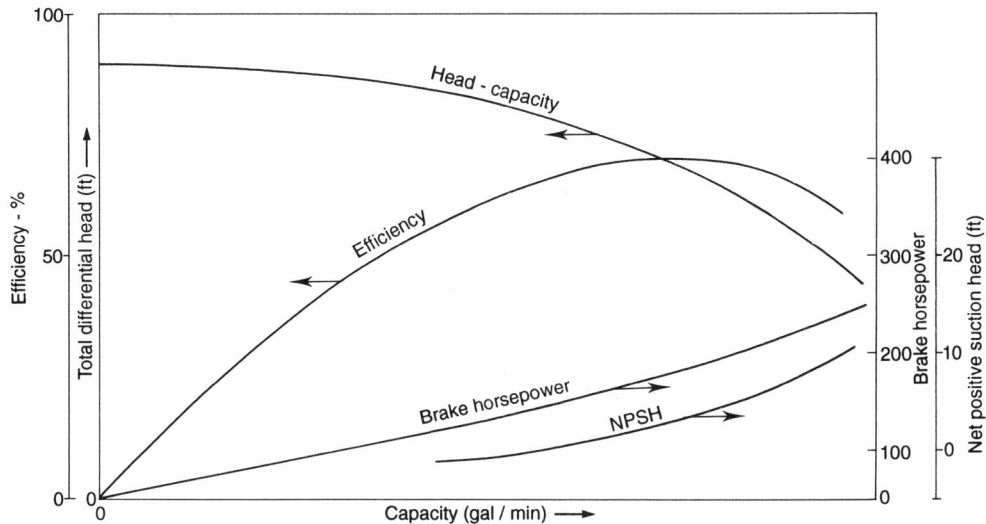

Figure 4.3. A typical characteristic curve for a centrifugal type pump

4.2 Pump Selection

Most industrial pumping applications favour the use of centrifugal pumps. The prominence of this type of pump stems from its ability to handle a very wide spectrum of fluids over a large range of pumping conditions. It is fitting that in considering pump selection the first choice has to be the centrifugal pump, and all others become a selection by exception. Centrifugal pumps are generally the simplest in construction, lowest first cost and easiest to operate and to maintain. This section therefore begins with the selection characteristics of the centrifugal pump.

THE CENTRIFUGAL PUMP

Before looking at the selection of the centrifugal pump it is necessary to define the following terminology associated with pumps in general. These are:

- Capacity
- Differential head
- Available NPSH
- Required NPSH

- *Capacity* This can be defined as the amount of fluid the pump can handle per unit time and at a differential pressure or head. This is usually expressed as gallons per minute at a differential head of so many pounds per square inch or so many feet.
- *Differential head* This is the difference in pressure between the suction of

the pump and the discharge. It is usually expressed as psi and feet in specifying a pump. The following formula is the conversion from psi to feet:

$$\Delta H = \frac{2.31 \times \Delta P}{SG}$$

where

H = the differential head in feet of fluid being pumped
P = the differential pressure of the fluid across the pump measured in pounds per square inch
SG = the specific gravity of the fluid at the pumping temperature

- *Available NPSH* The available net positive suction head (NPSH) is the static head available (in feet or metres) above the vapour pressure of the fluid at the pumping temperature. This is a feature of the design of the system which includes the pump.
- *Required NPSH* This is the static head above the vapour pressure of the fluid required by the pump design to function properly. The required NPSH must always be less than the available NPSH.

SELECTION CHARACTERISTICS

Selection of any pump must depend on its ability to handle a particular fluid effectively and efficiency of the pump under normal operating conditions. The second of these primary requirements can be determined by the pump's characteristic curves. These have already been described in Section 4.1 and a further discussion on these now follows.

Capacity range

Normal

Figures 4.4 and 4.5 show the normal capacity range for various types of centrifugal pumps in two different speed ranges, 3350 rpm 2950 rpm. These values correspond to motor full load speeds available with current at 60 and 50 cycles, respectively. Most process applications call for these speed ranges. Lower speeds are for low or medium head and high capacity requirements, and for special abrasive slurries or corrosive liquids. Low-capacity centrifugal pump applications may require special recirculation provisions in the process design to maintain a minimum flow through the pump. Because of practical consideration in impeller construction, the smallest available process type centrifugal pumps are rated at about 50 gpm (maximum efficiency point).

High and low capacity ranges

Pumps above the limits shown in Figures 4.4 and 4.5 will normally require large horsepower drivers. Special investigation of efficiency, speed, NPSH requirements, etc. will normally be justified. As an example, heads at or above the limits shown for multi-stage pumps at standard motor speed may be obtained by speed-increasing gears (motor drive) or turbines to give pump operating speeds above maximum motor speeds (NPSH requirements increase with speed).

In general, centrifugal pumps should not be operated continuously at flows less than approximately 20% of the normal rating of the pump. The normal rating for the pump is the capacity corresponding to the maximum efficiency point. The following table lists minimum desirable flow rates which should be maintained by continuous recirculation if the required process flow conditions are of lower magnitude:

Head range (ft)	Pump type	Minimum continuous capacity rating of pump (gpm)	Normal rating (gpm)
60 cycle speed (3550 rpm)			
To 100	1 stage	10	60
100–350	1 stage	15	75–100
350–650	2 stage	30	150
650–1100	2 stage	40	160
400–1200	Multi-stage	15	50
1200–5500	Multi-stage	40	100–120
50 cycle speed (2950 rpm)			
To 75	1 stage	10	50
75–250	1 stage	15	60–80
250–450	2 stage	25	120
450–775	2 stage	30	130
250–850	Multi-stage	10	40
850–3800	Multi-stage	30	80–100

Care must be exercised in the design of any recirculation system to ensure that the recirculated flow does not increase the temperature of pump suction and cause increased vapour pressure and reduction of available NPSH.

For low head pumps that can operate at 1750 or 1450 rpm, the above normal and minimum continuous capacities are reduced by 50%.

Effect of liquid viscosity

When suitably designed, centrifugal pumps can satisfactorily handle liquids containing solids, dirt, grit and corrosive compounds. Though fluids with viscosities up to 20 000 SSU (440 cSt) can be handled, 3000 SSU (650 cSt) is usually the practical limitation from an economical operating standpoint.

Effect of suction head

An important requirement is that there be sufficient net positive suction head at the eye of the first-stage impeller. This is static pressure above the vapour

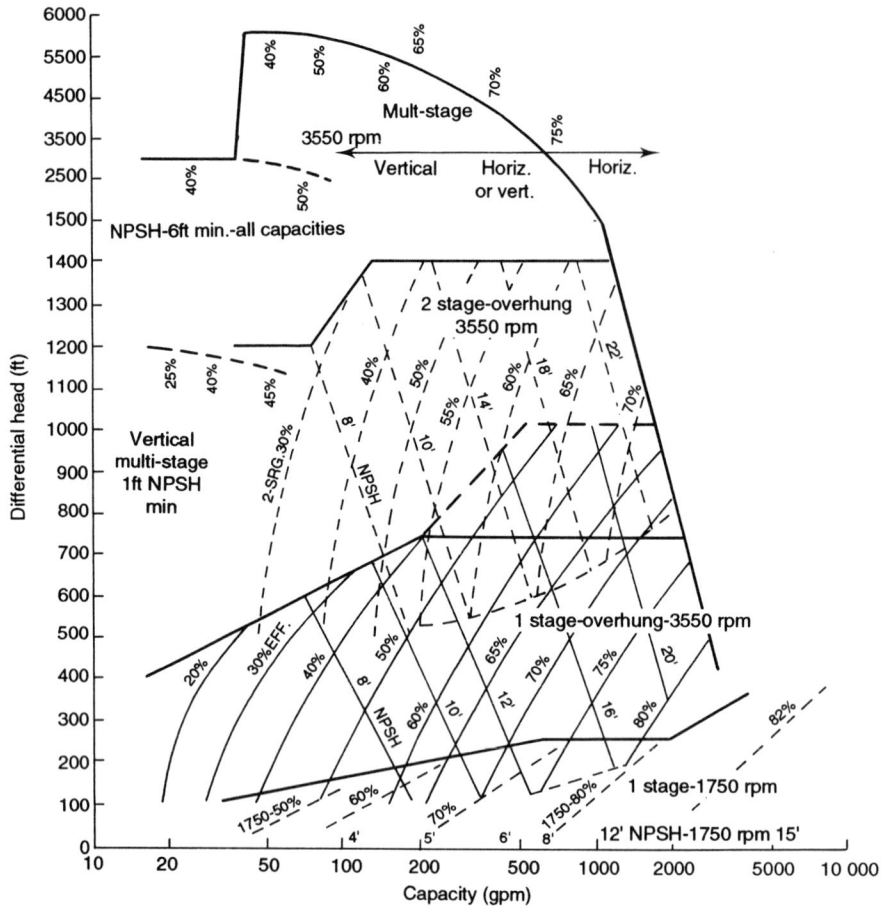

Figure 4.4. Centrifugal pumps at 3550 rpm. The % lines are for pump efficiencies and the curves labelled in feet are required NPSH

pressure of the fluid handled to prevent vaporization at the impeller eye. Flashing of the fluid produces a shock or cavitation effect at the impeller which results in metal loss, noise, lowered capacity and discharge pressure, and rapid damage to the pump. NPSH requirements for various centrifugal pumps will normally vary from 6 to over 20 (in feet of fluid) depending on type, size and speed. Vertical pumps can be built for practically no NPSH at all at the nozzle. These will have extended barrels in order to provide the required NPSH at the eye of the first-stage impeller.

Efficiency

The efficiency of centrifugal pumps varies from about 20% for low-capacity (< 20 gpm) pumps to a range of 70–80% for high-capacity (> 500 gpm) pumps.

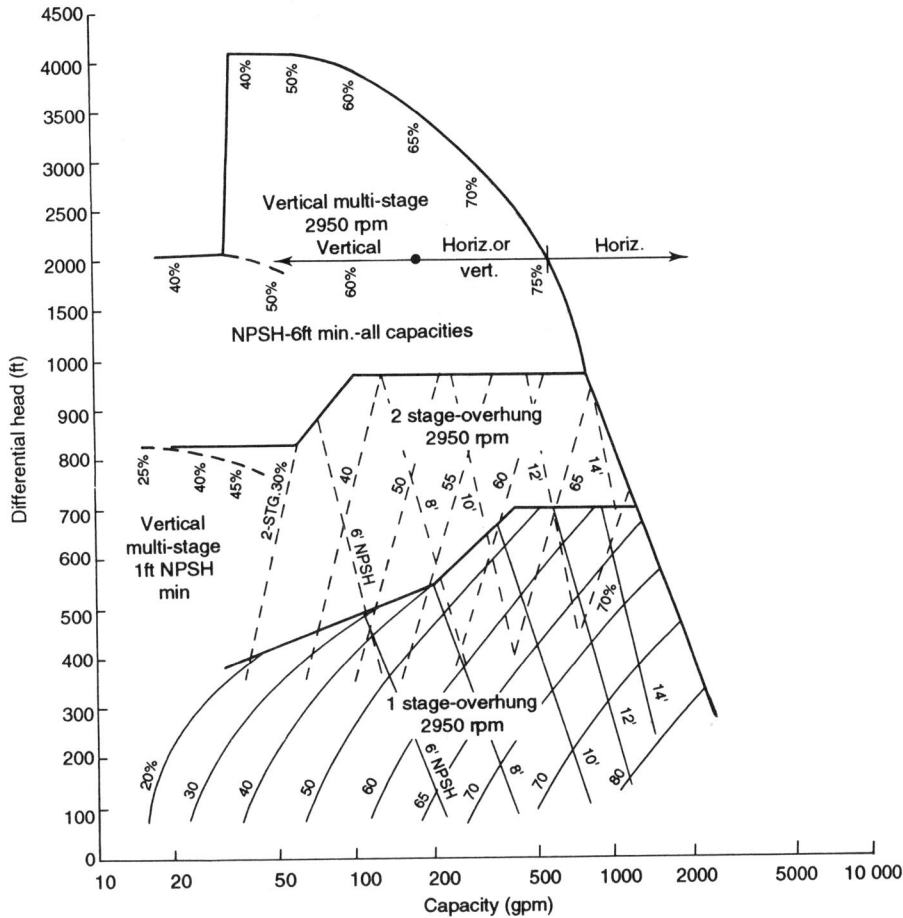

Figure 4.5. Centrifugal pumps at 2950 rpm

Extremely large capacity pumps (several thousand gpm) may have efficiencies of up to nearly 90%.

ROTARY PUMPS

Rotary pumps deliver constant capacity against variable discharge pressure. This is a feature of all positive displacement pumps. Rotary pumps are available for process application over a range of 1 to 5000 gpm capacity and a differential pressure up to about 700 psi. Displacement of the pump varies directly with the speed except in the case where the capacity may be influenced by the viscosity of the fluid. In this case thick viscous liquids may limit capacity because they cannot flow into the cylinder fast enough to keep it completely filled. Rotary pumps are used mostly in low-capacity service where the efficiency of a centrifugal pump would be very low.

RECIPROCATING PUMPS

The liquid discharge from reciprocating pumps is pulsating. The degree of pulsation is higher for *simplex* pumps than for the *duplex* type. The pulsation is also higher for direct-driven pumps than for those driven by motor or turbine through a gearbox.

Pulsation is generally not a problem in the case of small low-speed pumps of this type. It affects only the associated instrumentation which can be compensated for by local dampeners. However, as the pump speed is increased the pulsation effect becomes more serious, affecting the piping design of the system. Under these conditions of high pulsation instantaneous piping pressures may often exceed the design pressure of the piping. The piping must then be designed to meet this higher pressure requirement which invariably results in a higher cost. As an alternative discharge dampeners are often considered, but these to add to the cost of the pump installation.

Reciprocating pumps are used mostly in situations where their low piston speeds will withstand corrosive and abrasive conditions. They are ideal also for pumping low capacity against high differential head and where it is necessary to maintain a constant flow rate against a gradually increasing discharge pressure.

4.3   Evaluating Pump Performance

In general, process engineers are involved with the day-to-day operation of process plants. Many of their duties in this respect are concerned with maintaining plant efficiency, locating trouble areas and solving operational problems. This section is directed to these engineers and presents the calculation methods to check pump performance in terms of the pump horsepower and the available NPSH for a pump.

BRAKE HORSEPOWER EFFICIENCY

Actual running efficiency can be calculated from plant data. These may be compared with a typical expected efficiency to evaluate pump performance. The steps below are followed to arrive at this efficiency figure:

- *Step 1*   Obtain flow rate from plant readings. Also read discharge pressure and, if available, suction pressure. If this latter reading is not available calculate it from source pressure, height of liquid above pump suction and frictional loss.
- *Step 2*   Read stream temperature and obtain SG of stream from lab data. Calculate SG at flow conditions.
- *Step 3*   Calculate differential head which is discharge pressure psia − suction pressure psia. Convert to differential head in feet by

$$\frac{\Delta P \times 144}{62.2 \times SG} \text{ (at flow condition)}$$

- *Step 4* Convert feed rate to pounds per hour. Then calculate hydraulic horsepower from the expression

$$\frac{\text{Head (ft)} \times \text{rate (lb/h)}}{60 \text{ (min)} \times 33\,000 \text{ (hp/min)}}$$

- *Step 5* From motor data sheet obtain pump motor running efficiency from plant data read power usage in kW.
- *Step 6* Convert pump power to horsepower by dividing kW by 0.746. Multiply this by pump efficiency expressed as a fraction. This is the brake horsepower.
- *Step 7* Divide hydraulic horsepower by brake horsepower and multiply by 100 to give efficiency as a percentage.
- *Step 8* Check against Figures 4.4 or 4.5 to evaluate pump performance. Is the calculated efficiency within reasonable agreement with the expected efficiency given by Figures 4.4 or 4.5? If there is a large discrepancy then the appropriate mechanical or maintenance engineers should be informed.

CHECKING AVAILABLE NPSH

This needs to be done if the pump is showing signs of vibration and losing suction under normal operations.

- *Step 1* Obtain details of the fluid being pumped (temperature, SG, flow rate, source pressure and vapour pressure of fluid).
- *Step 2* Calculate the frictional pressure drop in the suction line (see Chapter 1).
- *Step 3* Calculate the suction pressure of the pump by taking source pressure and adding in the static head. For this calculation take static head as being from the bottom of the vessel (not from the liquid level).
- *Step 4* From the suction pressure calculated in step 3 take out loss through friction.
- *Step 5* Calculate NPSH available as being net suction pressure less vapour pressure. This is usually quoted in feet (or metres) so convert using:

$$\frac{\text{psi} \times 144}{62.2 \times \text{SG}}$$

- *Step 6* Check against manufacturer's data sheet for required NPSH. If the available NPSH is less than required the pump will continue to cavitate. Fill tank to a level until the vibration stops and maintain it at that level.

If the problem is really troublesome and maintaining a liquid level as suggested in step 6 above is not practical, contact the pump manufacturer. Very often the manufacturer will be able to make some minor changes to the pump design that will solve the problem.

Example calculation

(1) Pump brake horsepower efficiency

$$\text{Flow capacity of pump at flow conditions} = 200 \text{ gpm}$$
$$\text{(from plant data)} = 82\,632 \text{ lb/h}$$
$$\text{Suction pressure (plant data)} = 120 \text{ psig}$$
$$\text{Discharge pressure} = 450 \text{ psig}$$
$$\Delta P = 330 \text{ psi}$$
$$\text{SG at flow conditions} = 0.827$$

$$\text{Differential head (ft)} = \frac{330 \times 144}{62.2 \times 0.827} = 924 \text{ ft}$$

$$\text{Hydraulic horsepower} = \frac{82\,632 \times 924}{60 \times 33\,000}$$
$$= 38.56$$

From motor rating, motor efficiency is 92%

$$\text{Plant readings show motor power usage} = 48.1 \text{ kW}$$

$$\text{Then motor hp input} = \frac{48.1}{0.746} = 64.5 \text{ hp}$$

$$\text{Motor output} = 64.5 \times 0.92$$
$$= 59.3$$

This is brake horsepower.

$$\text{Then pump efficiency} = \frac{38.56}{59.3} = 65\%$$

Motor is 50 cycle speed from Figure 4.5 for multi-stage 2950 rpm. This efficiency figure is about right.

(2) Checking available NPSH

Fluid is gas oil from surge drum at 250 °F and 15 psig drum is 12 ft above grade (bottom of drum which is horizontal). Boiling point at 34 psia is 488 °F. Vapour pressure of gas oil at 250 °F = 0.8 psia (VP curves)

$$\text{Source pressure} = 29.7 \text{ psia}$$
$$\text{SG at conditions} = 0.815$$
$$\text{Head above grade} = 12 \text{ ft (to bottom of drum)}$$
$$\text{Pump centreline} = 2 \text{ ft above grade}$$
$$\text{Liquid head to pump} = 10 \text{ ft} = 3.5 \text{ psi}$$
$$\text{Friction pressure drop} = 0.4 \text{ psi (calculated from 0.2 psi/100)}$$
$$\text{Less friction} = 0.4$$
$$\text{Less vapour press} = 0.8$$
$$\text{Available NPSH} = 33.2 - (0.4 + 0.8) = 32 \text{ psia}$$
$$= 91 \text{ ft}$$

Most pumps only require 10 ft or less.

4.4   Specifying a Centrifugal Pump

Two disciplines are responsible for correctly specifying a pump. These are the mechanical engineer and the process engineer. Most pumps are designed and built in accordance with set and accepted industrial codes, such as the API codes. The mechanical engineer ensures that the mechanical data supplied to the manufacturer for a particular pump meet the requirements of the code and standards to which the pump is to be built. The process engineer develops and specifies precisely the performance required of the pump in meeting the process criteria of the plant. To accomplish this the mechanical engineer develops a mechanical specification and the process engineer initiates the pump specification sheet.

THE MECHANICAL SPECIFICATION

This specification is in a narrative form and will contain at least the following topics:

- *Scope*   Introductory paragraph which gives the code the pump manufacturer to which it is to conform (such as API 610). A list of other standards to which the pump shall conform if these are required.
- *Main body of the specification*   This covers all additions and any exceptions to the selected code. It provides for the type of drive shaft acceptable if different from code. Items such as impeller size as a percentage of maximum allowable by code is given. The needs for special bearing arrangements in the case of multi-stage pumps are detailed in this document.
- *Ancillary equipment and piping arrangements*   The specification describes in detail the type of cooling medium that shall be used. It provides a guide also to the piping requirements that are required to satisfy the cooling system(s).
- *Seal or packing requirements*   The mechanical specification details the type of seal or packing that will be installed. It also provides details of the seal arrangement required if this is different from the standard code.
- *Pump mounting*   Some installation guide is provided by this specification. The method by which the pump is mounted on the baseplate is detailed. It also outlines under what conditions the manufacturer is to provide pedestal cooling facilities.
- *Metallurgy*   Although the process engineer will specify the general material of construction for the pump (such as carbon steel, cast iron, etc.) it is the mechanical engineer who details this. This includes the specific grade of the material and in many cases its pre-operational treatment.
- *Inspection*   The mechanical specification will provide details of the inspection that the company will carry out during the manufacture of the pump and before its delivery. This will include dimensional checking during manufacture and some checks on the metallurgy. Prior to shipping the purchaser may require a running test of the pump and will witness this test. For this purpose, the pump is run in the workshop under specified process conditions.

The mechanical specification may continue to detail other requirements that the purchaser may wish. Its objective is to ensure that the pump when delivered is mechanically robust, is safe and easily maintainable. The mechanical specification must also be cognisant of the cost implications of the requirements on the pump and to keep them as low as possible.

THE PROCESS SPECIFICATION

The data provided by the process engineer must be sufficient to ensure that the pump delivered for the process purpose will meet the duty required of it. These data are supplied to the pump manufacturer in the form of a data sheet similar to the one shown as Figure 4.6. The data sheet collects the essential input from the process engineer, the mechanical engineer and, later, the manufacturer to describe fully what is required of the pump and what the manufacturer has supplied. All the data given here will be unique to this pump.

The process input to the pump specification shown on Figure 4.6 are those items marked with the 'P', while input by other disciplines and the manufacturer are as indicated on the form. The process engineer compiles much of these data from an 'Hydraulic Analysis' of the pump system similar to that described in Chapter 1, Section 1.10. A calculation sheet given as Figure 4.7 shows the development of this calculation using the data from Section 1.10 and is described as follows.

Compiling the pump calculation sheet

(1)   The pump number, title and service

This first section of the calculation sheet is important because it identifies the pump and what it is intended to do. The item number and service description will be unique to this item and will remain as its identification throughout its life. All the data below this section will refer only to this pump and to no other. The item number may contain the suffixes 'A', 'B', 'C', etc. This indicates identical pumps in parallel service or as spare or both. This section also shows how many of these pumps are motor driven and how many are turbine (steam) driven. Usually spare pumps in critical service will be turbine driven. The remark column in this section should give any information that will be of benefit to the pump manufacturer or future operators of the pump. For example, if the spare pump is turbine driven, the process engineer may require an automatic start-up of the turbine on a 'low flow' of the pumped stream. This should be noted here.

(2)   Operating conditions each pump

The details of the fluid to be pumped and a summary of the calculations given below are entered here. Starting on the left of this section:

● *Liquid*   This is a simple definition of the pumped material. In the example given here this will simply be 'vacuum gas oil'.
● *Pumping temperature*   This is the temperature of the gas oil at the pump. There are two temperatures called for: 'normal' and 'max'. The normal temperature is that shown on the process flow diagram, while the max.

SHEET NO. P   REV.
JOB NO. P   DATE P
BY P   CHK'D.
P.O. NO.

NOTE: ○ INDICATES INFORMATION TO BE COMPLETED BY

    □ BY MANUFACTURER

FOR _____ P _____   SITE _____ P _____

UNIT _____ P _____   SERVICE _____ P _____

NO. PUMPS REQ'D _____ P _____ NO. MOTORS REQ'D _____ P _____ ITEM NO. _____ P _____ PROVIDED BY _____ MTD BY _____

NO. TURBINES _____ P _____ ITEM NO. _____ P _____ PROVIDED BY _____ MTD BY _____

PUMP _____ SIZE AND TYPE _____ (P-Type _____ SERIAL NO. _____

OPERATING CONDITIONS, EACH PUMP

LIQUID _____ P _____ m^3/h at PT, NOR. _____ P _____ RATED _____ P _____

DISCH. PRESS., kg/cm^2g _____ P _____

PT,°C, _____ P _____ MAX. _____ P _____ SUCT. PRESS., kg/cm^2g MAX. _____ P _____ RATED _____ P _____

SP.GR. at PT _____ P _____ DIFF. PRESS., kg/cm^2 _____ P _____

VAP. PRESS. at PT, kg/cm^2a _____ P _____ DIFF. HEAD, m _____ P _____

VIS. at PT, Ssu _____ P _____ CP _____ P _____ NPSHA, m _____ P _____

CORR/EROS. CAUSED _____ P _____ HYD. HP(metric) _____ P _____

PERFORMANC

PROPOSAL CURVE NO. _____

RPM _____ NPSHR (WATER) m _____

EFF. _____ metric BHP RATED _____

MAX. metric BHP RATED IMP _____

MAX. HEAD RATED IMP m _____

MIN. CONTINUOUS m^3/h _____

ROTATION (VIEWED FROM CPIG END) _____

CONSTRUCTION

NOZZLES	SIZE	RATING	FACING	LOCATION
SUCTION				
DISCHARG				

CASE-MOUNT: □ CENTERLINE □ FOOT □ BRACKET □ VERT. (TYPE) _____

  -SPLIT: □ AXIAL □ RAD; TYPE VOLUTE □ SGL □ DBL □ DIFFUSER

  -PRESS: □ MAX. ALLOW _____ kg/cm^2g _____ °C □ HYDRO TEST _____ kg/cm^2g

  -CONNECT: □ VENT □ DRAIN □ GAGE

IMPELLER DIA.: □ RATED _____ □ MAX. _____ □ TYPE: _____

MOUNT □ BETWEEN □ OVERHUNG

BEARINGS- □ RADIAL _____ □ THRUST _____

  LUBE: □ RING OIL □ FLOOD □ OIL MIST □ FLINGER □ PRESSURE

COUPLING: □ MFR _____ □ MODEL _____

  DRIVER HALF MTD ○ PUMP ○ DRIVER ○ PURCHASER

PACKING: □ MFR & TYPE _____ □ SIZE/NO. OF RINGS _____

MECH. SEAL: □ MFR & _____ API CLASS. CODE _____

  □ MFR CODE _____

SHOP TESTS

○ NON-WIT. PERF.   ○ WIT. PERF.

○ NON-WIT. HYDRO   ○ WIT. HYDRO

○ NPSH REQ'D.   ○ WIT. NPS!4

○ SHOP

○ DISMANT. & INSP. AFTER

○ OTHE _____

MATERIALS

PUMP: CASE/TRIM CLASS ○ _____

BASEPLATE □ _____

VERTICAL

PIT OR SUMP DEPTH ○ _____

MIN. SUBMERGENCE □ _____

COLUMN PIPE: □ FLANGED □ THREADED

LINE SHAFT: □ OPEN □ ENCLOSED

BRGS: □ BOWL _____ □ LINE SHAFT _____

BRG. LUBE □ WATER □ OIL □ GREASE

FLOAT &   ○ C.S. ○ S.S. ○ BRZ ○ NON

FLOAT □ _____

PUMP THRUST, □ UP _____ □ DOW _____

AUXILIARY

○ C.W. PIPE PLAN _____ ○ CU; ○ S.S.; ○ TUBING; ○ PIPE _____

□ TOTAL COOLING WATER REQ'D, m^3/h _____ ○ SIGHT F.I. REQ'D _____

○ PACKING COOLING INJECTION REQ'D: □ TOTAL m^3/h _____ □ kg/cm^2g

○ SEAL FLUSH PIPE PLAN _____ ○ C.S. ○ S.S. ○ TUBING ○ PIPE _____

○ EXTERNAL SEAL FLUSH FLUID _____ □ m^3/h _____ □ kg/cm^2g _____

○ AUXILIARY SEAL PLAN _____ ○ C.S. ○ S.S. ○ TUBING ○ PIPE _____

  ○ AUX. SEAL QUENCH FLUID _____

MOTOR DRIVER

HP(metric) _____ RPM _____ P _____ FRAME _____ VOLTS/PHASE/CYCL _____ P _____

MFR _____ BEARINGS _____ LUBE _____

TYPE _____ INSUL _____ FULL LOAD AMPS _____

ENC _____ TEMP RISE, C _____ LOCKED ROTOR _____

○ VHS ○ VSS   VERT. THRUST CAP., _____ MTR. ITEM NO. _____

APPROX. WT, PUMP & BASE _____

MOTOR _____ TURBINE _____

API STANDARD 610 GOVERNS UNLESS OTHERWISE   APPLICABLE TO: ○ PURCHASE ○ AS BUILT ○

Figure 4.6. A centrifugal pump specification sheet (P = specified by process)

PUMP CALCULATION.

Item No *P103 - A+B* Unit *CRUDE VACUUM UNIT* Sheet NO *1* Rev *0*
Service *HGO PRODUCT AND BPA* Motor Drive *1*
Turbine Drive *1* Remarks *TURBINE TO HAVE AUTO START BY DSTT* App.*J.S.*

OPERATING	CONDITIONS (Each Pump)			TURBINE CONDITIONS
Liquid *VAC GAS OIL*	US GPM @Pt.Min ___ NOR *137* Rated *1585*			Inlet Steam psig *600*
PT P NOR *545* MAX *740*	Disch Press Psig *85.5*			Temp P *670*
SP GR @ PT *0.755*	Suct Press Psig Max *50* Rated *- 0.7*			Exhaust Psig *50*
Vap Press @ PT.Psig *0.29*	Diff Press Psi *86.2*			PUMP MATERIALS
Vis @ PT Cp *0.906*	Diff Head FT *264.3.*			Casing *C.S.*
Corrosion\Erosion *NONE*	NPSH Available, FT *40* Hyd HP *79.7* (1)			Internal Parts *C.S.*
ALTERNATES	**B.C.**			**SKETCH**
DESTINATION :				
Destination Press psig	*50*			
Static Head psi	*5.4*			
Line Loss psi	*10.7*			
Meter Loss psi	*0.2*			
11-E-9/10 ΔHt Exchangers psi	*14.0*			
Δ Control Valves psi	*5.20*			
TOTAL DISCHARGE PRESS psig	*85.50*			
SUCTION.				
Source Press psig	*-14.4*			
+ Static Head psi	*14.7*			
- System Losses psi	*1.0*			
TOTAL SUCTION PRESS psig	*- 0.7*			
NPSH AVAILABLE.				
Source Press psia	*0.29*			
- (Vap Press + suct losses) psia SUb total Psia\Ft	*-1.29* *-1.00*			
Elev of liquid - pump CL Ft	*45.00*			
NPSH Available Ft	*40.40*			

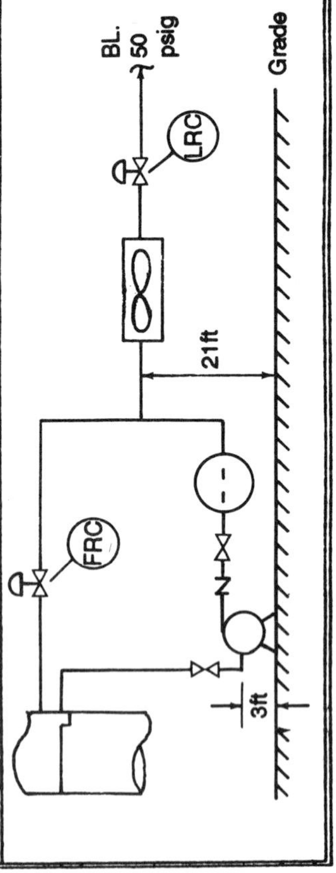

DATED *29.3.92* REV *0* Dated _____ REV ____ _____.

NOTES (1) BASED ON RATED FLOW.

Figure 4.7. A centrifugal pump calculation sheet

temperature is that used for the pump design conditions. It should be the same as the design temperature of the vessel from which the fluid is pumped.

- *Specific gravity at PT* This is self-explanatory. Note the item also calls for the SG at 60 °F.
- *Vapour pressure at PT* This is read from the vapour pressure curves given in the Figure A1.5 in Appendix 1. First locate the VP of the stream at atmospheric pressure (this is the material's normal boiling point). Follow the temperature line down or up to the PT and read off the pressure at that point.
- *Viscosity at PT* This too is self-explanatory. Note this calculation sheet requires the viscosity to be in *centipoise*. This is centistokes × specific gravity.
- *US gpm at PT* This is the pump capacity and three rates are asked for:
  —Minimum rate: The anticipated lowest rate the pump will operate at for any continuous basis. This rate sets the control valve range.
  —Normal rate: This is the rate given in the material balance and the basis for the hydraulic analysis.
  —Maximum rate: This is normally set based on the type of service that the pump will undertake. For example, pumps used only as rundown to storage will have a max. rate about 10% above normal. Those used for reflux to towers will have between 15% and 20% above normal.
- *Discharge pressure (psig)* This figure is calculated in the column below. It will also have been determined by the hydraulic analysis described in Section 1.10 in Chapter 1.
- *Suction pressure (psig)* Two pressures are asked for in this item. Rated pressure is that calculated in the column below and in the hydraulic analysis given in item 1. It is based on the 'norm' rate. The 'max.' suction pressure is based on a source pressure at the design pressure rating of the vessel from which the pump is taking suction.
- *Differential pressure (psi)* This is the discharge pressure minus the rated suction pressure.
- *Differential head (ft)* The head is determined from the differential pressure by

$$\frac{\text{Diff. press. psi} \times 144}{62.2 \times \text{SG at PT}}$$

- *NPSH (ft)* This is calculated in the column below. It is the suction head available greater than the fluid vapour pressure (at the PT) at the pump impeller inlet.
- *Hydraulic horsepower* This is calculated from the weight per unit time (usually minutes or seconds) of fluid being pumped times the differential head in feet divided by 550 ft-lb/s or 33 000 ft-lb/min. The differential head is always based on the rated suction pressure and the weight on the rated capacity (gpm) for this calculation.
- *Corrosion/erosion* The process engineer notes any significant characteristic of the fluid regarding its corrosiveness or abrasiveness here.

(3) Turbine conditions

Although this item is not strictly part of a pump definition it should be included for completeness in the case of turbine drives. The data required to complete this item are self-explanatory.

(4) Pump material

The process engineer indicates here the acceptable material for the pump in handling the fluid (for example, carbon steel, cast iron, etc.). It is not necessary to specify grade of steel, etc.

(5) The calculation columns

The objective of this section of the calculation sheet is to itemize all the data used to provide the figures given in the operating conditions described in item 2 above. The first column lists those items while the other three columns are available for entering the corresponding numbers. These three columns are provided to cater for alternative conditions that may need to be studied. A space is left on the right of the form to sketch the pumping system (it is very advisable to do a sketch). The first column starts with the destination pressure and continues down with the list of the pressure drops in the system to the pump discharge. This section of the column ends with the sum of the pressure drops giving the pump discharge pressure. The items that make up the pump suction pressure are listed next. This starts with the source pressure (usually a vessel) and its static head above the pump. All the pressure drops in the suction side are listed and deducted from the sum of the source pressure and static head to give the pump suction pressure.

The last section in the column itemizes the data that gives the *available* NPSH for the pump. The development of the NPSH is self-explanatory.

4.5   Centrifugal Pump Seals

A pump seal is any device around the pump shaft designed to prevent the leakage of liquid out of or air into a pump casing. All industrial pumps have shafts protruding through the casings which require sealing devices. Pump-sealing devices are usually either a 'packed box' with or without a lantern ring or a mechanical seal. Controlled leakage is a system sometimes used.

A flushing stream must be introduced into the pump seals for one or more of the following reasons:

- To effect a complete seal
- To provide cooling, washing or lubrication to the seal
- To keep grit from the seal
- To prevent corrosive liquid from reaching the seal

The facility for accomplishing this is called the 'flushing system' and there are two types of these in general use:

- A dead-end system
- A through system.

In a 'dead-end' system the flushing liquid enters the casing through the stuffing box and combines with the pumped fluid (see Figure 4.8(a)). A 'through' system is one in which the flushing liquid is recirculated between a double-seal arrangement and does not enter the pump (Figure 4.8(b)). The liquid source may be external to the pump or, as on most mechanical seals, is a self-flushing system in which the pumped liquid is used as the flushing fluid.

A description of each of the types of sealing devices is presented below and illustrated in Figure 4.9.

(1) Packed boxes (without lantern ring) (Figure 4.9(a))

This is the simplest type of pump seal. Its principal components are a stuffing box, rings of packing, a throat bushing, and a packing gland. A slight leakage through the packing is required at all times to lubricate the packing. A water quench is used at the packing gland if the packing 'leakage' is considered flammable or toxic.

(2) Packed box (with lantern ring) (Figure 4.9(b))

When a packed box pump seal is used in conjunction with a flushing oil system, a lantern ring is usually supplied. This metallic ring provides a flow path for the

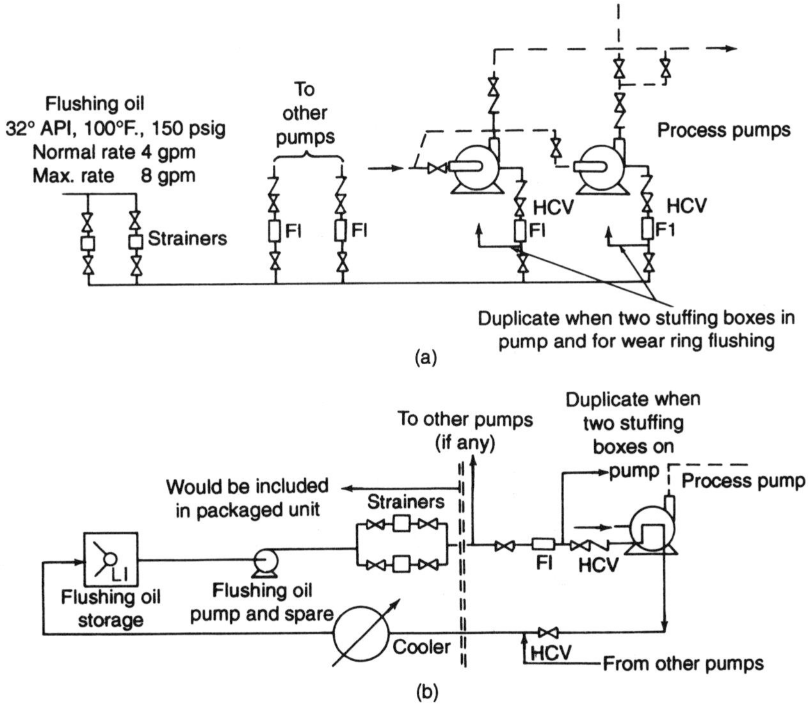

Figure 4.8. Typical flushing systems. (a) A dead-end system; (b) a through-recirculating system

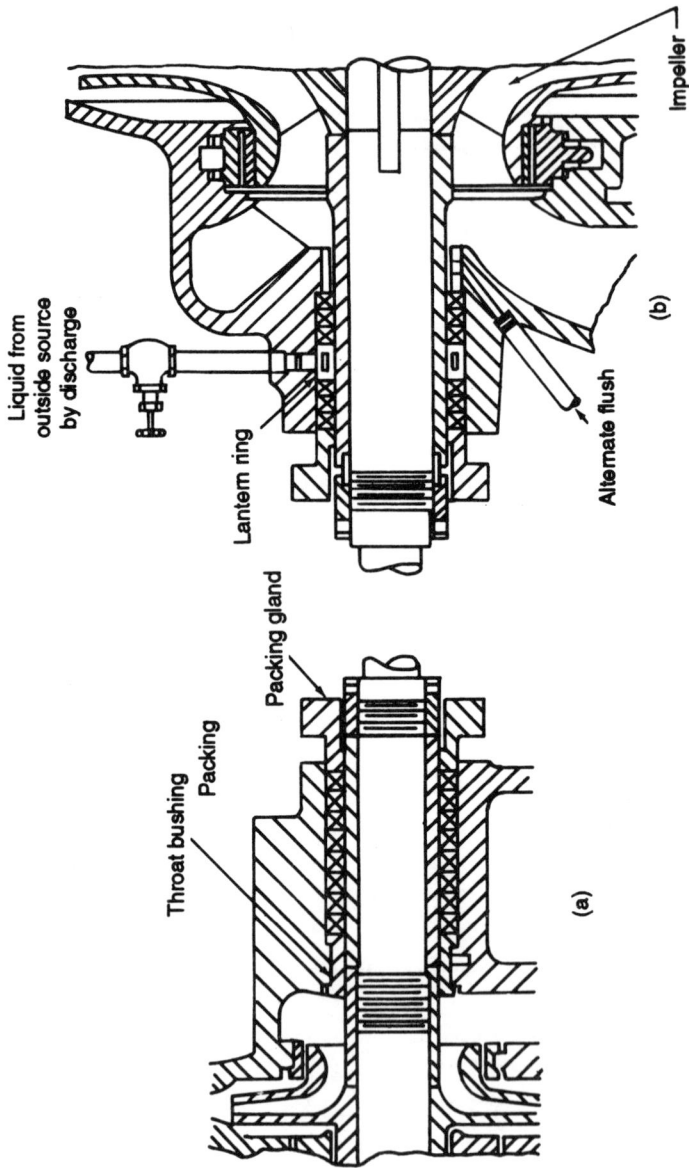

Liquid from
outside source
by discharge

Impeller

Lantern ring

Alternate flush

(b)

Packing gland

Throat bushing

Packing

(a)

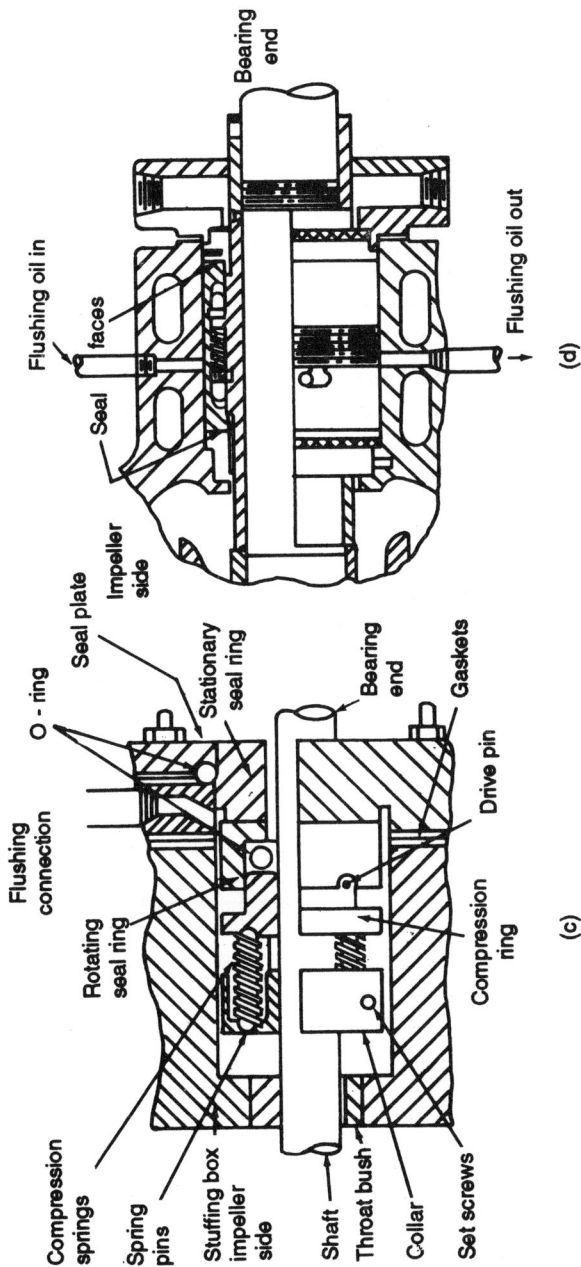

Figure 4.9. Pump shaft packing and seals. (a) 'Packed solid': stuffing box completely filled with packing, no lantern ring; (b) externally sealed stuffing box; (c) single mechanical seal; (d) double mechanical seal

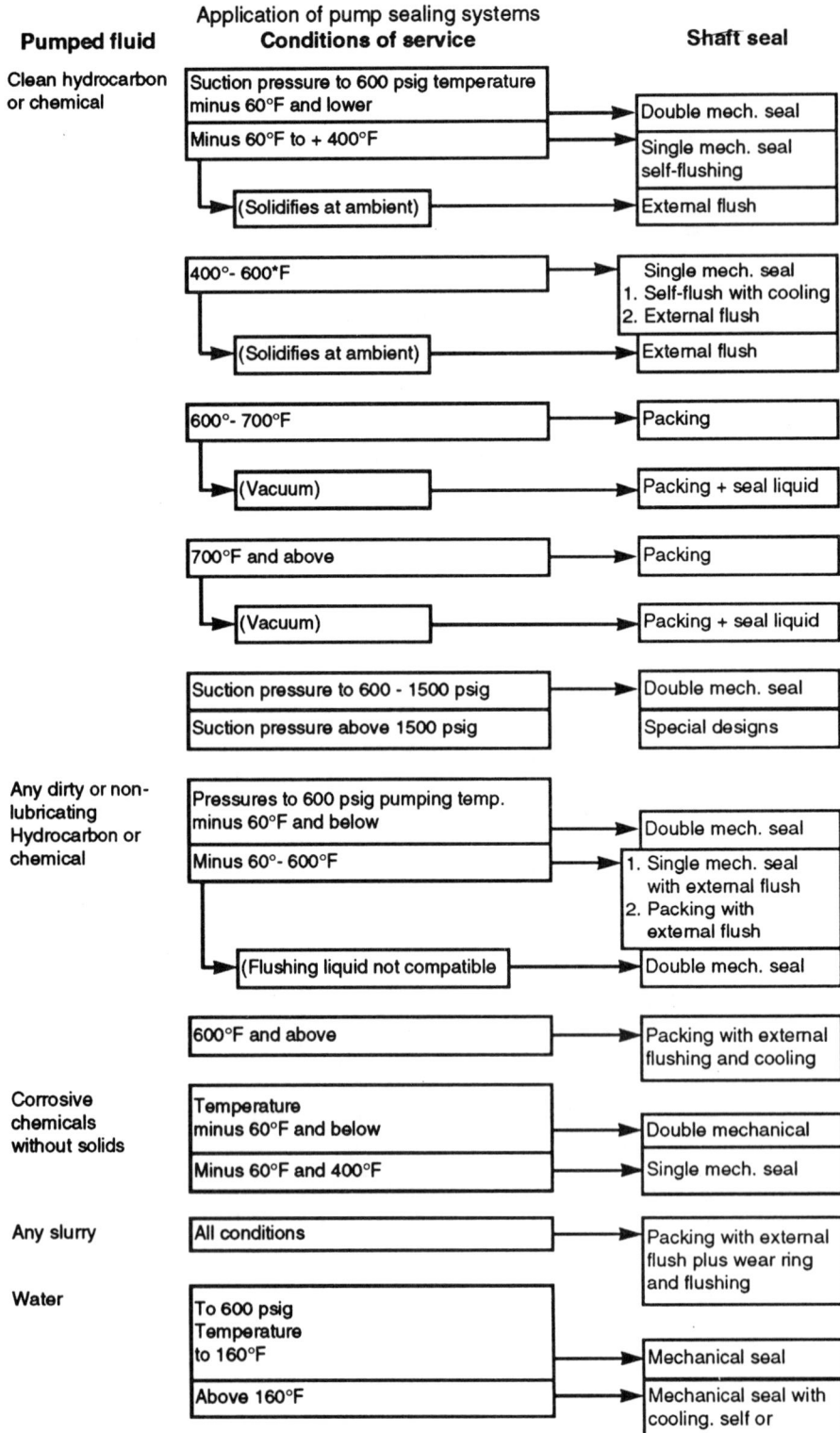

Application of pump sealing systems

Pumped fluid	Conditions of service	Shaft seal
Clean hydrocarbon or chemical	Suction pressure to 600 psig temperature minus 60°F and lower	Double mech. seal
	Minus 60°F to + 400°F	Single mech. seal self-flushing
	(Solidifies at ambient)	External flush
	400°- 600°F	Single mech. seal 1. Self-flush with cooling 2. External flush
	(Solidifies at ambient)	External flush
	600°- 700°F	Packing
	(Vacuum)	Packing + seal liquid
	700°F and above	Packing
	(Vacuum)	Packing + seal liquid
	Suction pressure to 600 - 1500 psig	Double mech. seal
	Suction pressure above 1500 psig	Special designs
Any dirty or non-lubricating Hydrocarbon or chemical	Pressures to 600 psig pumping temp. minus 60°F and below	Double mech. seal
	Minus 60°- 600°F	1. Single mech. seal with external flush 2. Packing with external flush
	(Flushing liquid not compatible	Double mech. seal
	600°F and above	Packing with external flushing and cooling
Corrosive chemicals without solids	Temperature minus 60°F and below	Double mechanical
	Minus 60°F and 400°F	Single mech. seal
Any slurry	All conditions	Packing with external flush plus wear ring and flushing
Water	To 600 psig Temperature to 160°F	Mechanical seal
	Above 160°F	Mechanical seal with cooling. self or

flushing oil to reach the pump shaft. For very erosive or corrosive services, the lantern ring is often located next to the throat bushing and a liquid is injected into the throat bushing to prevent the pumped fluid from reaching the packing area. For a pump operation with vacuum suction conditions, the lantern ring is installed at the middle of the box and liquid is injected to prevent air entering the system. This type also operates with positive leakage with the same comments as the packed box without lantern ring.

(3) Mechanical seals (Figures 4.9(c) and 4.9(d))

Typical basic elements of a single seal are shown in Figure 4.9(c). Sealing is affected between the precision-lapped faces of the rotating seal ring and stationary seal ring. The stationary seal ring is usually carbon, and is mounted in the seal plate by an O-ring. The two O-ring packings serve the dual purpose of sealing off any liquid tending to leak behind the seal rings and also to provide flexibility in allowing the seal faces to align themselves exactly so as to compensate for any slight 'wobble' of the rotating seal face caused by shaft whip.

The rotating seal ring is usually stainless steel with a Stellite face. The springs provide the necessary force to set the O-ring and hold the seal faces closed under low stuffing box pressures. Any pressure in the box exerts additional force on the rotating sealing ring. The seal is frequently 'balanced' so that the face pressure is in the correct ratio to the liquid pressure to ensure adequate sealing without excessive loading of the faces. Flushing oil enters the stuffing box through a connection in the seal plate.

A double seal consists of two single seals back to back (see Figure 4.9(d)). As a double seal is more expensive and requires a complicated seal-oil system, it is used only where single seals are not practical.

The diagram on pages 236 and 237 summarizes the application of the various sealing systems.

4.6   Pump Drivers and Utilities

Most pumps in the process industry are driven either by electric motors or by steam, usually in the form of steam turbines. This section deals with the calculation of the driver requirements and its specification.

ELECTRIC MOTOR DRIVERS

Electric motors are by far the most common pump drivers in industry. They are more versatile and cheaper than a comparable size of steam turbine. The electric motors used for pump drivers are the induction type motor. They range in size from fractional horsepower to 500 and higher horsepower. Sizing the required motor for a pump driver takes into consideration the pump brake horsepower, the energy losses occurring in the coupling device between the pump and the motor and a contingency factor of about 10%. These are expressed by

$$\text{Minimum driver bhp} = \frac{\text{Maximum pump bhp} \times 1.1}{\text{Mechanical efficiency of coupling}}$$

If the pump is driven through a direct coupling the efficiency will be 100%. With gears or fluid coupling the efficiency will be between 94% and 97%.

Specifying motor driver requirements

Process engineers are called upon very often to specify pump driver requirements or to check those already existing. In doing this two items of data need to be obtained or calculated:

● The actual required horsepower of the pump motor to drive the pump at its specified duty
● What is actually installed in terms of horsepower

These data are tabulated in terms of power load as follows:

● *Operating load (kW)*   This is power input to the motor at normal operating horsepower.
● *Connected load (kW)*   This is power input to the motor at motor-rated horsepower.

If the pump is spared by another motor-driven pump then the connected load will be the sum of both motors:

$$\text{Operating load} = \frac{\text{Minimum required driver hp} \times 0.746}{\text{Efficiency of the motor at its operating hp}}$$

$$\text{Connected load} = \frac{\text{Rated motor hp} \times 0.746 \times \text{number of motors}}{\text{Efficiency of the motors at 100\% full load}}$$

Table 4.1 gives the motor sizes, and efficiencies at % of full load.

Example calculation

Calculate the operating and connected loads for pump 11-p-3 A&B as specified in Section 4.4 of this chapter. From the pump calculation sheet the hydraulic horsepower is calculated:

$$\text{hhp} = \frac{\text{lb/min} \times \text{diff. head in feet}}{33\,000}$$

$$= \frac{8665 \times 264.3}{33\,000}$$

$$= 69.4 \text{ hp}$$

From Figure 4.4 and assuming 60 cycle pump speed pump efficiency is 79%, then:

$$\text{Brake horsepower} = \frac{\text{Hydraulic horsepower}}{0.79}$$

$$= 87.8 \text{ hp}$$

Table 4.1 Electric motors, size and efficiencies

Motor rating (hp)	Motor connected load (kW)	Motor efficiency at % of full load capacity		
		50	75	100
1	0.98	68	74	76
1.5	1.42	72	76.5	79
2	1.86	73	78	80
3	2.76	77.5	81.5	81
5	4.39	83	83	85
7.5	6.65	81	83.5	84
10	8.78	84	85	85
15	13.0	85	86	86
20	17.05	86.5	87.5	87.5
25	21.00	87.5	88.5	88.5
30	25.10	88	89	89
40	33.50	88	89	89
50	41.70	88	89.5	89.5
60	49.70	89.5	90	90
75	62.10	89	90	90
100	82.00	84	89	91
125	102.00	85	89.5	91.5
150	123.00	86	89	91
200	161.00	88	91	92.5
250	201.00	90.5	92.5	92.5
300	241.00	90.5	92.5	92.8
350	281.00	90.9	92.6	92.9
400	320.00	91.1	92.8	93.1
450	360.00	91.2	93	93.2

This will be a direct-driven pump. Thus coupling efficiency is 100%.

$$\text{Minimum motor size} = \text{bhp} \times 1.1 = 96.6 \text{ hp}$$

The closest motor size to this requirement is 100 hp (Table 4.1). This is a little too close so a motor size of 125 hp will be selected.

$$\text{Operating load} = \frac{\text{Rated hp} \times 0.746}{\text{Efficiency at \% of full load}}$$

$$\text{\% of full load} = \frac{87.8}{125} = 70.2\%$$

$$\text{Efficiency} = 89\% \text{ (from Table 4.1)}$$

$$\text{Operating load} = \frac{87.8 \times 0.746}{0.89}$$

$$= 73.6 \text{ kW}$$

$$\text{Connected load} = \frac{125 \times 0.746}{0.915}$$

$$= 101.9 \text{ kW (say 102 kW)}$$

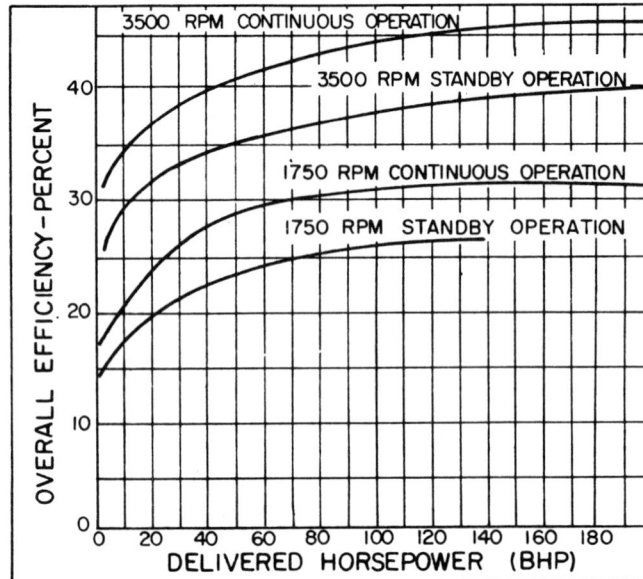

GRAPH I
EFFICIENCY OF TURBINES AT 110 PSIG SAT. STEAM
INLET WITH 20 PSIG STEAM EXHAUST

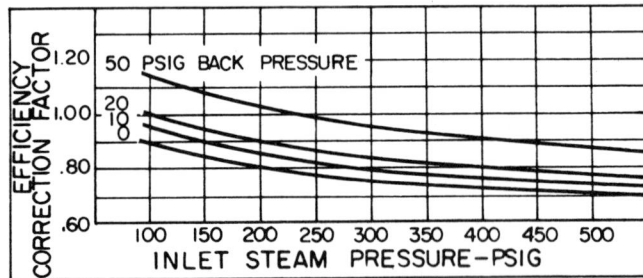

GRAPH 2
STEAM CONDITION CORRECTION FACTOR FOR GRAPH I

Figure 4.10. Steam turbine efficiencies

Note that if both regular and spare pumps were motor driven then the connected load would be $2 \times 102 = 204\,kW$.

Reacceleration requirement

To complete the specification for the motor requirement a degree of process importance of the pump must be established and noted. Voltage drops that can occur in any system may be sufficient to stop the pump. The process engineer must determine how important it is to the process and the safety of the process to

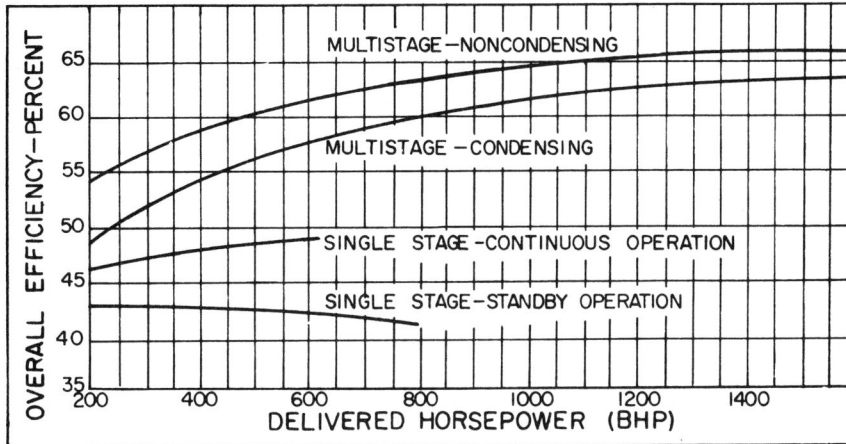

GRAPH 3
EFFICIENCY OF HIGH HORSEPOWER TURBINES

be able to restart and reaccelerate the particular pump quickly. The following code of importance has been adopted:

- Reacceleration absolutely necessary (A)
- Reacceleration desirable (B)
- Reacceleration unnecessary (C)

The 'A' category involves any pump critical to keeping the process onstream safely and with no possibility of equipment damage. The 'B' category applies to

APPROXIMATE CORRECTION FOR
SUPERHEATED INLET STEAM
ADD 9% TO THEORETICAL
HEAT AVAILABLE FOR EVERY
100 °F SUPERHEAT

APPROXIMATE CORRECTION FOR
UNSATURATED INLET STEAM
SUBTRACT 2% FROM
THEORETICAL HEAT AVAILABLE
FOR EACH 1% OF MOISTURE
PRESENT

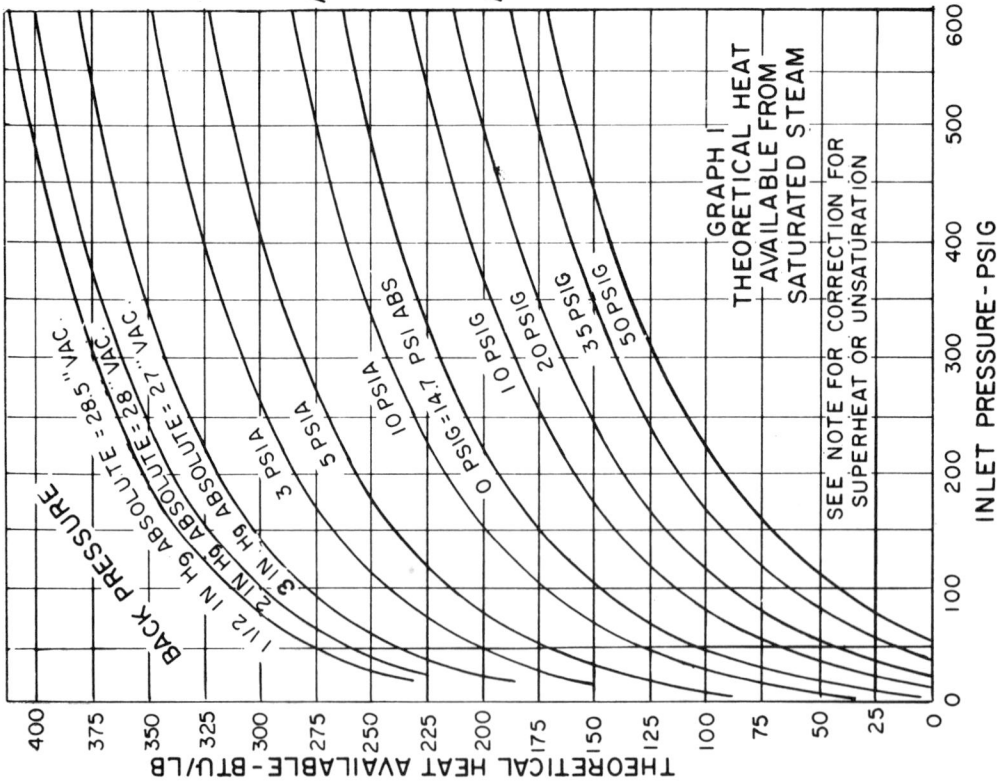

GRAPH II
THEORETICAL HEAT
AVAILABLE FROM
SATURATED STEAM

SEE NOTE FOR CORRECTION FOR
SUPERHEAT OR UNSATURATION

BACK PRESSURE

28.5"VAC.
28"VAC.
27"VAC.
3 PSIA
5 PSIA
10 PSIA
14.7 PSI ABS
10 PSIG
20 PSIG
35 PSIG
50 PSIG

1 1/2 IN Hg ABSOLUTE
2 IN Hg ABSOLUTE
3 IN Hg ABSOLUTE

0 PSIG =

THEORETICAL HEAT AVAILABLE - BTU/LB

INLET PRESSURE - PSIG

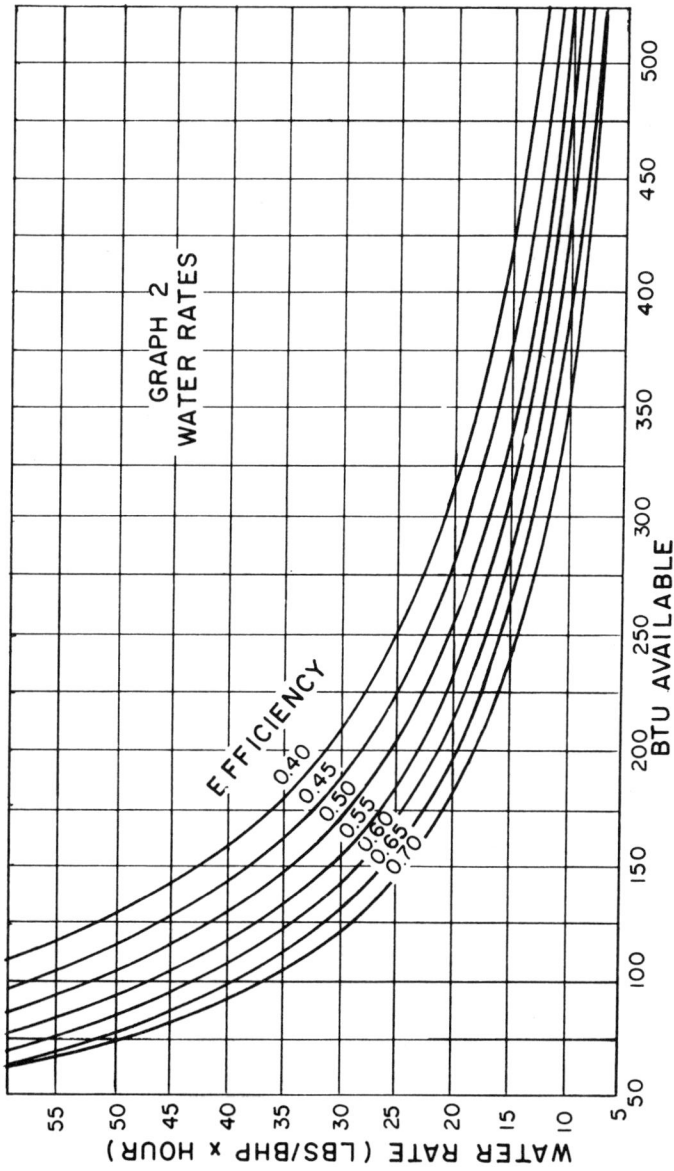

Figure 4.11. Water rates for condensing and non-condensing turbines

those pumps that in operation with the 'A' category will maintain the unit 'on spec'. The 'C' category refers to those pumps that can be started manually without any problems.

In the case of the example pump 11-P-3 A&B given here, the service required of the pump is so critical to the operation and orderly shutdown of the process in the case of power failure that the spare pump is turbine driven. Thus the motor-driven regular pump need only be coded 'C' for reacceleration.

STEAM TURBINE DRIVERS

Steam turbines are the second most common pump drivers in the modern process industry. Although more expensive than the electric motor they offer an excellent standby to retain the maximum process 'onstream' time. The one disadvantage with power-driven pumps is the reliability of power availability. Steam turbines therefore offer a good alternative in cases of power failure. Another alternative means of pump drivers is the diesel engine or gas engine, but these require their own fuel storage, etc. and are certainly not as reliable as the steam turbine. Most process plants therefore spare the critical pumps in the process with a turbine-driven unit which may be started automatically on low process flow.

The principle of the turbine driver

Turbines are the most flexible of prime movers in today's industry. Their horsepower output can be varied by the number and size of the steam nozzles used, speeds can be changed readily, and high speeds without gearing is possible. They have a very wide range of horsepower applications. The operation of the steam turbine is analogous to that of a waterwheel where buckets are attached to the wheel which collect the water. The wheel is moved downwards by the weight of the water collected thus causing the rotation of the wheel. In steam turbines the buckets are replaced by vanes which are impinged by the motive steam to cause the rotating motion. Turbines may consist of one set of vanes keyed to the shaft in the case of a single-stage machine or several sets of vanes in multi-stage machines. These sets of vanes are called simply 'wheels' and the number of stages are referred to as the number of wheels.

In multi-stage turbines, the steam leaving the first wheel is directed towards a set of stationary vanes attached to the casing. These stationary vanes reverse the steam flow and serve as nozzles directing the steam towards the second wheel attached to the same shaft.

Most turbines used on a regular basis in a process plant are single-stage. Multi-stage machines are more efficient but are also much more expensive. Their use therefore are for drivers requiring horsepower in excess of 300. The power industry is a good example for the use of large multi-stage turbine drivers. Single- or multi-stage turbines may be operated either condensing or non-condensing. However, pump drivers should not be made *condensing* without a rigorous review to see if other types of drives can be used. The complexity of condensing is hardly worth the small savings in utilities that are made.

The performance of the steam turbine

The salient factors in the performance of the steam turbine are:

- Horsepower output
- Speed
- Steam inlet and outlet conditions
- Its mechanical construction (e.g. number of wheels, size of the wheel, etc.)

These factors are interrelated and their effect on the performance of the turbine is reflected by a change in overall efficiency. The overall efficiency may be defined as the ratio of the energy output to the energy of the steam theoretically available at constant entropy as obtained from a Mollier diagram. This overall efficiency is the product of mechanical and thermal efficiencies. The losses in turbines are due partly to friction losses of the rotating shaft and partly to thermodynamic losses and turbulence. Figure 4.10 gives the overall efficiencies plotted against delivered horsepower.

The steam required by a turbine for a given horsepower application is called its 'water rate'. The actual water rate for a turbine is supplied by the manufacturer from test runs carried out on the machine in the workshop. Process engineers very often need to be able to estimate these water rates for their work. A typical situation arises when determining the best steam balance for a plant. Such estimates may be obtained from Figure 4.11. This and the accompanying notes are self-explanatory.

COOLING WATER REQUIREMENTS

Many pumps in process service require water cooling to various parts of the pump. This cooling water is applied to bearings, stuffing boxes, glands, and pedestals. The application of the cooling water is determined by the manufacturer in accordance with standards for the service and conditions that the pump must satisfy. Most of the cooling water may be recovered in a closed cooling water system. However, gland cooling water is never recovered but is routed to the waste water drain. The following lists the approximate cooling water requirements for pumps and steam turbines:

Pumps	To 1000 gpm	Above 1000 gpm
Up to 350 °F	0 gpm	0 gpm
350 °F to 500 °F	2 gpm	4 gpm
Above 500 °F	3 gpm	6 gpm
Steam turbines		
450 °F	0 gpm	0 gpm
Above 450 °F	3 gpm	3 gpm

5 COMPRESSORS

5.1 Types of Compressors and Selection

Compressors are divided into four general types:

- Centrifugal
- Axial flow
- Reciprocating
- Rotary

The name given to each type is descriptive of the means used to compress the gas and comparison of the different types of compressors and typical applications is shown in Table 5.1. A brief description of each of the types now follows.

CENTRIFUGAL

This type of compressor consists of an impeller or impellers rotating at high speed within a casing. Flow is continuous and inlet and discharge valves are not required as part of the compression machinery. Block valves are needed for isolation during maintenance.

Centrifugal compressors are widely used in the petroleum, gas and chemical industries due primarily to the large volumes of gas that frequently have to be handled. Long continuous operating periods without an overhaul make centrifugal compressors desirable for use for petroleum refining and natural gas applications. Normally they are considered for all services where the gas rates are continuous and above 400 acfm (actual cubic feet per minute) for a clean gas and 500 acfm for a dirty one. These rates are measured at the discharge conditions of the compressor. Dirty gases are considered to be gases similar to those from a catalytic cracker, which may contain some fine particles of solid or liquid material.

The slowly rising head capacity performance curves make centrifugal compressors easy to control by either suction throttling or variable-speed operation.

The main disadvantage of this type of compressor is that it is very sensitive to gas density, molecular weight and polytrophic compression exponent. A decrease in density or molecular weight results in an increase in the polytrophic head requirement of the compressor to develop the required compression ratio.

AXIAL FLOW

These compressors consist of bladed wheels that rotate between bladed stators. Gas flow is parallel to the axis of rotation through the compressor. Axial flow compressors become economically more attractive than centrifugal compressors in applications where the gas rates are above 70 000 acfm at *suction* conditions. The compressors are extremely small relative to capacity and have a slightly higher efficiency than the centrifugals. Axial flow compressors are widely used as air compressors for jet engines and gas turbines.

RECIPROCATING

Reciprocating compressors are widely used in the petroleum and chemical industries. They consist of pistons moving in cylinders with inlet and exhaust valves. They are cheaper and more efficient than any other type in the fields in which they are used. Their main advantages are that they are insensitive to gas characteristics and they can handle intermittent loads efficiently. They are made in small capacities and are used in applications where the rates are too small for a centrifugal. Reciprocating compressors are used almost exclusively in services where the discharge pressures are above 5000 psig.

When compared with centrifugal compressors, reciprocating compressors require frequent shutdowns for maintenance of valves and other wearing parts. For critical services this requires either a spare compressor or a multiple compressor installation to maintain plant throughput. In addition, they are large and heavy relative to their capacity.

ROTARY

Recent developments in the rotary compressor field have opened up areas of application in the process industry with the use of the following types of rotary compressors:

(1) *High-pressure screw* These compressors have been developed into heavy-duty type machines. They consist of two rotating helices that rotate in a casing without actual contact. Rotary compressors are lower in cost and have a higher efficiency than centrifugal compressors. They are not sensitive to gas characteristics since they are positive displacement machines. Parts are standardized production items so that a spare rotor is not generally required

Table 5.1. Comparison of compressor types and typical applications

Type and controllable range	Percent Availability	Operating speed, volumetric capacity, compression ratio per stage (6)	Compression efficiency	Advantages	Disadvantages	Usual drivers	Common applications
Centrifugal 70–100%	99.5–100% (1)	3000–15 000 rpm (2) 400–500 acfm minimum at discharge 150 000 acfm max. Suction volume (5) 80 000–100 000 ft polytrophic head/casing	70–78%	1. Long continuous operating periods 2. Low maintenance costs 3. Small size relative to capacity 4. Ease of capacity control	1. Pressures ratio is sensitive to gas density and molecular weight 2. Spare rotor required	Steam turbine Gas turbine Electric motor Waste gas expander	Large refrigeration system Cat cracker air Large catalytic reformer recycle gas
Axial flow 80–100%	99.5–100%	4000–12 000 rpm (2) 70 000 min. acfm 2–4 compression ratio per casing	75–82%	1. Very high throughputs possible 2. Extremely small size relative to capacity 3. Higher efficiency than centrifugals 4. Good for parallel operation with other axials or centrifugals	1. Capacity flexibility limited by steep head-capacity curve and short stable operating range except when variable-pitch stators are used 2. Performance and efficiency are sensitive to fouling 3. Spare rotor and spare stator blading are required	Steam turbine Gas turbine Electric motor Waste gas expander	Cat cracker air (large)

Type	Mechanical efficiency	Operating characteristics	Efficiency	Advantages	Limitations	Drivers	Applications
Reciprocating	98% clean gas (3) 95% dirty gas (3) 95% clean gas (4) 93% dirty gas (4)	300–1000 rpm 5 max. compression ratio or 330–380 °F max. discharge temp.	75–85%	1. Handles intermittent loads efficiently 2. Lower cost for small capacities 3. Used for very high discharge pressures (up to 50 000 psig) 4. Higher efficiency than centrifugal in lower capacity ranges 5. Insensitive to gas characteristics	1. Short continuous operating periods require spare or multiple machine installations if service is critical 2. Higher maintenance costs than centrifugal 3. Pulsation and vibration require engineered piping arrangement 4. Availability decreases when non-lubricated machines required to avoid lubricating oil in gas discharge	Synchronous motor coupled or integral electric motor Coupled or integral engine	Instrument air Refinery air Fuel gas Synthesis gas Crude gas Small catalytic reformer recycle gas Small refrigeration system Low mole. wt gas
Rotary high-pressure screw 55–100%	99–99.5%	2500–10 000 rpm 1000–20 000 acfm at suction 4 to 7 max. compression ratio per casing but not exceeding 100 psi ΔP	75–80%	1. Lower cost than centrifugals 2. Higher efficiency than centrifugals 3. Not sensitive to gas characteristics 4. Parts are standardised production items and no spare rotor required	1. Noisy, require inlet and discharge silencers 2. Sensitive to temperature rise due to close clearances 3. Not recommended for use where fouling produces hard deposits 4. Speed or bypass control are only type applicable	Electric motor Steam turbine Gas turbine Waste gas expander	Refinery air Fuel gas Cat. cracker air (small)

continued overleaf

Table 5.1. (*continued*)

Type and controllable range	Percent Availability	Operating speed, volumetric capacity, compression ratio per stage (6)	Compression efficiency	Advantages	Disadvantages	Usual drivers	Common applications
Low-pressure screw, lobe and vane type rotaries, fixed capacity	Not recommended for continuous service	1500–3600 rpm 100–12 000 acfm suction 2 compression ratio per stage 50 psi max. discharge pressure	75–80%	1. Low first cost 2. Low maintenance cost 3. Parts are standardized production items and no spare rotor is required	1. Limited life 2. Speed or bypass control are only type applicable 3. Very noisy	Electric motor Steam turbine	Low-pressure, light-duty, non-critical services

Notes:
(1) Clean service machines have the highest availability.
(2) Large machines run at lower speeds.
(3) Between turnarounds of 3 days every 8000–12 000 hours with electric drive. 95–98% includes 8-hour shutdowns every few months for valve maintenance.
(4) Between turnarounds of 2 weeks every 8000–12 000 hours with engine drive. 93–95% includes 8-hour shutdowns every month for maintenance checks on compressor valves and engine driver.
(5) Axial flow compressors should be considered at gas rate above 70000 acfm at suction conditions.
(6) Stages can be compounded in series for higher rates.

to be stocked for emergency replacement. This compressor is noisy and sensitive to temperature rise along the screws due to the close clearances involved. They are good for fouling services where the fouling material forms a soft deposit. This decreases the clearances and leakage along the screws and casing. They are not recommended for use in fouling services in which the deposits are hard. Variation in speed or a discharge bypass to suction are the only types of control that can be used.

(2) *Low-pressure screw, lobe and sliding vane*   These compressors should be used only for low-pressure, light-duty, non-critical applications. They operate on the same principle as the high-pressure screw type but have different mechanical designs. The same advantages and disadvantages apply as those for rotary high-pressure screw compressors. They are even lower in cost than the high-pressure screw compressors but contain parts having limited life, thus requiring more maintenance. Only centrifugal and reciprocating compressors will be discussed further in this book.

5.2 Calculating Horsepower of Centrifugal Compressors

Centrifugal compressors are used in process service where high-capacity flows are required. A typical example is the recycle compressor for handling a hydrogen-rich stream in some oil refining and petrochemical processes. The following table gives some idea of the centrifugal compressor's capacity range.

	Centrifugal compressor flow range		
Nominal flow range (inlet) acfm)	Average polytrophic efficiency	Average adiabatic efficiency	Speed to develop 10 000 ft head/wheel
500–7500	0.74	0.70	10 500
7500–20 000	0.77	0.73	8200
20 000–33 000	0.77	0.73	6500
33 000–55 000	0.77	0.73	4900
55 000–80 000	0.77	0.73	4300
88 000–115 000	0.77	0.73	3600
115 000–145 000	0.77	0.73	2800
145 000–200 000	0.77	0.73	2500

In general, the head or differential pressure levels served by a centrifugal compressor is considerably lower than that for reciprocal. The following diagram illustrates this feature.

Process engineers are often required to establish the capability of a centrifugal compressor in a particular service or to assess the machine's capability to handle a different service. In conducting these studies it is necessary to determine the machine's horsepower under the study conditions. This section provides a

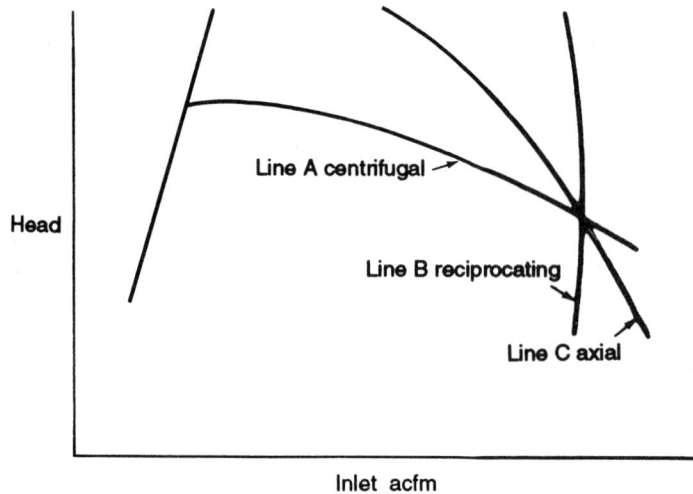

Inlet acfm

procedure where the gas horsepower (and thereafter the brake horsepower) of a compressor can be calculated. This procedure is as follows:

- *Step 1*   Establish the duty required from the compressor in terms of:
  Capacity in cf/min at inlet conditions
  Design inlet temperature
  Design inlet pressure
  The mole wt of the gas to be handled
  Compression ratio (P_2/P_1), P_2 being the discharge pressure and P_1 the inlet pressure.
- *Step 2*   Establish the K value for the gas. If this is a pure gas (such as oxygen) the K value can be read from data books. Otherwise the K value is the ratio (C_p/C_v). See Figure A1.3 in Appendix 1. (Note: Do not confuse this K factor with equilibrium constants.)
- *Step 3*   Calculate volume of the gas in scf/min. This is the inlet cfm times inlet pressure times 520 divided by 14.7 psia times inlet temperature in °R, thus

$$\text{scfm} = \frac{1\ \text{cfm} \times \text{inlet press.} \times 520}{14.7 \times \text{inlet temp. °R}}$$

- *Step 4*   Calculate number of moles gas/min by dividing scf/m by 378. Multiply number of moles by mole wt for lb/min of gas.
- *Step 5*   Read off the estimated discharge temperature from Figure 5.1 using this and the discharge pressure calculate the volume in cf/min at discharge.
- *Step 6*   Calculate density of gas at suction and discharge using the weight calculated in step 4 and the cfm for suction and the cf/min calculated in step 5 for discharge. This density will be in lb/ft^3.
- *Step 7*   The average value for Z is taken as Z at suction + Z at discharge divided by 2. Z (compressibility factor) is calculated by the expression

$$Z = \frac{MP}{T\rho_v \times 10.73}$$

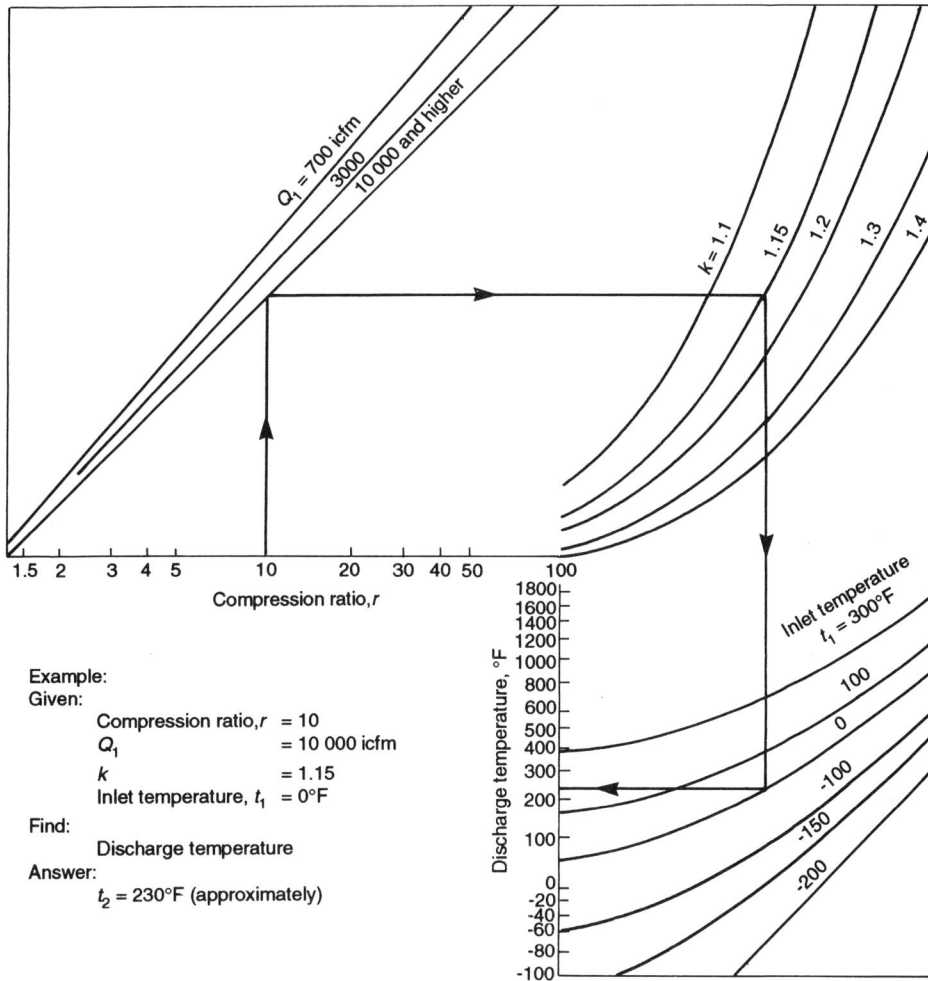

Figure 5.1. Estimated discharge temperatures of centrifugal compressors (Reproduced by permission of Gas Processors Suppliers Association)

where

M = mole weight
P = pressure at psia
T = °R (°F + 460)
ρ_v = density in lb/ft³

- *Step 8* Calculate the adiabatic head in ft lb/lb using the expression

$$H_{ad} = \frac{Z_{ave} \times R \times T_i}{(K-1)/K}\left[\left(\frac{P_2}{P_1}\right)^{(K-1)/K} - 1\right]$$

where

H_{ad} = adiabatic head (ft lb/lb)
Z_{ave} = average compressibility factor
R = gas constant = 1545/mole wt
K = adiabatic exponent C_p/C_v
P_1 = suction pressure (psia)
P_2 = discharge pressure (psia)
T = inlet temperature (°R)

● *Step 9*   The gas hp is obtained using the expression

$$\frac{W \times H_{ad}}{\eta_{ad} \times 33\,000}$$

where

W = weight in lb/min of gas
H_{ad} = adiabatic head (ft lb/lb)
η = adiabatic efficiency (0.7–0.75)

● *Step 10*   Check ghp using Figure 5.2.

Example calculation

To determine the gas HP of a centrifugal compressor assuming isentropic compression:

$$\text{Compression ratio} = 10.0$$
$$\text{Capacity (actual inlet cf/min)} = 10\,000$$
$$K_{ave} = 1.15$$
$$T_1\ (°F) = 100$$
$$P_1\ (\text{psia}) = 100$$
$$\text{Mole wt} = 30$$
$$\text{lb/min} = 5013$$

For isentropic compression

$$H_{ad} = \frac{ZRT}{(K-1)/K}\left[\left(\frac{P_2}{P_1}\right)^{(K-1)/K} - 1\right]$$

where

H_{ad} = adiabatic head (ft lb/lb)
Z = compressibility factor (ave)
R = gas constant = 1545/MW
K = adiabatic exponent C_p/C_v = 1.15
T = temperature in °R = °F + 460 °F
P_1 = suction pressure (psia)
P_2 = discharge pressure (psia)

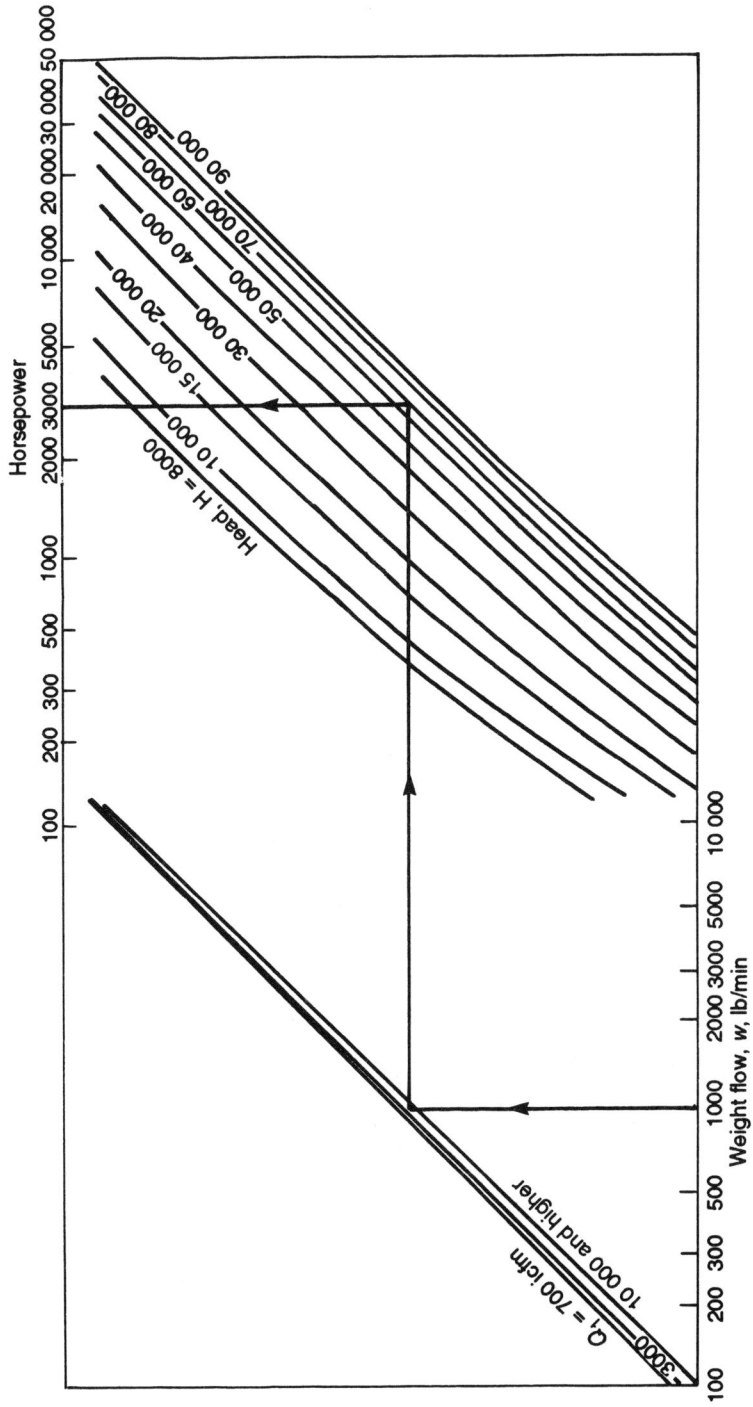

Figure 5.2. Determination of centrifugal compressor horsepower (Reproduced by permission of Gas Processors Suppliers Association)

Z at inlet conditions

$$\rho_v = 0.5 \text{ lb/ft}^3$$

$$Z = \frac{MP}{T\rho_v} \times 10.73 = \frac{30 \times 100}{560 \times 0.5} \times 10.73 = 0.998$$

Estimated discharge temp. (Figure 5.1) = 400 °F

$$Z_{dis} = \frac{30 \times 1000}{860 \times 3.26} \times 10.73 = 0.997$$

Use 0.998

$$H_{ad} = \frac{0.998 \times 51.5 \times 560}{\left(\dfrac{0.15}{1.15}\right)} \left[\left(\frac{1000}{100}\right)^{0.13} - 1 \right]$$

$$= 77\,004 \text{ ft lb/lb}$$

$$= 220\,664[1.349 - 1]$$

$$\text{gas hp} = \frac{W_1 \times H_{ad}}{\eta_{ad} \times 33\,000}$$

Let η_{ad} be 0.75

$$\frac{5013 \times 77\,004}{0.75 \times 33\,000} = 15\,598 \text{ ghp}$$

This compares well with the estimate based on Figure 5.2.

5.3 Centrifugal Compressor Surge Control, Performance Curve and Seals

Centrifugal compressors can be counted on for uninterrupted run lengths of between 18 to 36 months after the initial shakedown run. The 18-month run corresponds to a compressor handling dirty gas, such as furnace gas, and the 36-month run to a clean gas service, such as refrigerant.

Spare compressors are not usually provided. A spare rotor, however, is required to be stocked as insurance against an extended downtime. Since this rotor is part of the capital cost of the equipment, it is not accounted for as spare parts. Only reliable drivers such as an electric motor, steam or gas turbine can be used where long, continuous run lengths are required. In the case of steam and gas turbines, the drivers will probably dictate the maximum possible run length. The high operating speed of a centrifugal compressor also favours the selection of these type of high-speed drivers. The speed of these drivers can be specified to be the same as those of the compressor. For electric motor drive, a speed-increasing gear is normally required. Centrifugal compressors can be broadly classified with regards to head and capacity as follows:

	Speed (rpm)	Suction (acfm)	Polytrophic head (ft) #/#
Small standard multi-stage	3000–3600	100–1000	to 8500
Standard single-stage	3000–3600	700–60 000	1000–6700
Special single-stage	3000–15 000	1000–60 000	6700–11 500
Special multi-stage casing, uncooled	3000–15 000	1000–140 000	6700–100 000
Special multi-stage, multi-casing, intercooled	3000–15 000	2500–140 000	37 000 up

As a guide, the maximum head per impeller is about 10 000 ft. Normally, about eight impellers can be used in a casing.

The minimum allowable volume of gas at the compressor discharge is about 400 acfm for a clean gas and 500 acfm for a dirty one. Dirty gases are considered to be similar to the gas from a steam or catalytic cracking unit.

The discharge temperature is limited to about 250 °F for gases that may polymerize and 400 °F for other gases. Normally, intercoolers will be used to keep the discharge temperature within these limits. These temperature limitations do not apply to special centrifugal flue gas recirculators which can be obtained to operate at over 800 °F. There is also a temperature rise limitation of 350 °F per casing. This is the maximum temperature rise that can be tolerated due to thermal expansion considerations.

Use of cast iron as a casing material is limited to 450 °F maximum. Temperatures of −150 °F to −175 °F can be tolerated in conventional designs. Lower temperatures are not common and will require consulting on individual design features.

SURGE

A characteristic specific to centrifugal and axial compressors is a minimum capacity at which the compressor operation is stable. This minimum capacity is referred to as the surge or pumping point. At surge, the compressor does not meet the pressure of the system into which it is discharging. This causes a cycle of flow reversal as the compressor alternately delivers gas and the system returns it.

The surge point of a compressor is almost independent of its speed. It depends largely on the number of wheels or impellers in series in each stage of compression. Reasonable reductions in capacity to specify for a compressor are shown below.

Wheels/compression stage	% of normal capacity at surge–maximum
1	55
2	65
3 or greater	70

An automatic recirculation bypass is required on most compressors to maintain the minimum flow rates shown. These are required during start-up or when the normal load falls below the surge point. Cooling is required in the recycle circuit if the discharge gas is returned to the compressor suction.

PERFORMANCE CURVES

The rise of performance curves should be specified for a compressor. This is normally done by specifying the pressure ratio rise to surge required in each stage of compression. A continuously rising curve from normal flow rate to surge flow is required for stable control.

The pressure ratio rise to surge is largely a function of the number of impellers per compression stage. Reasonable pressure ratio rises to specify are shown below:

Wheels/compression stage	Minimum % of rise in pressure ratio from normal to surge flow
1	$3\frac{1}{2}$
2	6–7
2 or greater	$7\frac{1}{2}$

Frequently, the performance curves for a compressor have to be plotted to determine if all anticipated process operations will fit the compressor and its specific speed control. Three points on the head capacity curve are always known: normal, surge and maximum capacity. The normal capacity is always considered to be on the 100% speed curve of the compressor. The surge point and the compression ratio rise to surge have been specified. From this the head produced by the compressor at the surge point can be back-calculated using the head–pressure ratio relationship. The maximum capacity point is specified to be at least 115% capacity at 85% of normal head.

The head capacity curve retains its characteristic shape with changes in speed. Curves at other speeds can be obtained from the three known points on the 100% speed curve by using the following relationships:

(1)   The polytrophic head varies directly as the speed squared.
(2)   The capacity varies directly as speed.
(3)   The efficiency remains constant.

Figure 5.3 shows a typical centrifugal compressor performance curve.

CONTROL

Speed

Speed control is the most efficient type of control from an energy consideration. It requires, however, that a variable-speed driver such as a steam turbine or gas

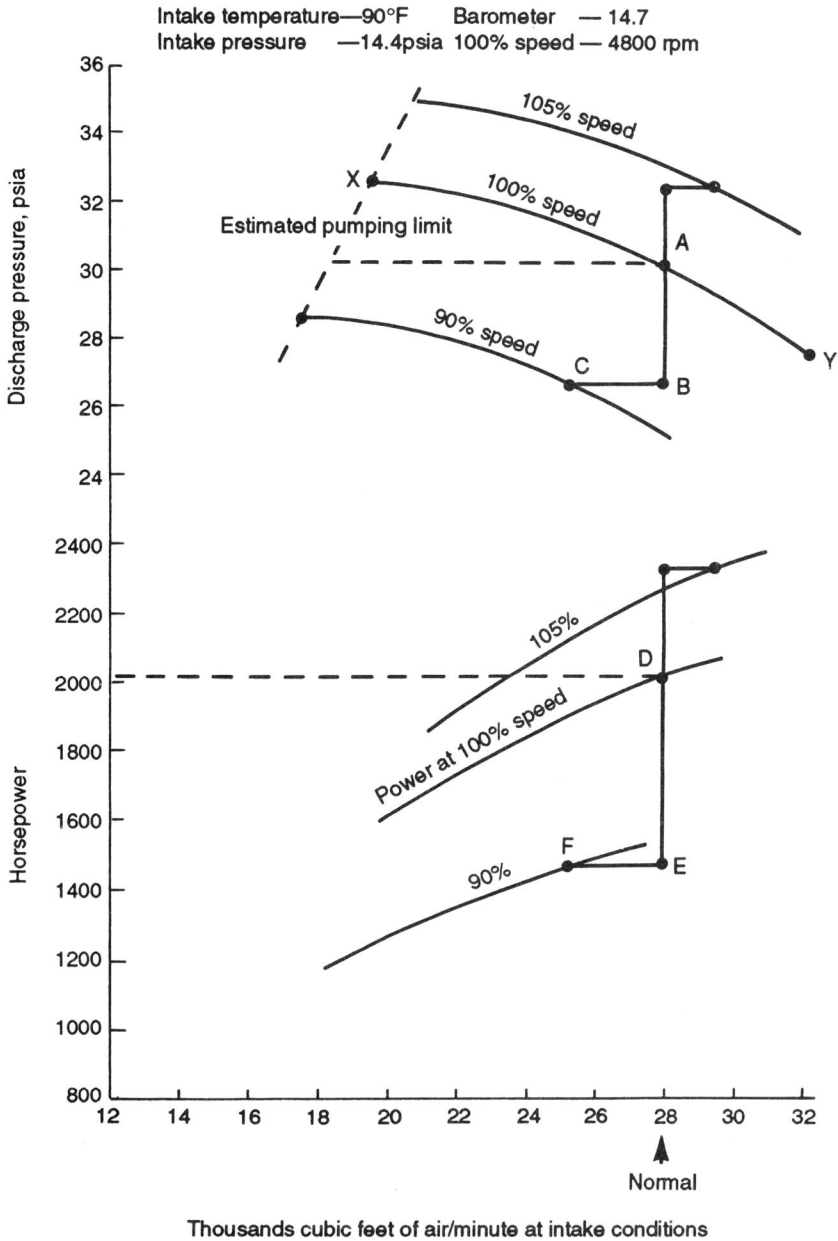

Figure 5.3. An example of a centrifugal compressor performance curve

turbine be used. The compressor is controlled by shifting its performance curve to match the system's requirement.

Suction throttling

- *Adjustable inlet guide vanes* This is the most efficient method of adjusting the capacity of a constant-speed compressor to match the system character-

istics. The guide vanes consist of a venetian blind device that is positioned by
a rack and pinion linkage. While the guide vanes do some throttling their
main effect is to change the velocity of the gas to that of the impeller vane by
changing the direction of flow. This changes the head produced and in effect
changes the characteristic of the machine.

- *Suction throttle*   This consists of a control valve located in the compressor
  suction which regulates the suction pressure to the compressor. The control
  valve results in a greater power loss compared to adjustable inlet guide vane
  control since it is a purely throttling effect. Suction throttle valves are lower in
  cost than adjustable inlet guide vanes.

- *Discharge throttling*   This consists of a control valve located in the compressor discharge. Discharge throttle valves are seldom used since they offer
  relatively little power reduction at reduced capacity. The effect is simply to
  'push' the compressor back on its curve.

SEALS

Table 5.2 shows the types of seals that are commonly used in centrifugal
compressors. The start-up as well as the operating conditions of the compressor
should be considered in selecting a seal. Often the system is evacuated when
hydrocarbons are handled prior to its start-up. This requires that the seal be good
for vacuum conditions.

5.4   Specifying a Centrifugal Compressor

The process specification must give all the information concerning the gas that is
to be handled, its inlet and outlet conditions, the utilities that are available and
the service that is required of the compressor. The process specification sheet
given here shows the minimum that a process engineer should provide in
approaching manufacturers. An explanation of this specification now follows
covering each of the items in the specification.

TITLE BLOCK

This requires the item to be identified by item number and its title. The number
of units that the specification refers to is also given here. For a centrifugal
compressor this will normally be just one, as very seldom is a spare machine
required.

NORMAL AND RATED COLUMNS

More often than not the conditions and quantities required to be handled will vary
during the operation of the machine. The two columns therefore will be
completed showing the average normal data in the first column and the most

Table 5.2. Centrifugal compressor seals

No.	Application	Gas being handled	Inlet pressure (psia)	Seal arrangement
1	Air compressor	Atmospheric air	Any	Labyrinth
2	Gas compressor	Non-corrosive Non-hazardous Non-fouling Inexpensive	Any	Labyrinth
3	Gas compressor*	Non-corrosive or corrosive Non-hazardous or hazardous Non-fouling or fouling	10–25	Labyrinth with injection or ejection of fluid being handled
4	Gas compressor	Non-corrosive Non-hazardous Non-fouling	All pressures	Oil seal combined with lube oil system
5	Gas compressor	Corrosive Non-hazardous or hazardous	All pressures	Oil seal with seal oil separate from lube oil system

*Where some gas loss or air induction is tolerable.

severe conditions and duty required by the compressor in the second. The severe conditions in column two are for a continuous length of operation, not instantaneous peaks (or troughs) that may be encountered.

GAS

The composition and gas stream identification must be included as part of the process specification. Usually the composition of the gas is listed on a separate sheet as shown in the example. Note in many catalytic processes that utilize a recycle gas the composition of the gas will change as the catalyst in the process ages. Thus it will be necessary to list the gas composition at the start of the run (SOR) and at the end of the run (EOR). The compressor may also be required to handle an entirely different gas stream at some time or other. This too must be noted. For example, in many petroleum-refining processes a recycle compressor normally handling a light predominantly hydrogen gas is also used for handling air or nitrogen during catalyst regeneration, purging and start-up.

VOLUME FLOW

This is the quantity of gas to be handled stated at 14.7 psia and 60 °F.

WEIGHT FLOW

This is the weight of gas to be handled in either lb/min or lb/h.

INLET CONDITIONS

- *Pressure*   This is the pressure of the gas at the inlet flange of the compressor in psia.
- *Temperature*   This is the temperature of the gas at the inlet flange of the compressor.
- *Mole weight*   The mole weight of the gas is calculated from the gas composition given as part of the specification.
- C_p/C_v This is the ratio of specific heats of the gas again obtained from the gas mole wt and Figure A1.3 in Appendix 1.
- *Compressibility factor (z)*   Use the value at inlet conditions calculated as shown in step 7 of Section 5.2.
- *Inlet volume*   This is the actual volume of gas at the conditions of temperature and pressure existing at the compressor inlet. Thus:

$$\text{acfm} = \frac{\text{scfm} \times 14.7 \times (\text{inlet temp. }^\circ\text{F} + 460)}{(60\,^\circ\text{F} + 460) \times \text{inlet press. psia}}$$

DISCHARGE CONDITIONS

- *Pressure*   This is the pressure at the compressor outlet flange and is quoted in either psia or psig.
- *Temperature*   This is estimated using Figure 5.1 in Section 5.2.
- C_p/C_v This will be the same as inlet.
- *Adiabatic efficiency*   This will be as given in Figure 5.4.

APPROXIMATE DRIVER HORSEPOWER

This item will include the adiabatic (or gas) horsepower as calculated in Section 5.2 plus the following losses:

Leakage loss—1% of adiabatic hp
Seal losses—35 hp for all hp ranges
Bearing loss—35 hp for all hp ranges

The remainder of the specification sheet contains all the essential data and requirements that may affect the duty and performance of the compressor. Much of this is self-explanatory, but there are some items that require comment:

(1)   Most compressor installations today are under an open-sided shelter with a small overhead gantry crane assembly for maintenance.
(2)   Usually the lube and seal oil assemblies have their own pump and control systems. Consequently even if the compressor itself is to be steam driven there may still be need to give details of utilities for the ancillary equipment.
(3)   Details of the gas composition is essential for any development of the compressor. This is listed on the last page of the specification together with any notes of importance concerning the machine and its operation.

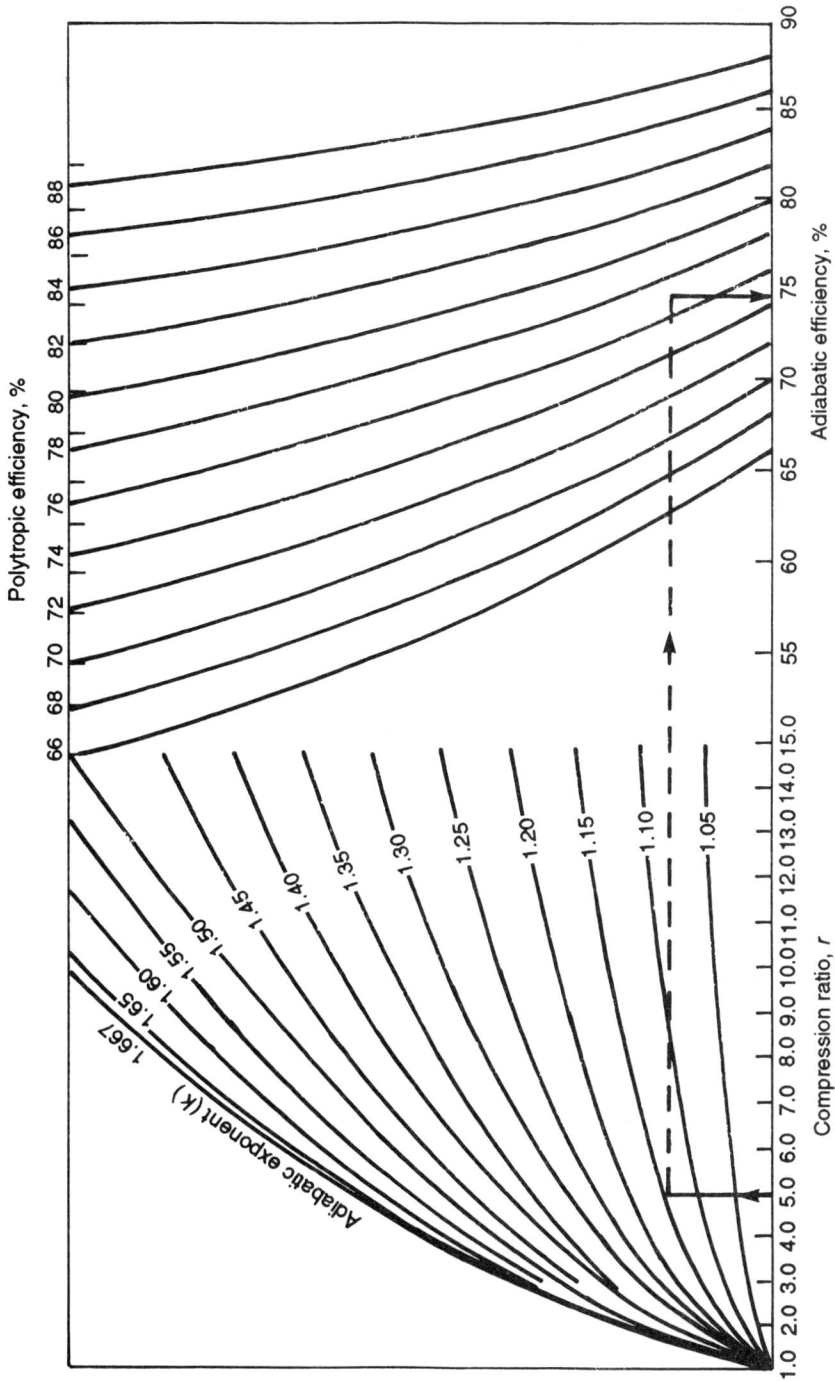

Figure 5.4. Adiabatic efficiencies for centrifugal compressors (Reproduced by permission of Gas Processors Suppliers Association)

Example calculation

Prepare a process specification sheet for a compressor to handle the hydrogen recycle stream in an O-xylene isomerization plant. Details are as follows:

Fresh feed rate	5000 bpsd of M-xylene
Recycle gas rate	7000 scf of hydrogen per bbl of fresh feed

Gas composition mole %	Start of run (SOR)	End of run (EOR)
H2	85.00	68.78
C_1	4.4	9.17
C_2	4.2	8.86
C_3	3.6	7.36
iC_4	0.78	1.62
nC_4	0.99	2.05
C_5s	1.03	2.16

Suction pressure	150 psig
Reactor pressure	500 psig
Suction temperature	100 °F

Step 1   Calculate the mole weight of the gas

Factor	SOR Mole %	MW	Wt	EOR Mole %	MW	Wt
H_2	85.0	2	170	68.78	2	138
C_1	4.4	16	70	9.17	16	147
C_2	4.2	30	126	8.86	30	266
C_3	3.6	44	158	7.36	44	324
iC_4	0.78	58	45	1.62	58	94
nC_4	0.99	58	57	2.05	58	119
C_5s	1.03	72	74	2.16	72	156
Total	100.00		700	100.00		1244
MW		7.0			12.44	

Step 2   Calculate volume flow of gas in scf/min

$$\text{Total volume of } H_2 \text{ required} = 5000 \text{ bpsd} \times 7000 \text{ scf}$$

$$= 35.00 \text{ mm scf/day}$$

$$= 24\,306 \text{ scf/min}$$

$$\text{For SOR volume gas flow} = \frac{24\,306}{0.85}$$

$$= 28\,595 \text{ scf/min}$$

$$\text{For EOR volume gas flow} = \frac{24\,306}{0.6878}$$

$$= 35\,339 \text{ scf/min}$$

Step 3   Calculate weight flow in lb/min

$$\text{Moles/min of gas} = \frac{\text{scf}}{\text{min}/378}$$

For SOR moles/min = 75.6

For EOR moles/min = 93.5

lb/min for SOR = 75.6 × 7.0 = 529 lb/min

lb/min for EOR = 93.5 × 12.44 = 1163 lb/min

Step 4   Calculate acfm at inlet conditions

Compressor inlet pressure = 165 psia

Compressor inlet temp. = 100 °F

$$\text{For SOR acfm} = \frac{28\,597 \times 14.7 \times 560}{520 \times 165}$$

$$= 2744 \text{ ft}^3/\text{min}$$

$$\text{For EOR acfm} = \frac{35\,339 \times 14.7 \times 560}{520 \times 165}$$

$$= 3391 \text{ ft}^3/\text{min}$$

Step 5   Estimate the C_p/C_v ratio

The molal proportions will be used for this purpose. The ratio for each component will be taken from Figure A1.3 in Appendix 1.

	C_p/C_v	SOR Mole %	C_p/C_v fact.	EOR Mole %	C_p/C_v fact.
H_2	1.40	85.0	119	68.78	96.29
C_1	1.30	4.4	5.7	9.17	11.92
C_2	1.22	4.2	5.12	8.86	10.81
C_3	1.14	3.6	4.10	7.36	8.39
iC_4	1.11	0.78	0.87	1.62	1.80
nC_4	1.11	0.99	1.10	2.05	2.28
C_5s	1.09	1.03	1.12	2.16	2.35
Total		100.00	131.9	100.00	133.84

Then C_p/C_v for the gas is

$$\text{SOR} = 1.319$$

$$\text{EOR} = 1.338$$

Step 6   Calculate compressability factors

$$Z = \frac{\text{MW} \times P_1}{T \times v \times 10.73}$$

For SOR flows

$$Z = \frac{7.0 \times 165}{560 \times 0.193 \times 10.73}$$

$$= 0.996$$

$$v = \frac{\text{wt lb}}{\text{min/acfm}}$$

For EOR flows

$$Z = \frac{12.46 \times 165}{560 \times 0.343 \times 10.73}$$

$$= 0.998$$

Step 7   Calculate outlet temperature
Approx. discharge temperature is read from Figure 5.1 in item 5.2 of this chapter using the following:

$$\text{acfm for SOR} = 2744$$

$$\text{acfm for EOR} = 3391$$

$$\text{Compression ratio} = \frac{515}{165}$$

$$= 3.12$$

$$\text{Inlet temp. °F} = 100$$

Then

$$\text{Discharge temp. for SOR} = 370\,°\text{F}$$

$$\text{Discharge temp. for EOR} = 340\,°\text{F}$$

Step 8   Calculate approx. driver hp
For isentropic compression

$$H_{\text{ad}} = \frac{ZRT}{(K-1)/K}\left[\left(\frac{P_2}{P_1}\right)^{(K-1)/K} - 1\right]$$

where

H_{ad} = adiabatic head (ft lb/lb)
Z = compressibility factor (ave.)
R = gas constant = 1545/MW
K = adiabatic exponent $C_{\text{p}}/C_{\text{v}}$ = 1.15
T = temperature in °R = °F + 460 °F
P_1 = suction pressure (psi)
P_2 = discharge pressure (psia)

Then for SOR:

$$H_{\text{ad}} = \frac{0.996 \times 221 \times 560}{(1.32-1)/1.32}\left[\left(\frac{515}{165}\right)^{(1.32-1)/1.32} - 1\right]$$

For EOR:

$$H_{ad} = \frac{0.998 \times 124 \times 560}{(1.34 - 1)/1.34}\left[\left(\frac{515}{165}\right)^{(1.34-1)/1.34} - 1\right]$$

Had for SOR conditions = 161 063

Had for EOR conditions = 91 546

Step 9 The gas HP is obtained using the expression

$$\text{Gas hp} = \frac{W \times H_{ad}}{\eta_{ad} \times 33\,000}$$

where

W = weight (lb/min) of gas
H_{ad} = adiabatic head (ft lb/lb)
η = adiabatic efficiency (0.7–0.75)

let η_{ad} be 0.73

$$\text{For SOR conditions gas hp: } \frac{529 \times 161\,063}{0.73 \times 33\,000} = 3536\text{ ghp}$$

Let η_{ad} be 0.73

$$\text{For EOR conditions gas hp: } \frac{1163 \times 91\,546}{0.73 \times 33\,000} = 4420\text{ ghp}$$

Step 10 The driver hp is as follows:

	SOR	EOR
Gas hp	3536	4420
Leakage losses	35	44
Bearing losses	35	35
Seal losses	35	35
Driver hp	3641	4534

These data are tabulated on the example specification sheet (Figure 5.5).

5.5 Calculating Reciprocating Compressor Horsepower

Reciprocating compressors are used extensively in the process industry. They vary in size from small units used for gas recovery (such as those on a crude distillation overhead system) to fairly large complex machines used for recycle gas streams and for transporting natural gas. Engineers are frequently required therefore to assess the horsepower of these machines and their capability to handle various streams. This section describes a method used to determine horsepower and proceeds with the following steps:

- *Step 1* Obtain the capacity and the properties of the gas to be handled. Fix the ultimate (discharge) pressure level.
- *Step 2* From the machine data sheet ascertain the number of stages.

ITEM No. <u>4-C-101</u>   TITLE <u>O-Xylene Isom Recycle Compressor</u>

Number of units required   <u>ONE</u>

		Normal	Rated
GAS (see attached composition)		Start of run	End of run
Vol flow	scf/min	28 595	35 339
Weight	lb/min	529	1163
INLET CONDITIONS:			
Pressure	psia	165	165
Temperature	°F	100	100
Mole weight		7.0	12.44
C_p/C_v		1.32	1.34
Compressibility factor		0.996	0.998
Inlet vol.	acf/min	2744	3391
DISCHARGE CONDITIONS:			
Pressure	psia	515	515
Temperature	°F	370	340
C_p/C_v		1.32	1.34
Adiabatic efficiency	%	73	73
APPROX. DRIVER HORSEPOWER		3641	4534

COMPRESSOR SERVICE REQUIRED:

Continuous ————————— <u>Yes</u>

Intermittent ————————— –

Standby ————————— –

CORROSIVENESS OF GAS ————————— None

Figure 5.5. An example of a centrifugal compressor process specification

ITEM No.   4-C-101   TITLE   O-xylene Isom Recycle Compressor

General data and requirements

Enclosure:

    Open to weather   No

    Under shelter   Yes

    In building   No

Ancillary equipment required:

    Lube oil assembly   Yes

    Seal oil assembly   Yes

Type of driver: Motor: None

    Steam turbine:
        Condensing   Yes
        Non-condensing —

Utilities:

    Power:
        Voltage   340
        Cycle   50
        Phase   3

    Steam:
        Inlet: Pressure psig   600
                Temperature °F   710
        Condensing exhaust:
                Pressure psia   3.0
                Temperature °F   —
        Non-condensing Exhaust:
                Pressure psig
                Temperature °F

    Cooling water: Pressure   60   psig: Temperature   40 °F
                         Allowable temperature rise   30 °F

Materials of construction:
    Casing   CS   Type   By mfr
    Internals   CS   Type   By mfr

ITEM No. _____   TITLE _____

General data and requirements (cont.)

Gas composition:

Remarks and notes:

Figure 5.5. (*continued*)

- *Step 3* Estimate the brake horsepower from the expression: bhp = 22 × (compression ratio/stage) × number of stages × capacity (in ft³/day- × 10⁶) × F
where

$$F = 1.0 \text{ for 1 stage}$$

$$1.08 \text{ for 2 stages}$$

$$1.10 \text{ for 3 stages}$$

Ratio/stage is $\sqrt{\text{ratio}}$ for two stages, $\sqrt[3]{\text{ratio}}$ for three stages

- *Step 4* Check the estimate with Figure 5.6. The use of these graphs is self-explanatory.
- *Step 5* Confirm actual suction conditions and compression ratio required (discharge pressure).
- *Step 6* Calculate compression ratio/stage.
- *Step 7* Calculate first-stage discharge pressure. This will be suction pressure times compression ratio per stage from step 6.
- *Step 8* Allow about 3% for interstage pressure drop, then calculate second-stage discharge pressure. Check that overall compression ratio/stage is close to that calculated for step 6.
- *Step 9* Calculate the K value of the gas. K value is C_p/C_v of the gas. If the gas is a mixture of components, the K value may be calculated as the sum of each component multiplied be each of their K values given in Table A1.3 in Appendix 1. Alternatively, for a good approximation data in Figure 5.7 may be used.
- *Step 10* Calculate discharge temperature from first-stage using Figure 5.8. Assume some intercooling (or calculate intercooling from plant data) and fix second-stage discharge temperature using also using Figure 5.8.
- *Step 11* Calculate the compressibility factor Z at suction and discharge from the expression

$$\rho_v = \frac{MP}{T \cdot Z \cdot 10.73}$$

where

ρ_v = gas density in lb/ft³ at condition
T = °Ranking (°F + 460 °F)
Z = compressibility factor
M = mole weight
P = pressure (psia)

Use average value at suction and discharge for each stage.

- *Step 12* Read off bhp/mm cfd at the compression ratio/stage (from step 7) and K from step 9 for each stage.
- *Step 13* Calculate bhp per stage from the expression:

$$\text{bhp} = (\text{bhp/mm cfd}) \times \left(\frac{P_L}{14.4}\right) \times \left(\frac{T_S}{T_L}\right) \times Z_{\text{ave}} \times \text{mm scf/D}$$

274   COMPRESSORS

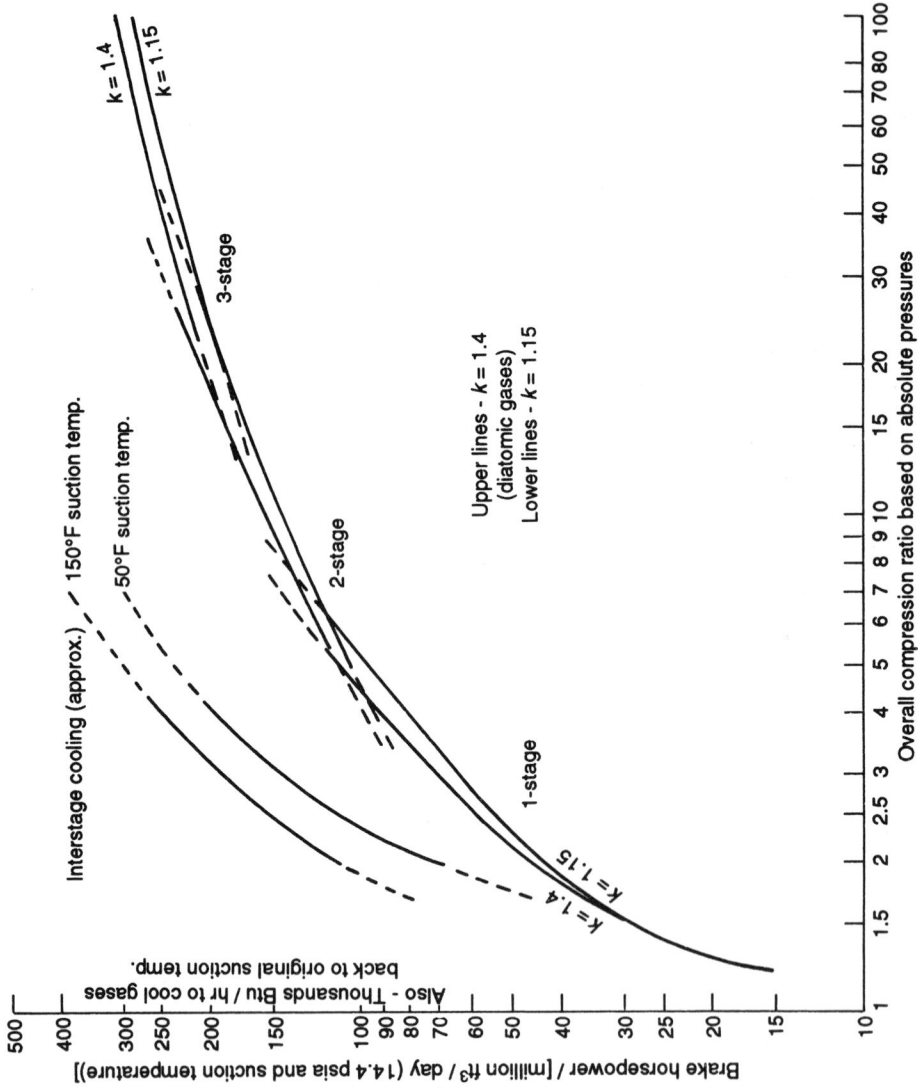

Figure 5.6. An estimate of brake horsepower per mm cfd for reciprocating compressors (Reproduced by permission of Gas Processors Suppliers Association)

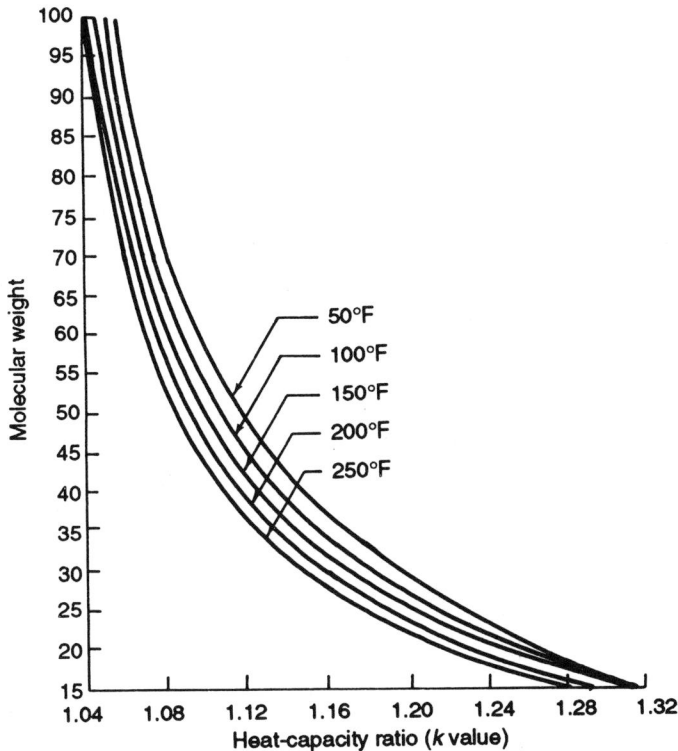

Figure 5.7. Approximation of K from mole weights (Reproduced by permission of Gas Processors Suppliers Association)

where

P_L = pressure base used in contract (psia)
T_S = intake temperature (°R)
T_L = temperature base used in contract °R (usually 520 °R)

- *Step 14* Brake horsepower for the machine is the sum of the bhp calculated for each stage in step 13 above.

Example calculation

(1) Reciprocating compressor

Determining the horsepower of a reciprocating compressor. Estimating the brake horsepower of a reciprocating compressor handling 3 mm cfd of 17 mole wt gas with two stages and a compression ratio of 12:1. Suction conditions are: 14.4 psia and 60 °F, $K = 1.25$.
By calculation:

$$\text{bhp} = 22 \times (\text{ratio per stage}) \times \text{no. of stages} \times \text{capacity} \times F$$

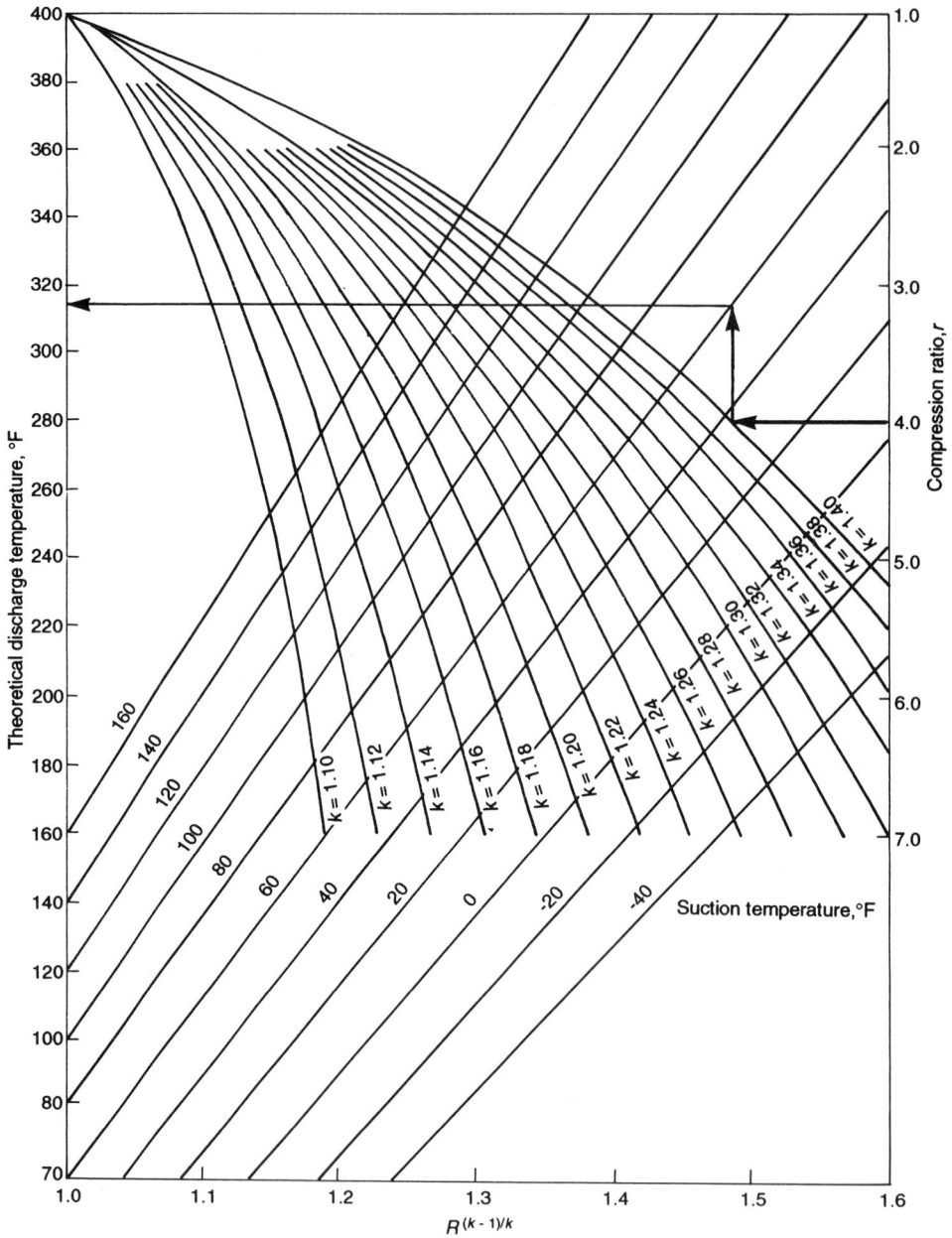

Figure 5.8. Determination of theoretical discharge temperature for reciprocating compressors (Reproduced by permission of Gas Processors Suppliers Association)

where

$$F = 1.0 \text{ for one stage}$$
$$= 1.08 \text{ for two stages}$$
$$= 1.10 \text{ for three stages}$$

$$\text{bhp} = 22 \times \sqrt{12 \times 2 \times 3 \times 1.08}$$
$$= 494$$

From Figure 5.6 bhp is read as 510, which is a reasonable concurrence.

(2) Detail calculation for horsepower

Using the above example but at a suction pressure of 100 psia and temperature of 100 °F. Compression ratio is 12 therefore discharge pressure is 1200 psia (12×100).

For a two-stage compressor compression ratio per stage $= \sqrt{12} = 3.46$

$$\text{First-stage discharge} = 100 \text{ psia} \times 3.46$$
$$= 346 \text{ psia}$$
$$\text{Suction to second stage} = 346 \times 0.97 \text{ (approx. } \Delta P)$$
$$= 336 \text{ psia}$$
$$\text{Compression ratio last stage} = \frac{1200}{336}$$
$$= 3.57$$

which is close enough

K value of gas:
Mole wt of gas $= 17$
From Figure 5.7 $K = 1.27$ (at 150 °F)

Discharge temperature of first stage:
From Figure 5.8 discharge temperature is 274 °F

Discharge temperature of second stage:
Assume intercooling to 120 °F for suction of second stage. Again using Figure 5.8 discharge temperature is 305 °F.

Compressibility factors Z:

$$\rho_v = \frac{MP}{TZ} \times 10.73$$

where

$$\rho_v = \text{gas density (lb/ft}^3)$$
$$T = {}^\circ\text{R}$$
$$Z = \text{Compression factor}$$
$$M = \text{Mole wt}$$
$$P = \text{Pressure}$$

First stage—suction:

$$1 \text{ mole at conditions} = \frac{378 \times 14.7 \times 560}{520 \times 100}$$

$$= 59.8 \text{ acf}$$

$$MW = 17 = \rho_v = 17/59.8 = 0.28$$

$$Z = \frac{17 \times 100}{0.28 \times 560 \times 10.73} = 0.996$$

$$\text{Ave.} = 0.998$$

Second stage—suction:

$$\rho_v = 0.92$$

$$Z = \frac{17 \times 336}{0.92 \times 580 \times 10.73} = 1.0$$

Discharge:

$$\rho_v = 2.5$$

$$Z = \frac{17 \times 1200}{2.5 \times 765 \times 10.73} = 0.99$$

$$\text{Ave.} = 0.997$$

BHP for compressor:
From Figure 5.6 bhp/mmcfd at 3.46 comp. ratio

$$\text{at } K = 1.27$$
$$= 74 \text{ first stage}$$

and for second stage

$$\text{at } K = 1.27 \text{ at comp. ratio} = 3.57$$

$$\text{bhp/mm cfd} = 78$$

$$\text{bhp} = (\text{bhp/mm cfd}) \times \left(\frac{P_L}{14.4}\right) \times \left(\frac{T_3}{T_L}\right) \times Z \text{ ave.} \times \text{mm scf/d}$$

For first stage:

$$\text{bhp} = 74 \times \left(\frac{14.7}{14.4}\right) \times \left(\frac{560}{520}\right) \times 3.0 \times 0.998$$

$$= 244$$

For second stage:

$$\text{bhp} = 78 \times \left(\frac{14.7}{14.4}\right) \times \left(\frac{580}{520}\right) \times 3.0 \times 0.997$$

$$= 266$$

Total bhp = 510, which compares well with the estimate.

5.6 Reciprocating Compressor Controls and Intercooling

A reciprocating compressor is a constant-displacement type compressor. It compresses the same volume of gas to the same pressure level without regard to whether the gas is hydrogen or butane. This characteristic makes it desirable for use in services where the gas will have a widely varying composition. In some cases when an extremely low density gas will be compressed, a reciprocating compressor may be more economical than a centrifugal compressor, even though the flow rate may be very high, due to the large number of stages required for the centrifugal.

Reciprocating compressors are widely used in process services where the flow rates are too small for centrifugal compressors. These units can be obtained with integral or coupled electric motors in sizes from a few hp to 12 000 hp and separate or integral gas drivers varying in sizes from 100 hp to 5500 hp.

A range of air-cooled light-duty compressors is available for intermittent service. They range in size from one-quarter to about 100 hp at pressures up to 300 psig and are usually single-acting. A primary process use of such equipment is for starting air compressors on gas engine-driven machines. Reciprocating compressors can be designed to handle intermittent loads efficiently. This is done by using cylinder unloaders such as clearance pockets or suction valve lifters. Power losses are low at part-load operation with these devices.

Reciprocating parts and pulsating flow present several engineering problems. The foundation and piping system must be constructed to withstand the vibrations produced by the compressor. The pulsating flow produced by the compressor must be dampened by the use of properly engineered suction and discharge bottles. These problems do not arise with the use of other types of compressors.

RECIPROCATING COMPRESSOR CONTROL

Control of the compressor to prevent driver overload can be accomplished with clearance pockets, suction valve lifters, a throttling valve in the suction line, or a control valve in a bypass around the compressor. A hand-operated bypass without cooler is usually provided inside the block valves for start-up purposes.

Clearance pockets

Of the above types of regulation, control by opening fixed clearance pockets gives the smoothest and most efficient control within its range of application. It has the following advantages:

- It minimizes the intake pulsations as the gas flow is not reversed in the intake lines to the cylinder.
- It results in lower bearing loads as all inertia loads are cushioned.
- It leads to very efficient part-load operation. When the gas compressed into the pockets is expanded, it follows the adiabatic line of compression and results in little power loss.

Clearance control has the following disadvantages which sometimes completely eliminate it from consideration:

- When low ratios of compression are combined with high suction pressure, clearance pockets of sufficient size to unload the compressor cannot physically be installed in the machine.
- Clearance control is designed for one set of pressure conditions, and any variation in either suction or discharge pressure affects the amount of unloading accomplished by a given pocket.
- Condensible corrosive gases sometimes cause corrosion and liquid slugging problems.

Suction valve lifters

Suction valve lifters are the other type of internal unloading devices for compressor cylinders and have characteristics that make them applicable when clearance control is not. They completely unload their end of the cylinder whenever they are opened, regardless of the pressure. They do result in increased bearing loads due to unbalanced inertia forces. Suction line pulsations may increase because the then single-acting cylinder may excite a different frequency in the gas.

Suction throttle valve

A throttle valve in the suction line should be considered only for small reciprocating compressors. For large machines, the suction valve cannot give tight enough shut-off to permit unloading the compressor for starting.

Bypass control

External bypass control around the compressor is applicable to all sizes of compressors. It results in a loss of power since the full compressor capacity must be compressed to and delivered at the full discharge pressure before being bled back to suction pressure. Care must be taken with this type of control to ensure that the bypassed gas is cooled sufficiently to prevent increasing the discharge temperature. This type of control is preferred for installations up to several hundred horsepower because of its smoothness and lack of complexity. Individual machines can be shut down for large process variations.

Variable-speed reciprocating compressors

With a variable-speed driver, cylinder control can usually be eliminated and speed control used to obtain desired process conditions. However, start-up unloading must be provided, usually consisting of a hand-operated bypass within the machine or cylinder block valves. On turbine-driven reciprocating compressors economics usually dictate that the compressor be run at constant speed and that cylinder controls or system bypasses be used to obtain the required control.

RECIPROCATING COMPRESSOR INTERCOOLING

Intercooling for multi-stage compressors is advisable whenever there is a large adiabatic temperature rise within the cylinder and the cylinder discharge temperature exceeds 350 °F. When intercooling is employed, the inlet temperature to the higher stage should be as close to the cooling water temperatures as practical. On standard commercial air intercoolers, approach temperatures of 15–20 °F are commonly used. Cooling to first-stage inlet temperature is usually economical on process gas compressors.

Intercooling is employed for two basic reasons:

(1) For mechanical reasons whereby discharge temperature must be limited to 350 °F for lubrication purposes.
(2) An economic reason as intercooling will save from 3% to 5% of the required bhp.

In general, on process compressors handling low n-value gases intercooling is not employed unless the temperature limitation is exceeded. On high n-value diatomic gas mixtures, such as air, intercooling is the rule above about a 4 compression ratio and ambient temperature at suction.

In general, cooling water for electric-driven compressors can be any water available, including salt water. (If the compressor is tied into a plant having gas engine-driven compressors, the electric-driven machine should be tied into the closed system). The cooling water should be available at a minimum pressure of 25 psig.

For estimating purposes, cooling water temperature rise across the cylinders can be taken as 15 °F and that across the inter- and after coolers as 15 °F cooling water requirements are as follows:

Jacket water cooling	500 BTU/bhp/h
Intercooler	1000 BTU/bhp/h
Aftercooler	1000 BTU/bhp/h

Use motor rating for number of horsepower required.

If the discharge temperature of the gas does not exceed 180 °F, it is common practice to eliminate cooling water on the cylinder and operate with cooling passages which are filled with oil. Any jacketed cylinder must be filled with some fluid to ensure even temperature distribution.

5.7 Specifying a Reciprocating Compressor

As in the case of the centrifugal compressor all data necessary to give a precis requirement for the duty and performance required of a reciprocating compressor must be given in the specification sheet (Figure 5.9). Many of these data are the same as those given in a specification for a centrifugal compressor (see Section 5.4). For completeness, however, all the items in a reciprocating compressor are included below.

ITEM No.   C101 A&B   TITLE Hydrotreater Recycle Gas Compressor

Number of units required   3 (2 + 1 spare)

	Normal				Rated			
GAS (see attached composition)	75% H_2				65.8% H_2			
VOL. flow scf/min	4337				4937			
WEIGHT lb/min	74.1				136.3			
INLET CONDITIONS (each stage)	stage 1	stage 2	stage 3	stage 4	stage 1	stage 2	stage 3	stage 4
Pressure psia	65	198*			65	198*		
Temperature °F	80	100*			80	100*		
Mol weight	6.44				10.44			
C_p/C_v	1.362				1.347			
Compressibility factor	0.994	0.996			0.996	0.997		
Inlet vol acf/min	1019	347			1159	394		
DISCHARGE CONDITIONS (each stage)								
Pressure psia	198	615			198	615		
Temperature °F	270	299			262	292		
C_p/C_v	1.362				1.347			
Compressibility factor	0.991	0.992			0.995	0.995		
APPROX. DRVIER HORSEPOWER	1055				1190			

COMPRESSOR SERVICE REQUIRED: *After intercooling
        Length of uninterrupted service: 8000   hours

           Type of compressor: Lubricated   Yes
                                      Non Lubricated

        Discharge RV setting (each stage)   250 and 750 psig

Figure 5.9. Process specification: reciprocating processor

ITEM No. C 101 A&B   TITLE   Hydrotreater Recycle Gas Compressor

General Data and Requirements

Enclosure:

Open to weather _____

Under shelter _____Yes_____

In building _____

Corrosiveness and remarks concerning gas _____

_____None_____

Type of driver: Motor: _____2 normal operating_____

Steam turbine:
Condensing   _____
Non-condensing   _____Spare machine_____

Utilities:

Power:
Voltage _____
Cycle   _____
Phase   _____

Steam:
Inlet: Pressure (psig)   _____600_____
              Temperature (°F)   _____710_____
Condensing exhaust
              Pressure (psia)   _____N/A_____
              Temperature (°F)   _____N/A_____
Non-condensing exhaust
              Pressure (psig)   _____50_____
              Temperature (°F)   ____(by vendor)____

Cooling water: Pressure ___60___ psig: Temperature ___40 °F___
                 Allowable temperature rise _____30 °F_____

Materials of construction:
Cylinder _____CS_____   Type ___By vendor___
Piston   _____Ni.Cr_____   Type ___By vendor___

Type of shaft seal: _____Labyrinth_____

ITEM No. C 101 A&B   TITLE   Hydrotreater Recycle Gas Compressor

General Data and Requirements (cont.)

Gas Composition

	Start of run	End of run
Mol %		
H_2	74.9	65.80
C_1	14.17	19.31
C_2	5.85	7.97
C_3	2.43	3.31
iC_4	1.13	1.54
nC_4	1.00	1.36
c_5s	0.52	0.71
Total	100.00	100.00

Remarks and notes:

1. Vendor to provide:

   Intercoolers: _____ Yes _____

   Aftercooler: _____ Yes _____

   Dampeners: _____ Yes _____

   Flushing and sealing oil systems: _____ Yes _____

2. If motor driven Re acceleration required class _____ A _____

3. Suction line: Size _____ 6" _____   RTG _____ 300# RF _____

   Discharge line: Size _____ 4" _____   RTG _____ 600# RF _____

Figure 5.9. (*continued*)

TITLE BLOCK

This requires the item to be identified by item number and its title. The number of units that the specification refers to is also given here. For a centrifugal compressor this will normally be just one, as very seldom is a spare machine required. This may not be so in the case of a reciprocating compressor.

NORMAL AND RATED COLUMNS

More often than not the conditions and quantities required to be handled will vary during the operation of the machine. The two columns therefore will be completed showing the average normal data in the first column and the most severe conditions and duty required by the compressor in the second. The severe conditions in column two are for a continuous length of operation, not instantaneous peaks (or troughs) that may be encountered.

GAS

The composition and gas stream identification must be included as part of the process specification. Usually the composition of the gas is listed on a separate sheet as shown in Figure 5.9. Note that in many catalytic processes that utilize a recycle gas the composition of the gas will change as the catalyst in the process ages. Thus it will be necessary to list the gas composition at the start of the run (SOR) and at the end of the run (EOR). The compressor may also be required to handle an entirely different gas stream at some time or other. This too must be noted. For example, in many petroleum-refining processes a recycle or make-up compressor normally handling a light predominantly hydrogen gas is also used for handling air or nitrogen during catalyst regeneration, purging and start-up.

VOLUME FLOW

This is the quantity of gas to be handled stated at 14.7 psia and 60 °F.

WEIGHT FLOW

This is the weight of gas to be handled in either lb/min or lb/h.

INLET CONDITIONS

In the case of multi-stage compressors the conditions for each stage must be shown. Where interstage cooling is used the effect must be reflected in the conditions specified.

- *Pressure* This is the pressure of the gas at the inlet of the compressor stage in psia. Note that if intercooling is used this pressure must include the intercooler pressure drop.
- *Temperature* This is the temperature of the gas at the inlet of the compressor stage—after the intercooler, if applicable.
- *Mole weight* The mole weight of the gas is calculated from the gas composition given as part of the specification.
- C_p/C_v This is the ratio of specific heats of the gas again obtained from the gas mole wt and Table A1.3 in Appendix 1.
- *Compressibility factor (z)* Use the value at inlet conditions calculated as shown in step 7 of Section 5.2.
- *Inlet volume* This is the actual volume of gas at the conditions of temperature and pressure existing at the compressor stage inlet. Thus:

$$\text{acfm} = \frac{\text{scfm} \times 14.7 \times (\text{inlet temp. }°F + 460)}{(60\,°F + 460) \times \text{inlet press. psia}}$$

DISCHARGE CONDITIONS

- *Pressure* This is the pressure at each stage outlet and is quoted in either psia or psig.
- *Temperature* This is estimated for each stage using Figure 5.8 in Section 5.5.
- C_p/C_v This will be the same as inlet.

APPROXIMATE DRIVER HORSEPOWER

The brake horsepower for the reciprocal compressor is calculated using the method in Section 5.5. This is brake horsepower and includes mechanical inefficiencies. The approximate minimum driver horsepower is $1.1 \times$ brake horsepower, but the approximate driver horsepower will be calculated using the inefficiencies for leakage, seals, etc. as in Section 5.4 for centrifugal compressors.

The remainder of the specification sheet contains all the essential data and requirements that may affect the duty and performance of the compressor. Much of this is self-explanatory but there are some items that require comment:

(1)   Most compressor installations today are under an open-sided shelter with a small overhead gantry crane assembly for maintenance.
(2)   Usually the lube and seal oil assemblies have their own pump and control systems. Consequently even if the compressor itself is to be steam driven there may still be a need to give details of utilities for the ancilliary equipment.
(3)   Details of the gas composition is essential for any development of the compressor. This is listed on the last page of the specification together with any notes of importance concerning the machine and its operation.

Example calculation

A hydrotreater make-up compressor is required to handle a gas stream such as to provide the unit with 260 scf per barrel of feed of pure hydrogen. The composition of the gas varies as follows:

Mole %	SOR	EOR
H_2	74.9	65.80
C_1	14.17	19.31
C_2	5.85	7.97
C_3	2.43	3.31
iC_4	1.13	1.54
nC_4	1.00	1.36
C_5s	0.52	0.71
Total	100.00	100.00

The fresh feed throughput is fixed at 30 000 bpsd (barrels per stream day). It is proposed to use $3 \times 60\%$ machines of which one will be standby and turbine driven. The inlet pressure of the gas is 50 psig at a temperature of 80 °F. The gas is to be delivered at a pressure of 600 psig and 100 °F. Prepare a process specification for reciprocating compressors to meet these requirements.

1   Calculating volume flows

$$\text{SOR conditions} \quad \text{Total flow required} = \frac{260}{0.749}$$

$$= 347 \text{ scf of gas per bbl of feed.}$$

$$= \frac{347 \times 30\,000}{24 \times 60} = 7229 \text{ scf/min}$$

$$\text{EOR conditions} \quad \text{Total flow required} = \frac{260}{0.658}$$

$$= 395 \text{ scf/bbl}$$

$$= \frac{395 \times 30\,000}{24 \times 60} = 8229 \text{ scf/min}$$

Volume flow per machine:

$$\text{SOR} = 7229 \times 0.6 = 4337$$

$$\text{EOR} = 8229 \times 0.6 = 4937$$

2   Calculating mole wt of gas

	MW	SOR Mol %	SOR Wt factor	EOR Mol %	EOR Wt factor
H_2	2	74.9	149.8	65.8	131.6
C_1	16	14.17	226.7	19.31	309.0
C_2	30	5.85	175.5	7.97	239.1
C_3	44	2.43	106.9	3.31	145.6
iC_4	58	1.13	65.5	1.54	89.3
nC_4	58	1.00	58.0	1.36	78.9
C_5s	72	0.52	37.4	0.71	51.1
Total		100.0	644.3	100.00	1044.6

SOR gas mol wt = 6.44   EOR gas mol wt = 10.44

3   Weight of gas (lb/min per machine)

One mol of any gas occupies 378 cf at 60 °F and 14.7 psia. Then

$$\text{For SOR conditions} \quad \text{moles/min of gas per machine} = \frac{4337}{378} = 11.5$$

$$\text{and lb/min} = 11.5 \times 6.44$$

$$= 74.06 \, \text{lb/min}$$

$$\text{For EOR conditions moles/min of gas per machine} = \frac{4937}{378} = 13.06$$

$$\text{and lb/min} = 13.06 \times 10.44$$

$$= 136.30 \, \text{lb/min}$$

4   Inlet conditions

$$\text{Inlet pressure} = 50 \, \text{psig} = 65 \, \text{psia}$$
$$\text{Required outlet pressure} = 600 \, \text{psig} = 615 \, \text{psia}$$
$$\text{Overall compression ratio} = \underline{615} = 9.46$$

This will be a two-stage compressor. Note: at this level of compression in reciprocating compressors the compression ratio should not exceed 4:1 for any stage.

$$\text{Compression ratio per stage} = \sqrt{9.46} = 3.07$$

$$\text{Discharge pressure stage 1} = 65 \times 3.07 = 199.6 \, \text{psia}$$

Allowing 2 psi for the pressure drop across the intercooler the suction pressure of stage 2 is 197.6 (call it 198 psia).

Check the compression ratio of stage 2:

$$\text{Required discharge pressure} = 615 \text{ psia}$$

$$\text{compression ratio} = \frac{615}{198} = 3.1$$

which is close to the originally predicted one of 3.07

5 Calculating C_p/C_v ratio

	C_p/C_v	SOR Mol %	SOR Factor	EOR Mol %	EOR Wt factor
H_2	1.4	74.9	1.049	65.8	0.921
C_1	1.3	14.17	0.184	19.31	0.251
C_2	1.22	5.85	0.071	7.97	0.097
C_3	1.14	2.43	0.028	3.31	0.038
iC_4	1.11	1.13	0.013	1.54	0.017
nC_4	1.11	1.00	0.011	1.36	0.015
C_5s	1.09	0.52	0.006	0.71	0.008
Total		100.0	1.362	100.00	1.347

$$C_p/C_v \text{ SOR gas} = 1.362$$

$$C_p/C_v \text{ EOR gas} = 1.347$$

6 Calculating inlet acfm per stage

SOR
Inlet volume for first stage:

$$\text{acfm} = \frac{\text{scf/min} \times 14.7 \times \text{inlet temp. }°R}{(60 + 460) \times \text{inlet press. psia}}$$

$$= \frac{4337 \times 14.7 \times 540}{520 \times 65}$$

$$= 1019 \text{ cf/min}$$

Inlet volume for second stage: (intercooled to 100 °F)

$$\text{acfm} = \frac{4337 \times 14.7 \times 560}{520 \times 198}$$

$$= 347 \text{ ft}^3/\text{min}$$

EOR
Inlet volume for first stage:

$$\text{acfm} = \frac{4937 \times 14.7 \times 540}{520 \times 65}$$

$$= 1159 \text{ ft}^3/\text{min}$$

Inlet volume for second stage:

$$acfm = \frac{4937 \times 14.7 \times 560}{520 \times 198}$$

$$= 488.5 \, ft^3/min$$

7 Calculating inlet compressibility factor (Z)

$$Z = \frac{MW \times P_i}{T_i \times \rho} \times 10.73$$

where

$$\rho = \frac{wt}{min/acfm}$$

SOR conditions

$$\text{First stage } Z = \frac{6.44 \times 65}{540 \times 0.0727 \times 10.73}$$

$$= 0.994$$

$$\text{Second stage } Z = \frac{6.44 \times 198}{560 \times 0.213 \times 10.73}$$

$$= 0.991$$

EOR conditions

$$\text{First stage } Z = \frac{10.44 \times 65}{540 \times 0.1176 \times 10.73}$$

$$= 0.996$$

$$\text{Second stage } Z = \frac{10.44 \times 198}{560 \times 0.345 \times 10.73}$$

$$= 0.997$$

8 Determining discharge temperature

From Figure 5.8

For SOR conditions: First stage   Comp. ratio = 3.07
$C_p/C_v = 1.362$
Suct. temp. = 80 °F
Dis. temp. read as 270 °F

Second stage   Comp. ratio = 3.01
$C_p/C_v = 1.362$
Suct. temp. = 100 °F
Dis. temp. read as 299 °F

For EOR conditions: First stage   Comp. ratio = 3.07
$$C_p/C_v = 1.347$$
Suct. temp. = 80 °F
Dis. temp. read as 262 °F

Second stage   Comp. ratio = 3.1
$$C_p/C_v = 1.347$$
Suct. temp. = 100 °F
Dis. temp. read as 292 °F

9   Compressibility factors for discharge conditions

SOR conditions:

First stage   acfm on discharge (before intercooler)

$$= \frac{4337 \times 14.7 \times 730}{520 \times 200 \text{ (neglect IC pressure drop)}}$$

$$= 447.5 \, \text{ft}^3/\text{min}$$

$$\rho = \frac{74.1}{447.5} = 0.166$$

$$Z = \frac{6.44 \times 200}{730 \times 0.166 \times 10.73}$$

$$= 0.991$$

Second stage   acfm on discharge

$$= \frac{4337 \times 14.7 \times 759}{520 \times 615}$$

$$= 151.3 \, \text{cf/min}$$

$$\rho = \frac{74.1}{151.3} = 0.490$$

$$Z = \frac{6.44 \times 615}{759 \times 0.49 \times 10.73}$$

$$= 0.992$$

EOR conditions:
These are calculated in the same way as those above and give the following results:

$$\text{First stage } Z = 0.995$$

$$\text{Second stage } Z = 0.995$$

10   Approximate driver horsepower

Use the expression bhp = 22 × (Comp. ratio/stage) × No. of stages × capacity × factor F (see also Section 5.5)

$$\text{Comp. ratio/stage} = 3.07$$
$$\text{No. of stages} = 2$$

Capacitypermachineinmmcf/dayatsuctiontemperature

$$\text{Factor for two-stage machine} = 1.08$$

For SOR conditions:

$$\text{bhp} = 22 \times 3.07 \times 2 \times (0.6 \times 10.81) \times 1.08$$
$$= 946$$

For EOR conditions:

$$\text{bhp} = 22 \times 3.07 \times 2 \times 7.38 \times 1.08$$
$$= 1077$$

Use the efficiency factors as given in Section 5.4. In this case there will be a gear assembly between compressor and driver. Use the efficiency of this as 97%. Thus:

	SOR	EOR
bhp	946	1077
Gear losses (5% of bhp)	29	33
leakage	10	10
Seal	35	35
Bearings	35	35
Driver hp	1055	1190

5.8   Compressor Drivers, Utilities and Ancillary Equipment

This section covers details on various compressor drivers, the utilities associated with operating the compressors and their ancillary equipment.

COMPRESSOR DRIVERS

Table 5.3 gives a listing of the more common types of compressor drivers. It provides some of the data that would influence the choice of the driver. The most common drivers by far in a process plant are the electric motor and the steam turbine. For very large machines as encountered in handling natural gas the gas turbine or gas engine becomes the more prominent prime mover.

Sizing drivers

As a basic rule, drivers are sized for the most severe duty required of the compressor plus a factor as an operating contingency. In general, the most severe

Table 5.3. Comparison of compressor drivers

Driver	HP range	Available speed (rpm)	Efficiency (%)	Common applications
Synchronous motor	100–20 000	3600	90–97	Reciprocating compressors
Induction motor	1–15 000	3600	86–94	All types of compressors
Wound rotor induction motor	–	–	–	Normally not used
Steam engine	10–4000	400–140	60–80	All types of rotary equipment
Steam turbine	10–2000	2000–15 000	50–76	Centrifugal, axial and reciprocating
Combustion gas turbine	3000–35 000	3600–10 000	19–24*	All types of compressors except reciprocating
Gas and oil engines	100–5000	300–1000	35–45	Reciprocating compressors

*The efficiency given here does not include for waste heat recovery. With WHR the efficiency can be increased to between 28% and 35%.

duty is that design case which has the highest suction temperature, the maximum ratio of specific heats, the lowest suction pressure, the highest required discharge pressure, and the gas molecular weight which gives the highest horsepower. The driver rated horsepower shall therefore be greater than:

$$\text{Driver brake hp} = \frac{\text{Max. compressor bhp at the most severe duty}}{\text{Mech. efficiency of the power transmission}}$$

The mech efficiency in this case includes energy losses for bearings, seals, lube oil, etc. in the case of centrifugal compressors and gears in the case of reciprocating compressors.

Electric motor drivers

Squirrel cage motors are preferred for this type of duty. These may be drip-proof open type where the location is not a fire or explosion hazard. Where it is required that the units must be explosion- or fireproof these motors must be totally enclosed. In sizing the motor efficiencies for squirrel cage motors up to 450 hp given in Section 4.6 for pumps may be used. Table 5.4 is used for motors above 500 hp.

The driver-rated brake horsepower is the compressor horsepower times a load factor divided by a service factor. Normally the load factor is 10% and a service factor for an enclosed squirrel cage motor is 1.0 and 1.15 for an open type.

Example calculation

Calculate the operating load and the connected load for the driver of a 4000 hp centrifugal compressor (includes leakage, seal, and bearing losses). A gear is used and this has a 97% efficiency. The load factor is 10% and the motor is open-type

Table 5.4. Motor efficiencies

Motor-rated hp	Motor efficiences at percentage of full load		
	50	75	100
500	91.4	93.1	93.4
1000	92.1	93.8	94.1
1500	92.4	94.1	94.4
2000	92.7	94.4	94.7
2500	92.9	94.6	94.9
3000	93.0	94.7	95.0
3500	93.0	94.7	95.0
4000	93.1	94.8	95.1
4500	93.1	94.8	95.1
5000	93.2	94.9	95.2

squirrel cage with a service factor of 1.15. There will be a normal operating unit and a spare, both motor driven.

$$\text{Minimum required driver hp} = \frac{4000 \times 1.1}{0.97}$$

$$= 4536$$

$$\text{Driver nameplate rating} = \frac{4536}{1.15} = 3944$$

$$\text{call it 4000 hp}$$

Connected load for the motor is: Motor nameplate rating $\times$ 1.15

$$= 4000 \times 1.15$$

$$= 4600 \text{ hp (rated hp)}$$

$$= \frac{4600 \times 0.746}{0.951} \text{ (at 100\% load)}$$

$$= 3608 \text{ kW}$$

There are two units: then total connected load

$$= 3608 \times 2$$

$$= 7216 \text{ kW}$$

Operating load for the motor is: $\dfrac{4000}{0.97} = 4124$ hp

$$\% \text{ load} = \frac{4142}{4600} = 90\%$$

$$\text{Operating load} = \frac{4124 \times 0.746}{0.948}$$

$$= 3245 \text{ kW}$$

Steam turbine drivers

Next to the motor drivers steam turbines are the most common form of drivers for rotary equipment in general and compressors in particular. The two most common types of these are turbines that exhaust to a lower pressure but the exhaust steam is not condensed and those in which the exhaust steam is condensed. Normally the latter are only used in the case of large driver horsepower (5000 and above). It is far more expensive than the non-condensing type as the exhaust is normally sub-atmospheric in pressure and, of course, the cost of the condenser must be included.

Steam turbine approximate efficiencies are listed in Table 5.5. These efficiencies are based on the exhaust pressure of the steam being 50 psig for the non-condensing type and 2-in. hg abs for the condensing type.

Determining the rated horsepower of the steam turbine driver follows closely the method for motor horsepower. First determine the minimum horsepower required of the turbine. Thus:

- *Step 1* Determine the *minimum* driver horsepower by multiplying the compressor bhp by 1.1.
- *Step 2* Now the turbine will deliver the normal hp at the normal speed. A contingency in the form of additional speed is added to the driver capability. This will be controlled in practice by a steam governor. This contingency is usually 5% above normal speed.
- *Step 3* Horsepower capability varies as the cube of the speed. Thus the rated horsepower of the turbine will be

$$\text{Rated hp} = \text{minimum hp} \times (105)^3$$

- *Step 4* The amount of steam that will be used is calculated by the change in enthalpy of the inlet steam to the outlet steam at constant entropy. The change in enthalpy for the two conditions is read from the steam Mollier diagram.
- *Step 5* The theoretical steam rate is

$$\frac{2544}{\text{Inlet enthalpy} - \text{Outlet enthalpy (in Btu/lb)}}$$

This figure divided by the turbine efficiency gives the steam rate in lb/bhp/h.

Table 5.5. Steam turbine efficiencies

Driver bhp	Adiabatic efficiencies (%) Inlet pressures (psig)	
	900	100
500	48	59
800	53	64
1000	56	67
1200	58	68
1500	60	71
2000	63	73
2500	65	74
3000 and up	67	76

Example calculation

Calculate the turbine horsepower requirements and the theoretical steam rates to drive a 4000 bhp centrifugal compressor. No gears are included in this case. Steam is available at 650 psig and 760 °F. The steam will exhaust into the plant's 125 psig header.

$$\text{Minimum driver horsepower} = 4000 \times 1.1$$
$$= 4400 \text{ bhp}$$
$$\text{Rated turbine hp at 105 \% speed} = 4400 \times (1.05)^3$$
$$= 5094 \text{ hp}$$
$$\text{Enthalpy of steam at 650 psig and 760 °F} = 1390 \text{ (entropy 1.62)}$$
$$\text{Enthalpy of steam at 125 psig} = 1225 \text{ (entropy 1.62)}$$
$$\text{Difference in enthalpy} = 165$$
$$\text{Efficiency of turbine (from Table 5.5)} = 67\%$$
$$\text{Theoretical steam rate} = \frac{2544}{165 \times 0.67}$$
$$= 23 \text{ lb/bhp/h}$$

Gas turbine drivers

These items of equipment are the most expensive and because they require a high capital investment their use can only be justified as compressor drivers where the continual load on the compressor is also very high. They therefore are met mostly in the natural gas industry and are used extensively in recompressing natural gas after treating for dew point control or desulphurizing.

The thermal efficiencies of gas turbines are low (about 16–20%) but it is common practice to use the exhaust gases which are usually at a temperature of above 800 °F in waste heat recovery. This involves exchanging the waste heat of the exhaust gases with boiler feed water to generate steam or to preheat a process stream (for example, distillation). Table 5.6 gives some turbine sizes and data. It should be noted that considerable development work is continuing in the field of gas turbines and consequently the data given here may be subject to revision or updating.

To obtain the gas turbine rated horsepower for a specific compressor duty follows closely the same calculation route as the steam turbine. Thus:

- *Step 1* Obtain the *minimum driver horsepower* by multiplying the compressor brake horsepower by 1.05 (bhp includes seals, leakage, etc.).
- *Step 2* The rated turbine horsepower is the minimum driver hp divided by the gear efficiency.
- *Step 3* The horsepower of the turbine selected must equal or slightly exceed

Table 5.6. Gas turbine sizes and data

HP rating at 80 and 1000 ft	Fuel consumption LHV lb/hp-h	Exhaust Flow #/s	Exhaust Temp °F	rpm
430	1.25	10.3	950	19 250
1 000	0.66	11.1	960	19 500
1 080	0.63	13.7	860	22 300
1 615	0.84	23.6	1000	13 000
2 500	0.76	43.0	795	9 000
3 800	0.75	53.0	900	8 500
5 500	0.76	77.3	945	5 800
7 000	0.8	101.0	938	5 500
8 000	0.65	102.0	935	5 800
9 000	0.70	130.0	850	5 000
10 000	0.59	123.5	805	6 000
12 000	0.65	160.0	720	4 750
13 500	0.62	187.0	800	4 860
15 000	0.61	188.0	835	4 860
24 000	0.61	258.0	850	3 600

the horsepower calculated in step 2. This hp must be corrected for site conditions as shown in step 4.

- *Step 4*   The horsepowers given in Table 5.6 are at an ambient temperature of 80 °F and at an elevation of 1000 ft. Correction for any specific site is given by the following expression:

$$\text{Site hp} = \text{quoted hp } (1.00 + A \times 10^{-2})(1.00 - B \times 10^{-2})$$

$$(1.00 - C \times 10^{-2}) \times \frac{\text{Site atmos. press.}}{14.7}$$

where

A = temperature adjustment of % per °F
B = inlet pressure loss % per inch water gauge
C = discharge pressure loss % per inch water gauge
    A will be plus for ambient temperatures above 80 °F and minus for ambient temperatures below 80 °F.

ANCILLARY EQUIPMENT

Reciprocating compressor dampening facilities

Dampening facilities are used in conjunction with reciprocating compressors to smooth out the pulsation effect of the compressor action. These facilities are simply in-line bottles sized larger than the gas line which cushion the gas motion. These are essential to minimize expensive piping designs that would be necessary without them. Calculating the size of these bottles is important in the design of the compressor facilities.

The following calculation technique is used to determine the size of new dampeners or to evaluate the adequacy of an existing facility. This calculation is described by the following steps:

- *Step 1* From a compressor data sheet obtain cylinder diameter and stroke dimensions.
- *Step 2* Calculate the swept volume per cylinder using the expression:

$$\frac{\pi}{4} D^2 \times S$$

where

D = cylinder diameter
S = stroke

- *Step 3* Knowing the suction and discharge pressures the pulsation bottle capacity (both suction and discharge) is obtained from Figure 5.10 in terms of a multiple of swept volume.
- *Step 4* Use the rule of thumb that pulsation bottle diameter equals one and a half times the compressor cylinder diameter. Calculate the suction and discharge bottle length.

Example calculation

Determine the dimensions of the compressor pulsation bottle of a reciprocating compressor having a 6-inch diameter cylinder and a stroke of 15 in. The compressor delivers 3.0 mm scf/d gas at a suction pressure of 100 psia and 100 °F and a discharge pressure of 1200 psia. The cylinder diameter is 6 in. and stroke is 15 in.

$$\text{Then swept volume} = \pi/4 \, 6^2 \times 15$$
$$= 424 \, \text{in}^3$$
$$\text{Capacity of machine} = 3 \, \text{mm scf/D}$$
$$\text{in a mmcf/D} = \frac{3 \times 14.7 \times 560}{520 \times 100}$$
$$= 0.475 \, \text{mmacfd}$$
$$= 330 \, \text{acf/min}$$
$$= 570\,240 \, \text{ac in/min}$$
$$\text{Machine speed} = \frac{570\,240}{424} = 1345 \, \text{rpm}$$

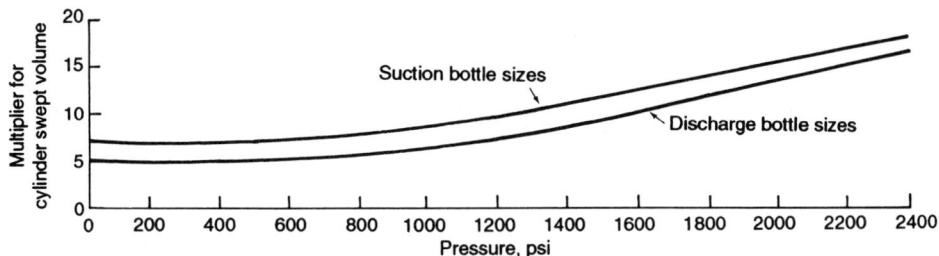

Figure 5.10. Dampener bottle capacity sizing

From Figure 5.10

Suction bottle size should be 7 × swept volume (at 100 psia)
discharge bottle size should be 7 × swept volume (at 1200 psia)

$$= 2968 \text{ in}^3 \text{ or } 1.718 \text{ ft}^3$$

As a rule of thumb diameter of bottle should be $1\frac{1}{2}$ × cylinder diameter = 0.75 ft (9 in.) length = 2968/63.6 = 47 in. or 4 ft.

6   HEAT EXCHANGERS

6.1   Types and Selection of Heat Exchangers

Heat exchange is the science that deals with the rate of heat transfer between hot and cold bodies. There are three methods of heat transfer:

- Conduction
- Convection
- Radiation

In a heat exchanger heat is transferred by conduction and convection, with conduction usually being the limiting factor. The equipment used in heat exchanger service is designed specifically for the duty required of it. That is, heat exchange equipment cannot be purchased as a stock item for a service but has to be designed for that service.

    The types of heat exchange equipment used in the process industry and their selection for use are as follows:

(1)   *The shell and tube exchanger*   This is the type of exchanger most commonly used in a process plant. It consists of a bundle of tubes encased in a shell. It is inexpensive and is easy to clean and maintain. There are several types of shell and tube exchangers and some of these have removable bundles for easier cleaning. The shell and tube exchanger has a wide variety of normal service. This includes vapour condensation (condensers), process liquid cooling (coolers), exchange of heat between two process streams (heat exchangers), and reboilers (boiling in fractionator service). Most of this chapter will be dedicated to the uses and design specification of the shell and tube exchanger.

(2)   *The double-pipe exchanger*   A double-pipe exchanger consists of a pipe within a pipe. One of the fluid streams flows through the inner pipe while

the other flows through the annular space between the pipes. The exchanger can be dismantled very easily and is therefore easily cleaned. The double-pipe exchanger is used for very small process units or where the fluids are extremely fouling. Either true concurrent or countercurrent flows can be obtained, but because the cost per square foot is relatively high it can only be justified for special applications. The following table gives the heat transfer area for various pipe lengths and diameters:

No. of tubes	Shell size (in.)	Tube size (in.)	Surface area (ft^2)		
			10 ft	20 ft	30 ft
1	2	1	5.8	11.0	16.3
1	3	1.5	10.9	20.9	30.9
1	4	2	13.7	26.1	38.5

(3) *Extended surface or fin tubes* This type of exchanger is similar to the double-pipe but the inner pipe is grooved or has longitudinal fins on its outside surface. Its most common use is in the service where one of the fluids has a high resistance to heat transfer and the other fluid has a low resistance. It can rarely be justified if the equivalent surface area of a shell and tube exchanger is greater than 200–300 ft^2.

(4) *Finned air coolers* These are the more common type of air coolers used in the process industry. Air cooling for process streams gained prominence during the early 1950s. In a great many applications and geographic areas they had considerable economic advantage over the conventional water cooling. Indeed, today it is uncommon to see process plants of any reasonable size without air coolers.

Air coolers consist of a fan and one or more heat transfer sections mounted on a frame. In most cases these sections consist of finned tubes through which the hot fluid passes. The fan located either above or below the tube section induces or forces air around the tubes of the section.

The selection of air coolers over shell and tube is one of cost. Usually, air coolers find favour in condensing fractionator overheads to temperatures of about 90–100 °F and process liquid product streams to storage temperatures. Air coolers are widely used in most areas of the world where ambient air temperatures are most times below 90 °F. At atmospheric temperatures above 100 °F humidifiers are incorporated into the cooler design and operation. The cost under these circumstances is greatly increased and their use is often than not justified.

In very cold climates the air temperature around the tubes is controlled to prevent the skin temperature of the fluid being cooled falling below a freezing criterion or, in the case of petroleum products, its pour point. This control is achieved by louvres installed to recirculate the air flow or by varying the quantity of air flow by changing the fan pitch.

(5) *Box coolers* These are the simplest form of heat exchange. However, they are generally less efficient, more costly and require a large area of the plant

plot. They consist of a single coil or 'worm' submerged in a bath of cold water. The fluid flows through the coil to be cooled by the water surrounding it. The box cooler found use in the older petroleum refineries for cooling heavy residuum to storage temperatures. Modern practice is to use a tempered water system where the heavy oil is cooled on the shell side of a shell and tube exchanger against water at a controlled temperature flowing in the tube side. The water is recycled through an air cooler to control its temperature to a level which will not cause the skin temperature of the oil in the shell and tube exchanger to fall below its pour point.

(6) *Direct contact condensers*   In this exchanger the process vapour to be condensed comes into direct contact with the cooling medium (usually water). This contact is made in a packed section of a small tower. The most common use for this type of condenser is in vacuum-producing equipment. Here the vapour and motive steam for each ejector stage is condensed in a packed direct contact condenser. This type has a low-pressure drop which is essential for the vacuum-producing process.

6.2   General Design Considerations

BASIC HEAT TRANSFER EQUATIONS

The following equations define the basic heat transfer relationships. These equations are used to determine the overall surface area required for the transfer of heat from a hot source to a cold source.

- *The overall heat transfer equation*   The usual heat transfer mechanism are conduction, natural convection, forced convection, condensation, and vaporization. When heat is transferred by these means the overall equation is as follows:

$$Q = UA(\Delta t_{\mathrm{m}})$$

where

Q = heat transferred in BTU/h
U = overall heat transfer coefficient (BTU/h/ft^2/°F)
A = heat transfer surface area (ft^2)
Δt_{m} = corrected log mean temperature difference (°F)

- *The overall heat transfer coefficient U*   This coefficient is the summation of all the resistance to the flow of heat in the transfer mechanism. These resistances are the resistance to heat transfer contained in the fluids, the resistance caused by fouling, and the resistance to heat transfer of the tube wall. The resistance to the flow of heat from the liquid outside the tube wall is measured by the film coefficient of that fluid. The resistance of the flow of heat from the fluid inside the tube is similarly the film coefficient of the inside fluid. These film coefficients are products of dimensionless numbers which include:

- The Reynolds number
- The Graetz number

- The Grashof number
- The Nusselt number
- The Peclet number
- The Prandtl number
- The Stanton number

The format of these numbers and their use are found in all standard textbooks on heat transfer (for example, Kern, *Process Heat Transfer*, and McAdams, *Heat Transmission*). These resistances are defined therefore by the following expression:

$$\frac{1}{U_o} = \frac{1}{h_0} + \frac{1}{h_i} \times \frac{A_o}{A_i} + \frac{1}{h_w} + (r_f)_o + (r_f)_i \times \frac{A_o}{A_i}$$

where

U_o = overall heat transfer coefficient based on outside tube surface (BTU/h/ft^2/°F)
h = The film coefficient (BTU/h/ft^2/°F)
r_f = fouling factors (1/BTU/h/ft^2/°F)
h_w = heat transfer rate through tube wall (BTU/h/ft^2/°F)
A = surface area (ft^2)

Subscripts o and i refer to outside surface and inside surface respectively.

FLOW ARRANGEMENTS

The two more common flow paths are concurrent and countercurrent. In concurrent flow both the hot fluid and the cold fluid flow in the same direction. This is the least desirable of the flow arrangements and is only used in those chemical processes where there is a danger of the cooling fluid congealing, subliming, or crystallizing at near-ambient temperatures.

Countercurrent flow is the most desirable arrangement. Here the hot fluid enters at one end of the exchanger and the cold fluid enters at the opposite end. The streams flow in directions opposite to one another. This arrangement allows the two streams, exit temperatures to approach one another.

LOGARITHMIC MEAN TEMPERATURE DIFFERENCE Δt_m

In either a countercurrent or a concurrent flow arrangement the log mean temperature difference used in the overall heat transfer equation is determined by the following expression:

$$\Delta t_m = \frac{\Delta t_1 - \Delta t_2}{\ln(\Delta t_1/\Delta t_2)}$$

The Δt's are the temperature differences at each end of the exchanger and Δt_1 is the larger of the two. In true countercurrent flow the Δt_m calculated can be used directly in the overall heat transfer equation. However, such a situation is not common and true countercurrent flow rarely exists. Therefore a correction factor needs to be applied to arrive at the correct Δt_m. These are given in Figure A1.6 in Appendix 1.

The equation given above for Δt_m is resolved to the following:

$$\Delta t_m = \frac{\Delta t_1 - \Delta t_2}{\log e(\Delta t_1/\Delta t_2)}$$

FLUID VELOCITIES AND PRESSURE DROPS

Film coefficients are a function of fluid velocity, density (vapour), and viscosity (liquids). Within limits, increasing the velocity of a fluid reduces its resistance to heat transfer (i.e. it increases its heat transfer coefficient). Increasing the fluid velocity, however, increases its pressure drop. An economic balance needs to be sought therefore between the cost of heat transfer surface and pumping cost. This exercise should be undertaken to find a payout balance of 2 to 4 years. This exercise has been done many times and the following data are considered a reasonable balance between velocity and pressure drop for some common cases:

	Tube side		Shell side	
	Velocity (ft/s)	Pressure drop (psi)	Velocity (ft/s)	Presure drop (psi)
Non-viscous liquids	6–8	10	1.5–2.5	10
Viscous liquids	6	20	3.0 max	15–20
Clean cooling water	6–8	10–15	–	–
Dirty cooling water	3 min	10+	–	–
Suspended solids in liquids	2–3 min	10*	1.5 min	15
Gases and vapours	$\frac{100}{\sqrt{Gas\,density}}$ Max	3–5	–	3
Condensing vapours	–	–	–	3–5

* Normally erosion by suspended solids in liquids occurs at velocities of above 6 ft/s.

For condensing steam pressure drop is usually not critical but a minimum steam pressure drop is desirable. Allowable steam velocities in tubes are as follows:

Pressure	Velocity (ft/s)
Below atmospheric	225
Atmos. to 100 psig	175
Above 100 psig	150

CHOICE OF TUBE SIDE VERSUS SHELL SIDE

There are no hard-and-fast rules governing which fluid flows on which side in a heat exchanger. Much is left to the discretion of individual engineers and their experience. There are some guidelines and these are as follows.

Tube side flow

- *Fouling liquids*   Tube cleaning is much easier than cleaning the outside of the tubes. Also, fouling can be reduced by higher tube side velocities.
- *Corrosive fluids*   It is cheaper to replace tubes than shells and shell baffles so, as a general rule, corrosive fluids are put tube side. There are exceptions and a major one are those corrosive fluids that become more corrosive at high velocities. An example of this are naphthenic acids which are present in some crude oils and their products.
- *High pressure*   Fluids at high pressure are usually put on the tube side as only the tubes, tube sheet, and channel need to be rated for high pressure in the units design. This reduces the overall cost of the exchanger.
- *Suspended solids*   Fluids containing suspended solids should, whenever possible, be made to flow tube side. Shell side flows invariably have 'dead spaces' where solids come out of suspension and build up to cause fouling.
- *Cold boxes*   These are exchangers used in cryogenic processes where condensing of a vapour on one side of the exchanger is accompanied by boiling of a liquid on the other side. The condensing fluid is preferred on the tube side. Better control of the refrigerant flow is accomplished by the level control across the shell side.

Shell side flow

- *Available pressure drop*   Shell side flows generally require a lower pressure drop than tube side. Therefore if a system is pressure drop limiting it should be routed shell side.
- *Condensers*   Condensing vapours should flow shell side wherever possible. The larger free area provided by shell side space permits minimum pressure drop and higher condensate loadings through better film heat transfer coefficients.
- *Large flow rates*   In cases where both streams are of a similar nature with similar properties the stream with the largest flow rate should be sent shell side where the differences in flow rates are significant. The shell side provides more flexibility in design by baffle arrangements to give the best heat transfer design criteria.
- *Boiling service*   The boiling liquid, as in the case of reboilers, waste heat recovery units, etc., should be on the shell side of the exchanger. This allows space for the proper disengaging of the vapour phase and provides a means of controlling the system by level control of the liquid phase.

TYPES OF SHELL AND TUBE EXCHANGERS

Figure 6.1 shows some of the more common arrangements in shell and tube exchangers design. The arrangements here are all one shell pass and one or two tube passes. Equipment with more than two tube passes (up to five) are also fairly common, particularly in petroleum refining. Shell arrangements are, however, left at one if at all possible. Where multi-pass shell side is required companies prefer to use complete exchangers in series or in parallel or both rather than making two or more shell passes using horizontal baffling in one exchanger.

6.3   Estimating Shell and Tube Surface Area and Pressure Drop

There are many excellent computer programs available that calculate exchanger surface area and pressure drops from simple input. The actual calculation when done manually is tedious and long. However, to understand a little of the importance of the input required by these programs it does well to at least view a typical manual calculation. The one given here is for a shell and tube cooler with no change of phase for either tube side or shell side fluids. The calculation follows these steps:

- *Step 1   Establish the following data by heat balances or from observed plant readings*
   —The inlet and outlet temperatures on the shell side and on the tube side.
   —The flow of tube side fluid and that for the shell side. It may be necessary to calculate one or the other from a heat balance over the exchanger.
   —Calculate the duty of the exchanger in heat units per unit time (usually in hours).
   —Establish the stream properties for tube side and shell side fluids. The properties required are SG, viscosity, specific heats, thermal conductivity.
- *Step 2   Calculate the log mean temperature difference* (ΔT_m)   Assume a flow pattern (i.e. either concurrent or countercurrent). Most flows will be a form of countercurrent. Then show the temperature flow as follows:

$$\begin{array}{c} \text{Shell in} \rightarrow \text{Shell out} \\ \text{Tube out} \leftarrow \text{Tube in} \\ \hline \text{Temp. diff. } \Delta T_1 \qquad \Delta T_2 \\ \hline \end{array}$$

The log mean temperature difference is then calculated using the expression

$$\Delta t_m = \frac{\Delta T_1 - \Delta T_2}{\log e \left(\Delta T_1 / \Delta T_2 \right)}$$

This temperature needs to be corrected for the flow pattern, and this is done using the correction factors given in Figure A1.6 in Appendix 1. The use of these are self-explanatory and are given in the figure.

- *Step 3   Calculate the approximate surface area*   From Table A1.1 in Appendix 1 select a suitable overall heat transfer coefficient U in BTU/h.ft^2.°F. Use the expression to calculate for A:

$$Q = UA\Delta T_m$$

where

Q = heat transfered in BTU/h (the exchanger duty)
U = overall heat transfer coefficient
A = exchanger surface area (ft^2)
ΔT_m = log mean temperature difference (corrected for flow pattern) (°F)

From the surface area calculated select the tube size and pitch, usually $\frac{3}{4}$ in. on a triangular pitch for clean service and 1 in. on a square pitch for dirty or fouling service. A single standard shell will hold about 4100 ft^2 of surface per pass. Now most companies do not use multi-pass shells and prefer sets of shells in series if this becomes necessary. The 'norm' therefore are single-pass shells each containing up to 4100 ft^2 of surface.

- *Step 4   Calculate the tube side flow and the number of passes*   If it cannot be read from plant data calculate tube side flow in ft^3/h by heat balance. Select the tube gauge and length. The tube data are given in Table A1.7 in Appendix 1 and standard lengths of tubes are 16 and 20 ft. Calculate the number of selected tubes per pass from the expression

$$N_p = \frac{\text{ft} \times 144}{3600 \times A_t \times V_t}$$

where

N_p = number of tubes per pass
F_t = tube side flow (ft^3/h)
A_t = cross-sectional area of one tube
V_t = linear velocity in tube (ft/s)

See Section 6.2 for recommended fluid velocities. The number of tube passes is arrived at by dividing the total surface area required by the total (external) surface area of the number of tubes per pass calculated above.

- *Step 5   Calculate tube side film coefficient corrected to outside diameter* (h_{io})   The tube side film coefficient may be calculated for water by the expression

$$h_{io} = \frac{300 \times (V_t \times \text{tube i/d (in.)})^{0.8}}{\text{tube o/d (in.)}}$$

where

h_{io} = inside film coefficient based on outside tube diameter (BTU/h.ft^2.°F)
V_t = linear velocity of water tube side (ft/s).

For fluids other than water flowing tube side use the expression[11]

$$h_{io} = \frac{K}{D_o}(C\mu/K)^{1/6}(\mu/\mu_w)^{0.14} \cdot \Phi(DG/\mu)$$

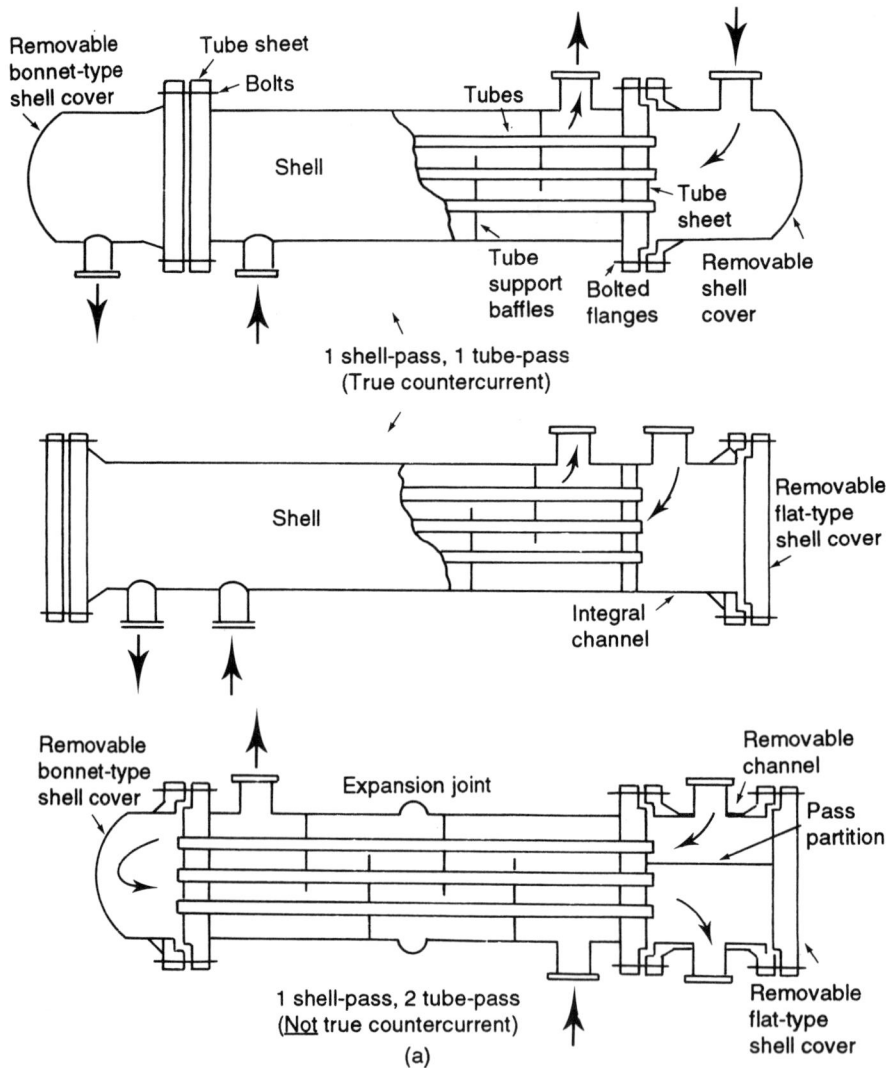

Figure 6.1. Some common types of shell and tube exchangers. (a) Fixed tube sheet; (b) removable tube bundle

where

h_{io} = inside film coefficient based on outside diameter (BTU/h.ft^2.°F)

K = thermal conductivity of the fluid in BTU/h.ft^2 (°F per ft). See Maxwell's *Data Book on Hydrocarbons* or Perry's *Chemical Engineers' Handbook*

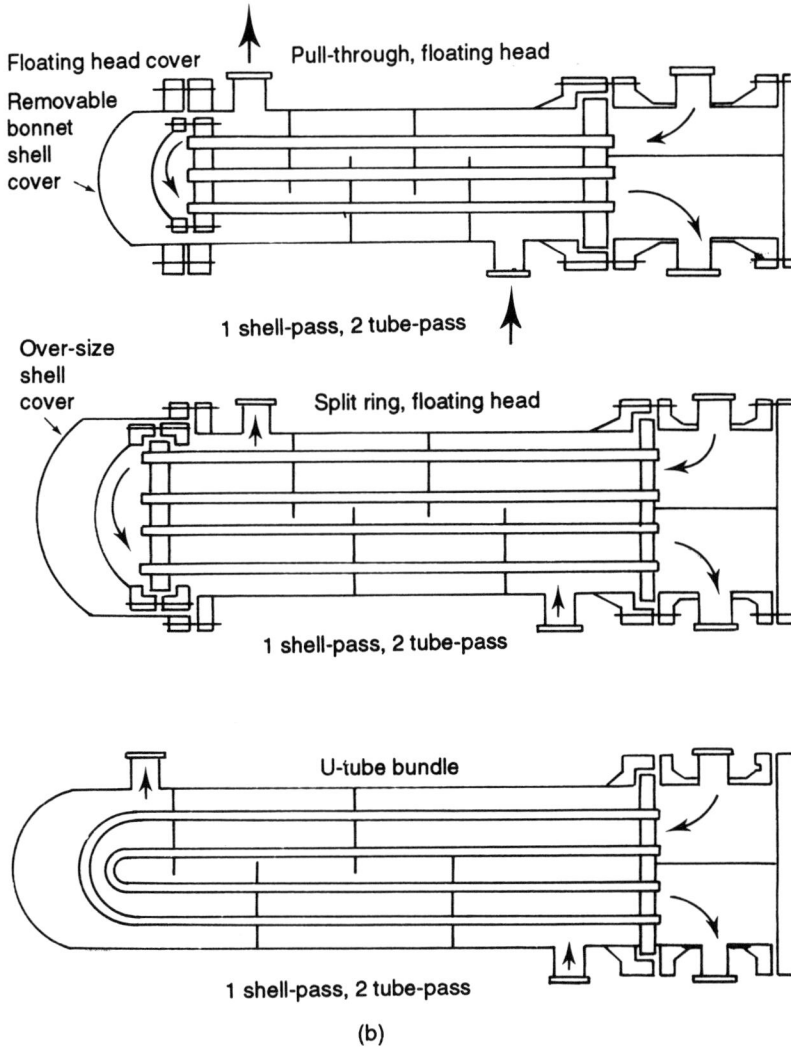

(b)

D = inside tube diameter (in.)

D_o = outside tube diameter (in.)

C = specific heat (BTU/lb/°F)

G = mass velocity (lb/s.ft²)

μ = absolute viscosity (eP) at average fluid temperature

μ_w = absolute viscosity (eP) at average tube wall temperature

$\Phi(DG/\mu)$ = from Figure 6.2

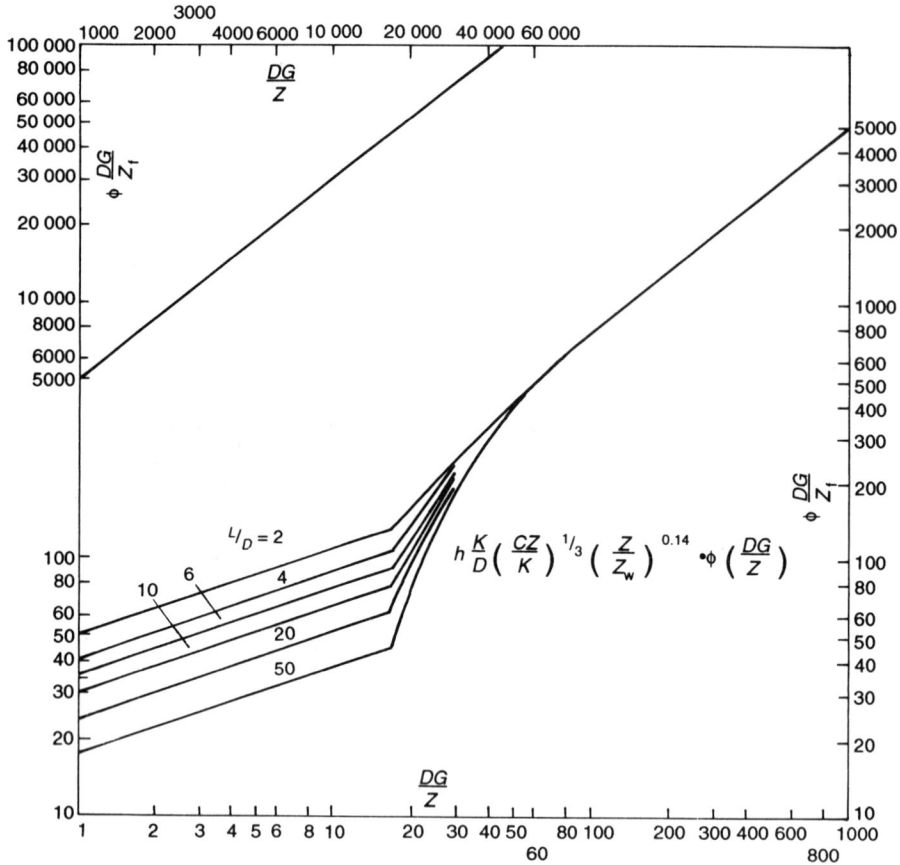

Figure 6.2. Heat transfer inside tubes (Reproduced by permission of Kreiger Publishing Company, Malabar, Florida, 1950, Maxwell, Data Book on Hydrocarbons)

- *Step 6   Calculate shell side dimensions*   First determine the shell side average film temperature as follows:

$$\text{Inlet ave.} = \frac{T_1 + T_2}{2} \quad \text{Outlet ave.} = \frac{T_3 + T_4}{2}$$

where

T_1 = shell fluid inlet temperature
T_2 = tube outlet temperature
T_3 = shell outlet temperature
T_4 = tube inlet temperature

Average shell side film temperature:

$$\frac{\text{Inlet ave. + outlet ave.}}{2}$$

Use this temperature to determine density and viscosity used in the shell side film coefficient calculations.

The shell diameter

Next calculate the diameter of the tube bundle and the shell diameter. For this use one of the following equations to calculate the number of tubes across the centreline of the bundle:

(1) For a square pitch tube arrangement:

$$T_{cl} = 1.19 \text{ (number of tubes)}^{0.5}$$

(2) For a triangular pitch tube arrangement:

$$T_{cl} = 1.10 \text{ (number of tubes)}^{0.5}$$

Note these are *total* number of tubes calculated in step 4 times number of tube passes. The minimum theoretical shell diameter is $T_{cl}/0.9$. The minimum actual shell diameter shall be at least an addition of 3 to the theoretical minimum.

Set number of baffles and their pitch

The type of baffles usually used are shown in Figure 6.3 disk and donut type baffles are used only where pressure drop available is very small and there is a pressure drop problem. Baffles on the bias are used in a square pitch tube arrangement and baffles perpendicular to the tubes are usual for triangular tube arrangements. The minimum baffle pitch should not be less than 16% of the shell diameter. Pitch in this case is the space between two adjacent baffles. Normally 20% of shell i/d is used for the baffle pitch. The number of baffles are calculated from

$$NB = \frac{10 \times \text{tube length}}{\text{baffle pitch \% } \times \text{ diam. of shell}}$$

Free area of flow between baffles

The space available for flow on the shell side is calculated as

$$W = D_i - (d_o \times T_{cl})$$

where

   W = space available for flow (in.^{2})
   D_i = shell inside diameter (in.)
   d_o = tube outside diameter (in.)
   T_{cl} = number of tubes across centreline

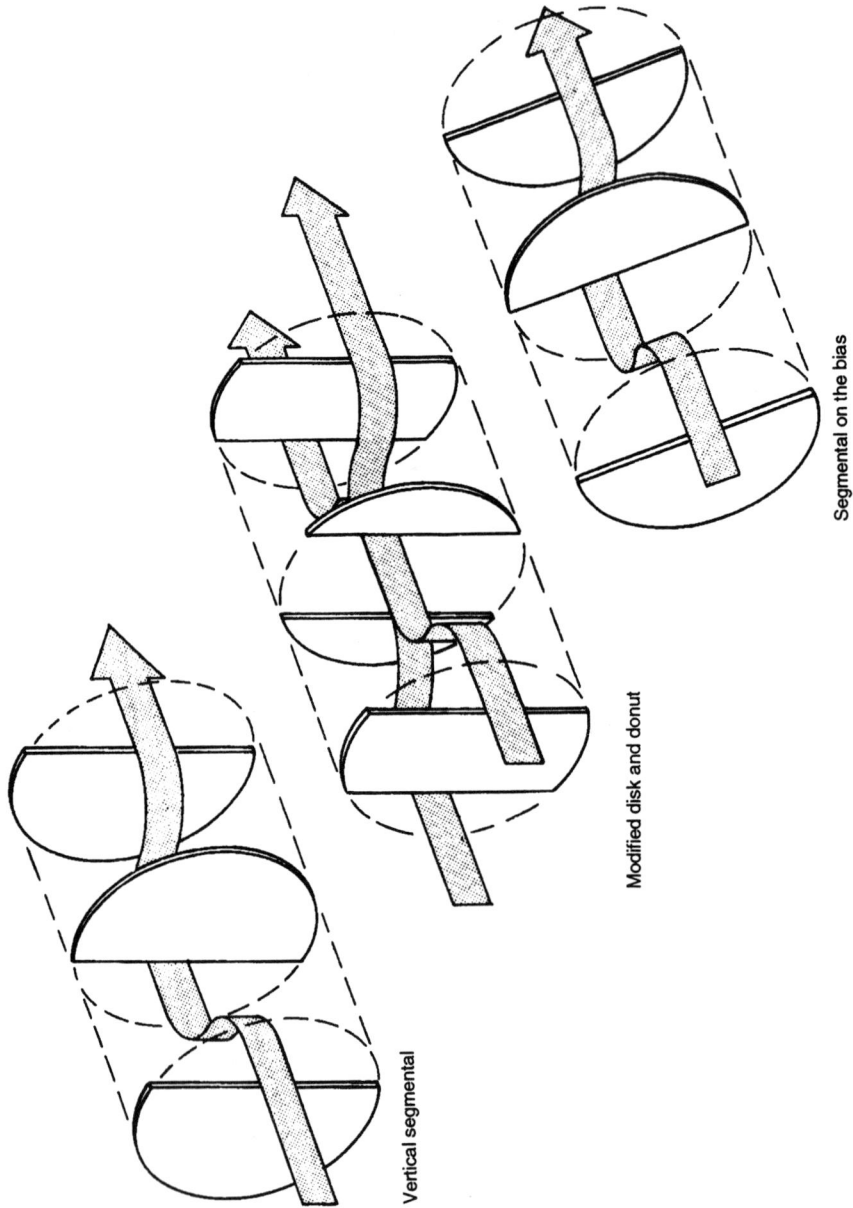

Vertical segmental

Modified disk and donut

Segmental on the bias

Figure 6.3. Types of baffle

The free area of flow between baffles is now calculated as follows:

$$A_f = W \times (B_p - 0.187)$$

where

A_f = free flow area between baffles (in.^{2})
B_p = baffle pitch (in.)

- *Step 7 Calculate the shell side film coefficient h_o* The following expession is used to determine the outside film coefficient:[12]

$$h_o = \frac{K}{d_o}(C\mu_f/K)^{1/3} \cdot \phi(d_o G_m/\mu_f) \cdot \frac{4P_b}{D}$$

where

h_o = outside film coefficient (BTU/h.ft^2.°F)
G_m = maximum mass velocity (lb/s.ft^2)
d_o = outside tube diameter (in.)
K = thermal conductivity
C = specific heat of fluid (BTU/lb/°F)
μ_f = viscosity at mean film temperature (cP)
P_b = baffle pitch (in.)
D = shell internal diameter (in.)

$\phi(d_o G_m/\mu_f)$ is a function of the Reynolds number read from Figure 6.4. The Reynolds number is

$$Re = \frac{d_o G_m}{\mu_f}$$

where G_m = lb/s.ft^2. This film coefficient is corrected for the type of baffle and tube arrangement by multiplying it with one of the following factors:

For square pitch vertical to tube rows	0.50
For square pitch on the bias	0.55
For triangular tube pitch	0.70

- *Step 8 Calculate the overall heat transfer coefficient U_o* The film coefficients calculated in steps 5 and 7 are now used in the expression

$$\frac{1}{U_o} = \frac{1}{h_{io}} + r_{io} + \frac{1}{h_o} + r_o + r_w$$

where

U_o = overall heat transfer coefficient (BTU/h.ft^2.°F)
r_{io} and r_o = tube side and shell side fouling factors respectively (h.ft^2°F/BTU). For clean tubes this is 0.001 as a sum of both factors.
r_w = tube wall resistance to heat transfer (h.ft^2°F/BTU), which is expressed as

$$r_w = \frac{t_w d_o}{12 \times K \times (d_o - 2t_w)}$$

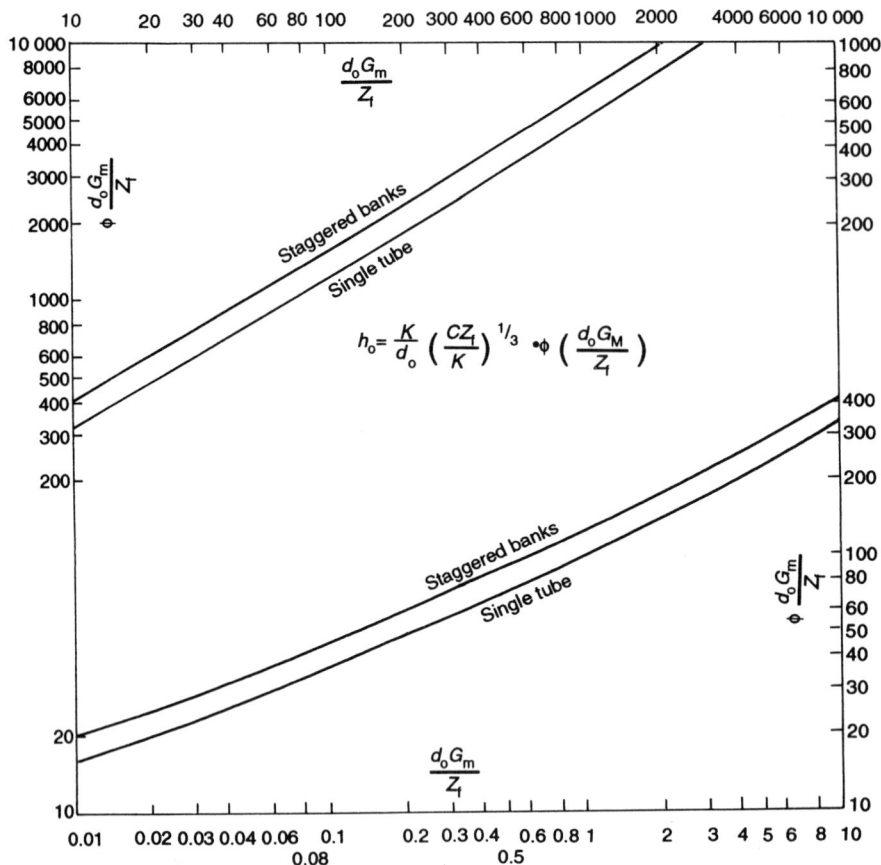

Figure 6.4. Heat transfer to fluids outside tubes. For tube and shell heat exchangers multiply h_0 by a 'bundle factor' as follows: 0.50 for square pitch tubes in line; 0.55 for square pitch tubes at 45°. G_m is evaluated at centre row of tubes (Reproduced by permission of Kreiger Publishing Company, Malabar, Florida, 1950, Maxwell, Data Book on Hydrocarbons)

where

t_w = tube wall thickness (in.)
d_o = outside tube diameter (in.)
K = thermal conductivity (BTU/h.ft^2°F/ft). See Table 6.1

Compare the calculated value of the overall heat transfer coefficient with the assumed one in step 3. If there is agreement within ±10% then the calculated one will be used for revising the calculation for surface area and the other dimensions. If there is no agreement repeat the calculation using a new value for the assumed U.

● *Step 9 Calculate tube side pressure drop* Using the adjusted dimensional values from the calculated U_o, calculate the tube side pressure drop using one of the following equations:

$$\Delta P_t = 0.02 F_t \times N_p \times ((V^2 + (0.158 L V^{1.73}/d_i^{1.27}))$$

Table 6.1. Thermal conductivity of tube metals

	K (BTU/h.ft^2.°F/ft)
Admiralty brass	64
Aluminium brass	58
Aluminium	117
Brass	57
Carbon steel	26
Copper	223
Cupronickel	41
Lead	20
Monel	15
Nickel	36
Red brass	92
Type 316 alloy steel	9
Type 304 alloy steel	9
Zinc	65

For water only. For fluids other than water use

$$\Delta P_t = F_t \times N_p \times (\Delta P_{tf} + \Delta P_{tr})$$

where

$$\Delta P_{tf} = F_3 \cdot \frac{L}{d_i} \cdot (\rho_m \times V^2/9270).(\mu_w)^{0.14}$$

$\Delta P_{tr} = 3 \times (\rho_m \times V^2/9270)$
F_3 = Factor based on Reynolds number (see Figure 6.5)
ρ_m = density (lb/ft^3) at mean fluid temperature
μ_w = viscosity of fluid at tube wall temperature (cP) (use mean film temperature)
V = linear velocity (ft/s)
F_t = pressure drop fouling factor as follows (dimensionless)

Tube o/d	Tube metal	ft
0.75	Steel	1.50
1.00	Steel	1.40
1.50	Steel	1.20
0.75	admiralty brass	1.20
1.00	Copper	1.15

The pressure drop figures calculated by these equations are for one unit. Where there are more than one shell in series multiply the figures by the number of shells.

• *Step 10   Calculate the shell side pressure drop*   Using the revised dimensions calculated in step 8 the total shell side pressure drop is calculated using the following equation:

$$\Delta P_s = F_s(\Delta P_{sr} + \Delta P_{sf})$$

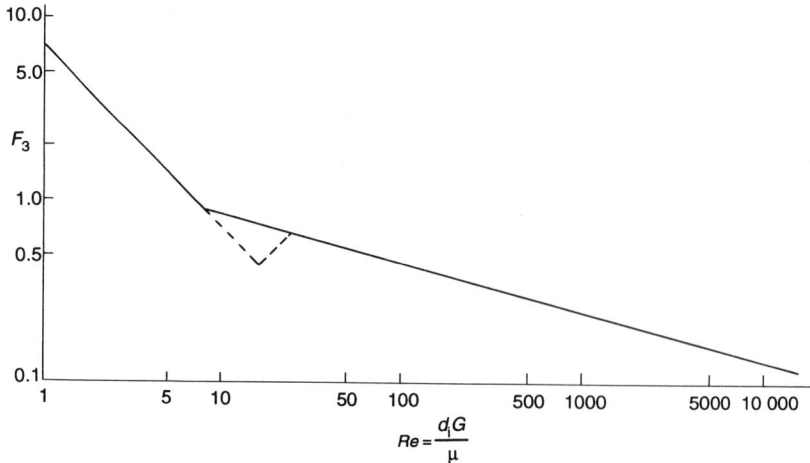

Figure 6.5. Pressure drop factor F_3 for flows inside tubes

where

$$\Delta P_{sf} = B_2.F_{sp}.N_{tc}.N_b.(m \times V^2/9270)$$
$$\Delta P_{sr} = \text{pressure drop due to turns given by}$$

$$(N_b + 1).(3.5 - 2P_b/D).\frac{(m \times V^2)}{9270}$$

B_2 = factor as follows:

Baffle position	Tube layout	B_2
Vertical	Square	0.30
Bias at 45°	Square	0.40
Vertical	Triangular	0.50

F_{sp} = factor based on Reynolds number (see Figure 6.6)
N_{tc} = number of tubes on centreline
N_b = number of shell baffles
P_b = space between baffles (in.)
D = shell i/d (in.)

The pressure drop calculated here is for one shell. If there are more than one shell in series then multiply these pressure drops by the number of shells.

Example calculation

A light oil is to be cooled by cooling tower water in a shell and tube exchanger. Calculate the surface area required, the overall exchanger dimensions, and the

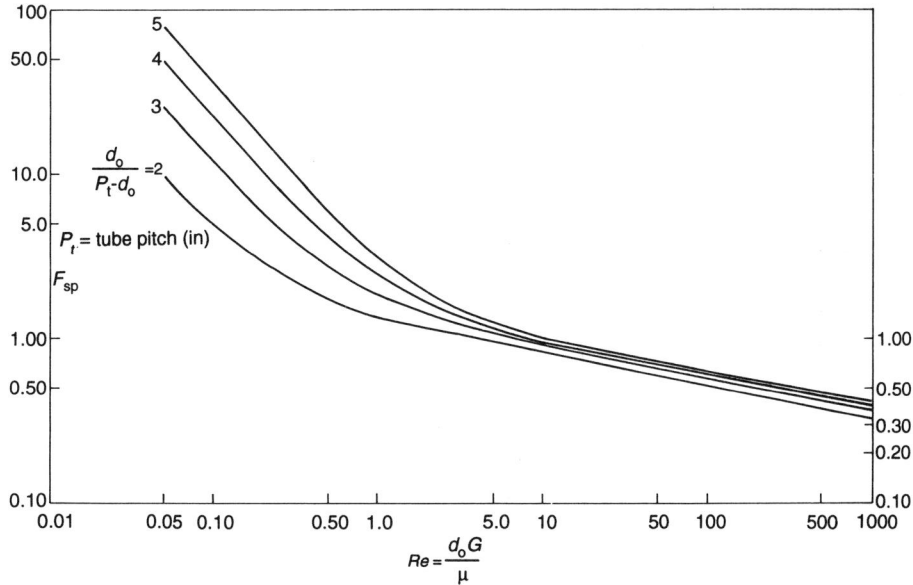

Figure 6.6. Pressure drop factor F_{sp} for flows across banks of tubes

pressure drops for shell side and tube side. The rate of flow for the light oil is 120 000 lb/h. Other data are as follows:

	Shell side Light Oil	Tube side Water
Temperature (°F)	215	60
Temperature out (°F)	100	90
Specific heat (BTU/lb.°F)	0.56	1.0
Specific gravity at 60 °F	0.850	1.0
Thermal conductivity (BTU/h.ft^2.°F)	0.082	0.36
Viscosity at 100 °F (cP)	2.0	0.68
Inlet pressures (psig)	175	60

1 Calculating the exchanger duty and flow of water

Heat removed from oil = lb/h × spht × temp. diff.
$$= 120\,000 \times 0.56 \times (215 - 100)$$
$$= 7\,728\,000 \text{ BTU/h}$$

$$\text{Water flow rate} = \frac{Q}{(\text{Temp.diff.})}$$

where Q = exchanger duty = 7.728 mm BTU/h

$$= \frac{7\,728\,000}{90 - 60} = 257\,600 \text{ lb/h.}$$

2   Calculate the log mean temperature difference ΔT_e

Assume countercurrent flow. Then temperature flow will be

$$T_1 \rightarrow T_2$$

$$t_2 \leftarrow t_1$$

where

T_1 = inlet temperature of oil shell side
T_2 = outlet temperature of oil shell side
t_1 = inlet temperature of water tube side
t_2 = outlet temperature of water tube side

$$215 \rightarrow 100$$
$$90 \leftarrow 60$$
$$\text{Diff. } \overline{125 \quad\quad 40}$$

$$\Delta T_e = 75\,°\text{F}$$

Using Figure A1.6 for correcting the calculated log mean temperature to flow pattern

$$J = \frac{t_2 - t_1}{T_1 - t_1} = \frac{90 - 60}{215 - 60}$$
$$= 0.194$$

$$R = \frac{T_1 - T_2}{t_2 - t_1} = \frac{215 - 100}{90 - 60}$$
$$= 3.83$$

Correction factor for one shell pass and two or more tube passes is 0.885. This value is above 0.8 and is acceptable.

Then corrected LMTD is $75\,°\text{F} \times 0.885 = \Delta T_m = 66.4\,°\text{F}$

3   Calculate approximate total heating surface area

Selected overall heat transfer coefficient from Table A1.1 in Appendix 1 = 80 BTU/h.ft^2.°F.

Using the expression

$$Q = UA\Delta T_m$$

$$A = \frac{7\,728\,000}{80 \times 66.4}$$
$$= 1455\,\text{ft}^2$$

This is fairly clean service so $\frac{3}{4}$ in. o/d tubes on a triangular pitch will be used. Maximum recommended surface area per shell is 4100 ft^2 for triangular pitch. Therefore one shell will easily accommodate this calculated surface area.

4   Tube side flow and the number of tube passes

Cooling water flows tube side. Set the linear velocity at about 2.5 ft/s.

$$\text{lb/h of water} = 257\,600$$
$$\text{Max. temp. } °\text{F} = 90$$
$$\text{lb/ft}^3 \text{ at } 90\,°\text{F} = 62.12$$
$$\text{ft}^3/\text{h} = 4147$$

Tubes are to be copper and 14 BWG. Use standard length of 16 ft.

Internal cross-sectional area of one tube = 0.302 in.^{2}

$$\text{Then number of tubes per pass} = \frac{4147 \times 144}{3600 \times 0.302 \times 2.5}$$
$$= 220 \text{ tubes/pass.}$$

External area per foot of tube = 0.193 (see Table A1.7 in Appendix 1)

Set tube length as a standard 16 ft.

$$\text{Then total ft}^2 \text{ per pass} = 16 \times 0.193 \times 220 = 691$$

$$\text{Number of passes} = \frac{1455}{691} = 2.1 \text{ (make it two passes)}$$

Based on a two-tube pass arrangement:

$$\text{ft}^2 \text{ per pass} = \frac{1455}{2.0} = 727.5$$

$$\text{Number of tubes per pass} = \frac{727.5}{16 \times 0.1963} = 232$$

$$\text{Velocity in tubes} = \frac{4147 \times 144}{3600 \times 0.302 \times 232} = 2.37 \text{ ft/s}$$

5   Calculate tube side film coefficient h_{io}

For water flowing tube side use

$$h_{io} = \frac{300 \times (V_t \times \text{tube i/d (in.)})^{0.8}}{\text{tube o/d (in.)}}$$

where

h_{io} = inside film coefficient based on outside tube diameter (BTU/h.ft^2.°F)
V_t = linear velocity of water tube side in ft/s = 2.37 ft/s

$$h_{io} = \frac{300}{0.75} \times (2.37 \times 0.584)^{0.8}$$
$$= 518.8 \text{ BTU/h.ft}^2.°\text{F}$$

6   Calculate shell side flow and dimensions

Shell mean film temperature is as follows:

$$\text{Inlet ave.} = \frac{T_1 + T_2}{2} \quad \text{Outlet ave.} = \frac{T_3 + T_4}{2}$$

where

T_1 = shell fluid inlet temperature. = 215 °F
T_2 = tube outlet temperature. = 90 °F
T_3 = shell outlet temperature. = 100 °F
T_4 = tube inlet temperature. = 60 °F

$$\text{mean shell side film temperature} = \frac{\text{inlet ave.} + \text{outlet ave.}}{2}$$

$$= 116.3 \,°\text{F}$$

Viscosity of light oil at 116.3 °F = 1.6 cP

Density of light oil at mean fluid temp. of 158 °F = 0.825 × 62.37

$$= 51.4 \, \text{lb/ft}^3.$$

To calculate the number of tubes across centreline of tube bundle and shell i/d use:

$$T_{cl} = 1.10 \times (2 \times 232)^{0.5}$$

$$= 23.7 \text{ tubes}$$

$$\text{Theoretical diameter of tube bundle} = \frac{23.7}{0.9} = 26.3 \text{ in.}$$

$$\text{For shell diameter use } 26.3 + 3.0 = 29.3 \text{ (make it 30 in. i/d)}$$

To calculate area of flow for fluid shell side:

$$W = D_i - (d_o \times T_{cl})$$

where

W = space available for flow (in.)^2
D_i = shell inside diameter (in.) = 30
d_o = tube outside diameter (in.) = 0.75
T_{cl} = number of tubes across centreline = 23.7
W = 12.23 in.

Baffles to be vertical to grade with a pitch of 20% of shell i/d.

$$\text{Number of baffles} = \frac{10L}{(0.2 \times 30)} = 27$$

Free area of flow between each set of baffles:

$$= W \times (\text{baffle pitch} - 0.187)$$

$$= 12.23 \times (6.0 - 0.187) = 71.1 \text{ in.}^2$$

7   Calculate shell side film coefficient h_o

$$h_o = \frac{K}{d_o}(C\mu_f/K)^{1/3} . \phi(d_o G_m/\mu_f).(4P_b/D_i)^{0.1}$$

where

h_o = outside film coefficient (BTU/h.ft^2.°F)
G_m = maximum mass velocity (lb/s.ft^2)
d_o = outside tube diameter (in.) = 0.75
K = thermal conductivity. = 0.0784 BTU/h.ft^2.°F/ft)
C = specific heat of fluid (BTU/lb/°F) = 0.85
μ_f = viscosity at mean film temperature (cP) = 1.3
P_b = baffle pitch (in.)
D_i = shell internal diameter (in.)

$\phi(d_o G_m/\mu_f)$ is a function of the Reynolds number read from Figure 6.4. The Reynolds number is

$$Re = \frac{d_o G_m}{\mu_f}$$

$$G_m = \frac{120\,000 \times 144}{3600 \times 71.1}$$

$$Re = \frac{0.75 \times 120\,000 \times 144}{3600 \times 71.1 \times 1.6} = 31.6$$

from Figure 6.4 $\phi(d_o G_m/\mu_f) = 780$

$$h_o = \frac{0.084}{0.75} \times (0.52 \times 1.6/0.0783)^{1/3} \times 780 \times 0.7^* \times 0.98$$

* Correction factor for baffles. See step 7.

$$h_o = 0.1045 \times 2.415 \times 700 \times 0.7 \times 0.98$$

$$= 121.9 \text{ BTU/h.ft}^2.°F$$

8   Calculate overall heat transfer coefficient U_o

$$K_m \text{ for admiralty brass} = 64$$

$$r_w = \frac{t_w \times d_o}{12 K_m \times (d_o - 2t_w)}$$

where

r_w = thermal resistance of wall to heat transfer (h.ft^2.°F/BTU)
t_w = wall thickness (in.)
$$r_w = \frac{0.083 \times 0.75}{12 \times 64 \times (0.75 - 0.166)}$$
$$= 0.00014 \text{ h.ft}^2.°F/BTU$$

Then

$$\frac{1}{U_o} = \frac{1}{518.8} + 0.001 + \frac{1}{121.9} + 0.0001 + 0.001$$

$$= 0.0122$$

$$U_o = 81.96$$

which is within acceptable limits.

Use calculated U_o to correct dimensions and exchanger arrangements as follows:

$$\text{Revised surface area} = 1420 \text{ ft}^2$$
$$\text{No. of tube passes} = 2$$
$$\text{No. of tubes per pass} = 226$$
$$\text{Velocity tube side} = 2.43 \text{ ft/s}$$
$$\text{Tube bundle diameter} = 26 \text{ in.}$$
$$\text{Shell i/d} = 30 \text{ in.}$$
$$\text{No. of baffles} = 27 \text{ on 6-in. pitch.}$$
$$\text{Free area of flow} = 72.8 \text{ in.}^2$$

9 *Calculate tube side pressure drop* ΔP_t

Fluid flowing is water. Then:

$$\Delta P_t = 0.02 \text{ ft} \times N_p \times ((V^2 + (0.158 L V^{1.73}/d_i^{1.27}))$$

where

$$F_t = \text{pressure drop fouling factor} = 1.2$$
$$N_p = \text{no. of tubes passes} = 2.0$$
$$V = \text{linear velocity} = 2.43 \text{ ft/s}$$
$$L = \text{tube length} = 16 \text{ ft}$$
$$d_i = \text{tube i/d} = 0.584 \text{ in.}$$
$$\Delta P_t = 0.02 \times 1.2 \times 2.0 \times (5.90 + 23.26)$$
$$= 1.40 \text{ psi}$$

10 *Calculate shell side pressure drop* ΔP_s

$$\Delta P_s = F_s(\Delta P_{sr} + \Delta P_{sf})$$

where

$$\Delta P_{sf} = B_2.F_{sp}.N_{tc}.N_b.(m \times V^2/9270)$$
$$\Delta P_{sr} = \text{pressure drop due to turns given by}$$

$$(N_b + 1).(3.5 - 2P_b/D).\frac{(m \times V^2)}{9270}$$

B_2 = factor for baffling = 0.5 (see step 10)
F_{sp} = factor based on Reynolds number. See Figure 6.6 = 0.64
N_{tc} = number of tubes on centreline = 24
N_b = number of shell baffles = 27.
P_b = space between baffles (in.) = 6
D = shell i/d (in.) = 30
V = linear velocity across tubes (call it 1.0 ft/s)

The Reynolds number is calculated as 31.6 (see calculation 7 above). Then F_{sp} is 0.64.

$$\Delta P_{sf} = 0.5 \times 0.64 \times 24 \times 27 \times (50.2 \times 1.0/9270)$$
$$= 1.12 \text{ psi}$$
$$\Delta P_{sr} = (20 + 1) \times ((3.5 - (2 \times 6)/30)) \times 0.0054$$
$$= 0.352 \text{ psi}$$

Total shell pressure drop $\Delta P_s = 1.12 + 0.352$
$$= 1.5 \text{ psi}$$

which is acceptable.

6.4 Air Coolers and Condensers

Air cooling of process streams or condensing of process vapours is more widely used in the process industry than cooling or condensing by exchange with cooling water. The use of individual air coolers for process streams using modern design techniques has economized on plant area required. It has also made obsolete those large cooling towers and ponds associated with product cooling. This section describes air coolers in general and outlines a method to estimate surface area, motor horsepower and plant area required by the unit.

As in the case for shell and tube exchangers, there are many excellent computer programs that can be used for the design of air coolers. The method given here for such calculations may be used in the absence of a computer program or for a good estimate of a unit. The method also emphasizes the importance of the data supplied to manufacturers for the correct specification of the units.

GENERAL DESCRIPTION OF AIR COOLERS/CONDENSERS

Figures 6.7 and 6.8 show the two types of air coolers used in the process industry. Both units consist of a bank of tubes through which the fluid to be cooled or condensed flows. Air is passed around the tubes either by a fan located below the tubes (Figure 6.7) forcing air through the tube bank or one located above the tube bank drawing air through the tube bank (Figure 6.8). The first arrangement is called 'forced draught' and the second 'Induced draught'.

Air in both cases is motivated by a fan or fans driven by an electric motor or a steam turbine or, in some cases, a gas turbine. The fan and prime driver are

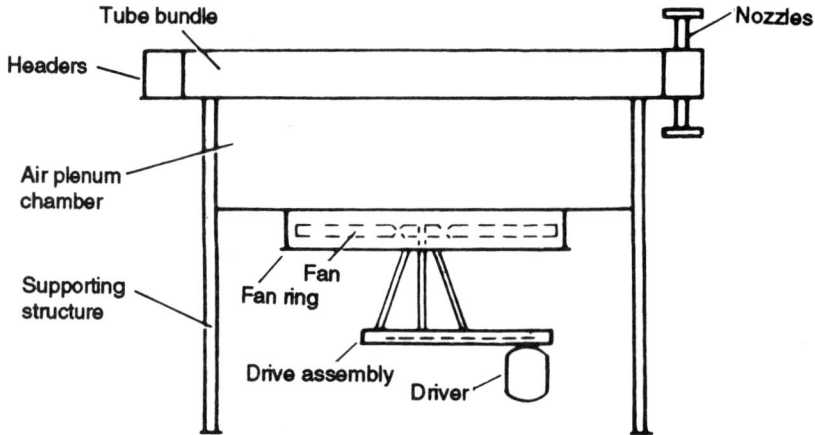

Figure 6.7. Air coolers—forced draught

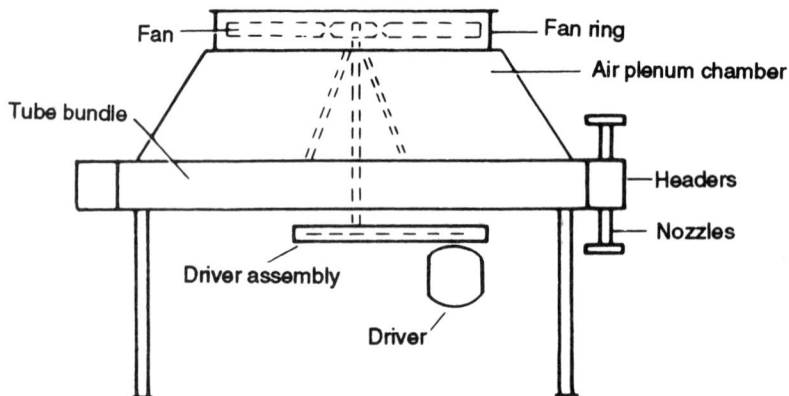

Figure 6.8. Air coolers—induced draught

normally connected by a 'V' belt or by a shaft and gearbox. Electric motor drives are by far the most common prime drivers for air coolers.

The units may be installed on a structure at grade or, as is often the case, on a structure above an elevated pipe rack. Most air coolers in condensing service are elevated above pipe racks to allow free flow of condensate into a receiving drum.

THERMAL RATING

Thermal rating of an air cooler is similar in some respects to that of a shell and tube described in Section 6.3. The basic energy equation

$$Q = U\Delta TA$$

is used to determine the surface area required. The calculation for U is different in that it requires the calculation for the air side film coefficient. This film coefficient is usually based on an extended surface area which is formed by adding fins to the bare surface of the tubes. Thermal rating, surface area, fan dimensions and horsepower are calculated by the following steps:

- *Step 1* Calculate the heat duty and the tube side material characteristics.
- *Step 2* Calculate the log mean temperature for the exchanger. Using the following equation determine the temperature rise for the air flowing over the tubes:

$$\Delta t_m = ((U_e + 1)/10)).((\Delta T_m/2) - t_1))$$

where

Δt_m = air temperature rise (°F)
U_e = overall heat transfer coefficient assumed (from Table 6.2)
ΔT_m = mean tube side temperature (°F)
t_1 = inlet air temperature (°F)

Calculate the log mean temperature difference (LMTD) as in step 2 of Section 6.3.

Table 6.2. Some common overall transfer coefficients for air cooling

Service	$\frac{1}{2}$ in. by 9 Fin ht by fin/in.		$\frac{5}{8}$ in. by 10 Fin ht by fin/in.	
	U_e	U_0	U_e	U_0
Process water	95	6.5	110	5.2
Hydrogen liquids				
Visc. at ave. temp. (cP)				
0.2	85	5.9	100	4.7
1.0	65	4.5	75	3.5
2.5	45	3.1	55	2.6
6.0	20	1.4	25	1.2
10.0	10	0.7	13	0.6
Hydrogen gases at pressures (psig)				
50	30	2.1	35	1.6
100	35	2.4	40	1.9
300	45	3.1	55	2.6
500	55	3.8	65	3.0
1000	75	5.2	90	4.2
Hydrocarbon condensers				
Cooling range 0 °F	85	5.9	100	4.7
10 °F	80	5.5	95	4.4
60 °F	65	4.5	75	3.5
100+ °F	60	4.1	70	4.2
Refrigerants				
Ammonia	110	7.6	130	6.1
Freon	65	4.5	75	3.5

U_e is transfer coefficient for finned surface.
U_0 is transfer coefficient for bare tubes.

- *Step 3* Determine an approximate surface area using the expression

$$AE = \frac{Q}{UE \cdot \Delta T_\mathrm{m}}$$

where

AE = extended surface area (ft^2)
Q = exchanger duty (BTU/h)
UE = overall heat transfer coefficient (based on extended surface from Table 6.2)
ΔT_m = log mean temperature difference corrected for number of passes (°F)

- *Step 4* Calculate the number of tubes from the expression

$$N_\mathrm{t} = \frac{AE}{A_\mathrm{f} \times L}$$

where

N_t = total number of tubes
AE = extended surface area (ft^2)
A_f = extended area per foot of fin tube read from Table 6.3
L = Length of tube (30 ft is standard).

- *Step 5* Fix the number of passes (usually three or four) and calculate the mass flow of tube side fluid using the expression

$$G = \frac{\mathrm{lb/h \ of \ tube \ side \ fluid} \times N_\mathrm{p} \times 144}{N_\mathrm{t} \times A_\mathrm{t} \times 3600}$$

where

G = mass velocity (lb/s/ft^2)
N_p = number of tube passes
A_t = inside cross-sectional area of tube (in.^{2})

Table 6.3. Fin tube to bare tube relationships (Based on 1 in. o/d tubes)

Fin ht by fin/in.	$\frac{1}{2}$ in. by 9		$\frac{5}{8}$ in. by 10	
Area/ft fin tube	3.8		5.58	
Ratio of areas fin/bare tube	14.5		21.4	
Tube pitch (in.)	2Δ	$2\frac{1}{4}\Delta$	$2\frac{1}{4}\Delta$	$2\frac{1}{2}\Delta$
Bundle area ft^2/ft*				
Three rows	68.4	60.6	89.1	80.4
Four rows	91.2	80.8	118.8	107.2
Five rows	114.0	101.0	148.5	134.0
Six rows	136.8	121.2	178.2	160.8

* Bundle area is the external area of the bundle face area in ft^2/ft.

- *Step 6*   Calculate the Reynolds number for tube side using the expression

$$Re = \frac{d_i \cdot G}{\mu}$$

where

Re = Reynolds number (dimensionless)
d_i = tube i/d (in.)
μ = tube side fluid viscosity at average temperature (cP)

- *Step 7*   Calculate the inside film coefficient from the expression

$$h_{io} = \frac{K}{D}(C\mu/K)^{1/3}(\mu/\mu_w)^{0.14} \cdot \phi(DG/Z)$$

where

h_i = inside film coefficient (BTU/h.ft^2.°F)
K = thermal conductivity of the fluid (BTU/h.ft^2(°F per ft). See Maxwell, *Data Book on Hydrocarbons*, or Perry, *Chemical Engineer's Handbook*
D = inside tube diameter (in.)
C = specific heat (BTU/lb/°F)
G = mass velocity (lb/s.ft^2)
μ = absolute viscosity (cP) at average fluid temperature
μ_w = absolute viscosity (cP) at average tube wall temperature
$\phi(DG/\mu)$ = from Figure 6.2

- *Step 8*   Calculate the mass velocity of air and the film coefficient on the air side, thus:

$$\text{Weight of air} = \frac{Q}{C_{air} \times \Delta T_{air}}$$

where

Q = exchanger duty (BTU/h)
C_{air} = specific heat of air (use 0.24)
ΔT_{air} = temperature rise of the air (°F)

Face area of tubes A_f is calculated as follows. Set the o/d of the tubes (usually 1 in.), length, fin size (usually 5/8 in. at 10 to the inch or $\frac{1}{2}$ in. at 9 to the inch), pitch (see Table 6.3), and number of tube rows (start with 3 or 4). Then face area is

$$A_f = \frac{\text{Total extended surface area } AE}{\text{External area per ft of bundle (from Table 6.3)}}$$

Mass velocity of air is calculated from the expression

$$G_a = \frac{\text{lb per hour of air flow}}{\text{Face area } A_f}$$

The film coefficient for the air side is read from Figure 6.9.

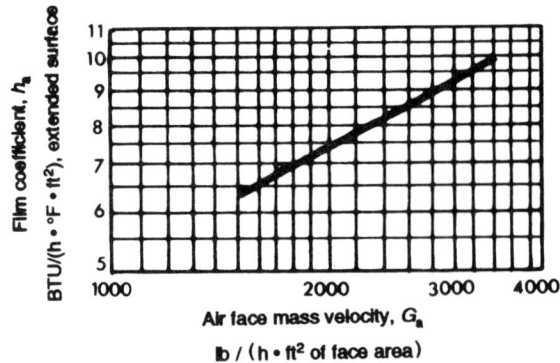

Figure 6.9. Air film coefficients

- *Step 9* Calculate the overall heat transfer coefficient as follows. Area ratio of bare tube outside to finned outside is read from Table 6.3. Then factor to convert all heat flow resistance to outside tube diameter basis is

$$F_t = \frac{A_r \times \text{tube o/d}}{A_t}$$

where

A_r = area ratio
A_t = inside tube cross-sectional area (in.^{2})

Then

$$\frac{1}{U_o} = \frac{1 + r_t \times F_t + r_w + 1}{h_i h_o}$$

where

r_t = inside fouling factor
r_w = tube metal resistance (normally ignored)

If the calculated U is within 10% of the assumed there will be no need to recalculate with a new assumed value for U. The dimensions and data are adjusted, however, using the calculated value for U.

- *Step 10* Calculate the required fan area and the fan diameter as follows:

$$\text{Fan area} = \frac{0.4 \times \text{face area } A_f}{\text{Assumed number of fans}}$$

Begin with assuming two fans and continue with multiples of 2 until a reasonable fan diameter (about 10–12 ft) is obtained. On very large units fans can be maximized at 16 ft.

Fan diameter = $\sqrt{(\text{fan area} \times 4/\pi)}$

- *Step 11* Calculate air side pressure drop and actual air flow in ft^3/min.

$$\text{Average air temperature} = \frac{t_1 + t_2}{2}$$

From Figure 6.10

$$D_r = \text{relative density factor for air at elevations of site}$$

From Figure 6.11

$$\Delta P_a = \text{pressure drop of air in inches of } H_2O$$

$$\Delta P_a \text{ corrected} = \frac{\Delta P_a \times \text{No. of rows}}{D_r}$$

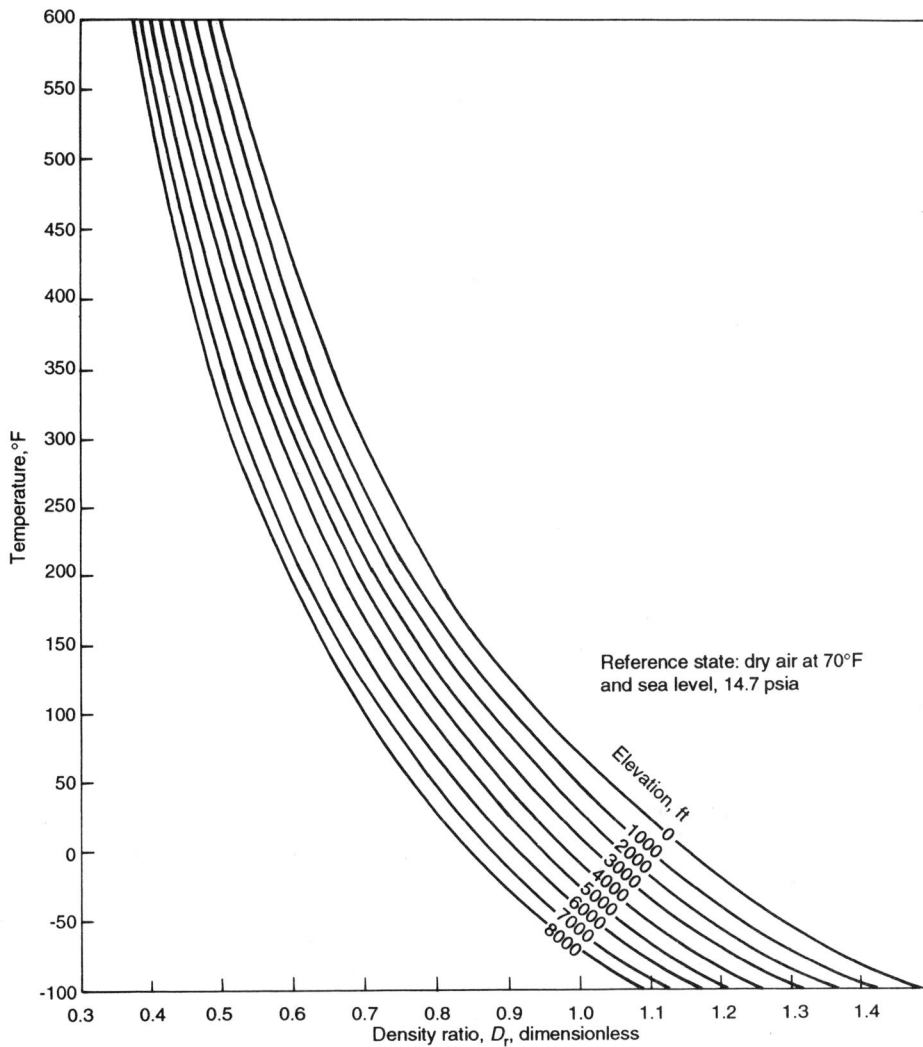

Figure 6.10. Relative density of air

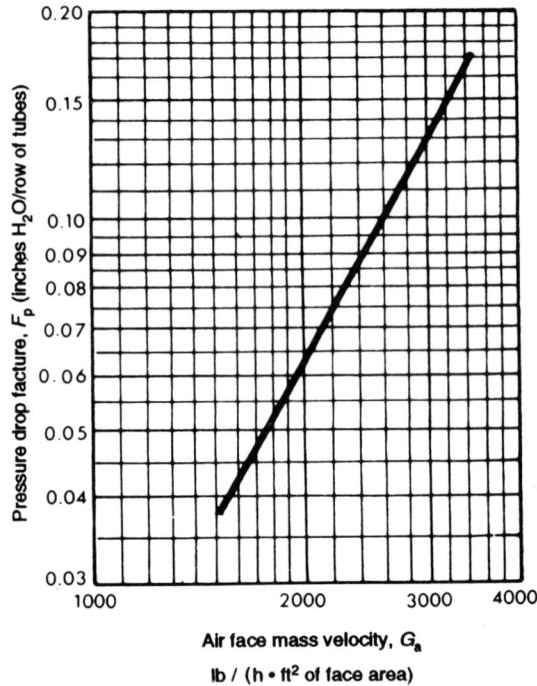

Figure 6.11. Pressure drop air side in inches of water

Density of air at corrected ΔP_a

$$\frac{29}{(378 \times 14.7 \times T_2)/((T_1 \times (\text{corr. } \Delta P_a + 14.7))}$$

where T_1 and T_2 are absolute temperatures. acfm of air is therefore

$$\frac{\text{lb/h of air}}{\text{Density} \times 60}$$

ΔP of air at the fan is obtained by the expression

$$\Delta P_a + \left(\frac{\text{acfm}}{4000(\pi d^2)/4}\right)^2$$

in inches of water gauge.

- *Step 11*   Calculate the fan horsepower as follows:

$$\text{Hydraulic hp} = \frac{\text{acfm} \times \text{density of air} \times \text{diff. head in ft}}{33\,000}$$

$$\text{Differential head} = \frac{\text{Total } \Delta P \text{ at fan in inches } H_2O \times 5.193}{\text{Density}}$$

$$\text{bhp} = \frac{\text{Hydraulic hp}}{\eta'}$$

Where η' is the fan efficiency (usually 70%).

Example calculation

The example given in Section 6.3 will be used here but in this case the light oil will be cooled by an air cooler. An example of a two-phase exchanger will be discussed and given later in Section 6.6.

$$\text{Ambient air temperature} = 80\,°F$$
$$\text{Elevation of site} = \text{at sea level}$$
$$\text{Tube length} = 30\,\text{ft}$$
$$\text{Fin arrangement} = \tfrac{5}{8}\,\text{in. at } 2\tfrac{1}{2}\,\text{in. } \Delta \text{ pitch}$$

Tube o/d standard 1-in. CS tubes 14 BWG

Average temperature of the light oil:

$$\frac{215 + 100}{2} = 158\,°F$$

Calculating temperature of the air out t_2

Assume overall heat trans coefficient $U_e = 3.6$ (Table 6.2)

$$\text{Then air temperature rise} = \left(\frac{U_e + 1}{10}\right) \cdot \left(\frac{(T_1 + T_2)}{2} - t_1\right)$$

where t_1 = air inlet temperature (°F).

Air temperature rise = 0.46 + 77.5 = 35.7 (say 36 °F) and air outlet temperature is 80 + 36 = 116 °F = t_2.

$$\Delta T_e = 215 \rightarrow 100$$
$$\underline{116 \leftarrow \quad 80}$$
$$99 \qquad 20$$

$$\Delta T_e = 49\,°F$$

Assume four tube passes. Then correction factor from Figure A1.6 in Appendix 1 is as follows:

$$J = \frac{116 - 80}{215 - 80} = 0.27$$

$$R = \frac{215 - 100}{116 - 80} = 3.2$$

$$F = 0.96$$

$$\Delta T_m = 0.96 \times 49 = 47\,°F$$

Approximate surface area is calculated from

$$A = \frac{Q}{\Delta T_m \times U_e}$$

$$A = \frac{7\,728\,000}{47 \times 3.6} = 45\,674\,\text{ft}^2$$

Use standard 1-in. tubes 30 ft long and $\frac{5}{8}$-in. fins (10 per inch). and four rows on $2\frac{1}{2}$-in. Δ pitch. Then from Table 6.3:

$$\text{External area per ft of bundle} = 107.2$$

$$\text{Face area} = \frac{45\,674}{107.2} = 426 \text{ ft}^2$$

Calculating the number of tubes:

$$N_{\text{t}} = \frac{\text{Approx. area}}{A_{\text{f}} \times 30}$$

$$A_{\text{f}} = \text{extended area of tube from Table 6.3}$$

$$= 5.58$$

$$N_{\text{t}} = 273 \text{ (call it 270 tubes)}$$

Calculating the mass velocity of the tube side fluid:

$$G = \frac{144 \times 120\,000 \times 4.0}{270 \times 0.5463 \times 3600} = 130.2 \text{ lb/s.ft}^2$$

Calculating the Re number:

$$Re = \frac{d_{\text{i}} \cdot G}{\mu}$$

$$= \frac{0.834 \times 130.2}{0.8} = 136$$

Calculating the inside film coefficient h_{i}:

$$h_{\text{i}} = \frac{K}{D}(C\mu/K)^{1/3}(\mu/\mu_{\text{w}})^{0.14} \cdot \phi(DG/\mu)$$

where

h_{i} = inside film coefficient in BTU/h.ft^2.°F
K = thermal conductivity of the fluid in BTU/h.ft^2 (°F per ft) (see Maxwell or Perry) = 0.082
D = inside tube diameter (in.) = 0.834
C = specific heat (BTU/lb/°F) = 0.56
μ = absolute viscosity (cP) at average fluid temperature = 0.8
μ_{w} = absolute viscosity (cP) at average tube wall temperature = 0.86
$\phi(DG/\mu)$ = From Figure 6.2 = 1060

$$\text{Average wall temperature: } \frac{T_1 + t_2}{2} = \frac{215 + 116}{2} = 166$$

$$\frac{T_2 + t_1}{2} = \frac{100 + 80}{2} = 90$$

$$\frac{166 + 90}{2} = 128 \text{ °F}$$

$$h_i = \frac{0.082}{0.834} \times \left(\frac{0.56 \times 0.8}{0.082}\right)^{1/3} \times \left(\frac{0.8}{0.86}\right)^{0.14} \times 1060$$

$$= 183.7 \text{ BTU/h.ft}^2.°\text{F}$$

Calculating the mass velocity of air and air side film coefficient h_0:

$$\text{Weight of air} = \frac{7\,728\,000}{0.24 \times (116 - 80)}$$

$$= 894\,444 \text{ lb/h}$$

$$G_a = \frac{\text{lb/h}}{\text{Face area}} = \frac{894\,444}{426}$$

$$= 2099 \text{ lb/h.ft}^2$$

From Figure 6.9 $h_0 = 7.4 \text{ BTU/h.ft}^2°\text{F}$

Calculating the overall heat transfer coefficient U_0:

$$\text{Area ratio of 1-in. fin tube (Ar)} = 21.4 \text{ for } \tfrac{5}{8}\text{-in. fin}$$

Then conversion factor to outside area is $\dfrac{A_r}{A_t} = \dfrac{21.4 \times 1.0}{0.834}$

$$= 25.66$$

$$\frac{1}{U_0} = \left(\frac{1}{(h_i)} \times 25.66\right) + (r_t \times 25.66) + \frac{1}{h_0}$$

$$(r_t = \text{tube-sidefoulingfactor} = 0.0005)$$

$$= 0.139 + 0.013 + 0.135$$

$$= 0.287$$

$$U_0 = 3.48 \text{ BTU/h.ft}^2°\text{F}$$

Assumed was 3.6 BTU/h.ft^2°F. Therefore calculated is acceptable being less than 10% difference.

Adjusting the data and dimensions to the calculated U_0:

$$\text{Surface area } A_e = 49\,229 \text{ ft}^2$$
$$\text{Face area} = 459 \text{ ft}^2$$
$$\text{No. of tubes} = 294$$
$$G_a = 1896 \text{ lb/h.ft}^2$$

Calculating the fan area and diameter:

$$\text{Fan area} = \frac{0.4 \times \text{face area}}{\text{No. of fans}}$$

$$= \frac{0.4 \times 459}{2} \text{ (two fans assumed)}$$

$$= 91.8 \text{ ft}^2$$

$$\text{Fan diameter} = \sqrt{(91.8 \times 4/\pi)}$$

$$= 10.8 \text{ ft (say 11 ft, which is acceptable)}$$

Calculating air side pressure drop and acfm of air:

$$\text{Average air temperature across tubes} = \frac{80 + 116}{2}$$

$$= 98 \,°\text{F}$$

From Figure 6.10, $D_r = 0.94$ for sea level

From Figure 6.11, pressure drop factor $= 0.067$ in. WG

$$\text{Pressure drop through tubes} = \frac{0.067 \times 4}{0.94}$$

$$= 0.285 \text{ in. WG}$$

Density of air at atmos. $+ 0.285$ in. WG and 98 °F or 558 °R

$$29 \div ((378 \times 14.7 \times 558) \div (520 \times 15.09))$$

$$= 0.073 \text{ lb/ft}^3$$

$$\text{acfm of air} = \frac{1\,200\,000}{0.073 \times 60} = 273\,972 \text{ ft}^3/\text{min}$$

There are two fans. Then load per fan $= 136\,986$ ft^3/min.

$$\text{Total } \Delta P \text{ at fan} = \Delta P_a + \left((\text{ft}^3 \text{ per fan} \div \frac{(4000 \times \pi d_f)^2}{4} \right)^2 \times D_r$$

$$= 0.285 + 0.407 = 0.692 \text{ in. WG}$$

Calculating hydraulic and brake horsepower:

$$\text{Hydraulic hp} = \text{acfm} \times \text{density} \times \text{diff. head (ft)}$$

$$\text{Diff. head} = \text{lb/ft}^2 \div \text{density}$$

$$0.695 \text{ in. WG} = 0.695 \times 5.193 \text{ lb/ft}^2$$

$$\text{Hydraulic hp} = \frac{136\,986 \times 0.692 \times 5.193}{33\,000}$$

Assume an efficiency of 70%. Then the brake hp will be

$$\frac{136\,986 \times 0.692 \times 5.193}{33\,000 \times 0.7} = 21.3 \text{ per fan}$$

Calculating exchanger width:

$$\text{Width} = \frac{\text{Face area}}{\text{Tube length}}$$

$$= 459/30 = 15.3 \text{ (say 15 ft 6 in.)}$$

Summary

Exchanger has 294 tubes 1 in. o/d with $\frac{5}{8}$ in. fins. There will be four passes and the tubes are on $2\frac{1}{2}$ in. Δ pitch. These will be standard length of 50 ft with a total width of $15\frac{1}{2}$ ft. The exchanger will be equipped with two fans each requiring 21.3 bhp supplied by two motors each 30 hp.

6.5 Condensers

In chemical process plants vapours are condensed in the shell side of a shell and tube exchanger, the tube side of an air cooler, or by direct contact with the coolant in a packed tower. By far the most common of these operations are the first two methods. In the case of the shell and tube condenser the condensation may be produced by cooling the vapour by heat exchange with a cold process stream or by water. Air cooling has overtaken the shell and tube condenser in the case of water as coolant in popularity as described in Section 6.4.

In the design or performance analysis of condensers the procedure for determining thermal rating and surface area is more complex than that for a single-phase cooling and heating described in Section 6.3 and 6.4. In condensers there are three mechanisms to be considered for the rating procedure:

- The resistance to heat transfer of the condensing film
- The resistance to heat transfer of the vapour cooling
- The resistance to heat transfer of the condensate film cooling

Each of these mechanisms is treated separately and along preselected sections of the exchanger. The procedure for determining the last two of the mechanisms follows that described in Section 6.3 for single-phase heat transfer. The following expression is used to calculate the film coefficient for the condensing vapour:

$$h_c = \frac{8.33 \times 10^3}{(M_c/L_c . N_s)^{0.33}} \times k_f \times \left(\frac{SG^2}{\mu_f}\right)^{0.33}$$

where

h_c = condensing film coefficient
M_c = mass condensed (lb/h)
L_c = tube length for condensation
    $= \dfrac{A_{zone}}{A} \times (L - 0.5)$
$N_s = 2.08\, N_t^{0.495}$ for triangular pitch
k_f = thermal conductivity of condensate at film temperature
SG = specific gravity of condensate
μ_f = viscosity of condensate at film temperature (cP)

Again there are many excellent computer programs that calculate condenser thermal ratings, and these, of course, save the tedium of manual calculation. However, no matter which method of calculation is selected there is required one

major additional piece of data over that necessary for single-phase heat exchange.
That is the enthalpy curve for the vapour.

Enthalpy curves are given as the heat content per pound or per hour contained
in the mixed phase condensing fluid plotted against temperature. An example of
such a curve is given in Figure 6.12. These enthalpy curves are developed from
the vapour/liquid or flash calculations described in Section 1.4. Briefly, the
calculation for the curve commences with determining the dew point of the
vapour and the bubble point of the condensate. Three or more temperatures are
selected between the dew and bubble points and the V/L calculation of the fluid
at these temperatures carried out. Enthalpy for the vapour phase and the liquid
phase are added for each composition of the phases at the selected temperatures.
These together with the enthalpy at dew point and bubble point are then plotted.

As in the case of the shell and tube exchanger (Section 6.3) and the air cooler
(Section 6.4), a manual calculation for condensers is described here. Again this is
done to provide some understanding of the data required to size such a unit and
its significance in the calculation procedure. Computer-aided design should,
however, be used for these calculations whenever possible.

The following steps describe a method for calculating the film coefficient of a
vapour condensing on the shell side of a shell and tube exchanger. The complete
rating calculation will not be given here as much of the remaining calculation is
simply repetitive.

- *Step 1* Calculate the dew point of the vapour stream at its source pressure.
  Estimate the pressure drop across the system. Usually 3–5 psi will account for
  piping and the exchanger pressure drop. Calculate the bubble point of the

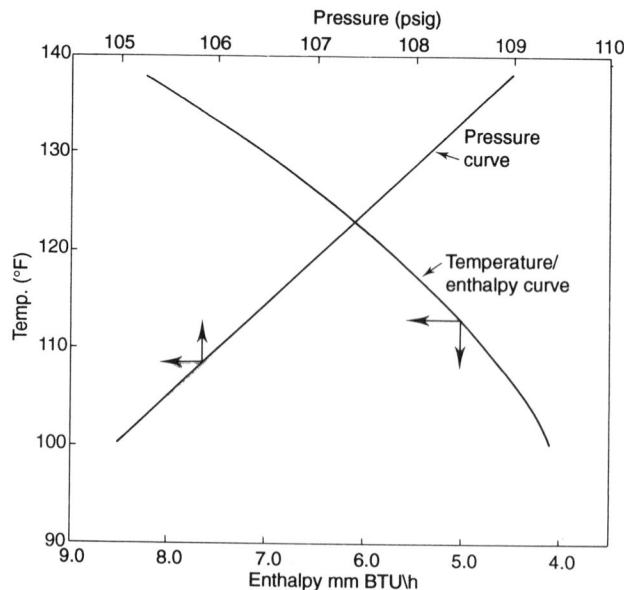

Figure 6.12. Enthalpy curve for a debutanizer overhead condenser

condensate at the terminal pressure. Selected three or more temperatures between dew point and bubble point and calculate the vapour/liquid quantities at these conditions of temperature and pressure.

- *Step 2* Calculate the enthalpy of the vapour and liquid at these temperatures. Plot the total enthalpies against temperature to construct the enthalpy curve. Establish the properties of the vapour phase and liquid phase for each temperature interval. The properties mostly required are SG, viscosity, mole wt, thermal conductivity and specific heats.
- *Step 3* In the case of a water cooler calculate the duty of the exchanger and the quantity of water in lb/h. Commence the heat transfer calculation by assuming an overall heat transfer coefficient (use the data given in Table A1.1 in Appendix 1). Calculate the corrected LMTD and the surface area.
- *Step 4* Using the surface area calculated in step 3 define the exchanger geometry in terms of number of tube passes, number of tubes on the centreline, shell diameter, baffle arrangement and the shell free flow area. Calculate also the water flow in feet per second.
- *Step 5* Divide the exchanger into three or four zones by selecting the zone temperatures on the enthalpy curve. Calculate the average weight of vapour and the average weight of condensate in each zone. Using these averages, calculate the average heat transferred for:

Cooling of the vapour Q_v

Cooling of the condensate Q_l

Condensing of the vapour which will be:

Total heat in the zone (from the enthalpy curve) less the sum of Q_v and Q_l.

- *Step 6* Calculate the film coefficient for the tube side fluid (see Section 6.3).
- *Step 7* Starting with zone 1 and knowing the outlet temperature of the coolant fluid, the total heat duty of the zone, and the shell side temperatures calculate the coolant inlet temperature. Using this calculate the LMTD for the zone. Assuming a zone overall heat transfer coefficient U calculate a surface area for the zone. Using this and the total exchanger area estimated in step 4 establish L_c in feet.
- *Step 8* Calculate the condensing film coefficient from the equation given earlier. This will be an uncorrected value for h_c. This will be corrected to account for turbulence by the expression

$$h_{c\,(corr)} = h_c \times (G_v/5)^{0.7}$$

where G_v = average vapour mass velocity (lb/h.ft^2)

- *Step 9* Calculate the value of G_v using the free flow area allocated to the vapour γ_v. The following expressions are used for this:

$$\gamma_v = 1 - \gamma_l$$

$$\frac{1}{\gamma_l} = 1 + \frac{\text{Ave. mass vapour}}{\text{Ave. mass liquid}} \times (\mu_v/\mu_l)^{0.111} \times (\rho_l/\rho_v)^{0.555}$$

$$G_v = \frac{\text{Ave. mass vapour}}{25 \times \text{free flow area} \times \gamma_v}$$

- *Step 10* Calculate the film coefficient h_v for the vapour cooling mechanism. This will be the procedure used for a single-phase cooling given in Section 6.3. This is corrected to account for resistance of the condensate film by the expression

$$\frac{1}{h_{v\,\text{corr}}} = \frac{1}{h_c} + \frac{1}{1.25h_v}$$

- *Step 11* Calculate the film coefficient for the condensate cooling mechanism. Again this is the procedure described in Section 6.3 for single-phase cooling on the shell side. This is corrected for drip cooling that occurs over a tube bank:

$$\text{Drip cooling } h_{dc} = 1.5 \times h_c$$

$$\text{and } h_{l\,\text{corrected}} = \frac{2 \times h_{dc} \times h_l}{h_{dc} + h_l}$$

- *Step 12* Calculate the total zone film coefficient h_0 using the following expression:

$$h_0 = \frac{Q_{\text{zone}}}{(Q_c/h_c) + (Q_v/h_v) + (Q_l/h_l)}$$

where Q_c, Q_v, Q_l are the enthalpies for condensing, vapour cooling, and condensate cooling respectively.

- *Step 13* Calculate the overall heat transfer coefficient neglecting the shell side coefficient from step 12. Thus:

$$\frac{1}{U_x} = r_0 + r_w + r_{i0} + R_{i0}$$

where r_0, r_w, r_{i0} are fouling factors for shell fluid, wall, and tube side fluid respectively, and R_{i0} is the tube side film coefficient calculated in step 6.

- *Step 14* Calculate the overall heat transfer coefficient U_{zone} for the zone using the expression

$$U_{\text{zone}} = \frac{h_0 \times U_x}{h_0 + U_x}$$

Check the calculated U against the assumed for the zone. Repeat the calculation if necessary to make a match.

- *Step 15* Calculate the zone area using the acceptable calculated U. Repeat steps 7 to 14 for the other zones. The total surface area, etc. is the sum of that for each zone.

Note that the following example calculation is carried out for zone 1 only. The remaining calculation is simple repetition.

Example calculation

A debutanizer tower operates at 123 psia and 138 °F at the tower top. The total overhead (reflux and product) is condensed in a shell and tube water cooled condenser to its bubble point at 120 psia and 100 °F. Calculate the surface area required by the condenser. The total mass flow is 27 027 lb/h with a mole wt of 52.3 and 4.58 lb/gal. The composition of the stream is as follows:

	Moles/h
C_2	24.96
C_3	169.65
iC_4	62.79
nC_4	250.38
iC_5	8.97

1 Calculating dew point and bubble point

Bubble point calculation—reflux drum conditions

	Mole fract.	1st trial at 120 psig		2nd trial at 110 psig	
		K	$Y = Kx$	K	$Y = Kx$
C_2	0.048	3.7	0.178	4.2	0.202
C_3	0.328	1.3	0.426	1.43	0.469
iC_4	0.122	0.62	0.076	0.65	0.079
nC_4	0.485	0.46	0.223	0.48	0.233
iC_5	0.017	0.20	0.003	0.22	0.004
Total	1.000		0.906		0.987

Second trial: use K for C_3:

$$\frac{1.3}{0.906} = 1.43$$

Corresponding pressure = 125 psia (110 psig)

$$\text{Actual pressure} = \frac{1.43}{0.987} = 1.45$$

corresponding to 120 psia or 105 psig

Debutanizer reflux drum will operate at 100 °F and 105 psig for the *total* condensation of overhead product and reflux.

Predicting the tower top pressure:

Reflux drum pressure	= 105 psig
Overhead line pressure drop =	1 psi
Condenser pressure drop	= 3 psi
Total tower top pressure	= 109 psig

	Mole fract. Y	Tower top temperature			
		1st trial at 165 °F		2nd trial at 138 °F	
		K	$x = Y \pm K$	K	$x = Y \pm K$
C_2	0.048	5.2	0.009	4.8	0.010
C_3	0.328	2.50	0.131	2.00	0.164
iC_4	0.122	1.38	0.088	1.00	0.122
nC_4	0.485	1.0	0.485	0.746	0.650
iC_5	0.017	0.52	0.033	0.360	0.047
Total	1.000		0.746		0.993

Second trial: use $NC_4 K$ value:

$$\text{New } K = 1.0 \times 0.746$$
$$= 0.746$$

Corresponding to 138 °F

This trial is close enough.

Tower top conditions
Pressure 109 psig 123.7 psia
Temperature 138 °F

2   Calculating flash curve points

At 123 psia and 130 °F

			Assume $V/L = 0.3$					
Comp.	Mol/h	K	$V/L.K$	$L = \dfrac{F}{1 + V/LK}$	Mol Vapour	Lb Vapour	Lb Liquid	gph Liquid
C_2	24.96	5.5	16.5	1.43	23.53	706	43	14
C_3	169.65	1.8	5.4	26.5	143.15	1299	1167	276
iC_4	62.79	0.95	2.85	16.31	46.48	2996	946	202
nC_4	250.38	0.67	2.01	83.18	167.2	9698	4826	993
iC_5	8.97	0.33	0.99	4.5	4.46	321	325	63
Total	516.75			131.93	384.02	19720	7307	1548

Calculated $V/L = 2.92$
MW vapour = 51.2
MW liquid = 55.4
SG liquid = 0.567

At 122 psia and 120 °F

Comp.	Mol/h	K	Assume $V/L = 0.5$		Mol Vapour	Lb Vapour	Lb Liquid	gph Liquid
			$V/L.K$	$L = \dfrac{F}{1 + V/LK}$				
C_2	24.96	5.1	2.55	7.03	17.93	538	211	71
C_3	169.65	1.69	0.845	91.95	77.70	3419	4046	957
iC_4	62.79	0.86	0.43	43.91	18.88	1096	2547	544
nC_4	250.38	0.60	0.30	192.6	57.78	3353	11171	2299
iC_5	8.97	0.27	0.35	7.90	1.07	77	569	109
Total	516.75			343.39	173.36	8483	18544	3980

Calculated $V/L = 0.503$
MW vapour = 49
MW liquid = 54
SG liquid = 0.56

At 120 psia and 110 °F

Comp.	F Mol/h	K	Assume $V/L = 0.1$		Mol Vapour	Lb Vapour	Lb Liquid	gph Liquid
			$V/L.K$	$L = \dfrac{F}{1 + V/LK}$				
C_2	24.96	4.8	0.48	16.86	8.10	243	506	
C_3	169.65	1.5	0.15	147.52	22.13	974	6492	
iC_4	62.79	0.73	0.073	58.52	4.27	248	3394	
nC_4	250.38	0.54	0.054	237.55	12.83	746	13778	
iC_5	8.97	0.24	0.024	8.76	0.21	15	631	
Total	516.75			469.21	47.54	2226	24801	

Calculated $V/L = 0.101$
MW vapour = 46.8
MW liquid = 52.9
SG liquid = 0.553

3 Calculating enthalpies at the flash intervals

At 130 °F

$$\text{Vapour} = 19\,720 \text{ lb/h} \times 300 \text{ BTU/lb} = 5\,916\,000 \text{ BTU/h}$$
$$\text{Liquid} = 7307 \text{ lb/h} \times 172 \text{ BTU/lb} = 1\,257\,000 \text{ BTU/h}$$
$$\text{Total} = 7\,173\,000 \text{ BTU/h}$$

At 120 °F

$$\text{Vapour} = 8483 \text{ lb/h} \times 320 \text{ BTU/lb} = 2\,715\,000 \text{ BTU/h}$$
$$\text{Liquid} = 18\,544 \text{ lb/h} \times 165 \text{ BTU/lb} = 3\,060\,000 \text{ BTU/h}$$
$$\text{Total} = 5\,775\,000 \text{ BTU/h}$$

At 110 °F

$$\text{Vapour} = 2226 \text{ lb/h} \times 315 \text{ BTU/lb} = 702\,000 \text{ BTU/h}$$
$$\text{Liquid} = 24801 \text{ lb/h} \times 160 \text{ BTU/lb} = 3\,968\,000 \text{ BTU/h}$$
$$\text{Total} = 4\,670\,000 \text{ BTU/h}$$

At dew point

$$\text{Vapour} = 27\,027 \text{ lb/h} \times 304 \text{ BTU/lb} = 8\,216\,000 \text{ BTU/h}$$

At bubble point

$$\text{Liquid} = 27\,027 \text{ lb/h} \times 150 \text{ BTU/lb} = 4\,054\,000 \text{ BTU/h}$$

These points are plotted in Figure 6.12 to give an enthalpy curve for the condenser. The total duty of the condenser is

$$8\,216\,000 - 4\,054\,000 = 4\,162\,000 \text{ BTU/h}$$

4   Calculating the flow of water coolant

Water inlet is 60 °F and outlet is fixed at 90 °F. Then

$$
\begin{aligned}
\text{lb/h of water} &= \frac{Q}{(90 - 60)} \\
&= \frac{4\,162\,000}{30} \\
&= 138\,733 \text{ lb/h}
\end{aligned}
$$

5   Calculating approximate surface area of condenser

From Table A1.1 in Appendix 1 let U be 90 and the exchanger LMTD corrected is as follows:

$$
\begin{array}{ccc}
\Delta T_e & 138 \rightarrow 100 & \\
& \underline{90 \leftarrow 60} & \\
& \underline{48} & \underline{40} \\
\end{array}
$$

$$\Delta T_e = \ 44\,°F$$

Assume two or more tube passes and one shell pass. Then

$$J = \frac{90 - 60}{138 - 60} = 0.385$$

$$R = \frac{138 - 100}{90 - 60} = 1.27$$

$$F = 0.89$$

$$\Delta T_m = 39\,°F$$

$$Q = UA\Delta T_{\mathrm{m}}$$

$$A = \frac{Q}{U\Delta T_{\mathrm{m}}} = \frac{4\,162\,000}{90 \times 39}$$

$$= 1186 \text{ sqft}$$

6 Establishing the condenser configuration

Tubes will be Admiralty brass $\frac{3}{4}$-in. 14 BWG on triangular pitch with standard 16-ft length:

$$\text{Number of tubes } NT = \frac{3.82 \times \text{total surface area}}{(L - 0.5) \times d_{\mathrm{o}}}$$

$$= \frac{3.82 \times 1186}{15.5 \times 0.75}$$

$$= 390$$

Assume four tube passes. Then tubes per pass = 98

$$\text{Cross-sectional area of tube side} = \frac{98 \times 0.286}{144}$$

$$= 0.182 \text{ ft}^2$$

$$\text{Water linear velocity } V = \frac{\text{lb/h}}{62.1 \times 3600 \times 0.182}$$

$$= \frac{138\,733}{40\,687} = 3.4 \text{ ft/s}$$

$$\text{Number of tubes across centreline} = NT_{\mathrm{C}}$$

$$= 1.10 \times (NT)^{0.5}$$

$$= 22.8$$

$$\text{Outer tube limit} = (NT_{\mathrm{C}} - 1)P_{\mathrm{T}} + d_{\mathrm{o}}$$

$$\text{Where } PT \text{ is tube pitch} = 1 \text{ in.}$$

$$= (21.8 \times 1) + 0.75$$

$$= 22.55 \text{ in.}$$

$$\text{Shell i/d} = 22.55/0.9 = 25 \text{ in. min.}$$

$$\text{Add 3 in. shell i/d} = 28 \text{ in.}$$

$$\text{Fix baffle pitch at 20\% of shell i/d} = 5.6 \text{ (say 6 in.)}$$

$$\text{Free width for vapour } W = D - (d_{\mathrm{o}} \times NT_{\mathrm{C}})$$

$$= 28 - (0.75 \times 21.8)$$

$$= 11.75 \text{ in.}$$

Total free area of flow between two
adjacent baffles
$$= W(P_{\mathrm{b}} - 0.187)$$

$$= 67.7 \text{ in.}$$

$$\text{Number of baffles } N_{\mathrm{B}} = 10L/P_{\mathrm{B}}$$

$$= 160/6 = 27$$

7   Calculating the tube side film coefficient h_{i0}

$$h_{i0} = \frac{300}{d_o}(V.d_i)^{0.8}$$

where

h_{i0} = tube side film coefficient (BTU/h.ft^2.°F)

d_o and d_i = outside and inside diameters = 0.75 in. and 0.584 in.

V = linear velocity of water in tubes = 2.85 ft/s

h_{i0} = 601 BTU/h.ft^2.°F

*8   Dividing the condenser into temperature zones establishing zone
     conditions*

Divide the enthalpy curve into four zones as follows:

Zone 1 = Dew point (138 °F) to 130 °F.
Zone 2 = 130 °F to 120 °F
Zone 3 = 120 °F to 110 °F
Zone 4 = 110 °F to 100 °F

Calculate the enthalpy attributed to:

- Cooling the vapour
- Condensing the vapour
- Cooling the condensate

for each zone. Average mass flow of liquid and vapour in each zone will be used
for this.

Zone 1:

$$\text{Average vapour: } \frac{\text{Vapour in} + \text{vapour out}}{2} = 23\,374 \text{ lb/h}$$

$$\text{Vapour cooling} = 23\,374 \times C_p \times (138 - 130)$$
$$= 23\,374 \times 0.43 \times 8$$
$$= 80\,405 \text{ BTU/h} = Q_{v1}$$

$$\text{Condensate cooling} = \frac{7307}{2} \times 0.78 \times 8$$
$$= 23\,000 \text{ BTU/h} = Q_{l1}$$

$$\text{Total to cooling mechanism} = 80\,405 + 23\,000$$
$$= 103\,405 \text{ BTU/h}$$

$$\text{From enthalpy curve: total heat in Zone 1} = 1\,043\,000 \text{ BTU/h}$$

$$\text{Enthalpy due to condensing} = 1\,043\,000 - 103\,405$$
$$= 940\,000 \text{ BTU/h} = Q_{c1}$$

The remaining zone enthalpies are similarly calculated to give the following
table:

Zone	Q_v	Q_1 mm BTU/h	Q_c	Q_{total}
1	0.08	0.023	0.94	1.043
2	0.061	0.101	1.236	1.398
3	0.024	0.169	0.912	1.10
4	0.005	0.202	0.409	0.616
Total	0.170	0.495	3.497	4.162

9   Zone 1 calculations to determine zone surface area

Calculating temperature of water into the zone:

$$\text{Temp. of water out} = 90\,°\text{F}$$

$$\text{Then temp. of water in} = \frac{1\,043\,000}{138\,733} = 90 - x$$

$$x = 82.5\,°\text{F}$$

Calculating LMTD across the zone:

$$\Delta T_{zl} = 138 \rightarrow 130$$
$$\phantom{\Delta T_{zl} =}\ 90 \leftarrow 82.5$$
$$\phantom{\Delta T_{zl} =}\ \overline{\ 48 \qquad 47.5\ }$$
$$= \text{say } 48\,°\text{F}$$

Calculating percentage of tube allocated to zone 1 (approx.):

$$\text{Assume } U_{\text{zone 1}} = 120\,\text{BTU/h.ft}^2.°\text{F}$$

$$\text{Then approx. area} = \frac{1\,043\,000}{120 \times 48}$$

$$= 181\,\text{ft}^2$$

$$\% \text{ of tube} = \frac{181}{1180} = 15.3\%$$

$$L_c = 15.3\%\ (L - 0.5)$$
$$= 2.37\,\text{ft}$$

For calculating condensing film coefficient h_{cl} use the following expression:

$$h_c = \frac{8.33 \times 10^3}{(M_c/L_c.N_s)^{0.33}} \times k_f \times \left(\frac{SG^2}{\mu_f}\right)^{0.33}$$

where

h_{cl} = condensing film coefficient
M_c = mass condensed in lb/h = 7307
L_c = tube length for condensation = 2.37 ft
N_s = 2.08 $N_t^{0.495}$ for triangular pitch = 27.61
k_f = Thermal conductivity of condensate at film temperature = 110 °F

SG = Specific gravity of condensate = 0.53
μ_f = viscosity of condensate at film temperature (cP) = 0.13.

$$h_{cl} = \frac{8.33 \times 1000}{5.14} \times 0.0785 \times 1.2928$$

$$= 177.3 \text{ BTU/h.°F.ft}^2 \text{ (uncorrected)}$$

Calculating G_{v1} average vapour mass velocity in lb/h.ft^2:
Fraction of free flow area for vapour is

$$\gamma_{v1} = 1 - \gamma_{l1}$$

where

$$\frac{1}{\gamma_{l1}} = 1 + \frac{\text{Ave. mass vap.}}{\text{Ave. mass liquid}} \times \left(\frac{\mu_v}{\mu_l}\right)^{0.111} \times \left(\frac{\rho_l}{\rho_v}\right)^{0.555}$$

$$\frac{1}{\gamma_{l1}} = 1 + \frac{23\,374}{3653} \times \left(\frac{0.014}{0.13}\right)^{0.111} \times \left(\frac{35}{1.0}\right)^{0.555}$$

$$= 36.89$$

Then $\gamma_{l1} = 0.027$

$$\gamma_{v1} = 1 - 0.027 = 0.973$$

$$G_{v1} = \frac{\text{Ave. mass vapour}}{25 \times \text{free flow area} \times 0.973}$$

$$= \frac{23\,374}{25 \times 67.7} \times 09973$$

$$= 14.19 \text{ lb/h.ft}^2$$

Correcting h_{cl}:

$$h_{cl\,corr} = h_{cl} \times (G_{v1}/5)^{0.7}$$

$$= 177.26 \times 2.075$$

$$= 368$$

Calculating film coefficient for vapour cooling h_{v1}:

$$h_{v1} \text{ (uncorrected)} = \frac{k}{d_o} (c\mu_f/k)^{0.33} \times \phi(d_o G_{v1}/\mu_f) \times 0.7$$

where

$$k = \text{thermal cond.} = 0.011 \text{ BTU/h.ft}^2.\text{°F/ft}$$
$$c = \text{specific heat} = 0.4 \text{ BTU/lb}$$
$$\mu_f = \text{viscosity at film temp.} = 0.014 \text{ cP}$$
$$d_o = \text{tube o/d} = 0.75 \text{ in.}$$
$$Re \text{ no.} = \frac{d_o.G_{v1}}{0.014} = 760$$

$$\phi(d_o.G_{v1}/\mu_f) = 5300 \text{ from Figure 6.4}$$

$$h_{v1} = \frac{0.011}{0.75} (0.4 \times 0.014/0.011)^{0.33} \times 5300 \times 0.7$$

$$= 62.2 \times 0.7 = 43.54 \text{ BTU/h.ft}^2.\text{°F.}$$

Calculating h_{v1} corrected:

$$h_{v1} \text{ corrected} = \frac{1}{h_{cl}} + \frac{1}{1.25 \times h_{v1}}$$
$$= 0.0028 + 0.0183$$
$$= 47.6$$

Calculating film coefficient for condensate cooling h_{11}:

Liquid mass velocity $G_{11} = \dfrac{M_{11}}{25 \times \text{free flow area} \times \gamma_{11}}$

$$= \frac{3653}{25 \times 67.7 \times 0.027}$$
$$= 79.9 \,\text{lb/h.ft}^2$$

$$Re \text{ no.} = \frac{0.75 \times 79.9}{0.13} = 461$$

$\phi(d_o.G_{11}/\mu_{1f}) = 4000$ from Figure 6.2

$$h_{11} = \frac{k_1}{d_o} (c\mu_{1f}/k_1)^{0.33} \times 4000 \times 0.7$$
$$= \frac{0.0785}{0.75} (0.79 \times 0.13/0.0785)^{0.33} \times 4000 \times 0.7$$
$$= 457 \text{ (uncorrected)} \times 0.7$$
$$= 320 \,\text{BTU/h.ft}^2.°F$$

Correcting h_{11} for drip cooling:

$$h_{dcl} = 1.5 \times h_{cl}$$
$$= 552$$

$$\text{Corrected } h_{11} = \frac{2 \times h_{dcl} \times h_{11}}{h_{dcl} + h_{11}}$$
$$= \frac{2 \times 552 \times 320}{552 + 320}$$
$$= 405 \,\text{BTU/h.ft}^2.°F$$

Calculating zone shell side film coefficient h_{01}:

$$h_{01} = \frac{Q_{zone}}{(Q_c/h_{cl}) + (Q_v/h_{v1}) + (Q_l/h_{11})}$$
$$= \frac{1\,043\,000}{(940\,000/368) + (80\,000/47.6)\,(23\,000/405)}$$
$$= 243 \,\text{BTU/h.ft}^2.°F$$

Calculating shell side overall heat transfer coefficient for zone 1:

$$\frac{1}{U_{zl}} = \frac{1}{h_{0l}} + \frac{1}{h_{j0}} + r_{i0} + r_0 + r_w$$

$$= \frac{1}{243} + \frac{1}{601} + 0.0013 + 0.001 + neg$$

$$= 0.0083$$

$$U_{zl} = 123 \text{ BTU/h.ft}^2.°F \text{ which is close to assumed of } 120$$
$$\text{BTU/h.ft}^2.°F.$$

$$\text{Zone 1 surface area} = \frac{1\,043\,000}{123 \times 48}$$

$$= 177 \text{ ft}^2$$

This concludes the calculation to determine the surface area for the first zone. The assumed area at the beginning of the calculation was 181 ft^2 which is within 10% of that calculated. Then the calculated area of 177 ft^2 is acceptable. The surface areas for the remaining zones are calculated in the same way. The results of these calculations are summarized as follows:

	ft^2
Zone 1	177.0
Zone 2	303.0
Zone 3	315.5
Zone 4	331.5
Total	1127.0

which is 5.2% more than the assumed area at the beginning of the calculation. This should not warrant a revision but the largest calculated area should now be used for any further work.

6.6 Reboilers

Reboilers are used in fractionation to provide a heat source to the system, and to generate a stripping vapour stream to the tower. They are operated by either the natural circulation of a fluid or forced circulation of the fluid to be reboiled. This section deals only with natural circulation reboilers. There are three common types of reboilers:

- The kettle
- The once-through thermosyphon
- The recirculating thermosyphon

THE KETTLE REBOILER

The following is a sketch of a typical kettle reboiler showing its plot layout relative to the fractionation tower. More details of a kettle reboiler are given in Figure 6.13.

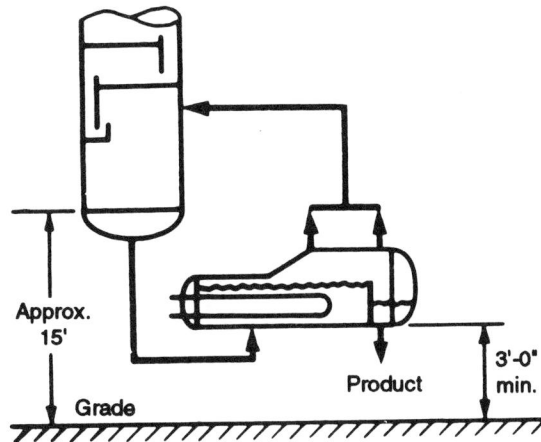

This type of reboiler is extremely versatile. It can handle a very wide range of vaporization loads (eg. when used as LPG vaporizer for fuel gas purposes it vaporizes 100% of the feed). The equipment consists of a large shell into which is fitted a tube bundle through which the heating medium flows. The liquid to be reboiled enters the bottom of the shell at the end adjacent to the tube inlet/outlet chamber. The liquid is boiled and partially vaporized by flowing across the tube bundle. The diameter of the shell is sized such that there is sufficient space above the tube bundle and the top of the shell to allow some disengaging of the liquid and vapour. A baffle weir is installed at the end of the tube bundle furthest from the inlet. This baffle weir establishes a liquid level over the tube bundle in the shell. The boiling liquid flows over this weir to the shell outlet nozzle, while the vapour generated is allowed to exit from the top of the shell through one or two nozzles.

The space downstream of the weir is sized for liquid hold-up to satisfy the surge requirements for the product. Thus it is not necessary to provide space in the bottom of the tower for product surge. If the heating medium is non-fouling it is permissible to use U-tubes for the tube bundle. Otherwise the tube bundle must be of the floating-head type. The kettle type reboiler should always be the first to be considered if there are no elevation constraints to pumping the bottoms product away.

ONCE-THROUGH THERMOSYPHON REBOILER

This type of reboiler and its location relative to the tower is given in the following sketch.

This type of reboiler should be considered when a relatively high amount of surge is required for the bottom product and when it is necessary to provide head for the product pump (NPSH requirement).

This reboiler takes the liquid from the bottom tray of the fractionator as feed. This stream enters the shell side of a vertical single tubepass shell and tube exchanger by gravity head to the bottom of the shell. The heating medium flows

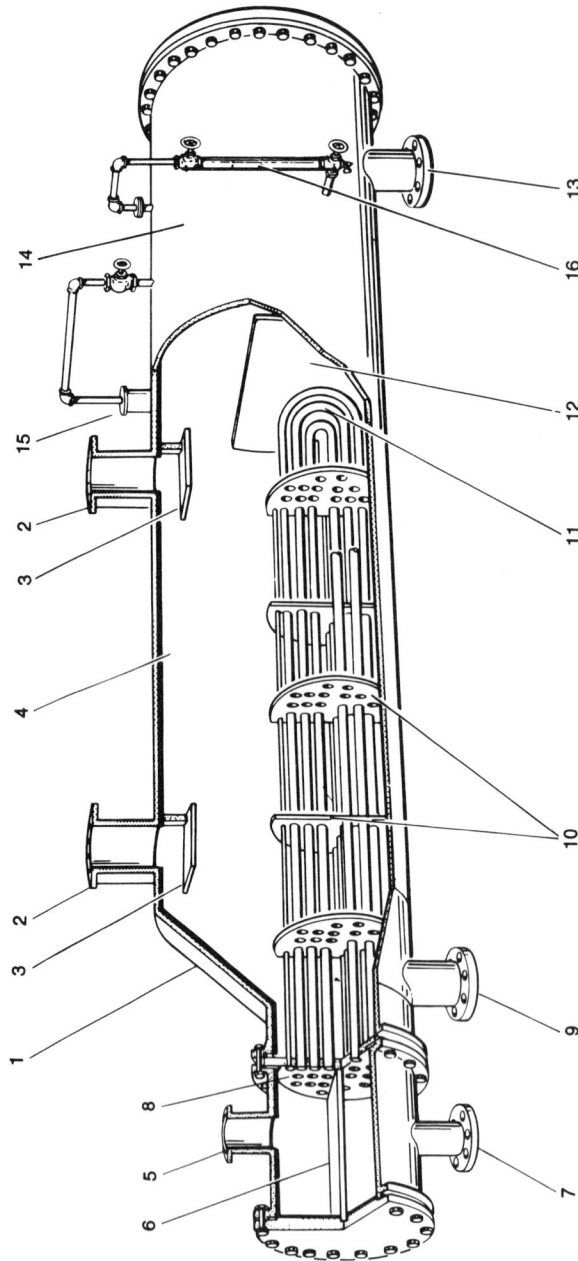

Figure 6.13. The components of a kettle reboiler. 1 Shell, 2 shell outlet nozzles (vapour), 3 Entrainment baffles, 4 vapour-disengaging space, 5 channel inlet nozzle, 6 channel partition, 7 channel outlet nozzle, 8 tube sheet, 9 shell inlet nozzle, 10 tube support plates, 11 U-tube returns, 12 weir, 13 shell outlet nozzle (liquid), 14 liquid hold-up (surge) section, 15 top of level—instrument housing (external displacer), 16 liquid level gauge

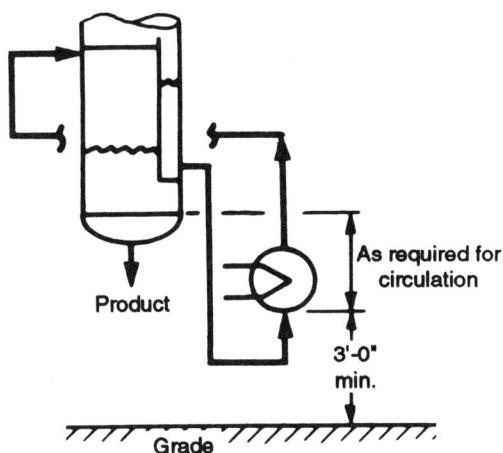

tube side to partially vaporize the liquid feed. A syphoning effect is caused by the difference in density between the reboiler feed and the vapour/liquid effluent. This allows the reboiler effluent to exit from the top of the shell side and re-enter the tower where the vapour disengages from the liquid phase. The liquid is the bottom product of the fractionator and is discharged from the bottom of the tower.

Both the kettle and the once-through thermosyphon types constitute a theoretical tray as regards fractionation. Unlike the kettle reboiler, the once-through thermosyphon is limited to a vaporization of not more than 60% of the feed. The low hold-up of the feed from a tray results in severe surging through the reboiler at high vaporization rates.

THE RECIRCULATING THERMOSYPHON REBOILER

When a vaporization rate higher than 60% of reboiler feed is required and a kettle reboiler is unsuitable a recirculating thermosyphon type reboiler should be considered. A sketch showing this type of reboiler is given below.

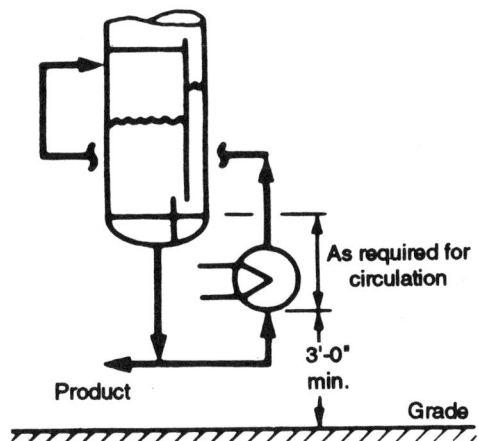

This reboiler is similar to the once-through thermosyphon in that it operates by flowing a liquid feed through the shell side of the vertical reboiler by the syphon mechanism. In the case of the recirculating reboiler, however, the feed to the reboiler is a stream of the bottom product from the fractionator. This is vaporized as described earlier and the liquid/vapour effluent returned to the tower. The vaporization by this reboiler can exceed 60% without danger of surging. However, vaporization in this type of reboiler should not exceed 80%. Its action is directed solely to inputting heat only to the tower and because it recycles the same composition stream to the tower bottom it cannot be considered as a theoretical fractionating tray (although some amount of fractionation does occur in this system).

Note in the description of both thermosyphon type reboilers the heating fluid is shown as flowing tube side. There may be cases where this stream will be routed shell side and the reboiler fluid directed tube side. Some guidance to this selection is provided by the following preference for tube side fluid:

(1)   Corrosive or fouling fluids
(2)   The less viscous of the two fluids
(3)   The fluid under the highest pressure
(4)   Condensing steam

REBOILER SIZING

As in the case of most heat exchangers, the sizing calculation is quite rigorous and complex. Normally, process engineers rarely need to compute this in detail. There will be a need, however, to estimate the size of these items for cost purposes or for plot layout studies. This sizing is greatly simplified by applying heat flux quantities to the predetermined reboiler duty. Heat flux is the value of heat transferred per unit time per ft^2 of surface. The following list gives a range of heat fluxes that have been used in design and observed in operating units.

	Design	Observed
	\multicolumn{2}{c}{(BTU/h.ft^2)}	
Kettle type	12 000	15 000–20 000
Once-through	15 000	17 500+
Recirculating	15 000	up to 20 000
Forced circulation	20 000	–

The duty of the reboiler is obtained by the overall heat balance over the tower. This is accomplished by equating the total heat out of a fractionating tower to the heat supplied, making the reboiler duty the unknown in the heat-supplied statement. Now the heat out of the fractionator is the total heat in the products leaving plus the condenser duty. The heat supplied to the tower is the heat brought in with the feed, and the heat supplied by the reboiler.

Example calculation

The feed to a fractionator is 87 960 lb/h of mixed hydrocarbons. It enters the tower as a vapour and liquid stream and has a total enthalpy of 15.134 mmBTU/h.

The overhead products are a distillate and a vapour stream at 95 °F. The vapour is 1590 lb/h with an enthalpy of 320 BTU/lb. The distillate is 8028 lb/h with an enthalpy of 170 BTU/h. The bottom product from the tower is 78 342 lb/h and leaves as a liquid at its boiling point at 440 °F. Its enthalpy is 370 BTU/lb. The overhead condenser duty is 4.278 mmBTU/lb. Calculate the reboiler duty.

Calculation

Calculate the reboiler duty from the overall tower heat balance as in the following:

	V/L	°API	°F	lb/h	Enthalpy	
					BTU/lb	MMBTU/h
In						
Feed	VL	–	300	87 960	–	15.134
Reboiler duty						x
Total in				87 960		$15.134 + x$
OUT						
Bottom prod.	L	–	440	78 342	370	28.986
O/head dist.	L	–	95	8 028	170	1.364
O/head vapour	V	–	95	1 590	320	0.508
Condenser duty						4.278
Total out				87 960		35.136

$$\text{Heat in} = \text{heat out}$$

Then

$$15.134 + x = 35.136$$
$$\text{Reboiler duty } x = 20.002 \text{ mm BTU/h}$$

Using a heat flux of 15 000 BTU/h.ft^2 the surface area for the reboiler becomes

$$\frac{20\,002\,000}{15\,000} = 1333.5 \text{ ft}^2$$

ESTIMATING THE LIQUID AND VAPOUR FLOW FROM THE REBOILER

It is necessary to know the vapour and liquid flow leaving the reboiler and entering the tower for the following reasons:

- To establish that there is sufficient vapour rising in the tower to strip the bottom product effectively
- To establish the vapour loading to the bottom tray for calculating the tray loadings
- To be able to calculate the driving force for flow through the exchanger in the case of thermosyphon reboilers.

The calculation of this flow is again based on a heat balance. In this case it is the heat balance across the reboiler itself. With the duty of the reboiler now established by the overall tower heat balance, as described above the balance over the reboiler can proceed as follows:

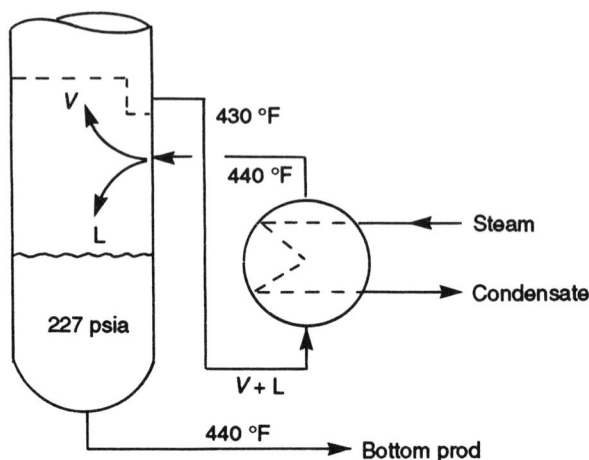

	V/L	°API	°F	lb/h	Enthalpy	
					BTU/lb	mmBTU/h
In						
Liquid from tray 1	L	–	430	$78342 + V$	369	28.908
						$+ 369V$
Reboiler duty						20.002
Total in				$7834 + V$		$48.910 + 369V$
OUT						
Bottom prod.	L	–	440	78342	370	28.986
Vapour to tray 1	V	–	440	V	458	$458 V$
Total out				$78342 + V$		$28.986 + 458V$

The temperature of the bottom tray (430 °F) is estimated from a straight-line temperature profile of the tower. As a rule of thumb, for a 30–40 tray tower the bottom tray will be about 10 °F lower than the bottom temperature:

$$\text{Again heat in} = \text{heat out}$$

$$\text{Then } 48.910 + 369V = 28.986 + 458V$$

$$V = 223\,865 \text{ lb/h}$$

Now the mole weight of the vapour is determined from the bubble point calculation of the bottom product used to determine the tower bottom temperature (see Section 1.3). In the case of the calculation example given above the bubble point calculation for the bottom product was as follows:

$$\text{Pressure at bottom of tower} = 220 + (30 \times 0.25)$$

$$= 227 \text{ psia}$$

	X_w	1st trial = 400 °F		2nd trial = 435 °F		MW	Weight factor	lb/gal	Vol. factor	
		K	$Y = XK$	K	$Y = XK$					
nC$_4$	0.017	2.7	0.046	3.1	0.053	58	3.1	4.86	0.64	
iC$_5$	0.047	1.9	0.089	2.2	0.103	72	7.4	5.20	1.42	°API 53.1
nC$_5$	0.055	1.7	0.094	1.9	0.105	72	7.6	5.25	1.45	
C$_6$	0.345	0.96	0.331	1.3	0.449	81	36.4	6.83	5.33	
C$_7$	0.322	0.44	0.142	0.59	0.190	102	19.4	6.84	2.84	
C$_8$	0.214	0.17	0.036	0.25	0.054	128	6.9	6.94	0.99	
Total	1.000		0.738		0.954	84.7	80.8	6.38	12.67	

Actual temperature = 440 °F

The mole weight of the vapour is that calculated for the Y column in the above table, which is 84.7.

CALCULATING THE PRESSURE HEAD DRIVING FORCE THROUGH A THERMOSYPHON REBOILER

The main advantage of thermosyphon reboilers is that there are no working parts, such as pumps, that can go wrong and cause failure. However, a major cause of failure in a thermosyphon reboiler is loss of driving force to move the fluid over or through the tube bundle. Much of this kind of problem occurs during commissioning when the reboiler has been incorrectly positioned relative to the tower nozzles, or during start-up where debris left after maintenance blocks one or other of the nozzles. In both these cases the problem is really the loss of pressure head that drives the fluid to be reboiled through the exchanger. The calculation to determine the theoretical driving force is based on the density of the incoming liquid, the head of that liquid to the inlet nozzle, the density of the out flowing liquid/vapour fluid, and its head. An example of a pressure driving force calculation based on a once through thermosyphon (as shown in the diagram below) is as follows. The flow data is based on the heat balance given earlier in this chapter.

The density of the liquid to the reboiler is 38.2 lb/ft^3 at 430 °F and the total flow is 302 207 lb/h:

$$\text{Hot ft}^3/\text{hr} = 7911$$
$$\text{Hot gpm} = 966$$

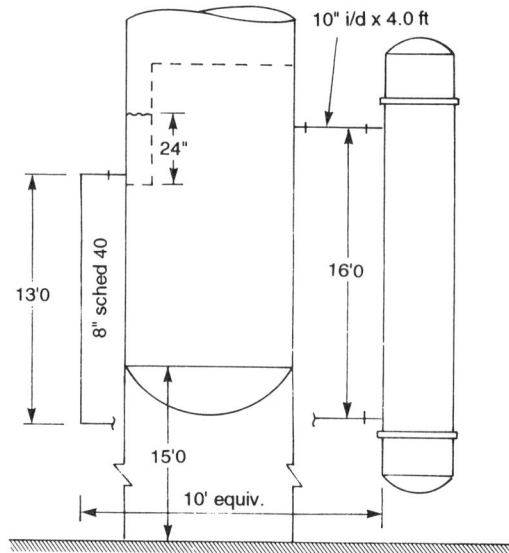

The transfer line from the tower to the bottom nozzle of the reboiler is a 8-in. schedule 40 seamless steel pipe. The head between the bottom of the tower draw-off pot and the reboiler nozzle is 13 ft. The equivalent horizontal line length including fittings is 15 ft.

From the friction loss tables in the appendix, head loss due to friction = 66.4 ft/100 ft of line (viscosity is taken as 1.1 cSt).

Total line length to the reboiler is 13 + 15 ft = 28 ft

$$\text{Head loss due to friction} = \frac{28 \times 66.4}{1000} = 1.85 \text{ ft}$$

$$\text{Head of liquid in draw-off pot} = 24 \text{ in.}$$
$$\text{Head of liquid to the reboiler inlet nozzle} = 13 + 2 \text{ ft}$$
$$= 15 \text{ ft}$$
$$\text{Pressure head at the reboiler inlet} = 15 - 1.85 \text{ ft}$$
$$= 13.15 \text{ ft}$$
$$= \frac{13.15 \times 38.2}{144} = 3.49 \text{ psi}$$

The density of the vapour/liquid stream leaving the reboiler is calculated as follows:

$$\frac{\text{Total mass of fluid}}{\text{ft}^3 \text{ liquid} + \text{ft}^3 \text{ vapour}}$$

$$\text{lb/ft}^3 \text{ of liquid (this is bottom product)} = 39.4 \text{ at } 440\,°F$$
$$\text{Mole wt of vapour (see bubble point}$$
$$\text{calculation above)} = 84.7$$
$$\text{ft}^3/\text{h of liquid} = \frac{78\,342 \text{ lb/h}}{39.9} = 1963.5$$

$$\text{ft}^3/\text{h of vapour} = \frac{223\,865\text{ lb/h}}{84.5} = 2643\text{ moles/h}$$

$$= \frac{2643 \times 378 \times 14.7 \times (460 + 440\,°\text{F})}{227 \times 520}$$

$$= 111\,974.6\text{ ft}^3/\text{h}$$

$$\text{Density of fluid from reboiler} = \frac{302\,207}{1963.5 + 111\,974.6}$$

$$= 2.65\text{ lb/ft}^3$$

In this case the fluid to be reboiled flows shell-side of the exchanger. The manufacturer's certified shell side pressure drop based on all vapour flow is 1.5 psi. The mixed phase pressure drop is calculated using Figure 6.14 thus:

$$m_v = \text{average mass of vapour} = \frac{223\,865}{2} = 111\,933\text{ lb/h}$$

$$m_l = \text{average mass of liquid} = \frac{302\,207 + 78\,342}{2} = 190\,275\text{ lb/h}$$

$$\rho_v = \text{average density of vapour} = \frac{2.0}{2} = 1.0\text{ lb/ft}^3$$

$$\rho_l = \text{average density of liquid} = \frac{38.8 + 39.9}{2} = 39.35\text{ lb/ft}^3$$

$$\text{Referring to Figure 6.14, } R_m = \frac{1}{(m_v/m_l) + (\rho_v/\rho_l)}$$

$$= \frac{1}{(111\,933/190\,275) + (1.0/39.32)}$$

$$= 1.63$$

From Figure 6.14

$$\alpha = 0.42$$

$$\Delta P_{\text{mixed phase}} = \alpha \times \Delta P_{\text{gas}}$$

$$= 0.42 \times 1.5\text{ psi}$$

$$= 0.63\text{ psi}$$

To calculate the total head of liquid from exchanger inlet to outlet nozzle:

Tube length is standard 16 ft
Assume bottom 20% of length is all liquid phase at a density of 38.8 lb/ft^3

Then head is $\dfrac{16 \times 0.2 \times 38.8}{144} = 0.862$ psi

Remaining head is mixed phase at a density of 2.65 lb/ft^3

Then head is $\dfrac{16 \times 0.8 \times 2.65}{144} = 0.24$ psi

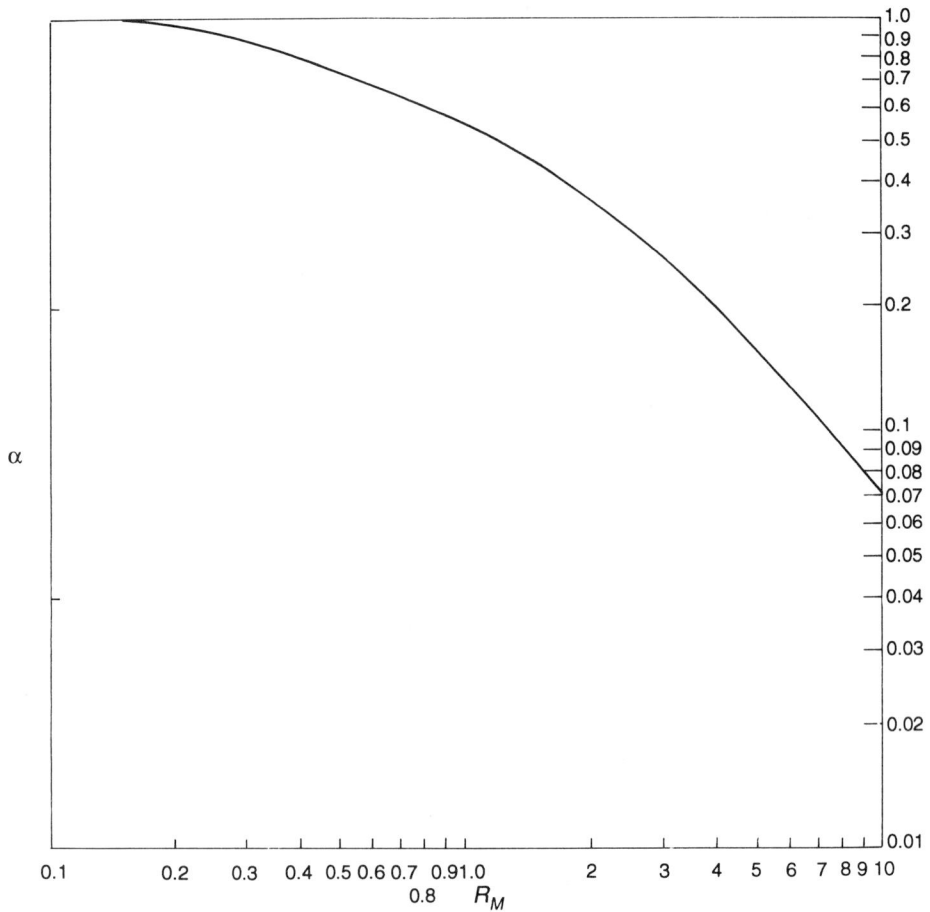

Figure 6.14. Two-phase pressure drop factor for flow across staggered tubes

Neglecting the small pressure drop due to friction in the 4 ft long return line to tower the total head to outlet nozzle =

Friction loss through exchanger = 0.630 psi
Lower section head = 0.862 psi
Upper section head = 0.240 psi
Total = 1.732 psi

Then driving force = pressure head available
— pressure head required.
= 3.49 psi − 1.732 psi
= 1.758 psi, which is satisfactory for good flow.

Note that the height of tower above grade is usually fixed by pump suction requirements in the first place. It may be adjusted upwards if necessary to accommodate a head to a reboiler. However, this necessity is quite rare. The transfer line to the reboiler should have its horizontal section at least 3.0 ft above grade to allow for maintenance, etc.

7   FIRED HEATERS

7.1   Types of Fired Heaters

This chapter provides some features and details of fired heaters. In this section the nature and purpose of fired heaters are explained. Most chemical plants contain a fired heater as a means of providing heat energy into a system. Because the equipment utilizes an outside source of fuel it is usually supported and enhanced by a heat exchange system to minimize the quantity of fuel required.

Generally, fired heaters fall into two major categories:

- Horizontal
- Vertical

The horizontal type heater usually means a box type heater with the tubes running horizontally along the walls. The vertical type is normally a cylindrical heater containing vertical tubes. Figures 7.1 and 7.2 show examples of these two types of heaters. These figures also give some nomenclature used in describing these items of equipment. Other terms used in connection with fired heaters are as follows:

- *Headers and return bends*   the fittings used to connect individual tubes.
- *Terminals*   the inlet and outlet connections.
- *Crossovers*   the piping used to connect the radiant with the convection section; usually external to the heater.
- *Manifold*   the external piping used to connect the heater passes to the process piping; may be provided with the heater.
- *Setting*   any and all parts that form:
  (a)   Coil supports
  (b)   Enclosure (housing)
- *Casing*   the steel shell which encloses the heater.

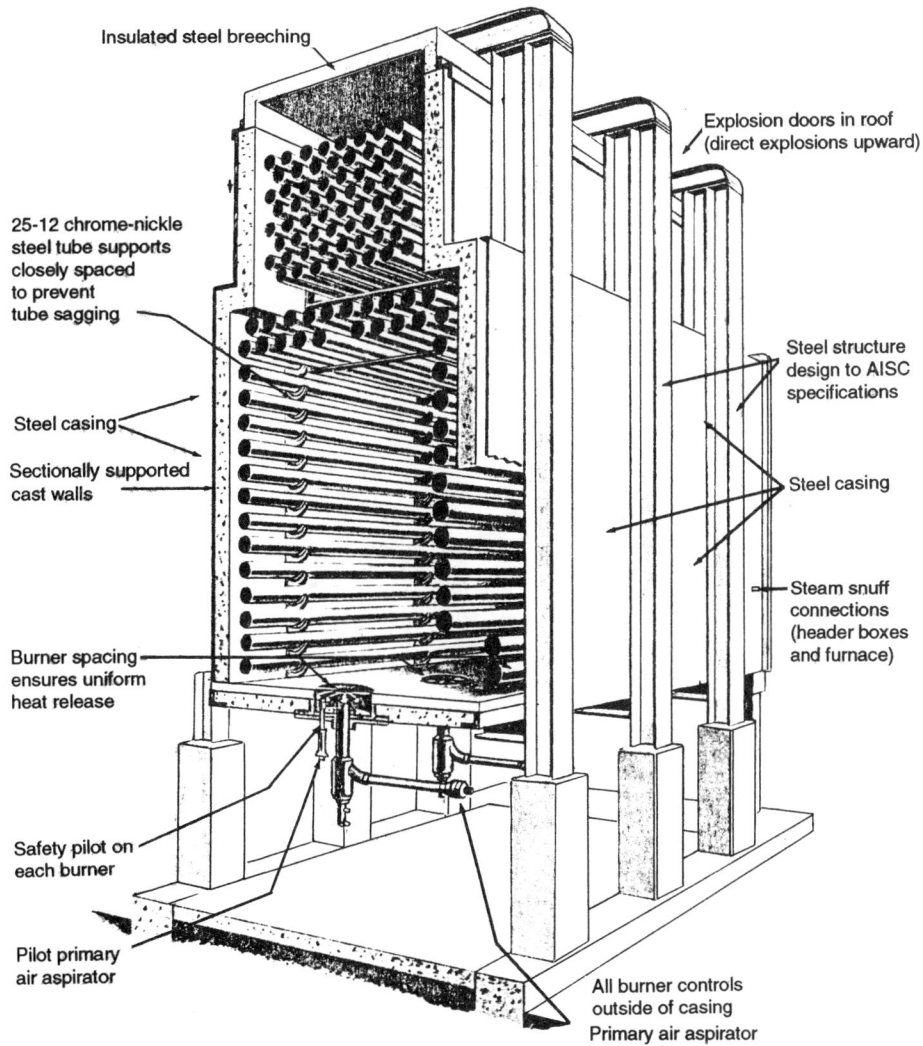

Figure 7.1. Horizontal type heater

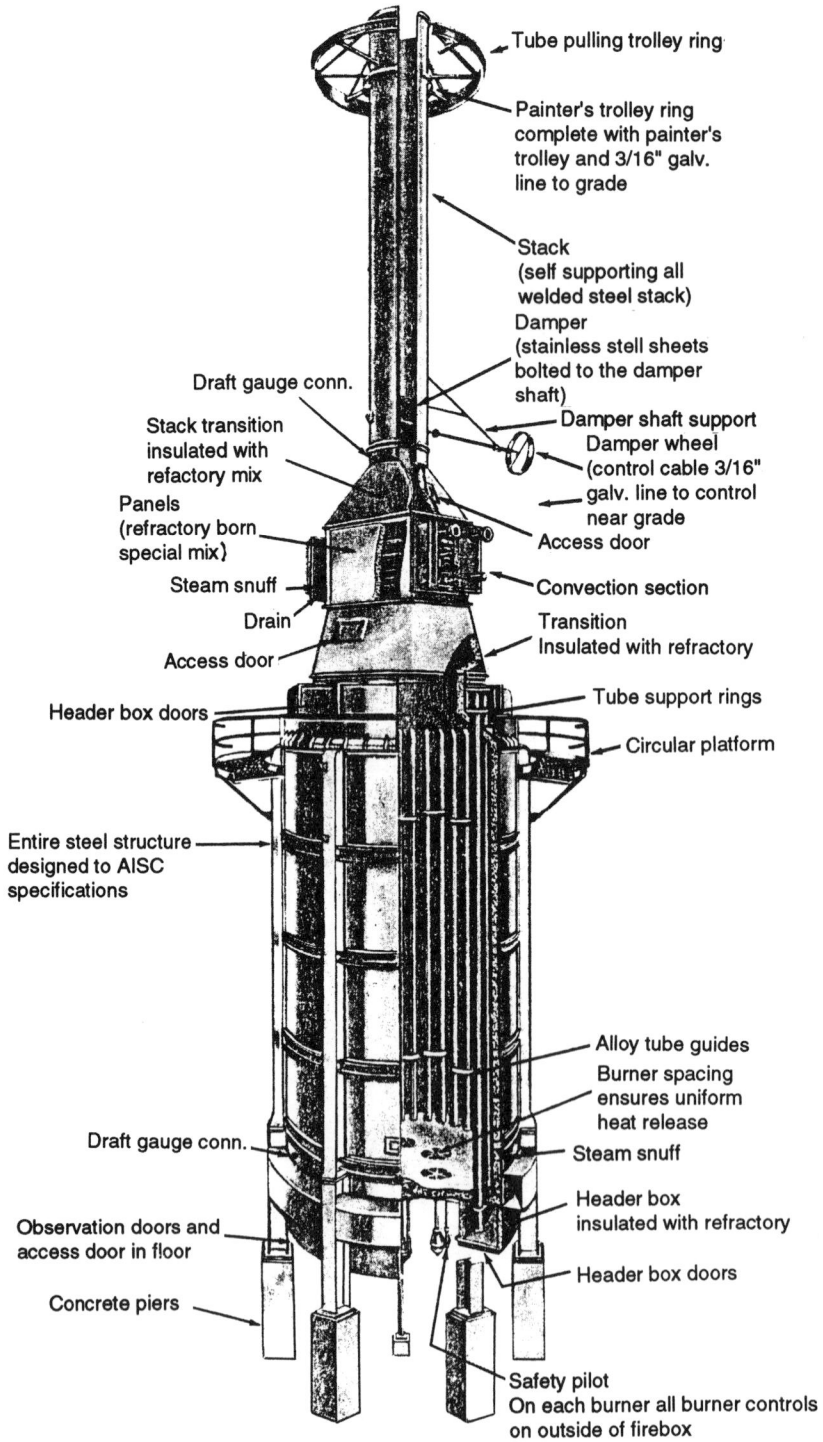

Figure 7.2. Vertical type heater

- *Bridge wall or partition wall* the refractory walls inside the heater which divide the radiant section into separately fired zones.
- *Shield tubes or shock tubes* the first two or three rows of tubes in the convection section. They protect or shield the convection tubes from direct radiant heat and must have same metallurgy as radiant tubes and have no fins.
- *Air plenum* the chamber enclosing burners under heater and having louvres to control air flow.

Cylindrical heaters require less plot space and are usually less expensive. They also have better radiant symmetry than the horizontal type.

Horizontal box types are preferred for crude oil heaters, although vertical cylindricals have been used in this service. Vacuum unit heaters should have horizontal tubes to eliminate the static head pressure at the bottom of vertical tubes and to reduce the possibility of two-phase slugging in the large exit tubes.

Occasionally, several different services ('coils') may be placed in a single heater with a cost saving. This is possible if the services are closely tied to each other in the process. A catalytic reforming preheater and reheaters in one casing are an example. A reactor heater and a stripper reboiler in one casing are another. This arrangement is made possible by using a refractory partition wall to separate the radiant coils. The separate radiant coils may be controlled separately over a wide range of conditions by means of their own controls and burners. If a convection section is used, it is usually common to the several services. If maintenance on one coil is required, the entire heater must be shut down. Also, the range of controllability is less than with separate heaters.

Each of these types may be shop fabricated if size permits, which reduces costs. However, shop fabrication should not be forced to the extent of getting an improperly proportioned heater.

7.2 Codes and Standards

Fired heaters have a 'live' source of energy. That is, they use a flammable material in order to impart heat energy to a process stream. Because of this the design, construction and operation of process fired heaters and boilers are strictly controlled by legislative and other codes and standards. This section outlines some of the more important of these codes and standards which need to be recognized by process engineers dealing with fired heaters in any way.

Codes and standards directly applicable to fired heaters are listed below. In addition, there are many codes and standards covering such factors as materials, welding, refractories, structural steel, etc. which apply to fired heaters. Process engineers usually need not concern themselves with these.

- *API RP-530 Calculation of Heater Tube Thicknesses in Refineries* This recommended practice sets forth procedures for calculating the wall thickness of heater tubes for service at elevated temperatures in petroleum refineries.
- *API Standard 630 Tube and Heater Dimensions for Fired Heaters for Refinery Service* This standard establishes certain standard dimensions for

heater tubes and for cast and wrought headers. Tube sizes and header centre-to-centre dimensions covered by the standard are as follows:

Tube o/d (in.)		Header c-to-c (in.)	
Primary	Secondary	Group A	Group B
2.375	–	4.00	4.75
2.875	–	5.00	5.25
3.50	–	6.00	–
4.00	–	7.00	6.50
4.50	–	8.00	7.25
–	5.00	9.00	7.75
5.563	–	10.00	8.50
–	6.00	11.00	9.00
6.625	–	12.00	10.00
–	7.625	14.00	12.00
8.625	–	16.00	14.00

Groove dimensions and tolerances for rolled headers are also given. Much of this standard is also used in chemical and petrochemical plants.

- *API RP-2002 Fire Protection in Natural Gasoline Plants* This contains a brief statement about the use of snuffing steam. This system provides the piping of a steam source to the heater firebox which in an emergency can introduce steam into the box to quench any uncontrolled fire. The system may be automatically controlled or activated manually.

- *API Guide for Inspection of Refinery Equipment, Chapter IX, Fired Heaters and Stacks* This reference gives a general description of fired heaters and describes how to inspect them, what damage to look for, and how to report the results of the inspection.

- *ASME Boiler Code and Boiler Codes of the USA* These are applicable in process plants if steam is generated, superheated or boiler feedwater preheated in the convection section. Special materials are required according to ASME Section I. The external piping and pressure relieving devices must also be in accordance with ASME Section I.

- *Contractors' standards* These will be covered in the Narrative Specification for the particular job and/or heater. The specification is written by the heat transfer engineer specialists in the Contractors' Mechanical Equipment Group. These specifications detail all the pertinent aspects required in the manufacture of the equipment and encompass all the requirements of the applicable codes.

7.3 Thermal Rating

Process engineers are seldom (if ever) required to thermal rate a fired heater or indeed check the thermal rating. This is a procedure that falls in the realm of specialist mechanical engineers with extensive experience in heater design and fabrication. Process engineers are, however, required to specify the equipment so that it can be designed and installed to meet the requirements of the process heat balance. To do this effectively it is desirable to know something about the mechanism of heater thermal rating.

A fired heater is essentially a heat exchanger in which most of the heat is transferred by radiation instead of by convection and conduction. Rating involves a heat balance between the heat-releasing and heat-absorbing streams, and a rate relationship.

Fuel is burned in a combustion chamber to produce a 'flame burst'. The theoretical flame burst temperature may vary from 4000 °F when burning refinery gases with 20% excess air preheated to 460 °F down to 2300 °F when burning residual fuel oils with 100% excess air at 60 °F. Heat is transferred from the flame burst to the gases in the firebox by radiation and mixing of the products of combustion. Heat is then transferred from the firebox gases to the tubes mainly by radiation.

The common practice is to assume a single temperature for the firebox gases for the purpose of radiation calculations. This may be the same as the exit gas temperature from the firebox to the convection section (bridge wall temperature) or it may be different due to the shape of the heater and to the effect of convection heat transfer in the radiant section. Experience with the particular type of heater is required in order to select the effective firebox temperature accurately.

While this chapter does not detail the rating procedure or give an example calculation, the following steps summarize the rating procedure:

(1) Calculate net heat release and fuel quantity burned from the specified heat absorption duty and an assumed or specified efficiency.
(2) Select excess air percentage and determine flue gas rates.
(3) Calculate duty in the radiant section by assuming 70% of total duty is radiant. This is a typical figure and will be checked later in the calculations. For very high process temperatures, such as in steam-methane reforming heaters, the radiant duty may be as low as 45% of the total.
(4) Calculate the average process fluid temperature in the radiant section and add 100 °F to obtain the tube wall temperature. The figure of 100 °F is usually a good first guess and can be checked later by using the calculated inside film coefficient and metal resistance.
(5) Calculate the radiant surface area using the average allowable flux. Convection surface is usually about equal to the radiant surface.
(6) Select a tube size and pass arrangement that will give the required total surface and meet specified pressure drop limitations.
(7) Select a centre-to-centre spacing for the tubes from the API 630 Standard or from dimensions of standard fittings, and calculate firebox dimensions. Long furnaces minimize the number of return bends and thus reduce cost. Shorter and wider fireboxes usually give more uniform heat distribution and lessen the probability of flame impingement on the tubes. For vertical cylindrical heaters, the ratio of radiant tube length to tube circle diameter should not exceed 2.7.
(8) The remainder of the calculation involves determining the firebox exit

temperature from assumption (3) above, applying an experience factor for the type of furnace to obtain the average firebox temperature, and then checking if this temperature will transfer the required radiant heat.

(9) The average heat flux (proportional to radiant surface) and the percentage of total duty of the radiant section (which affects average tube wall temperature) are varied until a balance is obtained. The convection section surface and arrangement can now be calculated.

(10) The heater is normally designed to allow adequate draught at the burners with at least 125% of the design heat release and an additional 10% excess air at the maximum and minimum ambient air temperatures.

HEAT FLUX

Although process engineers are not normally required to thermally rate a process heater they must often estimate the heater size, for preliminary cost estimates or plot layout and the like. This can be accomplished quite simply by the use of 'heat flux'. While this figure is quoted as $BTU/h.ft^2$, it is not, however, an overall heat transfer coefficient. It lacks the driving force ΔT for this. Heat flux is the rate of heat transmission through the tubes into the process fluid.

The maximum film temperature and tube metal temperature are a function of heat flux and the inside film heat transfer coefficient. The heat flux varies around the circumference of the tube, being a maximum on the side facing the firebox. The value depends upon the sum of the heat received directly from the firebox radiation and the heat reradiated from the refractory.

Single-fired process heaters are usually specified for a maximum *average* heat flux of $10\,000–12\,000\ BTU/h\,ft^2$. The maximum point heat flux is about 1.8 times greater.

Double-fired heaters are usually specified for about 13 500 to $18\,000\ BTU/h\,ft^2$ *average* heat flux with the maximum point flux being about 1.2 times greater.

The following are typical flux values for heaters in hydrocarbon service:

Horizontal, fired on one side	8 000–12 000
Vertical, fired on one side from bottom	9 000–12 000
Vertical, single row, fired on both sides	13 000–18 000

7.4 Heater Efficiency

The efficiency of a fired heater is the ratio of the heat absorbed by the process fluid to the heat released by combustion of the fuel expressed as a percentage. Heat release may be based on the LHV (lower heating value) of the fuel or HHV (higher heating value). Process heaters are usually based on LHV and boilers on HHV. The HHV efficiency is lower than the LHV efficiency by the ratio of the two heating values.

Heat is wasted from a fired heater in two ways:

- With the hot stack gas
- By radiation and convection from the setting

The major loss is by the heat contained in the stack gas. The temperature of the stack gas is determined by the temperature of the incoming process fluid unless an air preheater is used. The closest economical approach to process fluid is about 100 °F. If the major process stream is very hot at the inlet, it may be possible to find a colder process stream to pass through the convection section to improve efficiency, provided plant control and flexibility are adequately supplied. A more common method of improving efficiency is to generate and/or superheat steam and preheat boiler feedwater.

The lowest stack temperature that can be used is determined by the dew point of the stack gases. See Section 7.6 on stack emissions.

Figures 7.3 and 7.4 may be used to estimate flue gas heat loss. The loss to flue gas is expressed as a percentage of the total heat of combustion available from the fuel. These figures also show the effect of excess air on efficiency. Typically excess air for efficiency guarantees is 20% when firing fuel gas and 30% when firing oil.

Heat loss from the setting, called radiation loss, is about $1\frac{1}{2}$–2% of the heat release. The range of efficiencies is approximately as follows:

Very high   90% +. Large boilers and process heaters with air preheaters.
High   85%. Large heaters with low process inlet temperatures and/or
              air preheaters.

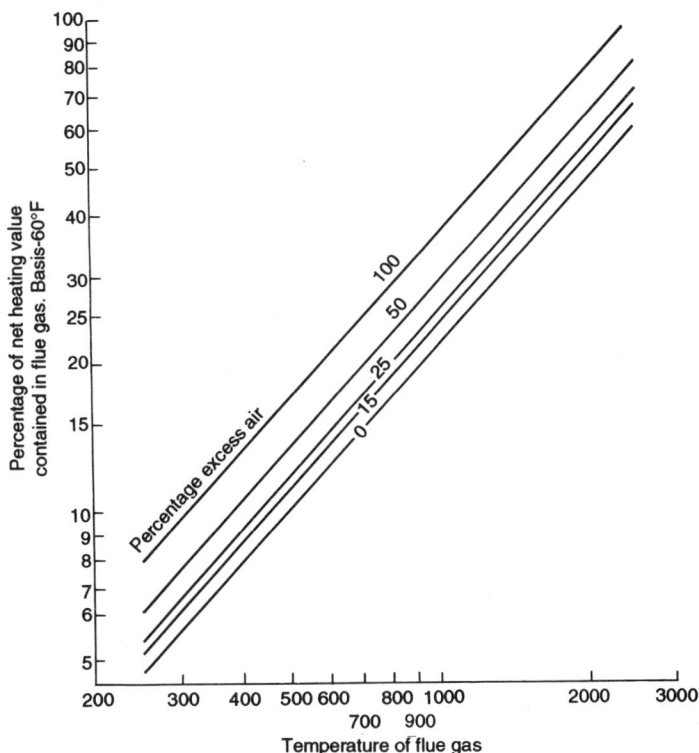

Figure 7.3. Percentage of net heating values contained in flue gas when firing fuel gas

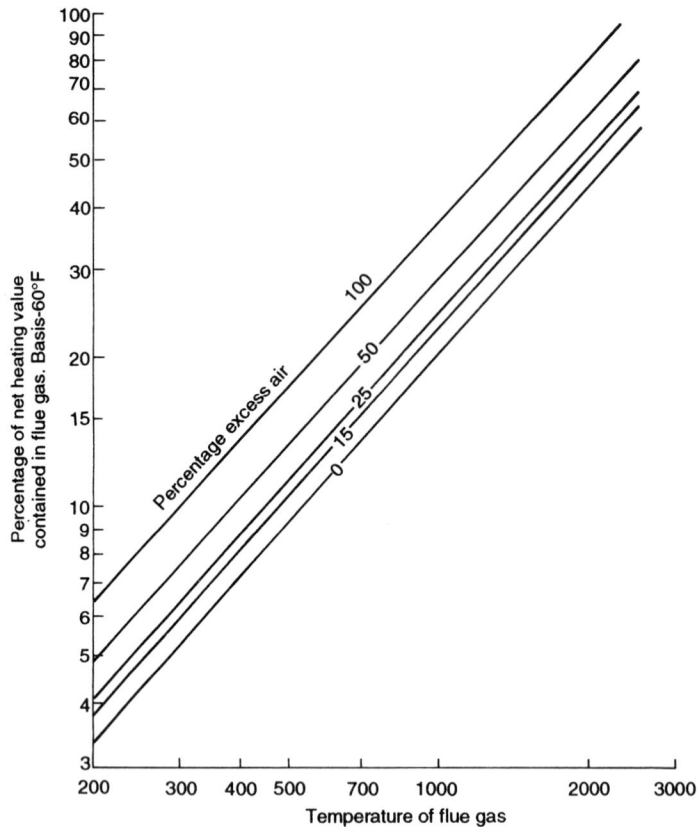

Figure 7.4. Percentage of net heating values contained in flue gas when firing fuel oil

Usual   70–80%.
Low   60% and less. All radiant.

Process engineers are often required to check the efficiencies of the process heaters on operating units assigned to them. This can be done using the heats of combustion given in Figures A1.1 and A1.2 in Appendix 1. The steps used to carry out these calculations are as follows:

● *Step 1* Obtain details of the heater from the manufacturer's data sheet or drawings. The data required are:
Tube area
Layout (is it vertical and how are the burners located?)
● *Step 2* From plant data obtain coil inlet and outlet temperature pressures and flow. Calculate the outlet flash (i.e. vapour or liquid or a mixture of vapour and liquid) condition, then calculate its enthalpy. Do the same for the inlet flow. Usually this will be a single phase, either liquid or vapour.
● *Step 3* Again from plant data obtain the quantity of fuel fired and its properties (API gravity in particular).

- *Step 4* The difference between the enthalpies calculated in step 2 is the enthalpy absorbed by the crude feed in BTU/h.
- *Step 6* Divide this absorbed enthalpy by the tube area to give the heat flux in BTU/h/ft^2. Heat fluxes are generally as follows:

Horizontal, fired one side	8000–12 000 BTU/h/ft^2
Vertical, fired from bottom on one side	9000–12 000 BTU/h/ft^2
Vertical, single row, fired on both sides	13 000–18 000 BTU/h/ft^2

  If the heat flux falls outside this range given above, there could be excessive fouling. Check the pressure drop—if this is far above manufacturer's calculated value then fouling is certainly present.
- *Step 7* Check the thermal efficiency of the heater by giving the fuel fired a heating value. This is provided by figures in Appendix 1. Use the LHV in BTU/lb and multiply it by the lb/h of the fuel.
- *Step 8* Divide the heat absorbed by the heat released calculated in step 7 to give the thermal efficiency. For most heaters this should be between 70% and 80%. If it falls below this range note should be taken of burner operation and the amount of excess air being used.

Example calculation

An atmospheric crude oil distillation heater handles 379 575 lb/h of crude oil at an inlet temperature of 515 °F (all liquid) and 720 °F outlet temperature having a flash of 51.1% wt vapour. Details of heater and fuel oil are as follows:

Heater details (from manufacturer)

- Tubular area total 5200 ft^2
- Vertical fired from the bottom on all sides
- Natural draught
- No steam generation

Plant data

- Flash zone temperature calculated—720 °F
- % vaporization of crude at calculated partial pressure as 51.1 % wt
- Total flow of crude
- Temperature of crude into heater
- Fuel oil fired lb/h and °API

(A) Degree of fouling

(see Section 1.12)   BTU/h $\times$ 10^6

Heat in crude vapour at 720 °F = 102.509
Heat in crude liquid at 720 °F = 73.430
Total = 175.939

Heat in with crude (all liquid) at 515 °F
$$= 379\,575 \times 272 = 103.24$$
$$\text{Heater duty} = \ \ 72.699$$

The heat flux under these conditions is:

$$\frac{72.699 \times 10^6}{5200 \text{ ft}^2} = 13\,980 \text{ BTU/h/ft}^2$$

This is within an acceptable limit and excessive fouling is not suspected.

Example calculation

Thermal efficiency
The fuel fired is 16 °API residuum

$$\text{lb/h of fuel fired} = 5440 \text{ (plant data)}$$

From Figure A1.1 in Appendix 1

$$\text{LHV (heat release)} = 17\,580 \text{ BTU/lb}$$
$$= 17\,580 \times 5440$$
$$= 95.685 \text{ BTU/h} \times 10^6$$
$$\text{Heat absorbed} = 72.699 \times 10^6 \text{ BTU/h}$$

$$\text{Thermal efficiency} = \frac{72.699 \times 10^6 \times 100}{95.635 \times 10^6}$$
$$= 76.02\%$$

which is within acceptable limits.

7.5   Burners

The purpose of a burner is to mix fuel and air to ensure complete combustion. The basic burner designs are:

- Direction—vertical up fired
  - vertical down dired
  - horizontally fired
- Capacity—high
  - low
- Fuel type—gas
  - oil
  - combination
- Flame shape—normal
  - slant
  - thin, fan-shaped
  - flat
  - adaptable pattern
- Hydrogen content—high

- Excess air—normal
            low
- Atomization—steam
              mechanical
              air-assisted mechanical
- Boiler types
- Low NO_x
- High intensity

Various combinations of the above types are available.

GAS BURNERS

The two most common types of gas burners are 'pre-mix' and 'raw gas'. Pre-mix burners are preferred because they have better 'linearity', i.e. excess air remains more nearly constant at turndown. With this type, most of the air is drawn in through an adjustable 'air register' and mixes with the fuel in the furnace firebox. This is called secondary air. A small part of the air is drawn in through the 'primary air register' and mixed with the fuel in a tube before it flows into the furnace firebox. A turndown of 10:1 can be achieved with 25 psig hydrocarbon fuels. A more normal turndown is 3:1.

OIL BURNERS

An oil burner 'gun' consists of an inner tube through which the oil flows and an outer tube for the atomizing agent, usually steam. The oil sprays through an orifice into a mixing chamber. Steam also flows through orifices into the mixing chamber. An oil–steam emulsion is formed in the mixing chamber and then flows through orifices in the burner tip and then out into the furnace firebox. The tip, mixing chamber and inner and outer tubes can be disassembled for cleaning.

Oil pressure is normally about 140–150 psig at the burner, but can be lower or higher. Lower pressure requires larger burner tips, and the pressure of the available atomizing steam may determine the oil pressure.

Atomizing steam should be at least 100 psig at the burner valve and at least 20–30 psi above the oil pressure. Atomizing steam consumption will be about 0.15–0.25 lb steam/lb oil, but the steam lines should be sized for 0.5.

COMBINATION BURNERS

This type of burner will burn either gas or oil. It is better if they are not operated to burn both fuels at the same time because the chemistry of gas combustion is different from that of oil combustion. Gases burn by progressive oxidation and oils by cracking. If gas and oil are burned simultaneously in the same burner, the flame volume will be twice that of either fuel alone.

PILOTS

Pilots are usually required on oil-fired heaters and are fired with fuel gas. They are not required when heaters are gas-fired only, but minimum flow bypasses around the fuel gas control valves are used to prevent the automatic controls from extinguishing burner flames.

EXCESS AIR AND BURNER OPERATION

The excess air normally used in process fired heaters is about 15–25% for gas burners and about 30% for oil burners. These excess air rates permit a wide variation in heater firing rates which can be effectively controlled by automatic controls without fear of 'starving' the heater of combustion air. There has been considerable work lately to reduce this excess air, mostly to minimize air pollution. This practice has not been used in process heaters to date. It has, however, been adopted in the operation of large power station type heaters with some success.

Normally companies specify that burners be sized to permit operation at up to 125% of design heat release with a turndown ratio of 3:1. This gives a minimum controllable rate of 40% of design without having to shut down burners.

BURNER CONTROL

Burner controls become very important from safety and operation considerations. Most systems include an instrumentation system with interlocks that prohibit:

- Continuing firing when the process flow in the heater coil fails
- The flow of fuel into the firebox on flame failure

Under normal operating conditions the amount of fuel that is burnt is controlled by flow controllers operated on the coil outlet temperature. With combination burners the failure of one type of fuel automatically introduces the second type. Such a switch-over can also be effected manually. This aspect is usually activated on pressure control of the respective the fuel system. That is, on low pressure being sensed on the fuel being fired the system automatically switches to the second fuel.

Most companies operate their own specific controls for the heater firing system. One such system is given in Figure 2.17.

HEATER NOISE

All heaters are noisy and this noise is the result of several mechanisms. Among these are the operation of the burners. Gas burners at critical flow of fuel emit a noise. This can be minimized by designing for low pressure drop in the system. Intake of primary and secondary air is another source of noise. Forced-draught burners are generally quieter than natural-draught if the air ducting is properly sized and insulated. The design of the fan can also reduce noise in this mechanism. A low tip-speed fan favours low noise levels.

7.6 Refractories, Stacks and Stack Emission

Refractories are used on the inside walls of the heater firebox, floor and through the convection side of the heater. The purpose of this refractory lining is to conserve the heat by limiting its loss to atmosphere by convection. It is also necessary for the personal safety of those working on or about the heater who may accidentally touch the heater walls.

Good insulation has the following qualities:

- It has good high-temperature strength.
- It is resistant to abrasion, spalling, chemical reaction and slagging.
- It has good insulating properties.

Among the most common refractories that meet some if not all of the above criteria are silica refractories, high alumina and fire clay brick. These have high resistance to spalling and to thermal shock. Their insulating qualities are also good.

Silica refractories tend to form slag with metal oxide dust and ashes. Compounds of sodium and potassium attack most refractories while refractories containing magnesium react with acids and acid gases. Carbon monoxide and other reducing chemicals that may be present in the firebox reduce the life of refractories, particularly fireclay brick and silica.

Dense refractories with low porosity are the strongest but have the poorest insulation qualities. Castable refractories containing a mixture of cement and refractory aggregate are the cheapest and the most easily to install. They are not very rugged however. Normally in process heaters conditions are such that the use of a light insulating refractory will satisfy all that is required from a refractory lining.

The ASTM standard part 13 gives more detail on refractories. This provides the classification of refractories and describes their characteristics and composition. It also offers a procedure for calculating the heat loss through the insulation and thus its thickness.

PREPARING REFRACTORIES FOR OPERATION

All new refractories need to be 'cured' after installation and before use. Refractories contain moisture, some due to the installation procedure and some in the form of water of crystallization. Curing the refractory means removing this moisture by applying a slow-heating mechanism. Heater manufacturers will usually provide details of the curing procedure they recommend. The following may, however, be used as a guide to refractory curing when manufacturers' procedures are not available:

(1) Raise the temperature of 50 °F per hour to 400 °F and hold for 8 hours at that temperature.
(2) Then increase the temperature again at 50 °F per hour from 400 °F to 1000 °F and hold for another 8 hours.

(3) If necessary, continue heating at 100 °F per hour to the operating temperature if higher than 1000 °F. Hold at the operating temperature for a further 8
hours.
(4) Cool at 100 °F per hour to about 500 °F and hold ready for operation.
(5) On start-up the heater can be heated up to its operating temperature at a
rate of 100 °F per hour.
(6) During the curing of the refractory it will be necessary to pass some fluid
through the heater coils to protect them from overheating. Steam or air may
be circulated through the coils for this purpose. In certain cases such as in
catalytic reforming of petroleum stock it may be necessary to circulate the
nitrogen or the hydrogen that was used to purge and pressure test the unit.
Air and steam in this process would not be desirable.

STACKS

Stacks are used to create an updraught of air from the firebox of a heater. The
purpose of this is to cause a small negative pressure in the firebox and thus enable
the introduction of air from the atmosphere. This negative pressure also allows
for the removal of the products of combustion from the firebox. The stack
therefore must have sufficient height to achieve these objectives and overcome
the frictional pressure drop in the firebox and the stack itself.

The height required for a stack to achieve good draught can be estimated from
the following equation:

$$D = 0.187 H (\rho_a - \rho_g)$$

where

D = draught in inches of water
H = stack height (ft)
ρ_a = density of atmospheric air (lb/ft^3)
ρ_g = density of stack gases (lb/ft^3 at stack conditions)

For stack gas temperature use 100 °F lower than gases leaving the convection
section. Stack gases have a molecular weight close to that of nitrogen. For this
calculation use 28 as the mole weight. Burner draught requirements range from
0.2 to 0.5 in. of water. Use 0.3 in. of water as a good design value.

Stacks must also be designed to handle and disperse stack emissions. This
usually results in having to build stack heights greater than that required for
obtaining draught. There are available specific computer programs relating plume
height of the stack gases above the stack outlet to the probable ground-level
fallout of the impurities in the gas. These programs also produce a map of the
relative concentrations of these impurities at ground level. Such data are usually
available from government offices in most countries. The predicted ground
concentration and map calculated are checked against local legal requirements.
The results usually form part of government approval to build and/or operate a
facility.

The stack diameter is based on an acceptable velocity of stack gases in the stack. This is generally taken as 30 ft/s. Some allowance must be made for frictional losses, i.e.:

1.5 velocity heads for inlet and outlet losses
1.5 velocity head for damper
1.0 velocity head for each 50 ft of stack height

STACK EMISSIONS

The obnoxious compounds in stack gas emissions arise from:

* Impurities in the fuel
* Chemical reactions resulting from the fuel combustion with air

The three major impurities in oil or gas fuels which produce undesirable emissions are:

* Sulphur
* Metals
* Nitrogen

Sulphur

All gas and oil fuels contain sulphur at some level of concentration. When these fuels are burned the sulphur reacts with air to form SO_2 and SO_3. These compounds are objectionable because:

* They cause air pollution in the form of smog.
* They contribute to the corrosion of heater tubes and stack.
* SO_3 lowers the dew point of the flue gases resulting in an objectionable visible plume at the stack exit.

Figure 7.5 shows the effect of sulphur in the feed on the flue gas dew point. This dew point also, of course, varies with the amount of excess air and the relative partial pressure of the combustion gases.

There is no precise way of determining the amount of SO_3 that is formed in the flue gas. Nor can it be determined with any degree of accuracy where the SO_3 is formed in the system. The total amount of the sulphur oxides is, of course, determined simply from the sulphur content of the feed. Because of this uncertainty it is important to restrict the minimum allowable convection side metal temperatures to 350 °F when firing fuels containing sulphur. Also the minimum temperatures to the stack should be 320 °F when firing fuel gas and 400 °F when firing fuel oil. In the case of metal stacks the use of a non-corrosive lining must be used in the colder section of the stack if the flue gas temperature falls below those stated above.

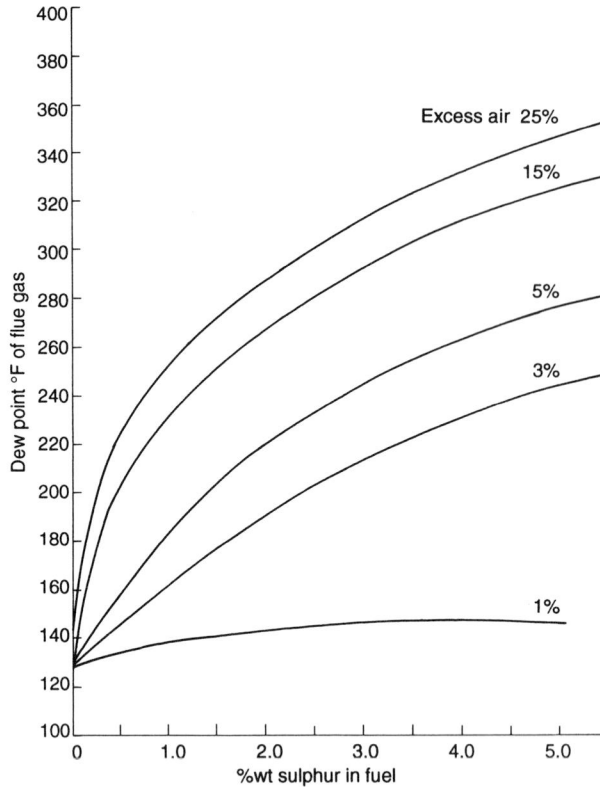

Figure 7.5. Dew point of flue gases versus fuel sulphur content

Metals

The most objectionable metal impurities in fuel oil are sodium and vanadium. These can cause severe corrosion of tubes and refractory lining. By high vanadium content is meant concentrations between 200 ppm and 400 ppm. Above 400 ppm is considered very high and should not be used as a fuel.

Vanadium in the presence of sodium and oxygen readily forms a corrosive compound:

$$Na_2O \cdot 6V_2O_5$$

This compound attacks iron or steel to form ferric oxides according to the equation

$$Na_2O \cdot 6V_2O_5 + Fe \rightarrow Na_2O \cdot V_2O_4 \cdot 5V_2O_5 + FeO$$

This vanadium product is oxidized back to the original corrosive compound according to the reaction

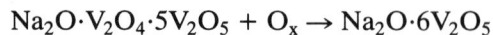

$$Na_2O \cdot V_2O_4 \cdot 5V_2O_5 + O_x \rightarrow Na_2O \cdot 6V_2O_5$$

These reactions occur at temperatures between 1070 °F and 1220 °F with the vanadium compound being continually regenerated to its corrosive state.

Sodium and sulphur also combine to form undesirable corrosive compounds with vanadium being present. This reaction forms sodium iron trisulphate $Na_3Fe(SO_4)_3$ at temperatures of around 1160 °F. The critical temperature for both vanadium and sodium corrosion is around 1100 °F. Vanadium is not a major problem at this temperature but becomes so at temperatures above this level.

Nitrogen

Some nitrogen oxides will be formed in combustion gases even if there are no nitrogen compounds in the fuel. The presence of the nitrogen compounds in the fuel significantly increases the nitrogen oxide content of the flue gas. There are six oxides of nitrogen present in flue gases but only two in any appreciable amount, i.e.

NO Nitrogen oxide.
NO_2 Nitrogen dioxide

Nitrogen oxide is a poisonous gas which readily oxidizes to nitrogen dioxide on entering the atmosphere. It is a yellowish gas which readily combines with the moisture in the air to form nitric acid. In the presence of sunlight and oxygen nitrogen dioxide also contributes to the formation of other air pollutants collectively labeled NO_x. The production of NO_x can be reduced by limiting the amount of excess air and by washing the flue gases with aqueous ammonia.

7.8 Specifying a Fired Heater

Some basic data concerning a fired heater must be made known before the equipment can be designed for fabrication or even costed. These data are provided by a specification sheet or sheets. In the case of a fired heater, as in the case for compressors, this specification can run into several sheets or forms. Such sheets will define the unit in terms of:

- Process requirement
- Mechanical detail
- Civil engineering requirement
- Operational requirement
- Environmental requirement

This section will deal only with the 'process requirement' and the duty of the heater. A typical process-originated specification sheet is shown as Figure 7.6. The data provided on this sheet are the minimum that will be necessary for a heater manufacturer or a heater specialist to begin to size the heater or price out the item. Usually these data would also be supported by a vaporization curve of the feed if there is a change of phase taking place in the heater coils. The process engineer developing this specification would also provide details of services required in addition to the main duty of the heater. For example, if it is intended that the convection side of the heater is to be used for steam generation, preheating, or steam superheating, the system envisaged must be properly described by additional diagrams and data.

The example specification sheet in Figure 7.6 is completed for a crude oil vacuum distillation unit heater with a steam superheater coil located in the convection side of the heater. The following describe the content on a line-by-line basis:

- *Line 1 Design duty* refers to the total duty required of the unit. In the case of the example given here it includes the duty required to heat and partially vaporize the oil in the radiant section and the duty required to superheat the steam. The data sheet will be split into two sections to reflect each of these duties.
  Service describes the main purpose for which the heater will be used. In this case it will be used to heat hydrocarbons.
- *Line 2 No. of heaters*. This is self-explanatory. In this case there is only one heater required. Should there have been more than one identical unit this would be reflected here.
  Unit This is the title given to this unit of equipment as it appears on an equipment list. It will correspond to the item number also given in the equipment list. In the case of the example it is 'Vac. column feed heater'.
- *Line 3 Item no*. This is the reference number given to the item in the equipment list. This reference number and unit title identifies the equipment on all drawings where it appears, and all documents used in its purchase, costing, maintenance, etc.
  Type The type of heater (if decided on) is given here. In the case of the example heater a cylindrical type has been selected for plot space considerations.
- *Line 4* This is the first line of the specification sheet proper. It commences with the service of the heater or section of the heater. In the case of this example only two columns have been provided. These are designated for the 'Radiant' section and 'Convection' section. On most preprinted specification forms there would be at least four columns.
- *Line 5 Heat absorption*. This line divides the duty of the heater into that required from the radiant coils and that required from the convection side. In the example the oil is routed through the radiant section only while saturated steam from a waste heat boiler is superheated in the convection coils. Both are measured in million BTU/h. Partial vaporization of the oil occurs in the radiant coil. Therefore a flash curve or a phase diagram of the oil must accompany this specification sheet. In the example the duty to the oil is calculated from data developed in the material balance and heat balance of the process as follows.
  Temperature of the feed into the heater is 554 °F and that of the coil outlet is 750 °F. From the flash curve at the outlet pressure of 35 mm Hg abs (0.68 psia) and the material balance the weight per hour of vapour is calculated to be

Design duty	40886000	BTU/h	Service	Hydrocarbons
No. of heaters	1		Unit	Vac unit preheater
Item no.	H 201		Type	Cylindrical

DESIGN DATA			Radiant section	Conv. section
Service			Red crude	Steam s. heat
Heat absorption		mmBTU/h	38.501	2.385
Fluid			Hydrocarbon	Sat steam
Flow rate		lb/hr	197 085	30 040
Allowable pressure drop		psi	300	30
Allowable average flux		BTU/h.ft^2	15 000	—
Maximum inside film temp.		°F	800	—
Fouling factor		°F.ft^2.h/BTU	0.004	0.001
Residence time		s	N/A	N/A
Inlet conditions				
Temperature		°F	554	368
Pressure		psig	270	155
Liquid flow		lb/h	197 085	nil
Vapour flow		lb/h	nil	30 040
Liquid density		lb/ft^3	48.4	—
Vapour density		lb/ft^3	—	Steam
Visc. liq/vap		cP	2.31/—	—/—
Specific heats liq/vap		BTU/lb	0.65/—	/
Thermal cond. liq/vap		BTU/h.ft^2.°F/ft	0.0671/—	/

Figure 7.6. Specification sheet: fired heater

127 444 lb/h and the liquid portion is 69 641 lb/h. The heat absorbed in the radiant section is

$$\text{Heat in with feed} = 197\,085\ \text{lb/h} \times 268\ \text{BTU/lb}$$
$$\text{(all liquid)} \quad = 52.819\ \text{mm BTU/h}$$

Heat out in feed:

$$\text{Liquid portion} = 69\,641\ \text{lb/h} \times 378\ \text{BTU/lb}$$
$$= 26.324\ \text{mm BTU/h}$$
$$\text{Vapour portion} = 127\,444\ \text{lb/h} \times 510\ \text{BTU/lb}$$
$$= 64.996\ \text{mm BTU/h}$$
$$\text{Total heat out} = 91.320\ \text{mm BTU/h}$$
$$\text{Duty of radiant section} = 91.320 - 26.324$$
$$= 38.501\ \text{mm BTU/h}$$

Outlet conditions			
Temperature	°F	750	500
Pressure (psia)	psig	(0.68)	125
Liquid flow	lb/h	69 641	nil
Vapour flow	lb/h	127 444	30 040
Liquid density	lb/ft³	47.3	
Vapour density	lb/ft	0.023	0.246
Visc. of liq/vap	cP	1.4/0.002	—/—
Specific heat liq/vap	BTU/lb	0.64/0.69	—/—
Thermal cond. liq/vap	BTU/h.ft².°F/ft	0.062/0.023	/
FUEL DATA			
Type (gas or oil)		OIL	GAS
LHV Gas / Oil	BTU/ft³ / BTU/lb	17 560	2320
HHV Gas / Oil	BTU/ft³ / BTU/lb	18 580	2520
Pressure at burner	psig	82	25
Temp. at burner	°F	176	68
Mol wt of gas			44
Visc. of oil at burner	cP	23.3	
Atomizing steam	Temp. °F / Press. psig	500 / 125	
Composition of gas	Mol %		
	H_2		0.2
	C_1		12.0
	C_2		28.2
	C_3		31.3

Figure 7.6. (*continued*)

For the convection section the duty is calculated as follows:

Weight of saturated steam at 155 psig is 30 040 lb/h

Temperature of 155 psig steam = 368 °F
From steam tables steam enthalpy = 1196.9 BTU/lb
Temperature of steam out = 500 °F
Pressure of steam out = 125 psig.
From steam tables steam enthalpy = 1275.3 BTU/lb
Duty of convection side = 30 040 (1275.3 − 1195.9)
= 2.385 mm BTU/h

	C_4		
	C_5		
	C_{6+}		
Properties of fuel oil			
° API		15.2	
Visc. at 100 °F	cSt	175	
Visc. at 210 °F	cSt	12.5	
Flash pt	°F	200	
Vanadium	ppm	12	
Sodium	ppm	32	
Sulphur	%wt	2.4	
Ash	%wt	< 1.0	

Remarks
(1) Flash curve of the reduced crude feed versus % volume distilled is attached for atmos. pressure and at 35 mm Hg abs.
(2) Gravity and mol wt curves for the reduced crude versus mid-volume % are attached.
(3) Soot blowers are to be considered in this package. Steam is available at 600 psig and 750°F.
(4) Studded tubes to be considered for the convection side.

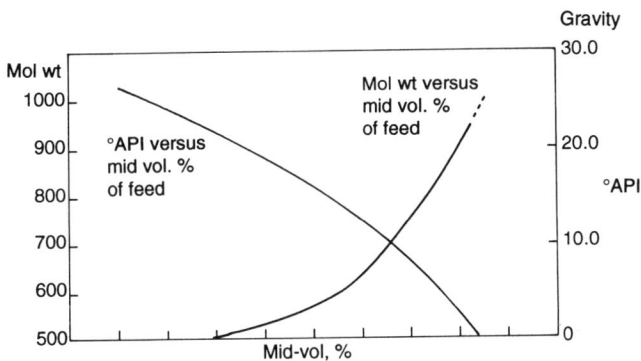

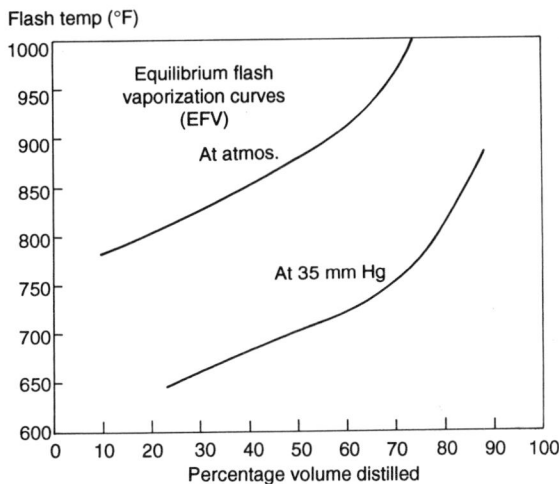

Specification Sheet Attachment 1: Flash curves, mol wt, and 5 g for feed

Figure 7.6. (continued)

- *Line 6   Fluid*. This refers to the material flowing in the coil. In the case of this example it will be hydrocarbons in the radiant coil and steam in the convection coil.
- *Line 7   Flow rate*. This is the total flow rate in lb/h entering the respective section of the heater. Thus for the radiant side the figure will be 197 085 lb/h and for the convection side 30 040 lb/h.
- *Line 8   Allowable pressure drop*. The process engineer enters the required pressure drop calculated from the hydraulic analysis of the system. This pressure drop is measured from the heater side of the inlet manifold downstream of the balancing control valves and the coil outlet downstream of the outlet manifold.
- *Line 9   Allowable average flux*. This is usually a standard set by the company for its various heaters. In the example here this value would be between 13 000 and 18 000 BTU/h.ft^2 (a vertical heater bottom fired). It is specified as 15 000 BTU/h.ft^2 for this example and refers only to the radiant section.
- *Line 10   Maximum inside film temperature*. It is important to notify the heater manufacturer of any temperature constraint that is required by the process. In the case of this example temperatures of the oil above 800 °F may lead to the oil cracking. Such a situation could adversely affect the performance of the downstream fractionation equipment and therefore high temperatures in excess of 800 °F must be avoided. There is no constraint on the convection coil.
- *Line 11   Fouling factor*. The fouling factors used in heat exchanger rating can be used here also. In this example therefore the radiant side would have a fouling factor of about 0.004 °F.ft^{2}h/BTU for the oil and 0.001 for the steam.
- *Line 12   Residence time*. This becomes important when a chemical reaction of any kind takes place in the heater tubes. In the case of this example this item does not apply. If the example were a thermal cracking heater or a visbreaker the appropriate kinetic equations and calculations would be attached to the specification sheet to support this item.
- *Lines 13–30*   These are self-explanatory. The only comment here is that the data are quoted at the inlet or outlet conditions of temperature and pressure.
- *Line 31   LHV*. The section of the specification sheet that follows deals with the characteristic of the fuel that will be used in the heater. This section is divided into oil and gas which are the usual fuels used in modern day processes. This item requires the 'Lower heating value' of the fuel. This can be read off charts such as Figures A1.1 and A1.2 in Appendix 1.
- *Line 32   HHV*. This is the other heating value data required by the heater designer. This 'higher heating value' data can also be read off charts such as Figures A1.1 and A1.2.
- *Line 33   Pressure at burner*. This normally refers to the oil fuel and is measured at the heater fuel oil manifold.
- *Line 34   Temperature at the burner*. This item too is self-explanatory and these measurements are also taken at the respective manifolds.
- *Line 35   Mol wt of gas*. This refers to the gas stream normally expected to be used. Obviously in practice this will vary with the process operation from day to day.

- *Line 36   Viscosity of the oil at the burner*. This viscosity is quoted at the burner temperature and may be arrived at from the two viscosity figures given later in the specification sheet. This is important for the best design or selection of the burner itself.
- *Line 37   Atomizing steam*. This item calls for the temperature and pressure of the steam that will be used for atomizing the oil fuel. This is also required for the best design or selection of the oil burner.
- *Line 38   Composition of gas*. This section requires the composition of the gas fuel in terms of mol %. This is the normal expected fuel gas that will be used in the heater. If there is likely to be a wide variation in the quality of the fuel gas that will be used this should be noted here as a range of two or even three compositions. This situation is particularly common in petroleum refining.
- *Line 46   Properties of fuel oil*. This final section of the specification sheet requires details of the fuel oil that will be used. These details are:

Gravity of the oil at 60 °F (in °API).

Viscosity of the oil at 100 °F and at 210 °F

Flash point in °F.

From the two viscosities the 'REFUTAS' graph can be used to determine the viscosity at any other temperature. This graph can be found in most data books that carry viscosity data. Note that the viscosity requested in this section is in centistokes (kinematic viscosity). This is because most suppliers quote in centistokes. To convert to centipoise multiply by the specific gravity (g/cm^3). The flash point required is that determined by the Pensky–Marten method and is a measure of the oil's flammability.

This completes the explanation of the main body of the specification sheet. It represents the minimum data required to commence the sizing of the item. The last part of the specification sheet is given to 'Remarks'. In this section the engineer should provide all the other data that may influence the design of the heater.

8 PROCESS STUDIES AND ECONOMIC ANALYSIS

Among the principal functions of a process engineer is to generate ideas for the enhancement of the company's production business. In carrying out this function the engineer is required to study various processing routes or changes to existing process routes and configurations. When satisfied that the technical aspect of any proposed changes or additions is feasible and sound he or she must now satisfy the company management that it is economically attractive. The proposal must also be shown to be the best of any possible alternatives that have been studied to achieve the same objective. This chapter sets out to describe some basic techniques used in carrying out this function.

8.1 The Study Approach

Process studies are usually initiated after a very precise definition by the company's management of the company's immediate and long-term production objectives. Apart from this definition being a premise of the company's annual operating budget it also aims to fulfil its marketing strategy. Often such a definition results in changes being required to be made to the process facilities in either the short term or long term to meet these objectives. The company will look to the process engineers to provide definition of these changes and to support the definition(s) with whatever technical and economic data necessary for management to make their 'go/no go' decisions.

Figure 8.1 shows the steps in a typical process study. This particular study route is to define an expansion of an existing process configuration by adding a new group of plants. The steps shown in Figure 8.1 are described as follows:

- *Step 1 Review management's requirements* This step is extremely important. The process engineer or engineers working on the study must understand completely what is required and what the end product of the study must fulfil. There is usually also a time element for the completion of the work. This too must be understood and planned for. Normally the study premise is presented in a written document with copies to all study participants. The contents of the document must be studied by all concerned. A 'kick-off' meeting is always beneficial where participants and management review the study premises together before commencing work on the study.

- *Step 2 Obtain stream flows from existing plants* In cases where existing plants are to form part of the study premise a detailed analysis of the performance of the plants must be made. This almost certainly will involve performance test runs under precalculated operating conditions. The test run results are evaluated and included in the analysis of products to meet the study's product slate.

- *Step 3 Develop product pattern from new plants* This will be executed in parallel with step 2. Here new processes that may be included in the study to meet the product requirements are examined and their product pattern developed. Process licensors are almost invariably involved at this stage to provide data on their proprietary processes.

- *Step 4 Evaluate surplus and shortfalls in meeting objective* Steps 2 and 3 are now used to determine the possible product slate achievable using plant yields and blending recipes. Any surplus or shortfalls in the product slate that may now exist must be eliminated by further work in steps 2 or 3 or both.

- *Step 5 Establish process options and constraints* With a firm product slate work may now commence to map processing routes that will achieve the product objective. This will include very rough plant configurations irrespective of cost or other constraints that will fulfil a processing route. This step should end with the elimination of the more obvious undesirable routes (for example, those routes which contain very small process units or unit processes that have undesirable byproducts). This step should provide five or six reasonable process options for further study.

- *Step 6 Commence screening studies* The purpose of this item is to reduce the number of possible options to two or, at most, three process configurations that appear most economically attractive. To arrive at the selected options those alternates arrived at in step 5 are allocated capital and operating costs. The capital costs in this case are very rough estimates arrived at by applying factors to the installed cost for similar processes. Details of this technique in cost estimating are given later in this chapter in Section 8.5. Licensors are usually requested to participate in the development of these cost items in the case of their proprietary processes. The operating costs for non-proprietary processes are also factored from similar processes. With these cost data in place an incremental payout time is calculated for each option. The option with the lowest capital cost is used as a base case and the others compared against it. Details of this calculation are given later in this chapter. The two (or three) options with the lowest positive incremental payout time against the base case are selected for further study.

- *Step 7 Prepare data for a more detailed cost estimate* The credibility of any

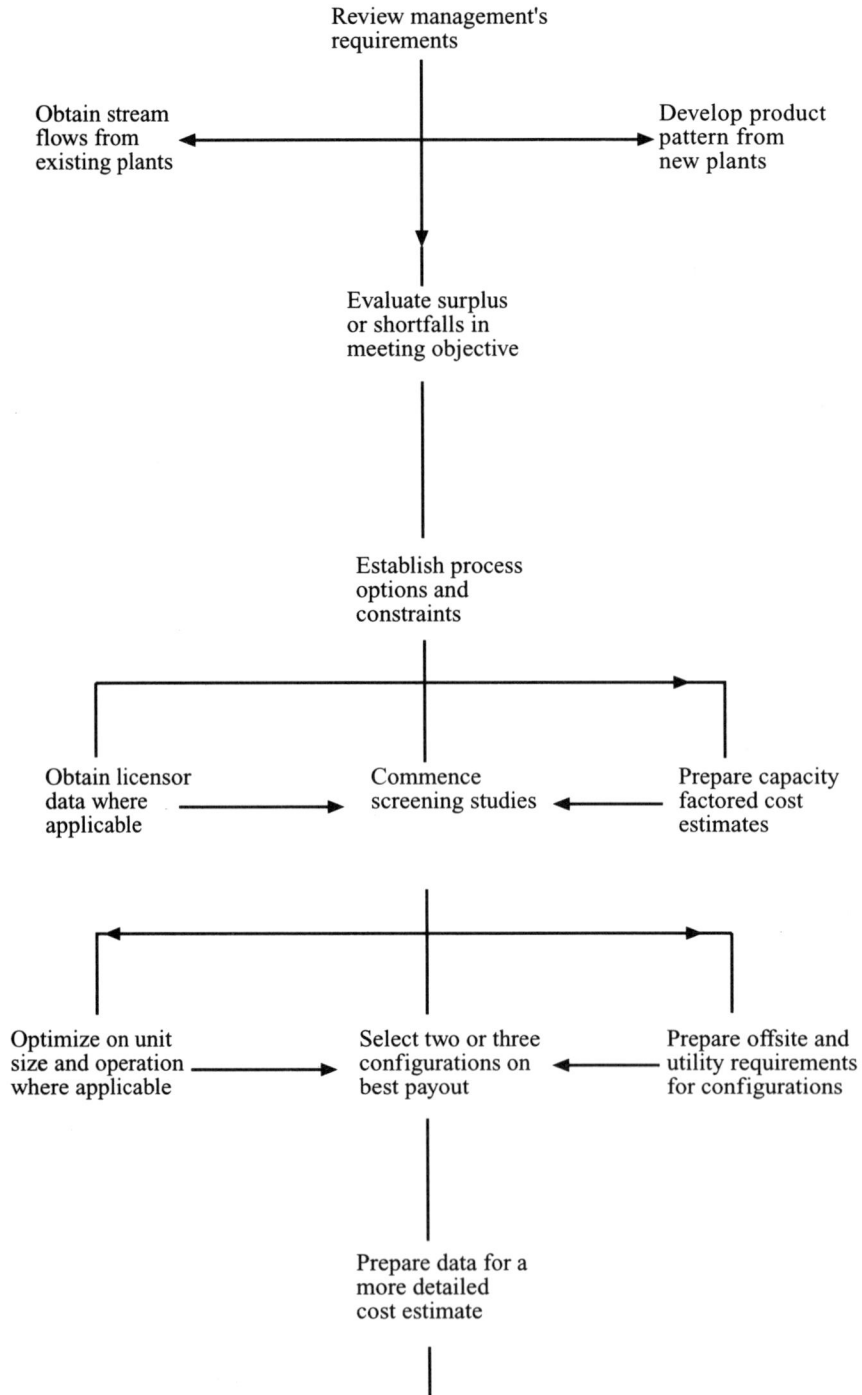

Figure 8.1. Steps in a typical process study

```
                              │
      ┌───────────────────────┼◄─────────────────────────┐
      │                       │                           │
      │               Calculate cash flow                 │
      │               for selected                        │
      │               configurations                      │
      │                       │                           │
      │                       │                           │
      │               Calculate ROI          Check and update
      │               on DCF basis for       cost estimate
      │               selected configurations             │
      │                       │                           │
      │                       │                           │
      │               Select process                      │
      │               route with best ◄───────────────────┘
      │               ROI
      │                       │
      │                       │
      │               Prepare report
      │                       │
      │                       │
   No go                 Make formal          For go decision
   decision ◄─────────── presentation to ───► prepare process
                         management            specification
```

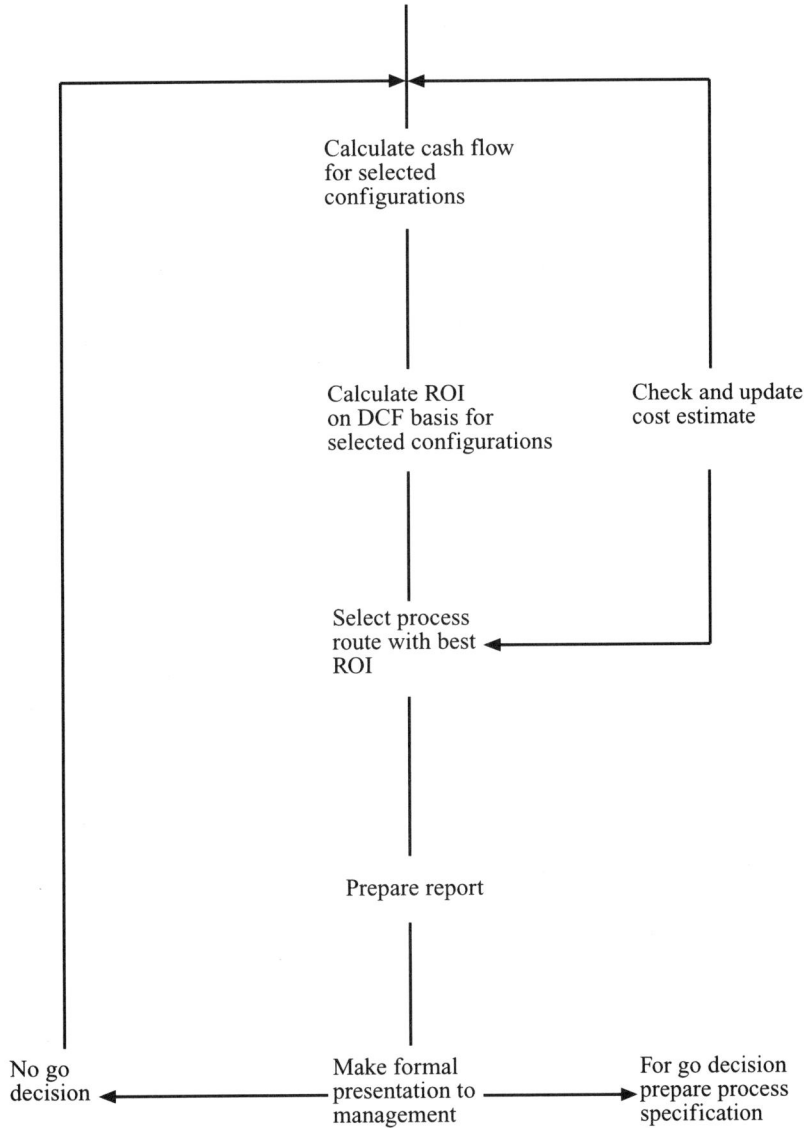

economic study depends to a large extent on the accuracy of the data used in predicting the various cost items. The principal cost item will be the capital cost of the processes to be installed and/or revamped. The next step in the study therefore is for the process engineer to provide basic engineering for a 'semi-definitive' capital cost estimate to be prepared. Should the project proceed from study stage to installation, this estimate will be used as the first project budget estimate for project control. Details of this type of estimate are given later in this chapter. Briefly, for this estimate process engineers will:

—Prepare process flowsheets (complete with mass balances)
—Prepare preliminary mechanical flowsheets (complete with line and instrument sizes with design conditions for temperature and pressure)
—Initiate all equipment data sheets sufficient for obtaining manufacturers' firm prices
—Review and select manufacturers' offers for equipment
—Using selected manufactures' equipment data, prepare preliminary utility balances and size by factoring new or expanded utilities facilities
—Factor additional offsite facilities to cater for the project objective
—Participate with others in preparing a preliminary plot layout for the new process facilities

● *Step 8   Calculate cash flow for selected configurations*   The selected manufacturer's equipment offer will also include a delivery time for the equipment item. Using these and other experience factors a preliminary schedule for the installation of the facilities is developed. This is executed by the experienced schedule engineers of the company or by accepted consultants. Using this schedule, the process engineer participates with others (usually members of the company's project management department or finance departments) to split the budget down to annual payments over the construction period as scheduled. The exercise continues to allocate payout of the capital costs on an annual basis over a prescribed period after plant start-up. This is also discussed later in this chapter.

● *Step 9   Calculate ROI on DCF basis for selected configurations*   The process engineer proceeds to develop the return on investment using the cash flow developed in step 8 for the selected configurations. A technique using the discounted cash flow principle is used for this purpose. This provides a more accurate and more credible result than a ROI arrived at by the simple method where no account is taken of the construction time and cash flow. Details of ROI calculations are given later in this chapter.

● *Step 10   Prepare the report and make formal presentation to management*   Much of the report should be written as the study project proceeds. In this way facts which are still fresh in one's mind are recorded accurately. The final report must be concise but complete with all salient points. Technique in report writing is not included as a subject in this book. It is sufficient to say that communicating the study and its results effectively is as important as any technique and good engineering 'know-how' used in the work. The same applies to the final oral presentation to management. The study, the methodology, and finally the results must be presented in a manner that maintains interest and stimulates the decision-making mechanism.

This completes a typical study approach. It needs to be pointed out that should the screening study reveal an option which far surpasses any of the others studied in terms of payout this could be the only option that then needs the further study as described in steps 7 to 10. Such a situation saves considerable time and effort.

8.2   Building Process Configurations and the Screening Study

From Section 8.1 it can be seen that the first major event that involves in-depth examination in a process study is the screening of options. Here a very simple and preliminary economic comparison of the options is carried out. This comparison is viable because the basis for the economic criteria is the same for each of the options. It must be borne in mind that a prerequisite for carrying out any of these process studies is a good working knowledge of the processes involved. In the example chosen here some knowledge of petroleum refining is a great advantage. This example has been chosen because the ultimate product slate is achieved through the addition of new processes, changing operating conditions of existing units, and, of course, blending base stocks. In many chemical or petrochemical plants only the first two of these mechanisms may be used and blending of base stocks would be unacceptable.

The example, then, is a requirement by a petroleum refinery to upgrade its fuel oil product to more financially lucrative lighter products. This is fairly common in this particular industry because of the fluctuating needs of the fuel market. The example calculation given here to illustrate the study is the base configuration published in an earlier book *Elements of Petroleum Processing* by this author. This may be found at the end of the first chapter of that book, and the block flow sheet is given here as Figure 8.2. The objective of this refinery configuration is to maximize gasolines and middle distillate but still retaining sufficient of the residue from the crude distillation as fuel oil.

In the example given here it is intended to reduce the amount of fuel oil and to produce as much gasoline and diesel as possible to give the best return on investment. For this purpose six well-proven processing routes have been selected for preliminary study:

(1)   Thermal crack all the atmospheric residue. This will be a single-stage unit.
(2)   Vacuum distill all the atmospheric residue. The waxy distillate will be cracked in a fluid catalytic cracker to give gas, LPG, gasoline naphtha, and light cycle oil as diesel blend components. The vacuum residue from the vacuum unit will be thermally cracked to give additional gas, LPG, naphtha (for gasoline production) and gasoil for diesel blending.
(3)   The same as the previous option except that there will be no thermal cracker for the vacuum residue.
(4)   Vacuum distill all the atmospheric residue and hydrocrack the waxy distillate.
(5)   Process the atmospheric residue through a delayed coking unit.
(6)   Finally use a proprietary fluid coking unit for the atmospheric residue that eliminates the coke by converting it into low-BTU gas.

The last two options are eliminated after a preliminary study because there is no economically attractive outlet for the coke and no effective use for the low-BTU gas in this refinery.

The example calculation that follows provides the detailed calculation for

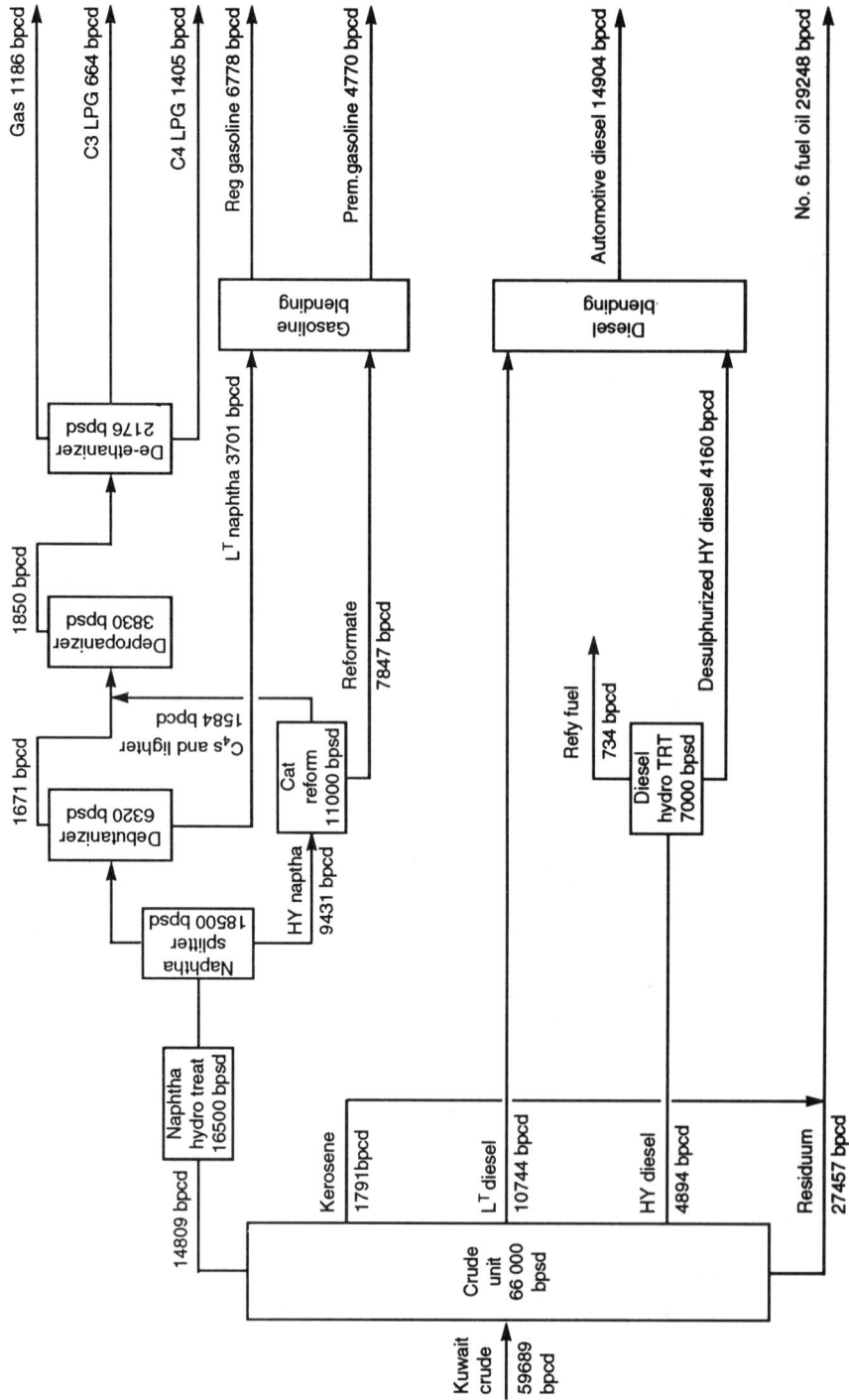

Figure 8.2. Existing refinery configuration

scheme (1) and provides the results of the same calculations for schemes (2)–(4). The configuration for scheme (1) is given in the block flowsheet (Figure 8.3). Similar block flowsheets would be prepared for the other study schemes but these are not shown here. However, a flow description is given for these three schemes.

Example calculation

It is required to develop a processing scheme in which the gasoline and diesel products are increased at the expense of the fuel oil product. The capacity of the present units and the flow configuration and flow rates are given in Figure 8.2. The refinery fence pricing of the products is as follows:

	$/bbl
LPG (C_3 and C_4)	20.2
Premium gasoline	35.2
Regular gasoline	27.6
Automotive diesel	28.6
No. 6 fuel oil	17.2

The cost of Kuwait crude oil is posted at $17.0/bbl delivered to refinery tankage.
    Cost of utilities and labour is as follows:

Power (purchased)	$0.042/kwh
Water	$0.5/1000 US gal
Steam	$0.83/1000 lb
Fuel	$0.01/lb
Labour	$15/h salary + 40% burdens. Three 8-hour shifts plus one shift off. Total hours per person per calender month = 185

The calculation

The first step is to determine the yields and the quality of the products that can be obtained by the thermal cracking of the atmospheric residuum. These are obtained by a method given in *Elements of Petroleum Processing*, Chapter 8, and are:

Stream	%wt	lb/h	lb/gal	gph	bpcd
Gas to C_5	9.02	35 079	4.67	7 506	4 289
Naphtha to 390 °F	20.29	78 899	5.34	14 775	8 443
Gas oil 390 °F to 622 °F	24.66	95 878	6.89	13 916	7 952
Residue +622 °F	46.03	179 013	7.73	22 870	13 069
Total	100.00	388 869	6.58	59 067	33 753

(Note: There is a volumetric increase in the total products over the feed. This is due to the changes in gravity of the products of cracking. The total weight and moles will still be the same between feed and products.)

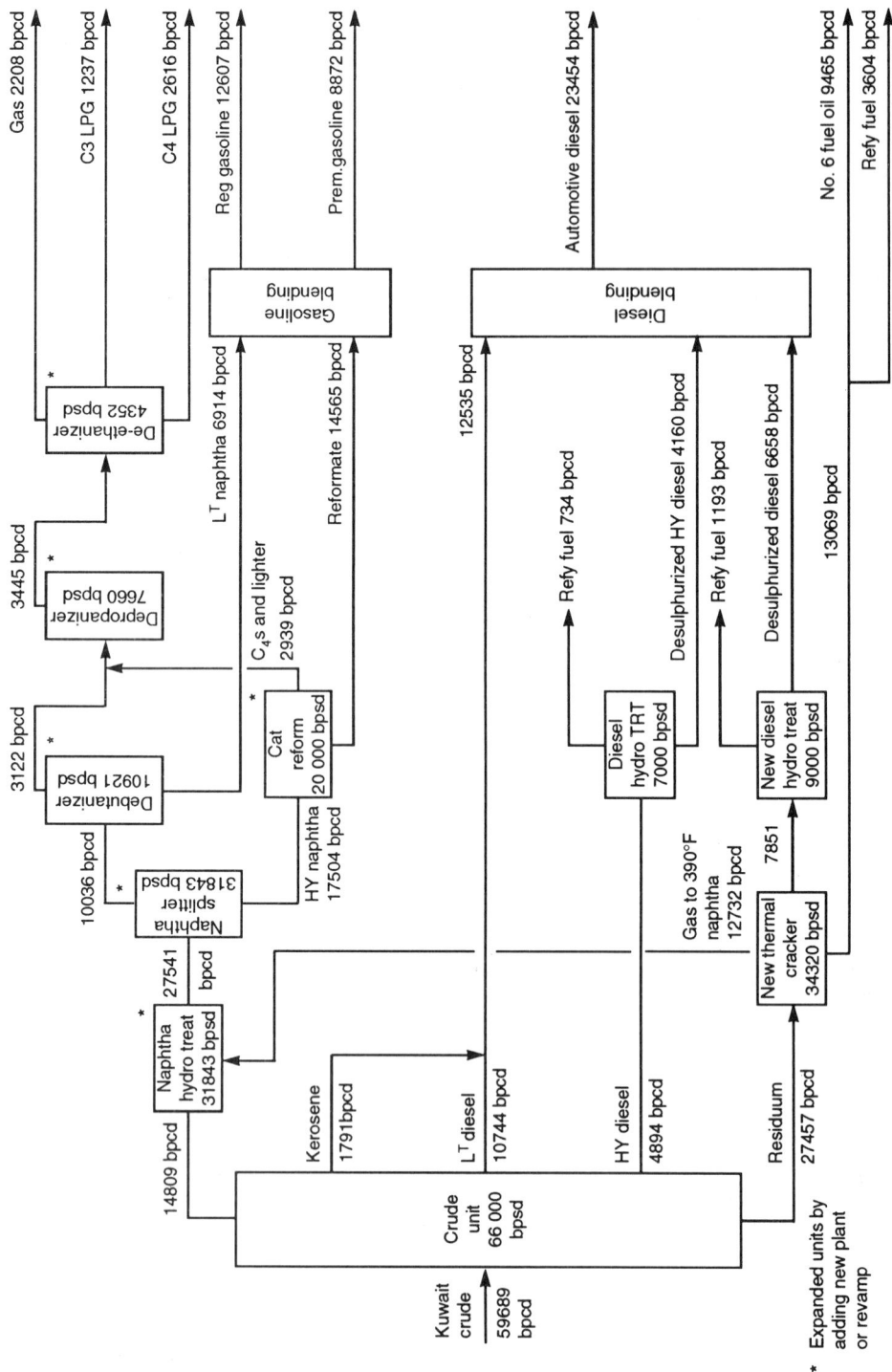

Figure 8.3. Scheme 1 – thermal cracking

HANDLING THE PRODUCTS

Figure 8.3 shows the routing of the product streams from unit to unit. The gas from the thermal cracker (gas to C_5) and the naphtha are routed to the existing naphtha hydrotreater which will be expanded to accommodate this. The hydrotreated material will subsequently go the splitter and debutanizer respectively. The existing reformer and light ends unit will also be expanded to cater for the thermal cracker light products. The gas oil from the thermal cracker will be hydrotreated in a new or expanded hydrotreater to meet diesel pool requirements. Thus new unit sizes will be:

- *Naphtha Hydrotreater*   Total new flow =

$$14\,809 \text{ bpcd} + 12\,732 \text{ bpcd} = 27\,541 \text{ bpcd}$$

New size at 86% service factor =

$$\frac{27\,541}{0.86} = 32\,024 \text{ bpsd}$$

which is an increase of 72% above the existing unit. This is too large to consider a revamp, therefore build a new unit 13 500 bpsd in parallel. The same applies to the splitter.
- *Catalytic reformer*   New feed to the reformer is 16 191 bpcd. At an 86% service factor the new size will be

$$\frac{17\,504}{0.86} = 20\,353 \text{ bpsd}.$$

Build a new cat reformer of 9000 bpsd.
- *Light ends*   New total feed to the debutanizer is 9283 bpcd. With a service factor of 90% this gives a new capacity of 10 000 bpsd, an increase of 58%. This unit will be revamped at a cost of 10% of a new unit. The depropanizer and de-ethanizer will be expanded with parallel units the same size as the existing ones.
- *The diesel hydrotreater*   A new hydrotreater of 9000 bpsd will be built to cater for the additional gas oil from the thermal cracker. The new total amount of gas oil to be treated is 4894 bpcd + 7851 bpcd = 12 745 bpcd. There is 706 bpcd spare capacity in the existing hydrotreater through improved catalyst and service factor. Taking this into consideration with a service factor of 80% then the additional capacity required will be 9000 bpsd.

OPERATING COSTS

The following table gives approximate factors for utilities and labour for some refinery plants. Note that the use of these is acceptable for the screening analysis given here. For detailed economic analysis of the selected configuration a more accurate operating cost using certified vendor and licensor data must be used.

| | Utilities requirement per bbl of feed | | | | |
	Steam (lbs)	Water (gals)	Power (kWh)	Fuel mm BTU	Labour Operation/shift
Light end dist.	25	12	0.4	—	1
Splitter.	0	9	0.4	0.1	
Cat reformer	50	15	12	0.35	3
Nap. hydrotreatment	—	4	6	0.12	1
Diesel hydrotreatment	30	8	1.5	0.12	1
Thermal cracker	70	20	3.0	0.5	2
FCCU	20	20	7.0	—	4
Wash plants	1.0	2.5	0.4	—	1
Vac. dist. units.	15.0	20	0.2	0.07	3

Looking at the configuration under study here, the operating cost for the extension of the refinery is as follows:

| Unit | | \$/calender day for utilities | | | Labour per ft^2 |
	Steam	Water	Power	Fuel	
Naphtha hydro (13 500 bpsd)	—	27	3 400	900	1
Naphtha splitter (1350 bpsd)	—	60	227	750	1
Lt ends (3000 bpsd)	60	18	50	—	
Thermal cracker (34 320 bpsd)	1 922	343	4 324	9 500	2
Diesel hydro (9000 bpsd)	216	36	567	600	1
Cat reformer (9000 bpsd)	360	68	4 536	263	3
Total	2 558	552	13 104	12 013	8

Total cost of utilities per stream year using 0.86 average service factor

$$= \$28\,227/\text{day} \times 0.86 \times 365 = \$8.840 \text{ million/year}$$

Labour salaries at \$15/h = 15 × 8 × 4 × 185 h/month × 1.4
Notes on labour:

(1)   The salaries are multiplied by 40% to take up government benefits, paid days off, vacation; etc.
(2)   There are four shifts: three on duty and one off every 24 h.
(3)   Hours worked per person per month is 185.

Then total labour cost per year = \$124 320/month × 12
                                                  = \$1.491 million/year
Total operating cost per year = \$ 10.331 million

INVESTMENT COSTS FOR THE NEW FACILITIES

There are two costs which make up the overall investment outlay:

- Capital cost of the plant
- Associated costs

The capital cost

For the screening study a capital cost of plant estimated on experience and factors are used. This and other types of estimates are described later in this chapter. A set of capacity cost data for plants of this size are listed below. Note that these numbers are pure fiction and bear no relationship to any company's cost data.

Type of unit	Capital cost factor ($/bpsd)
Light end units	412
Catalytic reformer	3120
Hydrotreater (naphtha)	780
Hydrotreater (diesel)	1200
Wash plants	170
Naphtha splitters	360
Thermal crackers	2000
Fluid catalytic crackers	3960
Vacuum distillation units	700

These factors increase or decrease with significant changes in plant capacity. They will increase with decrease in capacity and decrease to some extent with increase in capacity. The data here are assumed to be good for plant capacities under consideration.

Associated costs

To complete the full investment picture the following costs must be included as part of the investment:

- Cost of first catalyst and chemical inventory
- Cost of licensing fees for proprietary processes
- Cost of additional utility and offsite facilities. This cost is included because additional processing plants will either require additions to utilities and offsite facilities or take up available capacity. This will result at some future date in an expansion to these facilities which would not be necessary if this process expansion had not occurred.

For the purpose of a screening study a figure of 15% of the capital cost is used for these associated costs. In more definitive studies this item would be developed with more detail.

The capital cost of the process expansion considered here is as follows:

- *New naphtha hydrotreater*   Capacity 13 500 bpsd at $780/bpsd = $10 530 000

- *New naphtha splitter*   Capacity 13 500 bpsd at \$360/bpsd = \$4 860 000
- *Revamp debutanizer and new light ends*   Debutanizer revamp =

$$10\% \text{ of new plant} = \frac{2963 \times 412}{10}$$

$$= \$122\,000$$

$$\text{New light ends unit} = (3830 \text{ bpsd} \times 412) + (2176 \times 412)$$

$$= \$2\,474\,472$$

$$\text{Total} = \$2\,596\,500$$

- *New thermal cracker*   At a service factor of 80% the capacity of the unit is

$$\frac{27\,457 \text{ bpcd}}{0.8} = 34\,320 \text{ bpsd}$$

$$\text{Capital cost} = 34\,4320 \times \$2000$$

$$= \$68\,600\,000$$

- *New diesel hydrotreater*

$$9000 \text{ bpsd} \times \$1200 = \$10\,800\,000$$

- *New catalytic reformer*

$$9000 \text{ bpsd} \times \$3120 = \$28\,100\,000$$

$$\underline{\text{Total capital cost} = \$\,125.490 \text{ million}}$$

Associated costs:

This is taken as 15% of capital cost = \$18.823 million

$$\underline{\text{Total net cost} = \$\,144.313 \text{ million}}$$

THE PRODUCT SLATE

The new refinery product slate is summarized below (all in bpcd):

Product	Existing	Expanded	Difference
C_3 LPG	664	1237	+573
C_4 LPG	1405	2616	+1211
Reg. gasoline	6778	12607	+5829
Prem. gasoline	4770	8872	+4102
Auto diesel	15638	23454	+7816
Fuel oil	26323	9465	−16858
Refy fuel gas	1186	4135	+2949
Refy fuel oil	2925	3604	+679
Total	59689	65990*	+ 6301*

*Note difference in bpcd between input and output is due to changes in SG due to cracking. Mass (lb/h) in still equals mass out.

Gross income from increase in product slate

Product	Increase BPCD	$/bbl	$/day
LPGs	1 784	20.2	36 037
Reg. gasoline	5 829	27.6	160 880
Prem. gasoline	4 102	35.2	144 390
Auto diesel	7 816	28.6	223 538
Fuel oil	−16 858	17.2	−289 906
Refy fuel	−3 628	17.1	−62 039
Gross income			212 900

DEPRECIATION

In the economic analysis of the various processes the depreciation of the plant value must be taken into account. Details of depreciation will be discussed later in this chapter. For the purpose of the screening study the following depreciation factors may be used:

Crude distillation units	6.6% of capital cost per year
Cracking units	6.6% of capital cost per year
Light end units and gas plants	5.0% of capital cost per year
Hydrotreaters	6.0% of capital cost per year

The screening analysis based on a payout time and a return on investment (ROI) is given in Table 8.1. Note that ROI is given by

$$\frac{\text{Net income after taxes}}{\text{Net investment cost}} \times 100 = \%\,\text{ROI}$$

Table 8.1. Scheme 1—Addition of a thermal cracker: simple economic analysis

	$million
Capital cost of plant	125.490
Associated costs	18.823
Total net investment	144.313
	$million/year
Gross income	77.709
Less	
Operating cost	10.331
Depreciation	8.030
Net income before tax	59.348
Tax at 42.5%	25.223
Net income after tax	34.125
Payout time	3.4 yr
Return on investment	23.7%

and payout time is given by

$$\frac{\text{Net investment cost}}{\text{Net income after taxes} + \text{depreciation}} = \text{years}$$

DETAILS OF THE OTHER SCHEMES

- *Scheme 2* This calls for the vacuum distillation of the atmospheric residue to produce a waxy distillate and a heavy vacuum residue. The waxy distillate is cracked in a FCCU (fluid catalytic cracker) to give gas, LPG, naphtha for gasoline and gas oil for diesel. The heavy vacuum residue is also cracked in a thermal cracker to give gas, LPG, naphtha feed to the cat reformer, gas oil for diesel.

  The total capital cost for these additional facilities = $218.580 million
  Total additional product income = $108.202 million/year
  Total operating cost is an additional = $13.856 million/year

- *Scheme 3* This is the same as scheme 2 except the vacuum residue is not cracked. Details of this scheme are:

  Total capital costs for additional facilities = $194.780 million
  Total additional product income = $96.509 million/year
  Total operating cost is an additional = $8.122 million/year

- *Scheme 4* This also includes for the vacuum distillation of the atmospheric residue as in schemes (2) and (3). The wax distillate in this case, however, is routed to a hydrocracker where LPG, naphtha for reforming to gasoline and diesel are produced. An additional hydrogen plant is required for this scheme. Details of the scheme are summarized below:

  Total capital cost for additional facilities = $306.012 million
  Total additional income $119.502 million/year
  Total operating cost is an additional $12.6 million/year

The results of these three economic analysis are compared with scheme 1 in Table 8.2.

8.3 Using Linear Programs to Optimize Process Configurations

Linear programming is a technique to solve complex problems having multi-variable conditions by the use of linear equations. The equations are developed to define the interrelationship of the variables. Computers are used to solve these equations and to select from a matrix of these equations the solution or solutions to the problem.

This is not a new technique. It has been used for many years in industry and particularly the oil industry to plan and optimize its operation. Indeed, in oil refining today there is considerable development in process control using linear programming 'on-line' to optimize the units operation. In its use as a management tool it can refine the calculations described in Section 8.2 to a very fine degree. It

Table 8.2. Comparison of the economic analysis of the schemes studied

Scheme	1	2	3	4
		$million		
Capital cost	125.490	218.580	194.780	306.012
Associated costs	18.823	32.828	29.217	46.080
Total net investment	144.313	251.408	223.997	352.092
		$million/year		
Gross income	77.709	108.202	96.509	119.502
Less				
Operating cost	10.331	13.856	8.121	12.600
Depreciation	8.030	15.007	13.237	18.361
Net income before tax	59.348	79.339	75.150	88.541
Tax at 42.5%	25.223	33.719	31.939	37.630
Net income after tax	34.125	45.620	43.211	50.911
Payout time (yr)	3.4	4.15	3.97	5.1
Return on investment (%)	23.7	18.1	19.2	14.4

The obvious selection in this case is scheme 1 with the lowest payout time and the highest ROI.

is possible also by using this technique to examine many more options than in a manual operation.

The first objective using this technique is to build a mathematical model that fully defines the problem. The model itself will consist of many sub-models which will be interrelated by linear equations. In the case of an oil refinery configuration study the data for the model development will include:

- All the processing plants to be considered
- The yield from each plant
- The product and feedstream properties
- All the possible routing of the streams
- All the possible final blending recipes
- Utility and operating requirements for each process on a unit-throughput basis
- Investment cost for each process

These data are coded and the coded items used in equations which represent all the relationships to one another. The coded data items are shown graphically in Figure 8.4. as part of the graphical representation of the total model. This in it's entirety is too large to include here. The relationship equations representing physical and financial data are input to the computer program. Proprietary sub-routines included in the program resolve non-linear relationships to linear. Such routines as the DCF calculation are also added into the main program. The computer solves the many hundred equations, meeting the problem premises and any other applied constraints. The selected configurations are optimized within the program and the final printout shows product quantities, product quality, stream flow, unit capacities investment costs, operating cost and the ROI. The graphical representation of the model shown in Figure 8.4, contains 511 variables and over 400 equations. By present linear program standards it is not a large model.

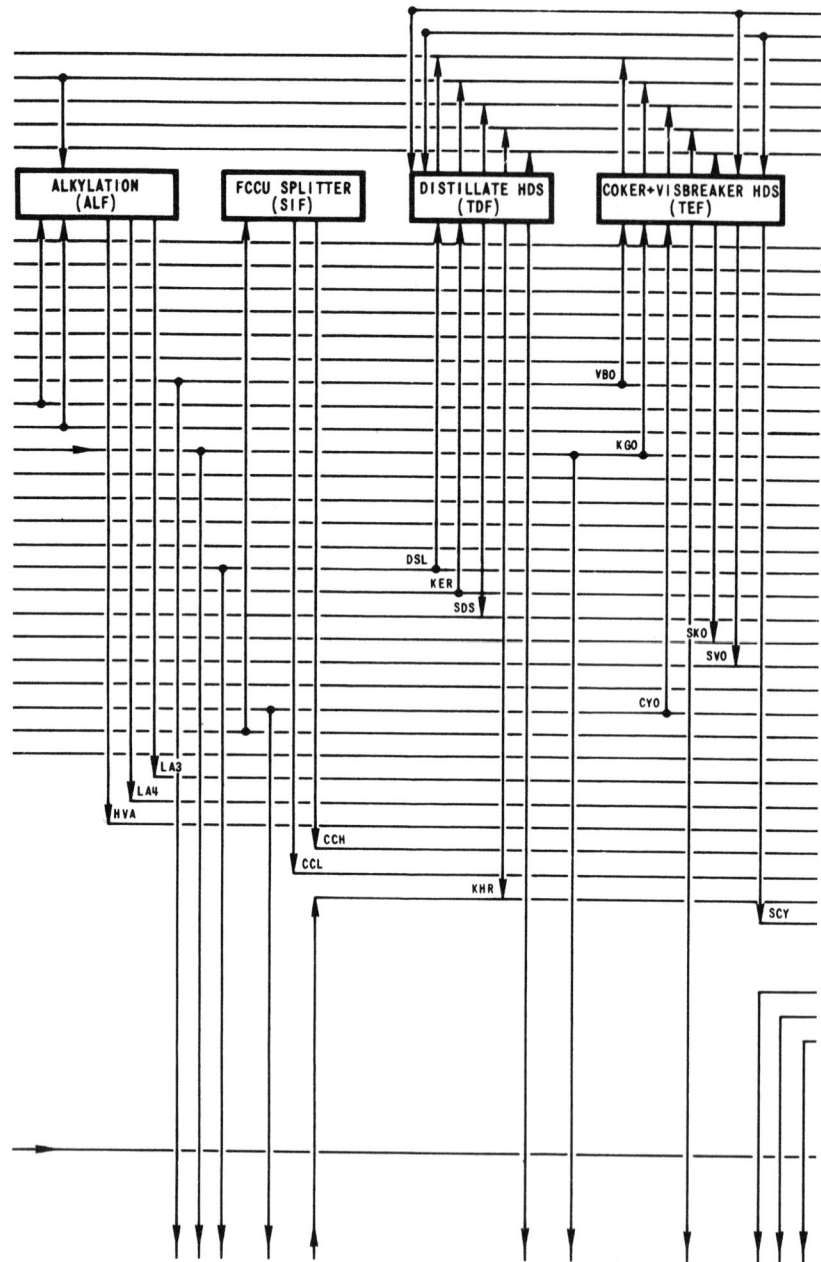

Figure 8.4. Economic study linear programming model

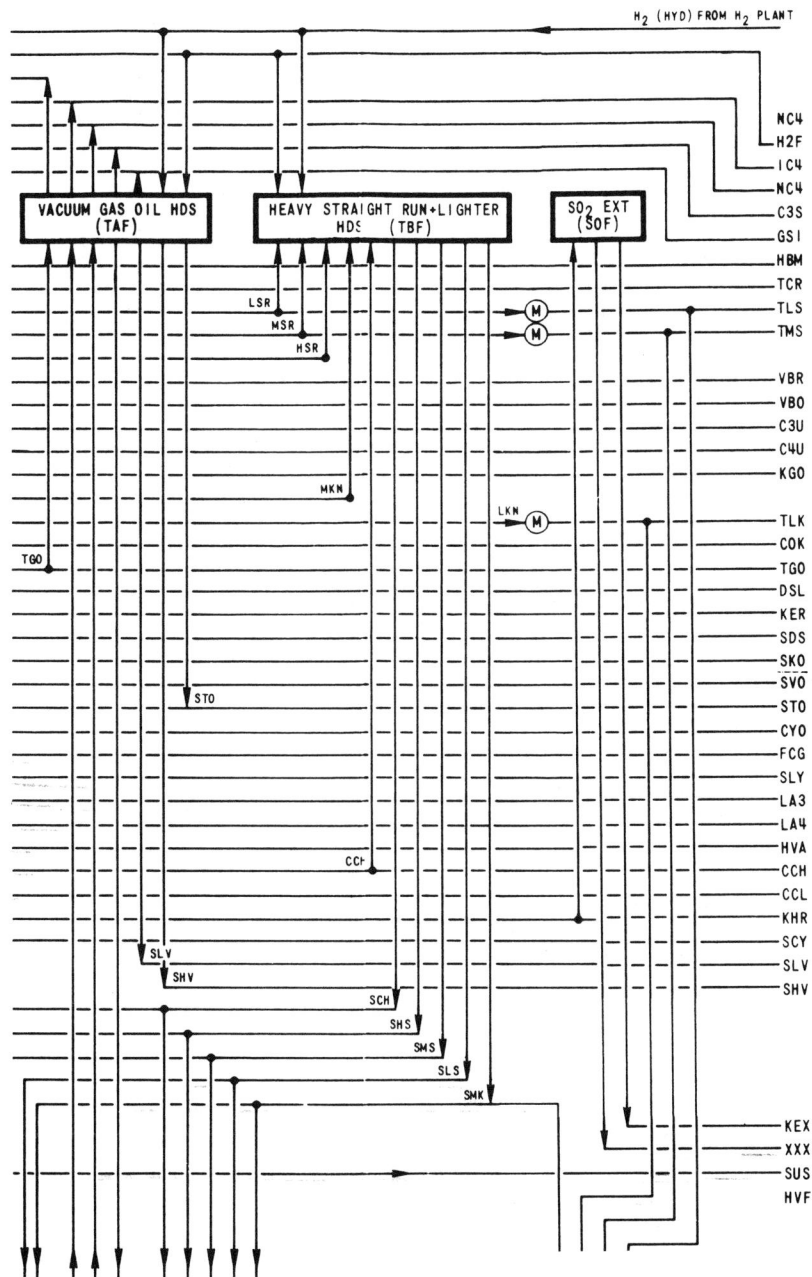

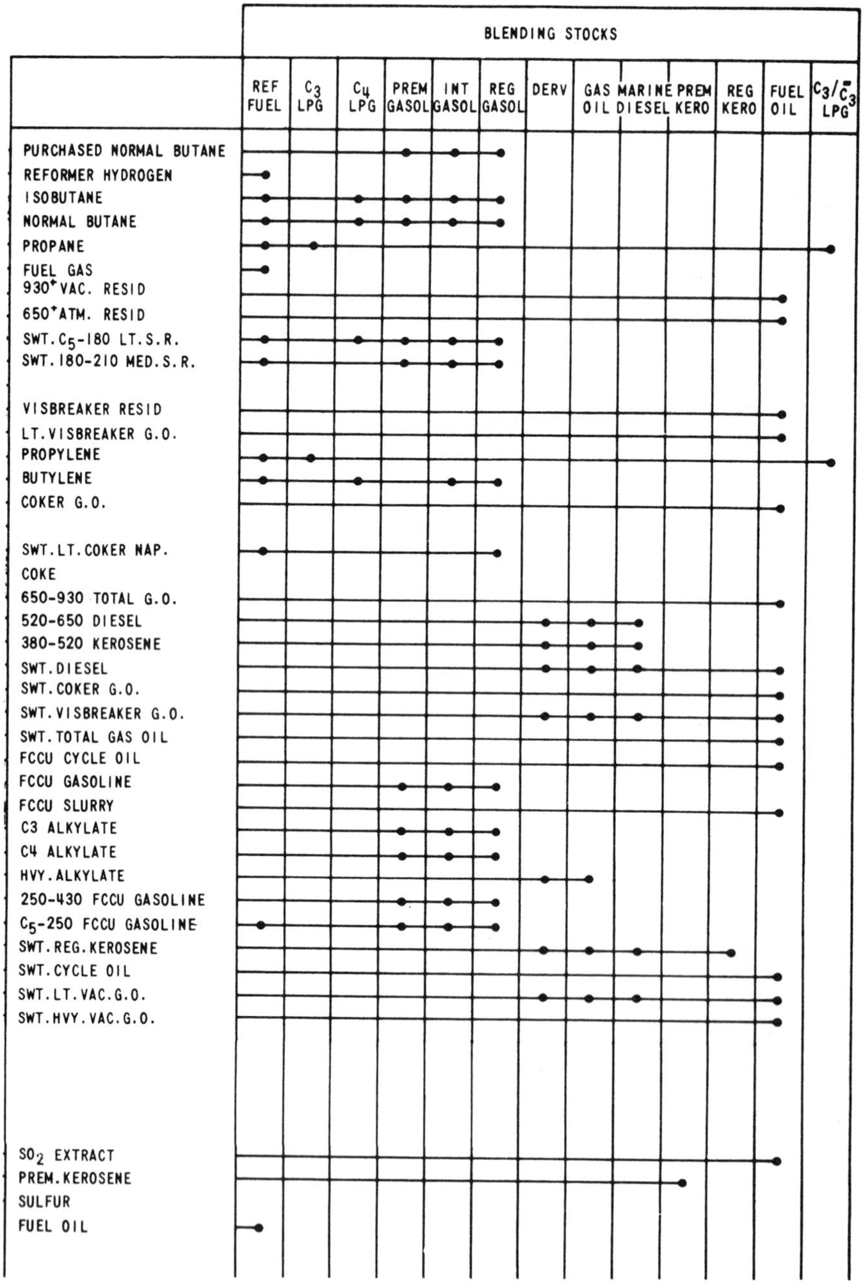

Figure 8.4. (*continued*)

Appendix 2 is a description of a process configuration study using a linear program. It was carried out for an American oil company who intended to build a refinery in Europe. The several solutions or options arrived at answered the constraints the company had in mind. Such an exercise would have taken a few years to accomplish by manual calculation with results probably not of the quality described here.

8.4 Preparing More Accurate Cost Data

Once a selected configuration or study case or cases have been selected from the screening study it is now necessary to firm up as much as possible the cost data used in the screening study. Three items of data fall into this category:

- The capital cost estimate
- The associated costs
- The operating cost

Process engineers supply the basic data from which an update can be obtained. This level of data is arrived at from a process design of the facilities making up the case studies. The process design will include a material balance and a firm process flow sheet. Sufficient data will be generated for all equipment to allow manufacturers to give a good and realistic equipment budget cost. Where applicable the manufacturers will also be requested to provide equipment efficiencies from which utility usage for the various items can be calculated. Where options exist within the process design itself these will be optimized for equipment sizing and layout. An example of such an optimization is the case of the process heat exchange system (see Section 1.7). Full and complete equipment data sheets need not necessarily be developed for this part of the study. Usually manufacturers are able to provide good budget costs from equipment summary sheets similar to those given in the example calculation below.

Where licensed processes are included in the case study the process engineer will need to develop a specification to a suitable licensor soliciting:

- A budget estimate for the plant (installed)
- The utility requirements and the operating cost (in the case of a catalytic process part of the operating cost will be the catalyst usage or loss in operation)
- First inventory of catalyst or chemicals where applicable
- The licensing fee

The specification must include the capacity of the plant and the required product quality. It must also, of course, give the licensor(s) details of the feedstock, utilities conditions, and in most cases local climatic conditions. The additional information concerning feed and product pricing together with utilities costs will enable the licensor to respond with firmer and optimized cost data.

The following example calculation follows the study screening given in Section 8.2.

H301 - Thermal cracker heater
Heating duty 121.789 mm BTU/h
Heat of reaction 249.106 mm BTU/h
Total duty 370.895 mm BTU/h

T301 - Syn tower
6'0" i/d and 13'5" i/d 108' T-T

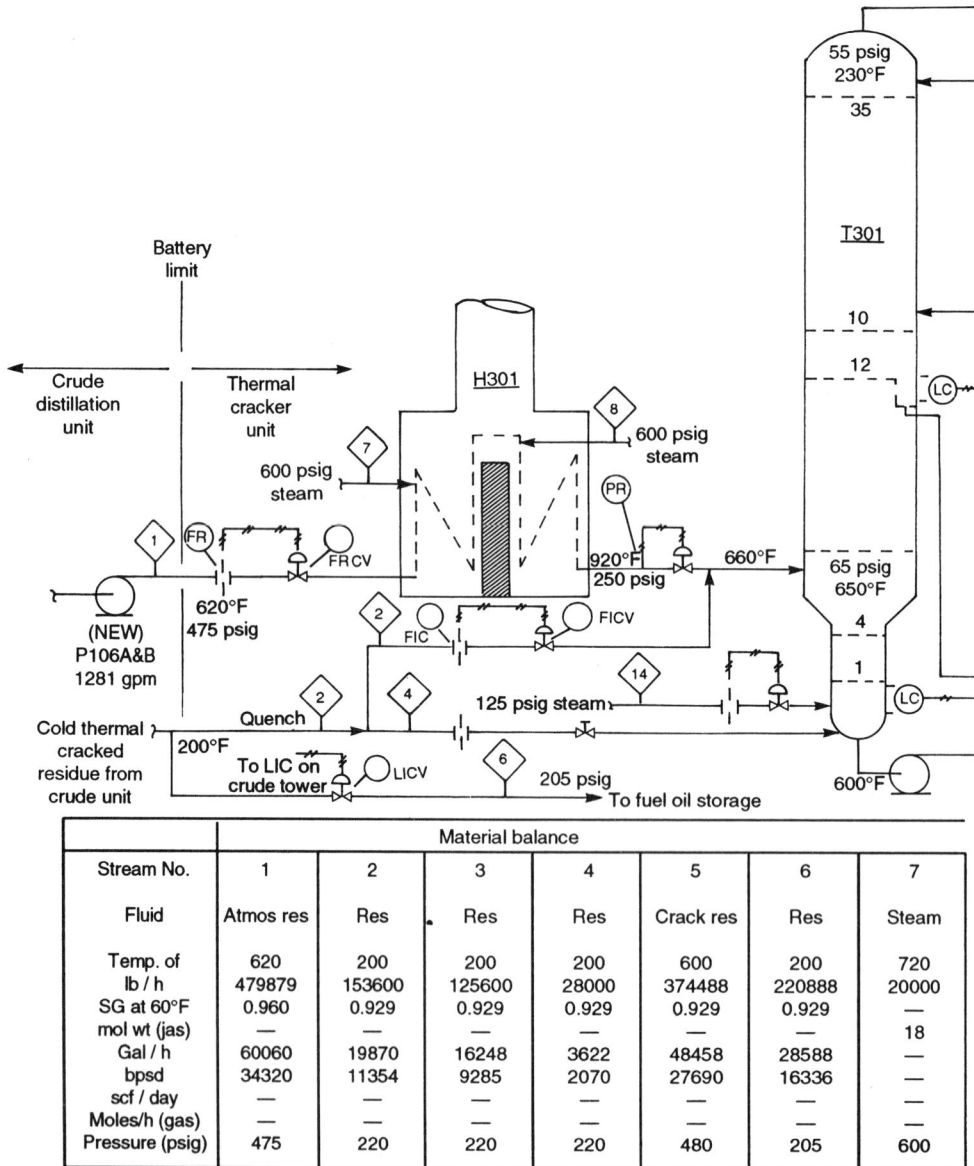

		Material balance					
Stream No.	1	2	3	4	5	6	7
Fluid	Atmos res	Res	Res	Res	Crack res	Res	Steam
Temp. of	620	200	200	200	600	200	720
lb / h	479879	153600	125600	28000	374488	220888	20000
SG at 60°F	0.960	0.929	0.929	0.929	0.929	0.929	—
mol wt (jas)	—	—	—	—	—	—	18
Gal / h	60060	19870	16248	3622	48458	28588	—
bpsd	34320	11354	9285	2070	27690	16336	—
scf / day	—	—	—	—	—	—	—
Moles/h (gas)	—	—	—	—	—	—	—
Pressure (psig)	475	220	220	220	480	205	600

Figure 8.5. Process flow diagram for a 34 320 bpsd thermal cracker

E-301 - Overhead condenser
Duty 157.373 mm BTU/h
Split range

D301 - Syn tower reflux drum
6'6" i/d x 20'0" T-T
and
4'0" i/d x 6'0" T-T
E-302 - Gas oil cooler
Duty 9.941 mm BTU/h

E-301

PR CV
To flare

PR CV
From fuel gas main

TR   PR

LCR

50 psig D301
90°F

LC   LCV   ⟨13⟩

To sour
water stripper

P301 A&B

LLV   ⟨9⟩

To new
naphtha
hydrotreater

⟨10⟩   FICV

P302 A&B

1420 gpm

⟨12⟩
210°F
FICV   FIC

E-302

TIC

415°F

⟨11⟩

P303 A&B
326 gph

To new diesel
hydrotreater
surge drum

⟨5⟩   LCV

To crude unit
existing heat
exchange system

P-304 A&B
1024 gpm

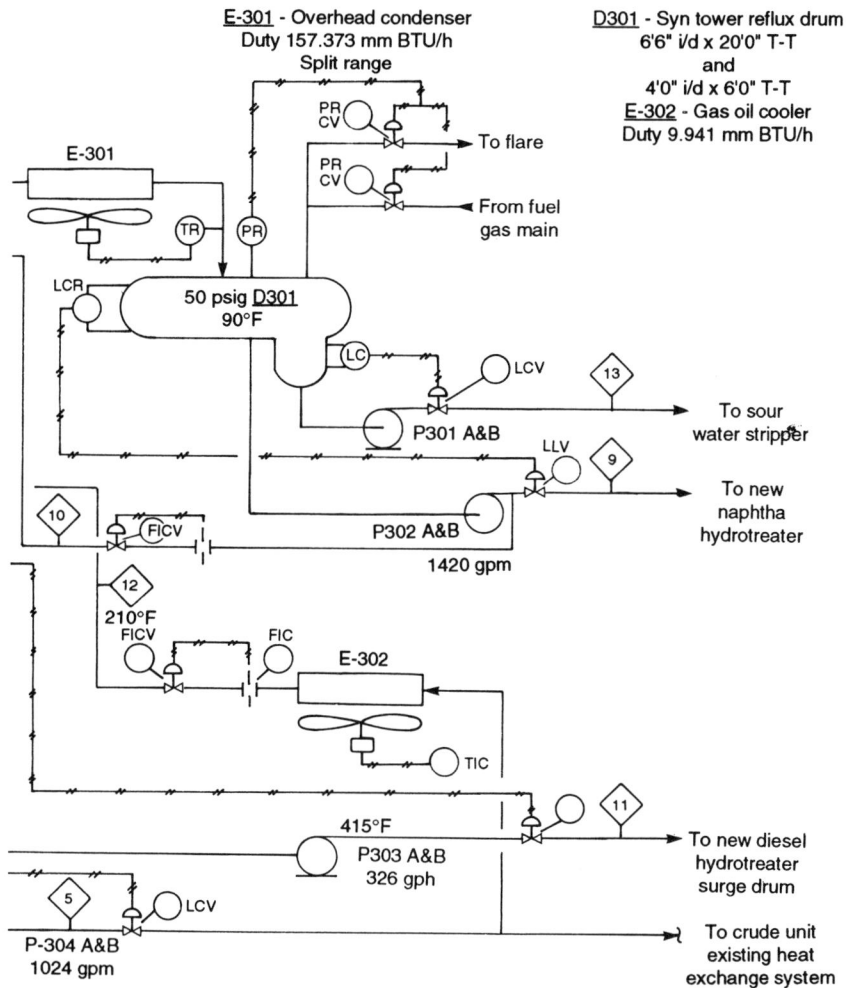

Material balance						
8	9	10	11	12	13	14
Steam	Naphtha	Naphtha	Gas oil	Pump around	Soap water	Steam
6720	90	90	415	210	90	380
5000	140653	281300	118338	88755	58200	33200
—	0.6070	0.6070	0.828	0.828	—	—
18	—	—	—	—	—	18
—	27852	55703	17175	12881	7217	—
—	15915	31830	9814	7361	—	—
—	—	—	—	—	—	—
—	—	—	—	—	—	—
600	120	75	135	89	152	125

Example calculation

From the screening study the new plants included in the selected case (which was case 1) were as follows:

- Naphtha hydrotreater (13 500 bpsd): licensed
- Naphtha splitter (13 500 bpsd)
- Catalytic reformer (9000 bpsd): licensed
- Depropanizer (3830 bpsd) and de-ethanizer (2176 bpsd)
- Diesel hydrotreater (9000 bpsd): licensed
- Thermal cracker (34 320 bpsd)

The existing debutanizer is to be revamped and a new sized equipment list will be prepared for this unit.

The unit's normally licensed processes are indicated as such in the above list. The thermal cracker may be licensed or can be designed by some major engineering companies as a non-proprietary process. It is a major item of cost and for this example the degree of process engineering required at this stage of the study is demonstrated for this unit.

THE PROCESS DESIGN OF THE THERMAL CRACKER

The process flowsheet for this unit is given in Figure 8.5. This contains the overall material balance for the unit together with the calculated reflux and quench flows. Major process stream routes are shown with stream temperatures where appropriate. The material balance table also gives the flow conditions (temp. and press.)

Table 8.3. Summary data sheet — centrifugal pumps

Item No.	P106 A&B	P301 A&B	P302 A&B	P303 A&B	P304 A&B
Service	Feed	S water	Ref + PR	PA/Prod	Bot. prod.
No. of units	1 + 1 spare	1 + 1 spare	1 + 1 spare	1 + 1 spare	1
Fluid	Resid	Water	Naphtha	Gas oil	Cr. Res.
Lb/h	479 879	58 200	421 953	207 093	374 488
Pump temp. (°F)	620	90	90	415	600
SG at PT	0.758	0.995	0.592	0.692	0.732
Visc. at PT (cSt)	1.8	0.76	0.33	0.25	1.2
Rate (norm) at PT (gpm)	1281	117	1420	599	1024
Rate (max) at PT (gpm)	1470	130	1700	690	1178
Norm. suct. press. (psig)	39	63	63	75	72
Max suct. press. (psig)	89	132	132	135	122
Discharge press. (psig)	475	152	85	135	480
Diff. press (psi)	436	89	22	60	408
Diff. head (ft)	1332	207	86	200	1290
NPSH avail. (ft)	>15	>15	>15	>15	>15
Hyd. horsepower	323	6.1	18.3	20.9	244
Corr/erros	Sulph.	H₂S	H₂S	Sulph.	Sulph.
Driver*					
R	Motor	Motor	Motor	Motor	Turb.
S	Turb.	Motor	Motor	Turb	Motor†

*Driver R = normally running, S = normally spare.
†May be common spared with P106 A&B.

for those major streams. The pump capacities are shown on the flowsheet next to the respective pump. These quantities are flows in gallon per minute at the pump temperature.

The major item in a thermal cracker is the furnace or heater. This has two duties, the first being to heat the feed to its reactor temperature and the second to maintain the feed at its reaction temperature for the prescribed period of time. Note therefore that the duty of this heater is shown on the flowsheet as the duty to heating the oil (which also includes the steam in this case) and the duty to supply the heat of reaction.

Data for the remainder of equipment are given in Tables 8.3–8.5. Sufficient

Table 8.4. Summary data sheet—vessels

Item No.	T301	D301
Service	Syn. tower	Ref. drum
Type	Vertical	Horizontal
Dimensions (ft):		
Top sect. I/D	13.5	6.5
Top sect. T-T	74.0	20.0
Bot sect. I/D	7.5	4.0
Bot sect. T-	31.0	6.0
Overall T-T	108.0*	20.0
Skirt	15.0	In struc.
Internals:		
Trays	Top sect. 30 sieve Bot. sect. 4 disk don.	None
Operating conditions:		
Press. (psig)	55/67	50
Temp. (°F)	230/650	90
Design conditions:		
Press. (psig)	125	125
Temp. (°F)	920	810
Vacuum (psia)	7.0	7.0
Min. temp. (°F)	90	90
Max. liquid level (ft)	8.0	N/A
Materials:		
Shell	CS[†]	CS
(corr. allow)	0.125 in.	0.125 in.
Internals	11/13 Cr	
(corr. allow)	0.125 in.	
Trays	Monel/Cr	None
(corr. allow)	0.125 in.	
Packing	N/A	
Insulation:		
Yes/no	Yes	No
Stress relieved:		
Yes/no	No	Yes

*Swage sect. 3.0 ft.
[†]Clad with 11/13 Cr.

Table 8.5. Summary data sheet—air condensers/coolers

Item No.	E301		E302	
Service	O/head cond.		Gas oil cooler	
Type	Forced		Forced	
Duty mmBtu/h	157.373		9.941	
Fluid	Nap. + H_2O		Gas oil	
Temp. (°F)	In 230	Out 90	In 415	Out 210
Liquid:				
M lb/h HC		422	88.8	88.8
H_2O		58.2	nil	nil
SG at 60 °F HC		0.607		0.828
Visc. at 60 cSt HC		0.12		0.52
Therm Cond HC		0.07		0.06
Specific heat (BtU/lb °F HC)				0.80
Vapour:				
M lb/h HC	422			
H_2O	58.2			
Mol. wt HC	96			
Therm. cond.	0.009			
Specific heat (BtU/lb °F)	0.41			
Fouling HC factor		—		
Ambient air temp. (°F)		60		
Allowable ΔP_{si}		5.0		10.0

Table 8.6. Process specification for a thermal cracker heater/reactor

Item number: H301 Thermal cracker heater

Overall heat duty: 370.895 mm BTU/h

Feed rate lb/h: 479 879

Temperature of feed in: 620 °F

Temperature of effluent: 920 °F

Specific gravity of feed at 60 °F: 0.960 (Kuwait Res. + 650 °F)

Effluent TBP and vaporization curves are attached (Fig 1)

Effluent SG curves versus mid boiling point curve is attached (Fig 2)

Feed conversion to gasoline of 390 °F cut point = 30% vol

Coil temperature profile (oil + steam) versus coil volume given as Fig 3 (attached)

Steam (600 psig) injected at coil volume of 50 ft³ and 150 ft³ respectively. Total steam to be injected is 74 000 lb/h

Outlet pressure of effluent to be 250 psig. This pressure to be maintained by downstream pressure control valve. (Supplied by others)

Coil pressure drop to be 225 psi max.

Table 8.6. (*continued*)

Coil tubes should be standard 4 in. 11/13 Cr with wide return bends. The firebox should be constructed to heat the incoming feed to a temperature of about 870 °F in the first 150 ft³ of coil. This should constitute the *heater section*. The *soaker section* and crossover should be designed to heat the oil to 920 °F and retain it at this temperature according to the profile given in Fig 3. The sections should be divided by a firebrick wall so that temperatures on both sides of the wall can be controlled independently by burner adjustments.

Vendor may recommend that the convection side of the heater be used for additional heater capacity or for steam generation or to remain unused if better temperature control may result.

Figure 4 attached shows a proposed layout of the heater coil with the temperature profile.

Note: The figures referenced in this text are not attached but are mentioned to illustrate the degree of process engineering required for this part of a process study.

data for obtaining a cost and efficiency for the heater are provided in a 'process specification' shown in Table 8.6. A special specification sheet is required for this item because, besides being an oil heater it is also a cracking reactor. In this function the design of the unit must be such as to allow the oil to be retained for a specific period of time at the reaction temperature. Table 8.7 gives a summary of accepted prices.

Table 8.7. Summary of accepted prices

The following are accepted vendor prices for equipment delivered to site.
Pumps and drivers

				Price $/unit	
	hhp	bhp	Pump	Motor	Turbine
P 106 A	323	430	606 826	489 082	—
106 B			606 826	—	503 903
P 301 A	6.1	10.1	14 253	19 563	—
301 B			14 253	19 563	—
P 302 A	18.3	28.1	39 655	39 126	—
302 B			39 655	—	40 312
P 303 A	20.9	32.2	45 441	48 908	—
303 B			45 441	48 908	—
P 304 A	244	325	458 648		
TOTAL			1 870 998	665 150	997 728

Total pumps and drivers = $3 533 884

Vessels

T 301 Syn. tower

	$
Shell and heads	1 766 810
Trays	1 590 129
Internals	176 681
Total	3 533 620

D 301 Reflux drum

Shell & Heads	561 222
Lining	37 415
Internals	24 943
Total	623 580

Total vessel account = $4 157 200

Coolers and condensers

E 301 Overhead condenser:

Unit including fans and structure = $1 627 540

E 302 gas oil cooler:

Unit including fans = $243 196 (unit located in pipeway, no standalone structure)

Total cooler and condenser account = $1 870 740

Heater/reactor

H 301 thermal cracker heater:

This unit will be field erected. Included in this account is the cost of fabricated material delivered to site.

Unit material cost = $8 314 400.
Total equipment material account summary.

	$
Pumps and drivers	3 533 884
Vessels	4 157 200
Coolersandcondensers	1 870 740
Heater/reactor	8 314 400
Total	17 876 224

8.5 Capital Cost Estimates

Capital cost estimates are usually prepared by specialist engineers who are fully conversant with their company's estimating procedures and material and labour cost records. Process engineers are involved in cost estimating only so far as they develop and provide the basic technical data used to prepare the capital cost of a plant or process. To execute this part of their duty properly, however, process engineers should know about cost estimates and their significance. This section therefore sets out to define in broad terms the various levels of estimating, their degree of accuracy and the scope of the engineering development that is required for each of them.

The accuracy of cost estimates will vary between that based on comparing similar plants or processes on a capacity basis to the actual cost of a plant when all the bills have been paid and the plant is operational. Obviously the accuracy based on similarity of the process of plant is very low while that based on project end cost must be 100%, that is, it is no longer an estimate but is now a fact. In between these two levels, however, estimates may be prepared giving an increasing degree of accuracy as engineering and construction of the process advances. The accuracy of estimates are enhanced as more and more money is committed to the project. The progress of a project uses milestone cost estimates coupled with completion schedules for its measurement and control. In the life of a construction project therefore four 'control estimates' may be developed. For want of better terms these may be referred to as

- The capacity factored estimate
- The equipment factored estimate
- The semi-definitive estimate
- The definitive estimate

Companies have their own terms for these estimates but the important point is to develop these estimates at a time in the project when they can be most useful and when sufficient information is available for their best accuracy. Figure 8.6 shows phases of a typical project when these estimates may be best developed. The following paragraphs now describe these estimates in more detail and emphasize the process engineer's input.

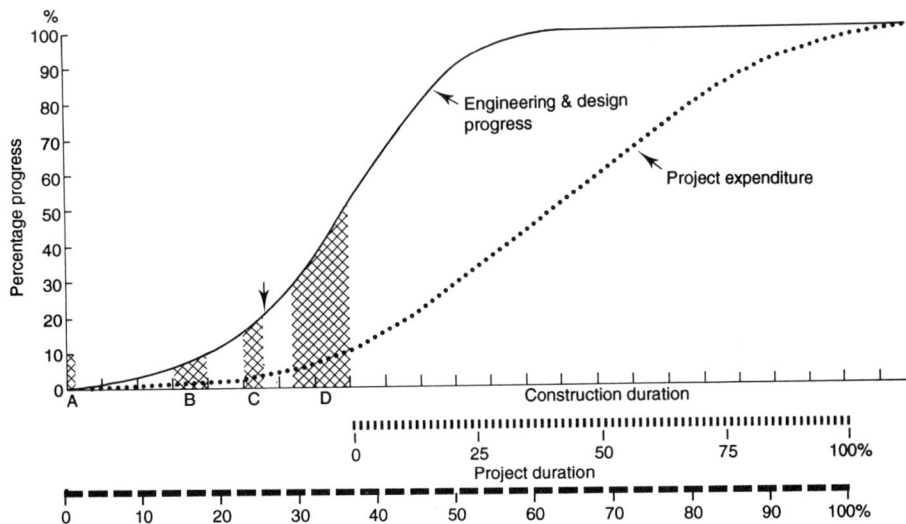

Figure 8.6. Time phasing of estimate types. A—capacity factor, B—machinery and equipment, C—semi-definitive (semi-detailed), D—definitive

THE CAPACITY FACTORED ESTIMATE

In most process studies this is the first estimate to be prepared. It is the one that requires the minimum amount of engineering but is the least accurate. This is because the plant that is used to factor from will not exactly match the plant in question. It is good enough, however, when comparing different processes where the estimates are on the same basis.

Past installed costs of similar plants (definitive estimate) are used coupled with some experienced factors to arrive at this type of estimate. The cost of a higher- or lower-capacity plant is obtained from the equation

$$C = K(A/B)^b$$

where

C = cost of the plant in question
K = known cost of a similar plant of size B
A = the capacity of the plant in question (usually in vol/unit time)
B = capacity of known plant
b = an exponential factor ranging between 0.5 and 0.8

The estimator's experience and records of the type of plant determines the exponent to be used. Usually this is 0.6. A listing of these factors are given in Appendix 1 in Table A1.8. Table 8.8 shows an example of a capacity factored estimate for the thermal cracking unit used in Section 8.4.

In this example the direct field cost is used as the cost figure to be capacity factored. This allows the cost estimator to adjust between the direct field cost and the total cost using current experience for field indirects, office costs, escalation and contingencies. In this case a multiplier of 1.5 times the direct field cost is used to collect these items.

THE EQUIPMENT FACTORED ESTIMATE

This requires a substantial amount of process engineering to define the specific plant or process that is to be estimated. Details of the degree of engineering to be performed is given in Section 8.4. Briefly, the following process activity is required:

- Development of a firm heat and material balance
- An acceptable process flowsheet to be developed
- An approved equipment list
- Equipment summary process data sheet for all major equipment
- Process specification for specialty items
- A detailed narrative giving a process description and discussion. (This will be required for the management review and approval of the estimate.)

This estimate is normally developed by the process engineer and the cost estimator only. There may be some participation from equipment vendors and licensors in the case of licensed processes. There is seldom any further input required from other engineering disciplines for this level of estimating.

Table 8.8. Capacity factored estimate

COST ESTIMATE SUMMARY

Plant: Description: Date:

			Estimated cost ($ million)			
A/C No.	Item	M/h	Labour	Sub-cont.	Material	Total
	Excavation		0.6			
	Concrete	$C =$	$K(A/B)$			
	Structural steel	where	$K = 28.93$ m	$ million		
	Buildings		$A = 34\,320$	bpsd		
	Machinery/equip.		$B = 16\,000$	bpsd		
	Piping					
	Electrical	C	$= 1.57 \times$	$28.93 million $= 45.733$		
	Instruments					
	Painting/scaffold					
	Insulation					
	DIRECT FIELD COST					45.733
	Temp. const. facil.					
	Constr. expense					
	Craft benefits & burdens					
	Equip. rental					
	Small tools					
	Field 0/heads					
	INDIRECT FIELD COSTS					
	TOTAL FIELD COSTS					
	Engineering salaries					$\times 1.5$
	Purchasing salaries					
	Gen office salaries					
	Office expense					
	Payroll burdens					
	Overhead costs					
	TOTAL OFFICE COSTS					
	TOTAL FIELD & OFFICE COSTS					
	Fee (contractor)					
	Escal/Contingency					
	TOTAL COSTS					68.596

The equipment summary process data sheets and any process specifications that may be required are used to solicit equipment prices from preselected vendors. In addition to the pricing data that will be used for the estimate development vendors should also be asked for delivery time required and sufficient details to enable the operating cost of the item to be calculated. The delivery time required is important to management to establish a project schedule and lifespan for the engineering and construction of the facilities. Later process engineers will also require this schedule to develop cash flow curves and data.

On receipt of the vendor's pricing data the cost estimator will apply factors to each of the equipment account data to arrive at a direct field cost for each group of equipment. This factored cost will be the multiplier of the equipment price which will include the materials and labour to install those items on-site. In Table 8.9 this factor is shown as a composite for the *whole* equipment prices to direct field cost. In this case the figure developed was a multiplier of 2.5. These factors are proprietary to respective companies but most major companies do have the statistics to develop their own factors.

The total cost is again obtained by factoring the DFC (direct field cost) as in the previous 'capacity factored' estimate. Note in the case of the example of an equipment factored estimate a factor of 1.53 has been used (slightly higher than that in Table 8.8). This is to account for the field erection of the heater, which was not known previously. This would impact on future home office engineering and field indirects to some extent.

THE SEMI-DEFINITIVE AND THE DEFINITIVE ESTIMATES

It is not proposed to describe these two estimates in detail here. It is sufficient to say that these estimates will include input from all engineering disciplines that are involved in the engineering and construction of the plant or process. Indeed after the equipment factored estimate the input from process engineers is merely an update of the process design as it develops. However, Tables 8.10 and 8.11 show the increasing amount of actual data as opposed to factored data that are included in these two estimates.

The numbers for these estimates are not included in these tables but the crosshatched items show where normally firm cost data would be included. The factored data are also indicated for both tables.

Although major manufacturers such as large oil and chemical companies have sufficient statistical data to develop semi-definitive and definitive estimates, these are normally left to the engineering and construction companies. These develop the estimates and use them during the course of the installation project as cost control tools for project management.

ACCURACIES OF ESTIMATES AND CONTINGENCIES

For an estimate to be a meaningful basis of a process study it's accuracy needs to be established and sufficient contingency be added to make the final cost figure as realistic as possible. Unfortunately this is easier said than accomplished. Most companies have, however, developed analysis programs to meet this requirement and many of these are quite sophisticated. Engineering and construction companies whose daily work depend on good cost estimates for project control are among the front-runners in this exercise. Most of these programs revolve around principles of risk analysis which is outside the scope of this book. Figure

Table 8.9. Equipment factored estimate

COST ESTIMATE SUMMARY

Plant: Description: Date:

A/C No.	Item	M/h	Estimated cost ($ million)			
			Labour	Sub-cont.	Material	Total
	Excavation					
	Concrete					
	Structural steel					
	Buildings					
	Machinery/equip.				17.876	
	Piping					
	Electrical					× 2.5
	Instruments					
	Painting/scaffold					
	Insulation					
	DIRECT FIELD COST					44.690
	Temp. const. facil.					
	Constr. expense					
	Craft benefits & burdens					
	Equip. rental					
	Small tools					
	Field 0/heads					
	INDIRECT FIELD COSTS					
	TOTAL FIELD COSTS					
	Engineering salaries					
	Purchasing salaries					× 1.53
	Gen office salaries					
	Office expense					
	Payroll burdens					
	Overhead costs					
	TOTAL OFFICE COSTS					
	TOTAL FIELD & OFFICE COSTS					
	Fee (contractor)					
	Escal/Contingency					
	TOTAL COSTS					68.377

8.7 has been developed from a collection of statistical values over a long period of time and does provide some quick guidance to judging the accuracy of the estimate and then setting a contingency value.

Figure 8.7 shows two sets of curves. The first is a positive to negative range of accuracy against the type of estimate. This type of estimate scale is based on Figure 8.6 as the progress of the project. Thus the widest accuracy range is for a

Table 8.10. Semi-definitive estimate

COST ESTIMATE SUMMARY

Plant: Description: Date:

A/C No.	Item	M/h	Estimated cost ($ million)			
			Labour	Sub-cont.	Material	Total
	Excavation				xxxxxxxx	
	Concrete				xxxxxxxx	
	Structural steel				xxxxxxxxx	
	Buildings				xxxxxxxxx	
	Machinery/equip.				17.876	
	Piping					xxxxxxxxx
	Electrical			xxxxxxxxx		
	Instruments				xxxxxxxxx	
	Painting/scaffold					
	Insulation					
	DIRECT FIELD COST	xxxxxxxx	xxxxxxxx		xxxxxxxxx	xxxxxxxxx
	Temp. const. facil.					
	Constr. expense					
	Craft benefits & burdens					
	Equip. rental			Factor		
	Small tools					
	Field 0/heads					
	INDIRECT FIELD COSTS					xxxxxxxx
	TOTAL FIELD COSTS					xxxxxxxx
	Engineering salaries					
	Purchasing salaries					
	Gen office salaries					
	Office expense					Factor
	Payroll burdens					
	Overhead costs					
	TOTAL OFFICE COSTS					xxxxxxxx
	TOTAL FIELD & OFFICE COSTS					xxxxxxxx
	Fee (contractor)					xxxxxxx
	Escal/Contingency					xxxxxxx
	TOTAL COSTS					xxxxxxxx

capacity factored estimate at the beginning of a project. The narrowest range therefore is at definitive estimate stage around 30–40% of project duration. The second set of curves gives a range of contingency to be applied to the various estimate types. This is a plot of percentage of total field and office cost against the estimate type used for the accuracy curves.

Table 8.11. Definitive estimate

COST ESTIMATE SUMMARY

Plant: Description: Date:

A/C No.	Item	M/h	Labour	Sub-cont.	Material	Total
			Estimated cost ($ million)			
	Excavation		xxxxxxxx	xxxxxxx	xxxxxxxxx	xxxxxxx
	Concrete		xxxxxxxx	xxxxxxx	xxxxxxxxx	xxxxxxx
	Structural steel		xxxxxxxx	xxxxxxx	xxxxxxxxx	xxxxxxx
	Buildings		xxxxxxxx	xxxxxxx	xxxxxxxxx	xxxxxxx
	Machinery/equip.		xxxxxxxx	xxxxxxx	17.876	xxxxxxx
	Piping		xxxxxxxx	xxxxxxx	xxxxxxxxx	xxxxxxx
	Electrical		xxxxxxxx	xxxxxxx	xxxxxxxx	xxxxxxx
	Instruments		xxxxxxxx	xxxxxxx	xxxxxxxxx	xxxxxxx
	Painting/scaffold			Factored		
	Insulation			Factored		
	DIRECT FIELD COST	xxxxxx	xxxxxxxx	xxxxxxxx	xxxxxxxxx	xxxxxxxx
	Temp. const. facil.		xxxxxxxx		xxxxxxxx	xxxxxxxx
	Constr. expense					xxxxxxx
	Craft benefits & burdens					xxxxxxx
	Equip. rental				xxxxxxxx	xxxxxxx
	Small tools				xxxxxxxx	xxxxxxx
	Field 0/heads					xxxxxxx
	INDIRECT FIELD COSTS		xxxxxxxx		xxxxxxxx	xxxxxxx
	TOTAL FIELD COSTS					xxxxxxx
	Engineering salaries	xxxxxxx	xxxxxxx			
	Purchasing salaries	xxxxxxx	xxxxxxx			
	Gen. office salaries	xxxxxxx	xxxxxxx			
	Office expense				xxxxxxxx	
	Payroll burdens				xxxxxxxx	
	Overhead costs				xxxxxxxx	
	TOTAL OFFICE COSTS xxxxxxxx		xxxxxxxx		xxxxxxxx	xxxxxxxx
	TOTAL FIELD & OFFICE COSTS					xxxxxxx
	Fee (contractor)					xxxxxxxx
	Escal/Contingency					xxxxxxxx
	TOTAL COSTS					xxxxxxxx

8.6 Discounted Cash Flow and Economic Analysis

Throughout this chapter a major role of a process engineer in conducting economic studies has been described and discussed. The purpose of these studies is to provide the company's management with sound technical data to enable

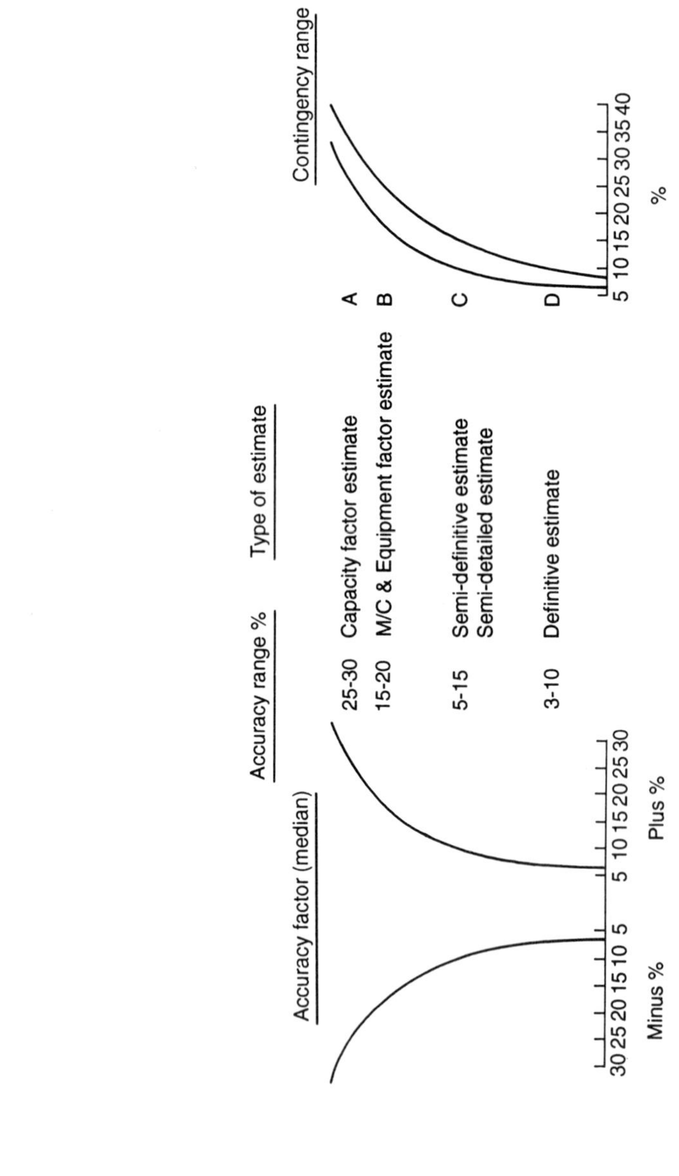

Figure 8.7. Accuracies and contingencies

them to improve the company's profitability at minimum financial risk. Up to now these studies have covered

- Identifying the viable options that will meet the study objectives
- Short-listing and screening the options using simple return on investment techniques
- Providing process input into the preparation of a more detailed capital cost estimate of those options selected for further detailed economic evaluation

The cost estimates based on firm equipment costs are the most accurate possible without committing to more detailed design and capital expenditure. Using these estimates the detailed and more reliable prediction of the profitability of the selected options can now proceed. This prediction is based on a return on investment calculated from a projected cash flow of the process over a prescribed economic life for the facility.

There are several methods of assessing profitability based on discounted cash flow (DCF) but the most reliable is a return on investment method using the present worth (or net present value) concept. This concept equates the present value of a future cash flow as a product of the present interest value factor and the future cash flow. *Based on this concept, the return on investment is that interest value or discount factor which forces the cumulative present worth value to zero over the economic life of the project.*

Whereas the development of capital cost of plants is usually a combined effort between the process engineer and cost estimator so is the development of a DCF return on investment a combined effort between the process engineer and the company's finance specialist. The process engineer provides the technical input to the work such as operating costs, type of plants, construction schedules and cost, yield and refinery fence product prices, etc. The financial specialist provides the financial data based on statutory and company policies, such as the form of depreciation, tax exemptions, tax credits (if any), items forming part of the company's financial strategy, etc. The calculation itself is in two parts:

- Calculation of cash flow
- Present worth calculation

These are described in the following paragraphs.

CALCULATING CASH FLOW

Figure 8.9 is a graph of the cumulative cash flow of a typical project in relation to its project life. Initially there is a financial loss when the company has to purchase land and equipment, pay contractors to erect equipment, etc. To do this, not unlike most private individuals, companies borrow money on which, of course, they have to pay interest. In addition to the cost of construction the company must keep in hand some capital with which it can buy feedstock and chemicals, and pay the salary of its staff during the commissioning of the plant. This working capital must also be considered as debt at the project initiation.

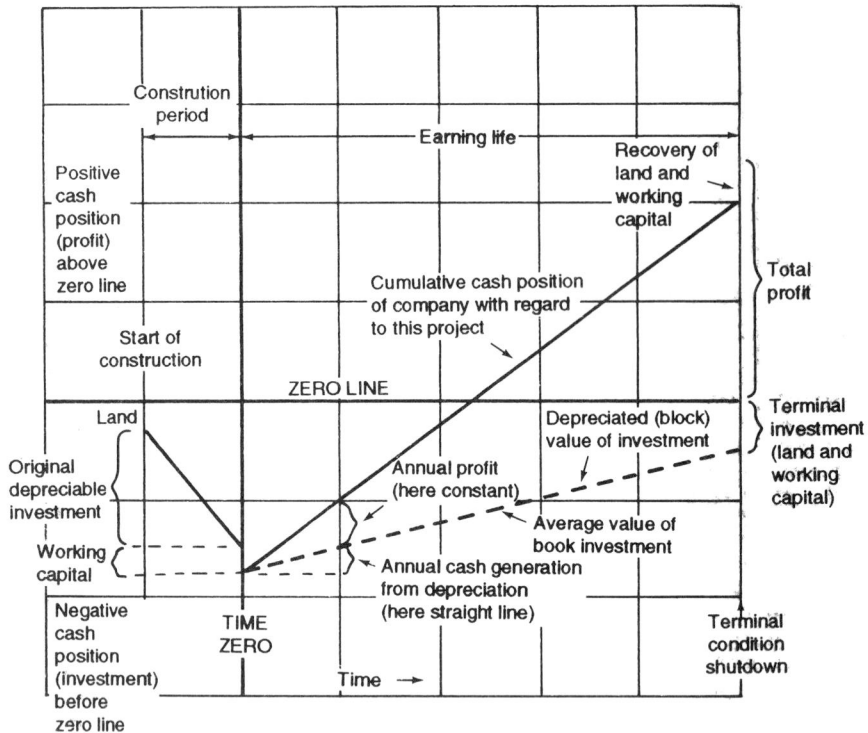

Figure 8.8. Graph of cumulative cash flow

At the end of construction and commissioning the cash flow into the system moves upward towards a positive value and after a prescribed number of years moves into a positive value. During this period of positive cash flow the money recovered from the operation must be sufficient to:

- Repay the loan and its interest
- Pay all associated taxes
- Pay all the operating costs of the project
- Make an acceptable profit

When these conditions have been met over the prescribed period of time called the 'earning life' or 'economic life' there still remains the plant hardware, the land, and the working capital. These are considered as the project's terminal investment and are added to the final project's net cost recovery.

The cash flow calculation recognizes the financial milestones shown in Figure 8.8 and the contents of a typical cash flow calculation are discussed below.

Economic life of the project

This is the number of years over which the project is expected to yield the projected profit and pay for its installation. These are the number of years starting at year 0, which indicates the end of construction and the commissioning of the facilities. The last year (usually year 10) is the year in which all loans and other project costs are repaid and the 'terminal investment' released.

Construction period

This is the period before year 0 during which the plant is constructed and commissioned. Assume this period is three years, then this is designated as end of year -2, -1, 0. During this period the construction company will receive incremental payments of the total capital cost of the plant with final payment at the end of year 0. The construction cost may be paid from the company's equity alone or from equity and an agreed loan or entirely from a loan. In the case of a loan to satisfy this debt the payment of the loan interest commences in this period. The interest payment over this period, however, is usually capitalized and paid over the economic life of the project.

Net investment

The net investment for the project includes the capital cost of the plant, which is subject to depreciation, and the associated costs, which are not subject to depreciation. The capital cost of the plant is the contractor's selling price for the engineering, equipment, materials and the construction of the facilities. In a process study using a DCF return on investment calculation the capital cost should be an estimate with an accuracy at least that based on an equipment factored type.

The associated costs include the following elements:

- Any licensor's paid-up royalties
- Cost of land
- First inventory of chemicals and catalyst
- Cost of any additional utilities or offsite facilities incurred by the project
- Change in feed and product inventory
- Working capital
- Capitalized construction period loan interest

Revenue

This is the single source of income into the project. In most cases it is the income received for the sale of the product(s). This is calculated from projected process yields of products multiplied by the marketed price of the products. A market survey should already have been completed to ensure that the additional products generated by the project are in demand and the price is in an acceptable range. Later a sensitivity analysis of the DCF return on investment may be conducted, changing the revenue recovered by price escalation or other means.

Expenses

This is the major cost of carrying out the project. It consists of the following items:

- *Plant operating cost* This includes the cost of utilities used in the process, such as power, steam and fuel. It also includes the cost of plant personnel in salaries, burdens, and indirects and the cost of chemicals and catalyst used.
- *Maintenance* There are two kinds of maintenance costs included in this item. These are the preventive maintenance carried out on a routine basis and those costs associated with incidental breakdowns and repair.
- *Loan repayment* The loan principal is paid back in equal increments over the economic life of the plant. This item includes the payback increments and the associated interest on a declining basis.

These are incurred running costs and as such are considered tax free.

Depreciation

The second cost to the project which is considered as a deductible from the gross profit for tax purposes is the depreciation of the plant value. This is calculated over the *plant life* as the plant capital cost divided by the plant life. The term 'plant life' is the predicted life of the facility before it has to be dismantled and sold for scrap. Usually this is set at 20 years and indeed all specifications relating to engineering and design of the facilities will carry this requirement, so that all material and design criteria such as corrosion allowances associated with the plant will meet this plant life parameter.

Ad valorem tax

This is the fixed tax levied in most countries payable to local municipal, provincial or state authorities to cover property tax, municipal service costs, etc.

Taxable income

Taxable income is revenue less operating cost, depreciation, and ad valorem tax. This, of course, is simply put, as in most countries, states or provinces there will probably be certain local tax relief principles and tax credits that will affect the final taxable income figure. The company's financial specialist will be in the position to apply these where necessary.

Tax

This is quite simply the tax rate applied to the 'taxable income' figure. This will vary from location to location but will be taken as one rate over the economic life of the project for the purpose of a process study, unless there are legislative changes already in place.

Profit after tax

This is the gross profit less the tax calculated in the previous item.

Net cash flow

This item is calculated for each year of the project's economic life. It commences with year 0 with the net investment shown as a negative net cash flow item. Then for each successive year until the end of the last year of the economic life, the net cash flow is calculated as the sum of profit after tax *plus* the depreciation. The depreciation is added here because it is not really a cost to the project. It is a 'book' cost only and is used specifically for tax calculation.

The cash flow item for the last year of the economic life must now include the 'terminal investment item'. This item is the sum of the net scrap value of the plant (scrap value less cost of dismantling), the estimated value of the land and the working capital initially used as part of the 'associated costs'. Thus the final cash flow item will be the sum of profit after tax, plus depreciation, plus terminal investment.

With the net cash flow in place the second part of the calculation, which is the determination of the return on investment based on the present worth concept for the project, can be carried out. Table 8.12 is an example of a completed cash flow calculation and Table 8.16 is an example of the present worth calculation.

Table 8.12. Net cash flow consolidation

| | $million | | | | | | | | | | |
End of year	0	1	2	3	4	5	6	7	8	9	10
Investment											
Capital cost	*719.2*										
Association cost	*22.2*										
Net investment	*141.4**·										18.7[†]
Revenue at 0% esc.		77.7	77.7	77.7	77.7	77.7	77.7	77.7	77.7	77.7	77.7
Expenses*											
Operating		11.0	11.3	11.7	12.0	12.4	12.7	13.1	13.5	13.9	14.4
Maintenance		4.8	4.8	4.8	4.9	4.9	5.0	5.1	5.1	5.2	5.2
Loan repayment		10.0	9.5	9.1	8.6	8.2	7.8	7.3	6.9	6.4	6.0
Total expense*		25.8	25.6	25.6	25.5	25.5	25.5	25.5	25.5	25.5	25.6
Depreciation*		7.5	7.5	7.5	7.5	7.5	7.5	7.5	7.5	7.5	7.5
Ad valorem tax*		2.4	2.4	2.4	2.4	2.4	2.4	2.4	2.4	2.4	2.4
Taxable income		42.0	42.2	42.2	42.3	42.3	42.3	42.3	42.3	42.3	42.2
Tax at 40%		16.8	16.9	16.9	16.9	16.9	16.9	16.9	16.9	16.9	16.9
Profit after tax		25.2	25.3	25.3	25.4	25.4	25.4	25.4	25.4	25.4	25.3
Net cash flow	*141.4**	32.7	32.8	32.8	32.9	32.9	32.9	32.9	32.9	32.9	51.5

*These are costs and therefore negative values in cash flow (shown here in italic).
[†]These are cash recoveries and therefore have positive values in cash flows.

Table 8.13. Schedule of operating and maintenance costs
Operating costs are escalated at a rate of 3.0% per year
Maintenance costs are escalated at 1.0% per year

End of year	Operating costs ($ million)	Maintenance ($ million)
1	11.01	4.76
2	11.34	4.81
3	11.681	4.56
4	12.031	4.904
5	12.392	4.953
6	12.764	5.003
7	13.147	5.053
8	13.541	5.103
9	13.947	5.154
10	14.366	5.206

CALCULATING THE CUMULATIVE PRESENT WORTH

This is an iterative calculation using three or more values as the discounting percentage. The result of the cumulative values of the present worth at the end of the economic life from these calculations are plotted against the discount percentages used for each case. The discount value when the cumulative present worth is zero is the return on investment for the project.

For each value of the discount percentages selected first calculate the discount factor for each year of the economic life starting at year 0, where the discount factor is always 1.0. Then divide the discount factor for year 0 which is 1.0 by $(1.0 + d)$ where d = discount percentage divided by 100. Do the same for each successive year. Thus if the discount percent selected is 10% then the discount factor for year $0 = 1.0$, for year $1 = 1/1.1 = 0.909$, and for year $2 = 0.909/1.1 = 0.826$, and for year $3 = 0.826/1.1 = 0.751$, and so on until the last year.

The present worth value for each year is then calculated as the net cash flow value for that year multiplied by the discount factor for that year. Remember, the cash flow value for year 0 is always negative. Next, determine the cumulative present worth value by adding the value for each successive year to the value of the year before. Starting with that value for year 0 which is negative add that value for year 1 which is positive to give a positive or negative cumulative value for year 1. Continue by adding the present worth value for year 2 to the cumulative value for year 1 to give the cumulative value for year 2, and so on through the last year. The cumulative value of the present worth for the last year in each of the discount percentages selected is plotted linearly against the discount percentage. This cumulative value can be either positive or negative. Indeed, to be meaningful the discount percentages selected must be such that the calculated present worth values for the last economic years be a mixture of negative and positive values. In this way the resulting curve plotted must pass through zero.

The following example calculation is based on scheme 1 in Section 8.2. Similar calculations may also be carried out on one or more of the other schemes that were screened to confirm their comparative profitability. In this case it is important to remember that the economic parameters used for each case are identical to enable a proper comparison and analysis to be made.

Example calculation

This scheme includes the addition of the following units into an existing oil refinery:

- New naphtha hydrotreater (13 500 bpsd)
- New naphtha splitter (1350 bpsd)
- Revamped debutanizer and new light end units (3830 bpsd)
- New thermal cracker (34 320 bpsd)
- New diesel hydrotreater (9000 bpsd)
- New catalytic reformer (9000 bpsd)

The capacity factored capital cost estimate used in the screening studies for this scheme was $125.490 million. A subsequent estimate based on a more definitive design and equipment definition (equipment factored estimate) gave a capital cost figure of $119.216 million for this configuration. This latter figure will be used as the capital cost in this example calculation.

Net investment cost

This will consist of:

- The capital cost
- Associated costs
- Capitalized construction loan interest

Associated costs:

(1) Paid-up royalties. This is a once-off licensing fee paid to the licensors of the hydrotreater processes and the catalytic reformer process. There will also be a running licensing fee for these processes but this will be included in the operating cost.
Paid-up royalties = $2.026 million
(2) First catalyst inventory = $4.052 million
(3) Cost of land = $1.0 million
(4) Cost of incremental utility/offsite facilities = $2.501 million
(5) Cost of increased product/feed inventory (only product inventory is considered here as there is no change in crude oil throughput). Statutory requirements for this refinery location is a mandatory inventory of 14 days' feed and product.

$$\text{Additional inventory cost} = 14 \times ((564\,845 + 162\,798) - (452\,756))$$
$$= \$3.848 \text{ million (see Section 8.2)}$$

(6) Working capital. This is taken as 5.0% of the capital cost = $5.835 million.

Total associated costs = $ 19.262 million

Construction cost and payment:

End of year	−2	−1	0
Construction schedule of payments ($ million)	11.922	47.686	59.608

The construction costs will be paid out of equity up to the limit of the equity. The remainder will be paid by a loan at 8.0% interest. Initial equity is 60% of capital + associated costs =

$$0.6 \times (119.216 + 19.262) = \$83.087 \text{ million}$$

The loan to complete the construction schedule of payments is raised during year 0 and interest on this is $2.89 million. This is capitalized but will be paid out of an increased equity.

Net investment:

	$million
Capital cost of plant	119.216
Associated costs	19.262
Construction loan interest	2.89
Net investment	141.368

Operating and maintenance costs

Operating cost for year 1 is made up as follows:

	$million
Operating labour	= 1.49
Utilities	= 8.32
Chemicals, catalyst, running royalties	= 1.20
Total	= 11.01

Operating costs escalate at a rate of 3.0% per year. The yearly operating cost schedule is given in Table 8.13.

The maintenance cost for year 1 is taken as 4% of capital cost which is 0.04 × 119.216 = $4.76 million. Maintenance costs are escalated at a rate of 1.0% per year. The annual schedule for this item is also given in Table 8.13.

Loan repayments and interest

The total loan for the project is $55.390 million and is repaid over 10 years at an interest rate of 8.0% per annum discounted annually. The schedule of repayments and interest is given in Table 8.14.

Table 8.14. Schedule of debt repayments and interest

Interest rate is 8.0% per year

End of year	Principal	Principal repayments ($ million)	Interest ($ million)	Total payments ($ million)
0	55.390			
1	49.851	5.539	4.431	9.970
2	44.312	5.539	3.988	9.527
3	38.773	5.539	3.545	9.084
4	33.234	5.539	3.102	8.641
5	27.695	5.539	2.659	8.198
6	22.156	5.539	2.216	7.755
7	16.617	5.539	1.772	7.311
8	11.078	5.539	1.329	6.868
9	5.539	5.539	0.886	6.425
10	0	5.539	0.443	5.982

Revenue

There is a single source of revenue which is from the sale of all products at the refinery price given in Section 8.2 of this chapter. For the base case given in this example there is no escalation of this figure which remains at $77.7 million per year. Sensitivity analysis performed later gives the change in ROI for escalated product pricing of 3% and 4% per year respectively. Schedules of increased pricing are given in Table 8.15.

Depreciation

Depreciation is normally the capital cost of the plant divided by the plant life. The plant life in this example is 20 years but the capital cost in this case has been taken as the original capacity factored estimate of $125.490 million. This allows for such

Table 8.15. Schedule of escalated revenue (to be used for sensitivity analysis)

End of year	Escalated at 3.0% ($ million/yr)	Escalated at 4.0% ($ million/yr)
1	77.709	77.709
2	80.040	80.817
3	82.441	84.050
4	84.915	87.412
5	87.462	90.901
6	90.086	94.545
7	92.789	98.327
8	95.572	102.260
9	98.439	106.351
10	101.392	110.601

Table 8.16. Present worth calculation

$ million

End of year	0	1	2	3	4	5	6	7	8	9	10
Net cash flow	141.4	32.7	32.8	32.8	32.9	32.9	32.9	32.9	32.9	32.9	51.5
Discounted at 10%											
Discount factor	1.0	0.909	0.826	0.751	0.683	0.621	0.564	0.513	0.467	0.424	0.385
Present worth	141.4	29.7	27.1	24.6	22.5	20.4	18.6	16.9	15.4	13.9	19.8
Cumulative PW	141.4	111.7	84.6	60.0	37.5	17.1	1.5	18.4	33.8	47.7	67.5
Discounted at 15%											
Discount factor	1.0	0.870	0.756	0.658	0.572	0.497	0.432	0.376	0.327	0.284	0.247
Present worth	141.4	28.4	24.8	21.6	18.8	16.4	14.2	12.4	10.8	9.3	12.7
Cumulative PW	141.4	113.0	88.2	66.6	47.8	31.4	17.2	4.8	6.0	15.3	28.0
Discounted at 25%											
Discount factor	1.0	0.800	0.640	0.512	0.410	0.328	0.262	0.210	0.168	0.134	0.107
Present worth	141.4	26.2	21.0	16.8	13.5	10.8	8.6	6.9	5.5	4.4	5.5
Cumulative PW	141.4	115.2	94.2	77.4	63.9	53.1	44.5	37.6	32.1	27.7	22.2
Discounted at 30%											
Discount factor	1.0	0.769	0.592	0.455	0.350	0.269	0.207	0.159	0.123	0.094	0.073
Present worth	141.4	25.1	19.4	14.9	11.5	8.9	6.8	5.2	4.0	3.1	3.8
Cumulative PW	141.4	116.3	96.9	82.0	70.5	61.6	54.8	49.6	45.6	42.5	38.7

Note: Figures in *italics* are negative.

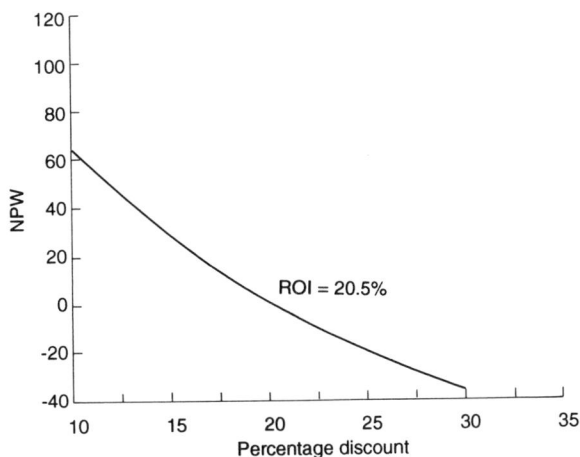

Figure 8.9. Plot of net present worth versus percentage discount

items in the associated costs such as the precious metal content of catalysts which are subject to depreciation. For this example therefore depreciation is taken as $7.7 million per year throughout.

Ad valorem tax

This item includes plant insurance and is set at 2.0% of capital cost per year.

Tax

This is 40% of taxable income. For the purpose of this study it is assumed there are no tax credits.

RESULTS

Figure 8.9 gives a plot of the cumulative net present worth versus percentage discount from the results of the calculations given in Table 8.16.

From Figure 8.9 the DCF return on investment for this scheme is 20.5% per year. Based on the plant location and the current investment environment, return on investment above 18% DCF makes the venture economically attractive. Similar calculations for scheme 3 will also be conducted to verify its ROI on a DCF basis for comparison with scheme 1.

The sensitivity of scheme 1 to escalation of refinery fence product costs are as follows:

- At an escalation rate of 3.0% per year the ROI becomes 23.5%
- At an escalation rate of 4.0% per year the ROI becomes 24.8%

9 THE PROCESS ENGINEER'S ROLE IN PROJECT MANAGEMENT

Invariably as a process engineer's experience increases he or she will become more and more involved in the company's management. More often than not, this involvement will take the engineer into the realms of project management in the installation of new facilities or the revamp of existing plants. In Chapter 8 the role of the process engineer in an operating company's management was discussed up to the point where a study becomes an approved appropriation for a new or revamped process. This chapter now continues with the activities of the senior process engineer in furthering the company's interests in the engineering and construction of the facility.

These activities begin with the development of the operating company's 'Project Duty Specification' and the process engineer's input to this most important document. Following this topic the chapter moves to the role of a senior process engineer in the office of the company contracted to engineer and build the facility. These sections describe the interface of the process engineering discipline with other disciplines in the contractor's organization and with the operating company's project team. They also discuss the major contractual process requirements and process milestones met during the project life.

9.1   Compiling the Project's Duty Specification

As soon as a project has been approved for implementation and the funds for its installation appropriated, work must begin to appoint an engineering and construction contractor. Before this can be accomplished, however, it is necessary to complete the following:

(1)   Prepare the duty specification for the plant.
(2)   Develop the contractor's scope of work.

(3) Develop the contractual terms and the contract itself.
(4) Assemble the document inviting contractor bids.
(5) Decide on the selection procedure for the contractor appointment.

Normally the operating company will assemble a team of people who will be dedicated to carry out these activities. This team will usually be headed by a member of the company's middle management who will be responsible for these precontract activities. The team itself will consist of senior personnel from its engineering disciplines, particularly process, and other parties such as the company's purchasing and legal departments.

Process engineer's may be invited to participate in some or all of the activities listed above. Their main participation, however, will be in the development of the duty specification. The importance of this document is that it forms the basis of the selected contractor's scope of work. It also establishes the standard of quality required by the operating company as the client. A well-produced project duty specification will contain at least the following sections and information:

- The process specification
- General design criteria
- Any preliminary flowsheets (duly labelled 'preliminary')
- Utilities specification
- Basis for economic evaluations
- Materials of construction
- Equipment standards
- Instrument standards

THE PROCESS SPECIFICATION

This document is developed entirely by the senior process engineer assigned to the team. It gives in precise terms the plant required, the number of units, its throughput, the product yields and quality, the required test standards, and any salient process requirements. An example of a typical process specification is given in Table 9.1.

GENERAL DESIGN CRITERIA

This section of the project duty specification is usually compiled by the project engineers assigned to the team. Some input is required from the process engineer to ensure that technical documents and data developed by others will conform to the company's usual format. This section, as the name implies, supplies the general data associated with any work done on the company's plant site. The following topics make up this section:

(1) *Scope* This is a brief statement covering the objective of the project. It is followed by a list of the company's standards that are to be used in the implementation of the work.
(2) *Climatic data* A list of the following data is given. These are the data that

Table 9.1. An example of a process specification

XYZ REFINERY PROJECT

Process specification for a thermal cracker

Number of units
one
Capacity
The unit shall have an input capacity of 34 500 barrels per stream day of a residue from the atmospheric distillation of Kuwait crude oil.
Charge
The normal feedstock will be an atmospheric residuum boiling above 700 °F TBP cut point on Kuwait crude.
Duty
The duty required from this unit will be to thermally crack the feedstock to produce gas, naphtha distillate, gas-oil, and fuel oil.
Yields
The unit shall be designed to make the required products in the following relative proportions from the feedstock specified above. The process licensor shall confirm these proportions by pilot plant tests on samples of the feedstock provided by the owner.

	wt% on feed
Conversion to 340 °F TBP cut point in cracker furnace and transfer line	25.0
Products:	
Gas to C_5	9.0
Naphtha distillate to 390 °F cut point	20.3
Gas oil 390 °F to 622 °F TBP cut	24.7
Fuel oil +622 °F cut point	46.0

Products

(1) The gas shall include C_3 and C_4 and shall be routed to the crude unit overhead distillate drum.
(2) The overhead naphtha distillate product shall be routed to the crude unit overhead naphtha hydrotreater. The naphtha product shall have an ASTM distillation end point of not more than 387 °F.
(3) The gas-oil sidestream product shall have an ASTM 90% vol distilled at a temperature no higher than 645 °F.
(4) The fuel oil as residue from the cracker primary tower shall have a minimum Pensky–Marten flash point of 200 °F and shall be thermally stable.

Process conditions

(1) *Thermal cracker furnace*   It is required that the furnace and transfer line be capable of effecting a crack of 25% wt on feed of gas and naphtha to a TBP cut point of 340 °F. The transfer line outlet temperature from the furnace shall not be higher than 920 °F at a pressure of 250 psig. The injection of HP steam into the furnace coils may be considered to increase turbulence and minimize the laydown of coke. Should steam be used its volume under the furnace conditions must be accounted for in the design of the furnace coil(s). The temperature of the feed to the inlet of the furnace soaking section must be controllable.

Table 9.1. (*continued*)

Process conditions

(2) *The main fractionator* The feed to the fractionator from the furnace must be quenched to a suitable temperature to meet the residue cut point requirement. This temperature must also be sufficient to produce enough overflash for proper fractionation between the distillate streams. The column will be operated at a pressure in the overhead distillate drum of 50 psig. Naphtha distillate stream shall be maximized and steam stripping of the gas-oil sidestream should be considered in this respect. Fractionation criteria in terms of gaps (and overlaps) between the distillate streams shall be ASTM distillate 95% temperature of naphtha and 5% temperature of gas-oil to be not less than +15 °F. The residue leaving the bottom of the tower shall be steam stripped for flash point control. It shall also be quenched in the well of the tower to prevent further cracking. A cold residue stream is recommended for this quench.

will be used in the various specifications and calculations developed during the course of the project:

- Dry and wet bulb temperatures
- Winter design dry bulb temperature
- Temperature extremes
- Barometric pressure

(3) *General design considerations* This section lists *all* the legislative criteria associated with the building of an industrial plant in the area. It will include the environmental requirements, safety and quality standards to be adhered to by contractors and licensor.

(4) *Units of measurement* The units of measurement that will be used in all calculations, flowsheets, and specifications are given in this item. It begins by stating the units in terms of:

- English units or
- SI units or
- Any others.

The next section then defines these units in more detail such as:

- Linear—millimetres (mm)
- Mass—kilograms (kg)
- Flow gas—normal cubic metres per hour (n m^3/h)
- Flow liquid—(large) cubic metres per hour (m^3/h)
            —(small) litres per minute (l/min).

There then follows a list of conversion units that should be used on the project.

(5) *Engineering flow diagrams* Many companies include the flow diagrams they have developed during the appropriation stage of the project. These are usually the process flow diagram and the mechanical flow diagram. The inclusion of these adds to the description of the work's scope by supporting the process specification. In most contracts, however, the client company will expect the contractor and/or the licensor to take responsibility for the process performance and to guarantee it. Under this circumstance these engineering diagrams are released into the project duty specification as

'preliminary' issues. The contractor is expected to check and, if necessary, revise them. Thereafter the contractor must accept responsibility for the technical content of the diagrams as a basis for future normal design development.

(6) *Utilities specification* Full details of the utility streams available in the client's plant is given in this item. This is generally prepared by the process engineers and must include data concerning all steam and condensate systems, water systems, air, and fuel. Such data will be at least the temperature and pressure of these streams. In the case of circulating systems such as fuel oil and cooling water both supply and return conditions will be required. Raw water available should also have a complete analysis of its impurities. The analysis of the normal fuel oil and fuel gas should be included. These would contain at least the following:

- Fuel oil: Supply temperature.
  SG at supply temperature
  Viscosity at 120 °F (cSt)
  Viscosity at 210 °F (cSt)
  Supply and return pressure
- Fuel gas: Supply temperature and pressure
  Mole weight
  Approximate molal analysis

The dew point of instrument air required at the air supply header must be stated. If the client has any preference as to the type of desiccant to be used in the drying of the air this too must be stated. All other preferences or standards that the client company requires to be utilized in the utilities design must be given in this section. For example, most companies have a standard burner control system for their plants. This should be fully described here together with some appropriate sketches. The electrical engineers usually add details of the plant power supply and distribution in this section. If some existing switchboards and sub-stations are to be utilized this should be noted together with a list of drawings that should be referenced. Other systems, although not strictly utilities, may be included here. Among these would be:

- The fire main
- The flare(s)
- Water effluent treating and disposal
- Other environmental protection systems
- Boiler blowdown systems

(7) *Basis for economic evaluations* The process engineer completes this section with the criteria used in the evaluation studies and the appropriation design. The section should begin with a statement that contractors are encouraged to review all flow and equipment arrangements so that all possible alternatives are considered. The incremental cost of any alternative arrangements must, however, yield a minimum of the company's stipulated return on investment. This return on investment is stated here. This section continues with the detailed costs of labour, utilities, feedstock, and products to be used in any such economic analysis.

(8) *Materials of construction*   This section is usually compiled by the team's project engineers with some input by the process engineers. It should begin with a paragraph on the references to be used in selecting the materials used in handling the corrosive streams. For example:
- Selection of steels exposed to hydrogen service shall be based on API 941, 1970 edition.
- Corrosion allowances for steels subject to hydrogen sulphide environment shall be in accordance with 'Computer correlations to estimate corrosion of steels' by Couper and Gorman, NACE paper No. 67.

The next paragraph or sub-section should detail the client's requirement for corrosion allowances in terms of equipment life. For example:

Equipment	Life in years
Columns, drums and reactors	20
Heat exchanger—shell	20
Heat exchanger—tubes	5
Pumps	10
Compressors	10
Atmospheric tankage	20
Piping material	10
Heater tubes	100 000 h

These data may also be supplemented by a table giving the company's accepted minimum corrosion allowance in millimetres or inches for the various equipment items.

(9) *Equipment standards to be used for sizing and design*   There follows in this section data to be used for the sizing and the specifying of the major equipment items which are:
- Vessels and columns
- Heat exchangers
- Pumps
- Compressors
- Heaters

Examples of such data included in this section are as follows:
Vessels and columns
Columns to be sized on a specified type of tray (e.g. valve trays)
Design for a percentage of flood (usually 80%)
Design pressure and temperature criteria for all vessels
Minimum diameter for trayed columns
Heat exchangers
Design standards to be used (e.g. TEMA, ASME, etc.)
Rating procedures
General design criteria (e.g. use of fixed- and floating-head tube bundles, expansion bellows, tube side flows where special considerations are required, use of kettle or thermosyphon reboilers, etc.)
Tube sizes and pitch relative to fouling factors
Table of fouling factors to be used

Allowable velocities and pressure drops
Approach temperatures to be used
Design temperature and pressures
Pumps
The design standard to which the pumps are to be designed (for example,
API 610, Centrifugal Pumps for General Refinery Services)
Rated capacity of the pump over normal. For example:

Service	Rated capacity over normal (%)	
	Flow control	Level or temp. control
Feed	5	10
Product	10	15
o/hd reflux.	15	20
Inter reflux.	20	20
Reboiler feed	15	20

Sparing requirement.
NPSH calculation criteria—source pressure
                              —Rate for suction line losses
                              —Level for static head (e.g. bot tan line for
                                columns).
                              —Pump centreline, etc.
Pump selection preferences
Minimum flow criteria
Casing design Conditions
Type of drivers
Piping hook-ups (company standard)
Compressors
Standard to be used for the design of compressors (e.g. centrifugal
compressors designed to API 617)
Compressor type selection preferences
Rating of compressors (includes % over normal capacity)
Sparing requirements
Mode of control
Type of drivers
Type of cooling
Piping hook-ups
Fired heaters
Types and selection preferences
Any company standards
Heater design criteria, e.g.:
● Burners and flues to be designed to 125% of normal heat release
● Environmental constraints
● Acceptable average flux, coil pressure drops, and mass velocities
● Surge volumes and acceptable % vaporization for fired reboilers
● The rated duty of the heater as a % of normal duty

Allowable maximum coil film temperature (where coking is a problem)
Coil design temperature and pressure criteria
Fuel system details (company standards)
Burner types preferred and burner control (company standards)

(10) *Instrument standards required* This section of the specification is usually compiled by the instrument and the process engineers. Its objective is to convey to the contractors the instrument, safety and control philosophy required for the company's plants. It may begin with the sizing criteria for control and relief valves that the company wishes to use. For example:

- Criteria to set the control valve pressure drop as a % of line losses, etc.
- Design capacities for control valves
- Design standards required for relief valve sizing (e.g. API 520 and 521).

This item should then continue to define the company's preferences in piping design for control valve hook-up, relief valve venting, and relief valve locations. Some basic criteria concerning the relief header and its condition should be described. Finally, the section should give a list of instrument drawing symbols that the company prefers or in the case of a plant being built in an existing complex, the symbols already in use.

This completes a review of a typical project duty specification as it applies to the process engineering discipline and interface. This document which usually forms part of a request for a contractor's quotes may be smaller and less detailed than described here. Sometimes the client company may wish to depend solely on the contractor's standards and criteria. However, to ensure good competitive quotations a document such as described here or even more detailed is desirable. It should be noted too that the complete project duty specification would contain details and criteria for other disciplines that have no or very little process engineering input. Among these would be civil, structural, electrical and piping.

9.2 Project Initiation and Subsequent Activities

This chapter continues now with a discussion of process engineering in the successful contractor's organization as it relates to the management of the project. First, however, it is desirable to understand a little of what is required to execute successfully the engineering and installation of a process plant. This section briefly discusses some of these basic requirements.

There is no doubt that the most complex and large process units can be built by a few people with limited technical knowledge and skills provided there is no limit on time, there are limitless funds available, and quality is not a requirement. The moment, however, any one or all of these constraints are applied, then order, method and discipline in the execution of the project is mandatory to meet the desired results. As all engineering and construction projects in the real world have these constraints then the art and management skills to work within them are required for a successful product. This is provided by project management and project control techniques.

Project management starts with the forming of a team of engineers and others specializing in the various disciplines necessary to engineer, design, procure and erect the facility. This team led by a designated project manager is supplemented by project schedulers and cost estimators. The disciplines represented on the team will be led by senior members of the respective disciplines. These engineers have the authority to direct the work of their particular discipline group and are responsible to the project manager for the quality of the work, its timeliness and cost. Details of a typical project organization are given later in this section.

The project proceeds with the application of the disciplines to fulfilling their role in a prescribed sequence and within time and budget limits. Some of these activities in which process engineers in particular participate are described below.

PROJECT INITIATION

The kick-off meeting

One of the most important conferences held during the course of the project life is the 'kick-off' meeting. This takes place as soon as possible after the award of the project by the client to the contractor. As the name implies, this meeting formally releases the contractor to begin work on the client's plant and describes again in detail the client's requirements. Its purpose is to communicate to the project team the scope of work, the time span required, the budget (approximate or firm), and details of quality requirements and specifications as outlined in the client's 'duty specification'.

This first or formal kick-off meeting conducted usually by the client is invariably followed by a second, less formal, meeting of the contractor's project team. The contractor's lead process engineer assigned to the project usually commences this meeting with a detailed description of the plant that is to be built. In this description the process engineer will draw from experience with the installation of similar processes to highlight any pitfalls that face the other disciplines in their work. For example, in the thermal cracker project used as an example in this book the process engineer may highlight some of the piping problems associated with coke formation and would certainly mention some particular requirements in furnace design to the mechanical engineer, and so on.

Other discipline lead engineers now follow with detailed descriptions of their discipline requirements as provided by the client's duty specification. Their descriptions and discussion complete the technical side of the kick-off meeting. Discussion should then follow on the project management aspect of the work. The completion date required for the project is tabled together with major milestones to be met during the project life. These milestones usually indicate when cost estimates (and schedules) are to be updated or when critical overall project decisions are to be made.

At the end of the kick-off meeting each key member of the project team should be absolutely sure of what the project requirements are and what role they must play in achieving them.

The project team organization

Perhaps even before a meaningful kick-off meeting can be conducted the team of key engineers, schedulers, purchasing staff, etc. who will carry out the work needs to be established. This and how the disciplines relate to one another is developed by the project manager designated to lead the team. A typical project organization that could be developed to execute the example 'Thermal Cracker Project' is shown in Figure 9.1 and a brief description of the various principal functions now follows.

The project manager has the overall responsibility and the authority for the execution of the project. He or she interfaces directly with the client's project manager and is the sole contact with the client's organization for all matters concerning the project. Reporting directly to the project manager are:

- A procurement coordinator—responsible for the purchasing, expediting and shop inspection activities required for the project.
- Area project engineers—who assist in the interdisciplinary activities and communication.
- The project control team of cost engineers, estimators, and schedulers—this team retains the progress reporting activities necessary to keep the project manager informed for his or her decision-making function.
- An engineering manager—in moderate to large projects an engineering manager is often included in the organization. His or her function is to maintain communication and interface between the engineering disciplines and project management. The engineering manager is also the single authoritative source conveying management directives into the engineering functions.

Reporting directly to the engineering manager therefore are the lead engineers of the various engineering disciplines. In the case of most chemical process type projects these disciplines are:

- Process engineering
- Instrument engineering
- Electrical engineering
- Mechanical engineering
- Vessel, civil, piping and structural engineering

It is not the intention here to describe the functions of these various disciplines as they are quite well known. However, the function of the lead engineers of these disciplines does warrant some description and comment. Lead engineers are responsible for the quality of their discipline's work, its timeliness, and the cost of the function in terms of person-hours expended. They prepare their discipline's execution plan for approval by the project manager. They also develop their function's person-hour budget and schedule to meet the approved plan's implementation. Commensurate with this responsibility they have the authority to direct the work within their discipline to meet the approved objectives.

Much of the total person-hours and the project schedule is utilized by the detailed design functions of the various disciplines. Indeed, the only products of the home office engineering function are equipment data sheets and detailed design drawings (including flow diagrams). The design function that produces

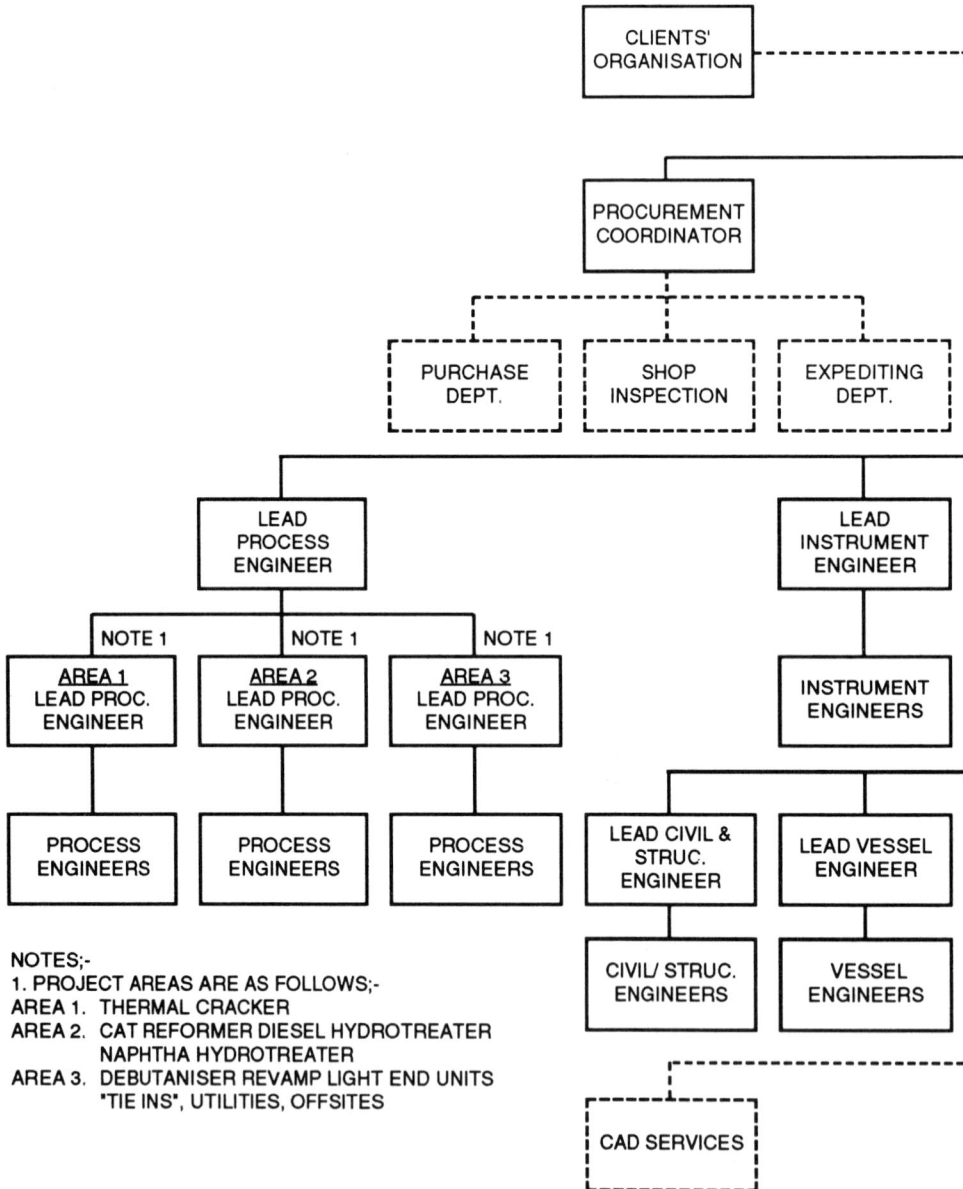

NOTES;-
1. PROJECT AREAS ARE AS FOLLOWS;-
AREA 1. THERMAL CRACKER
AREA 2. CAT REFORMER DIESEL HYDROTREATER
         NAPHTHA HYDROTREATER
AREA 3. DEBUTANISER REVAMP LIGHT END UNITS
         "TIE INS", UTILITIES, OFFSITES

Figure 9.1. A typical project team organization

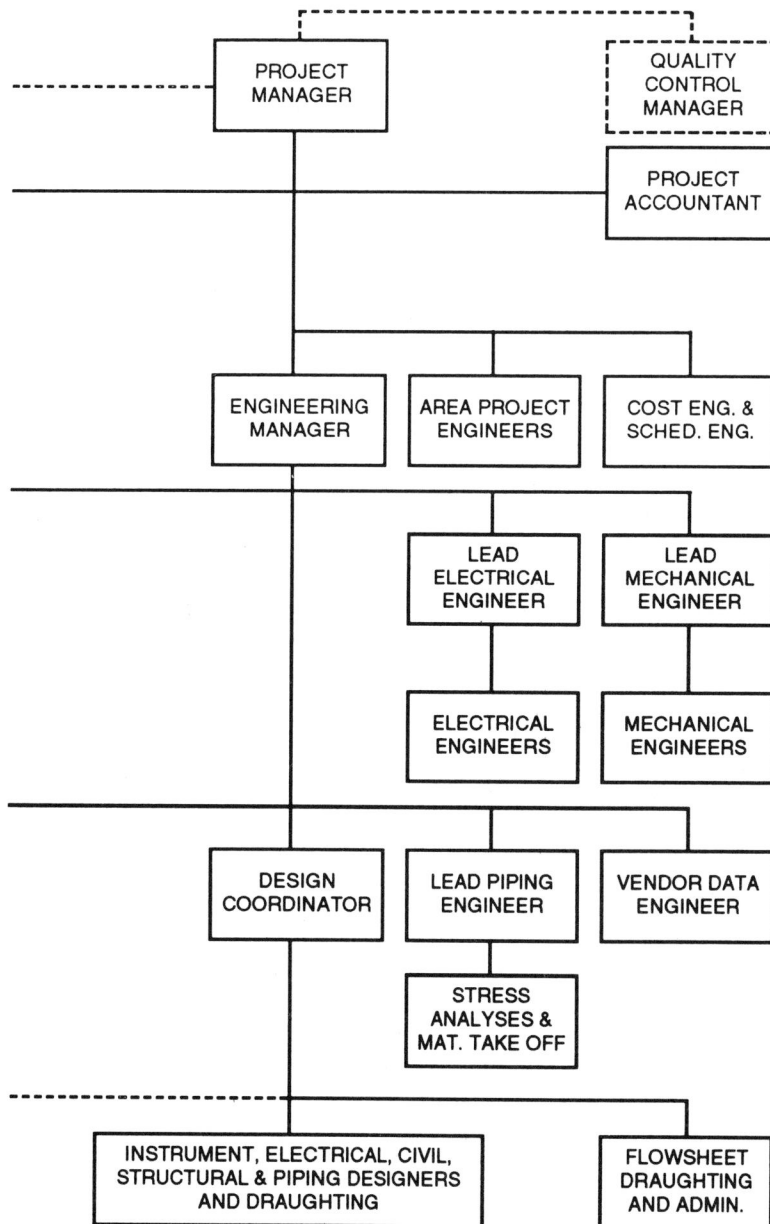

these drawings demands precise management controls and interface. This responsibility is often that of an assigned design coordinator. To meet this responsibility, they are given the same authority as any of the lead engineers. Normally they become a single source of communication between the engineering and the respective design functions. Thus all data from engineers to their designers flow through the design coordinators and, similarly, all design drawings, material take-off data and the like flow back into the project via the coordinators. The project organization therefore brings 'order' into the execution of the project—a necessary element to meet the project constraints.

The project procedure manual

One of the earliest activities of the project team is to initiate a project procedure manual. This is a document which compiles as many data as possible on the project and how it is to be executed. Needless to say, some of the data that will be included in the document are added piece by piece as the project develops. However, the basic requirements of project management is compiled at this very early stage. At this time many of the project team key people, although assigned, cannot begin work in their field until the process front-end work has been completed, approved by the client, and released into the project. This is an ideal time therefore for these key people to contribute to the project procedure manual.

A typical project procedure manual is divided into several parts, for example:

- Introduction
- Project organization and directory
- The master schedule
- Person-hour budgets (restricted issue)
- The cost code of accounts
- Project control procedures
- Correspondence and communication
- Conferences and meetings
- Filing index
- Individual engineering, procurement and design interface procedures
- Drawing index
- Field organization and directory
- Hand-over procedures and close-out reports

The items given above may be increased or decreased depending on the wishes of the project manager. After all, this is the project manager's document, although it is compiled by the members of the project team. Brief descriptions of the contents of these sections now follow.

Introduction

This section begins with the general information as to who the client is, the official address of the client and phone, fax or telex numbers, etc. It continues with a short history of the project up to the award and concludes with a synopsis of the process(es) involved. This section should not be more than three or four pages in length.

Project organization and directory

The project organization chart for the project, similar to that given in Figure 9.1, is included in this section. For this purpose, however, the names of the key people are included in their respective organization slots. Following the chart is a table of the key positions (and those of the client's project team) with the individual's name and office phone number or extension.

The master schedule

This is in bar chart form and shows the scheduled progress by discipline and activity. Initially, and of necessity, this will be preliminary but will indicate critical milestones dated clearly. It may be followed by a list of these milestone dates with explanatory notes. This will be subject to updates as the project progresses.

Person-hour budgets

This includes the allocated budgets by discipline for the project. This item is restricted to the copies of the manual issued to key personnel who are responsible for their discipline's budget control. As in the case of the master schedule, initially these estimates will be preliminary, and will be subject to updates as the project proceeds.

The cost code of accounts

This is a summary of the cost coding that will be used on the project to identify all the various cost centres. This coding will also be used to identify and code such items as purchase orders, client billing, change orders, etc. More often than not the contractor's normal cost codes are used for this purpose.

Project control procedures

This section outlines the reporting procedures the project manager will adopt to control the project in terms of progress and budget. It will detail the data required from the key members of the project and the timing of the data for each reporting period.

Correspondence and communication

All correspondence leaving the project does so under the project manager's signature. Similarly, all correspondence originating outside the project is addressed to the project manager. All correspondence is given a coded reference number and this reference number also identifies the origin of the correspondence. This reference code is given in this section, together with instructions on the routing of the correspondence. Detailed procedures as to the communication by other means, such as telephone and fax, are also given here. A list giving the required distribution of all correspondence to the project team personnel and possibly others outside the project is included in this section.

Conferences and meetings

Routine meetings and conferences are scheduled in this section. This will also include a list of permanent attendees at these meetings. As time taken up in meetings is a large utilizer of person-hours, the project manager may elect to outline certain recommended procedures for unscheduled meetings in this section. All meetings on the project are minuted and the minutes distributed within the team and other interested parties according to the project's correspondence distribution list.

The filing index

The project secretary organizes the project filing system. The files generated on the project often become legal documents which may be used for such purposes as 'job close-out' negotiations, any possible litigation, or enquiries at later dates into industrial accidents, etc. Maintaining the filing therefore is an important function and to accomplish this effectively a filing index is initiated and developed. This is given in the manual to enable members of the project team to utilize the files effectively when required.

Individual engineering, procurement and design interface procedures

During the life of a project there will a considerable amount of data and information generated and distributed to the various disciplines in the project team. To ensure the correct movement of these data each discipline develops its interface procedures. For example, a process engineer who has now completed the data sheet for a pump needs to ensure that this document is sent to the correct discipline for further work and ultimately purchasing. In this particular project the mechanical engineer is designated to be the interface between process and the purchasing department. Then this section of the procedure manual will detail such an interface with instructions on how the transmittal of the data sheet is to be done.

Drawing index

This is initiated and maintained by the design coordinator. In developing the index as a list of all the drawings the coordinator also demonstrates what the items in the drawing reference numbers signify. This is updated as the project proceeds.

The remaining items of the manual are self-explanatory. These are usually added only in later editions, just before or immediately after field 'move-in', etc.

Preliminary schedules and budgets

Before the project manager can apply project control measures, he or she must have some yardstick or criterion for control. Therefore very early in the project the project manager must establish the project execution plan and measure the cost, initially in terms of person-hours, and a preliminary schedule to implement the plan.

The project manager compiles the master project plan, budget, and schedule from input supplied by the various lead engineers and coordinators. The first of these input requirements will be that for the process 'front-end' activities. The lead process engineer therefore examines the scope of work needed to complete over, say, the first six months of the project. This time frame will allow the bulk of the process engineering to be completed. After this time it is expected that the process effort will resolve to advising other disciplines on process interface problems. After this time also the process engineer will be involved in completing operating manuals, compiling data for process guarantees and other activities that will not impact on other disciplines.

After examining the 'scope of work' the lead process engineer lists the activities that need to be completed during this time frame. From experience, these activities will be sequenced in a manner that:

- Utilizes the person-hours most effectively
- Provides the best early start to other disciplines' work
- Best meets the critical dates of the project

These activities and their sequence form the process engineering execution plan. With this in place, and approved by the project manager the lead process engineer takes each of the activities and, again from experience, adds in bar chart form time lengths for each activity. This completes the schedule for the plan, and by adding the number of engineers to be committed over the activity period the lead engineer can develop the person-hour estimate. Figure 9.2 is an example of a 'front-end' bar chart and person-hour spread.

Using the process engineering plan and schedule other disciplines who receive data from the process function can now proceed with their plan. The project manager receives and approves all these plans, and together with the assigned cost estimators and schedulers develops the initial overall plan, master schedule and budget for the project.

The mechanical flowsheet conference and approval

One of the major objectives of the process engineer's early activities is to provide sufficient data to enable other disciplines to commence effective work as early as possible. A major milestone in achieving this objective is to conference the mechanical flow diagram and have it approved by the client. This approval allows a considerable amount of design work (which is the major user of time on most projects) to commence. The development of the equipment data sheets has already allowed the mechanical engineering and the procurement functions to

XYZ Refinery Co. Ltd	Process engineering — Thermal cracker project area 1										Contract No. 2345					Page 1 of 2			

Week number — Number of engineers

Total man weeks this page = 32½
1 man week = 40 hours
Man hours this page = 1300

Planned activities	1	2	3	4	5	6	7	8	9	10	11	12	13	14	15	16	17	18	19	20
1. Review licensor data	1	½																		
2. Complete heat & material balance		1	½																	
3. Develop process Flow diagram			1	1																
4. Carry out prelim hydraulic analysis					½															
5. Size and specify process equipment																				
• Towers					1	1														
• Drums					½	½														
• Pumps					½	½	1													
• Heat exchangers						1	1	½												
• Heaters							½	½												
6. Develop mech. flow diagram						1	1	½	1											
7. Line size and spec						½	½	1			½	½	½	½						
8. Prepare instrument																				
• Flow elements						½	0	½												
• Control valves							1	1	½		0	½	0	½						
• Others								½	½	½										
9. Prepare safety relief valve specs										½	1	½	1	1						
Total man weeks this page	1	1½	1	1	3½	4½	5	4	3	1½	1½	1½	1½	2						(Continued on page 2)

Flow sheet conference

Continued on page 2

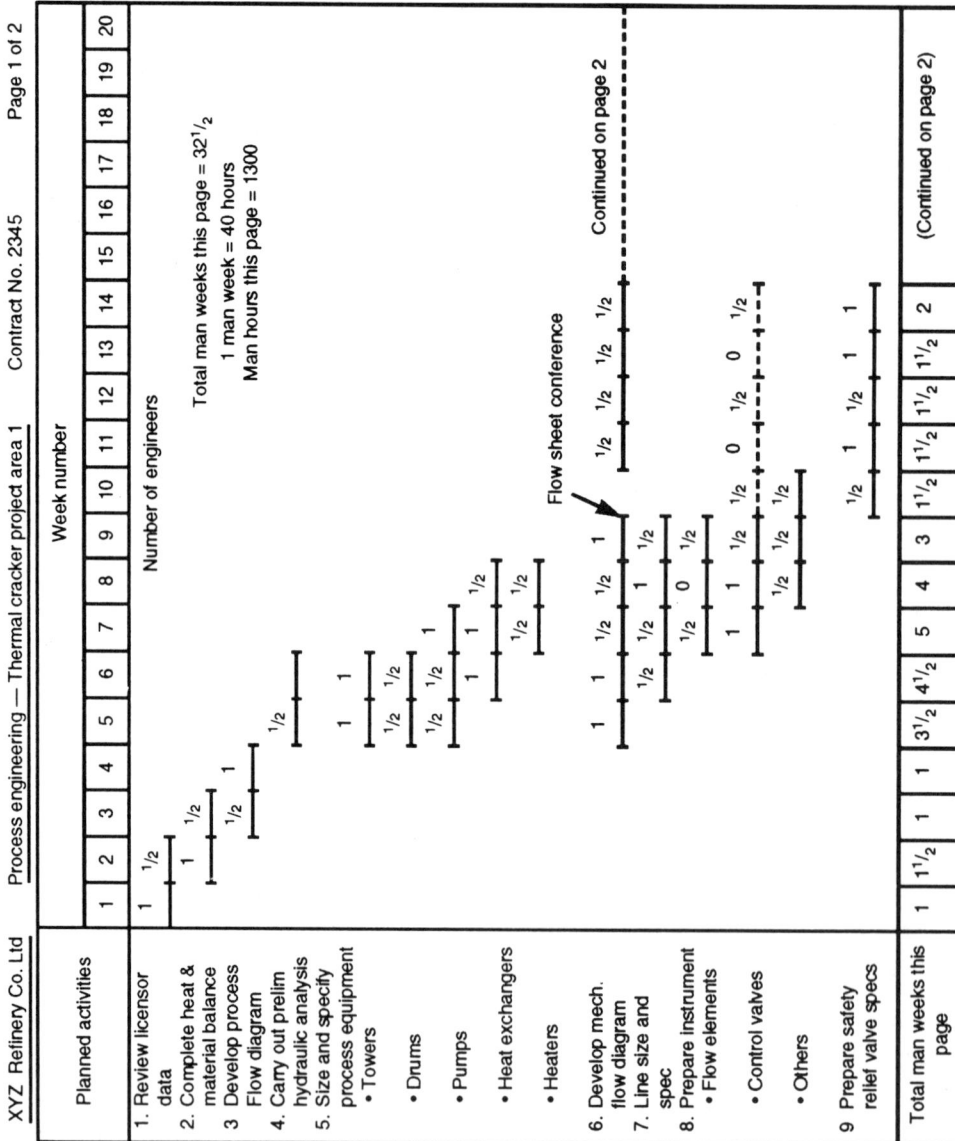

Figure 9.2. A section of process engineering front-end schedule

commence work, but it is the release of the approved mechanical flow diagram that really puts the activities of the project into top gear.

Participation of the client company in the flowsheet conference depends on the type of contract and the relative location of the client's offices to that of the contractor. Nevertheless, whether there is client participation or not, the flow sheet conference follows some basic guidelines:

- *Preparing for the conference* The lead process engineer is responsible for organizing and conducting the flowsheet conference. The flowsheet that will be used in the conference will be 'clean' and free from any pencil or other marks on the diagram itself. It will be given the revision number '0', and clearly marked 'flowsheet conference master'. The diagram will be complete with all lines sized and specified. Instrumentation will be correctly shown with control valves sized. Relief valves will also be shown but not necessarily sized at this time. All equipment will be properly titled and labelled in accordance with the agreed format for flowsheet production. Accompanying the 'flowsheet conference master' will be the current revision of the process flow diagram. This will also be 'clean' with no pencil marks other than the label 'Flowsheet conference copy'. The lead process engineer will also ensure that copies of all the equipment data sheets pertaining to the equipment shown on the flowsheet is available to the conference. The current equipment list will also be made available at the conference. Finally, the lead process engineer or the process engineer assigned to the particular process that is to be conferenced should have available a copy of the process calculations developed. This often helps to resolve minor issues that may arise.

- *Location and attendees* After notice of the conference has been issued by the lead engineer the project manager or assignee will designate the location where the conference is to be held. The manager will also issue directives to attendees giving the date, time and location of the conference. In cases of major conferences where client representatives are present professional secretaries may be employed to take notes at the meeting. Otherwise project engineers will be assigned this task. The notes taken at this meeting are most important to the final approval mechanism. The attendees are usually:
  —Client's representatives (if applicable)
  —Project manager (usually part-time)
  —Lead process engineer(s)
  —Responsible process engineer(s)
  —Engineering manager
  —Area project engineer(s)
  —Lead discipline engineers (except civil and structural—called for as required)
  —Design coordinator
  —Design section heads (piping, instruments, electrical)
  —Others as required, such as environmental engineer, cost and schedule control engineers, etc.

- *Conducting the conference* The process engineer conducting the conference commences with a brief outline of the process and its objective and proceeds to describe the process using the process flowsheet to illustrate the process flow and the equipment. The engineer then commences with the presentation of the mechanical flow diagram which is attached to a wall or suitable frame from which it can be easily read by the attendees. Now most companies have a colour code depicting 'additions to', 'deletions from', etc. Coloured pencils are used therefore to cover all such points made during the conference by marking up the flowsheet as the conference proceeds. Assume the colour coding in this case is as follows:

—Yellow: shows all items that have been presented
—Red: all changes and additions
—Blue: all deletions
—Green: numbers in green refer to note items taken during the conference

The process engineer conducting the conference starts at a suitable point on the left of the flowsheet, selects a major line, checks and 'yellows out' the source label, and traces the line using the yellow pencil to the first equipment item. In tracing the line the engineer highlights the line number, size, and specification shown on the line. The design coordinator or piping designer checks this against a line list. It must match exactly or it is marked as requiring further review. All valves and instrumentation along the line are similarly checked against piping lists or the instrument register. The equipment item into which the line is connected is checked for correct labelling and data. These shown on the flowsheet are compared with equipment list, data sheets and the process flow diagram for an exact match. The process engineer continues with a second line in a similar manner, and may elect to go back to the left side of the flowsheet to select this line or take a line leaving the equipment item. However, the same exercise is carried out for all lines and equipment until the flow diagram is completely yellowed out with the discussed notes marked and with all deletions and changes shown.

The flowsheet conference is a necessary activity but is nevertheless a high consumer of time. The process engineer should therefore use experience and knowledge to minimize the time taken in this activity. For example, where there is some considerable contention or where there is an area of concern to the client highlighted in the conference, the process engineer should attempt to defer a solution until after the conference. Appropriate notes should be made, however, of the problem and the item clearly marked with a 'hold' sign. Clients will normally approve flowsheets with some 'holds' provided that there are not too many. In further development of the flow diagram, priority is given to resolving the issue(s) and removal of the 'hold'.

Immediately after the conference the lead process engineer and the project manager formally solicit the client's approval and the release of the project to detailed design work. The approved flow sheet is marked as 'Rev. 1, Approved for Construction'.

SUBSEQUENT PROCESS ACTIVITIES

With the approval of the mechanical flow diagram by the client most of the process engineer's development work is now complete. Indeed, after the flow sheet conference the project team may be in a position to release some of the process engineers to other projects. The process activity is now directed towards checking vendor offers for equipment, participating in equipment selection, checking selected vendor's data, and monitoring continued mechanical flow diagram development on a routine basis. Two other major activities, however, face the contractor's process engineering effort:

• The compiling of the operating manual
• The development of the process and utility guarantees

Both these items are subject of the following two sections 9.3 and 9.4.

Participation in equipment selection

Of the subsequent routine activities that are carried out by the process engineer among the most important is participation in the equipment selection process. Process engineers must satisfy themselves and clients that their selection of any item of equipment will:

• Satisfy the heat and material balance
• Be safe and operable
• Be economically the best offered

In evaluating the first two items the process engineer will check the data supplied by the vendor against the data sheet originally prepared and issued for consistency. In evaluating the third item in the selection process the process engineer will calculate the payout incentive in terms of operating efficiency against incremental cost for all technically accepted items. This type of calculation has been discussed in earlier chapters of this book. Obviously, the lowest incremental payout time will influence the process engineer's recommendation for purchase.

Participation in plot layout development

Normally, plot layout studies and development are carried out by senior piping or other designers experienced in this kind of activity. With the approved mechanical flow diagram in place and the major items of equipment committed to respective vendors, plot layout studies can commence. The development and approval of a plot layout releases more definitive work into the design functions. For example:

• Piping layout design can commence, allowing piping material take-offs to be made.
• Similar activities are also extended to electrical cable, structural steel and underground piping.
• Civil engineering and design can commence with road layouts, effluent pond design, drainage, etc.

Process engineers participate actively in these layout studies, lending their knowledge of the process to the locating of the equipment items relative to one another. Finally, they participate in the plot layout conference and submission for client approval.

Subsequent development of the mechanical flow diagram

Although approved be the client this diagram remains a major document for the transmission of technical data throughout the engineering of the project. Thus it is in a continual state of updating and revision. The process engineer remains, however, responsible for the technical input to the diagram, and whatever is included on it and approved by the process engineer will be designed and installed. Maintaining control of the technical contents of the diagram is a routine duty for the process engineer.

The approved for construction issue of the diagram is usually located in an area accessible to all engineers and designers assigned to the project. These engineers and designers constantly utilize the 'piping master', as this copy of the diagram is sometimes called, adding data they have developed or taking note of developed data for further detailed design. These marks when added are identified by date and origin and are countersigned by the design coordinator when they have been noted and acted upon. Once per week on a fixed time schedule all the additions and actions made during the previous time interval are reviewed by the participating engineers, the design coordinator and the process engineer. If satisfied the process engineer will sign approval for the implementation of the changes.

The remaining routine activities

During the remaining engineering and design phase of the project the process engineer's routine activities are concerned with checking vendor data, answering clients' questions, acting in an advisory capacity to the project team as required, and responding to any client changes that may be requested. At the end of the project the process engineer will develop his or her part of the 'end of job' report. This is a concise summary of events that occurred during the project, their effect on the end result and the financial outcome of the project.

9.3 Developing the Operating Manual and Plant Commissioning

Among the final activities required of process engineers in the execution of a project is the development of an operating manual for the process. Very often the process engineers who have been most closely associated with particular process plant(s) on the project may be required to assist in the commissioning of those plants. This section describes the activities associated with both these functions.

DEVELOPING THE OPERATING MANUAL

The operating manual is a compilation of instructions and data reflecting recommended procedures for:

- Pre-start-up conditioning of the plant
- Plant start-up
- Normal operation and troubleshooting
- Emergency action and shut-down
- Normal shut-down
- Catalyst regeneration and decoking (where applicable)

These procedures include manufacturers' recommended handling, conditioning and operation of their items of equipment and the experience of the process engineer who designed the plant. To the process engineer the writing of the operating manual is the final 'in-depth' review of the mechanical flow diagram of the plant. During the course of writing the manual the engineer will use the diagram continually to make reference to the operating procedures proposed. Anomalies or missing valves and piping become obvious as the process engineer develops the procedure logic.

A typical operating manual has the following table of contents:

- *Introduction and process description*   This will include a brief statement as to the contents of the manual and its purpose. This will be followed by a process description which will include the 'Approved for Construction' revision of the process flow diagram as reference.
- *Pre-start-up conditioning*   This item describes the cleaning out of the plant after construction and its reassembly following the clean-out. For example, it will designate those lines and equipment that will be flushed out by water and those that will be blown out by air. It will also describe the equipment and materials that will be used for this clean-out such as piping spool pieces to replace control valves and nozzles and lines that will be blanked off during this activity. Finally, this item will draw attention to the manufacturer's procedures for various equipment clean-out and conditioning. Copies of the actual manufacturer's documents relating to this activity should be included in appendices to the manual. It is important in order to maintain the warranties that accompany most equipment that the manufacturer's instructions are carefully followed.
- *Plant start-up*   This section of the manual presupposes that the plant has been properly cleaned, all utilities to the plant have been commissioned, all equipment is ready for operation, all drains are free, and firefighting equipment and procedures are in place. The process engineer then begins a detailed description of the activities to be carried out in their proper sequence to bring the plant on-line and producing the products intended. To accomplish this the engineer refers to the mechanical flow diagram. Using this document the engineer describes each action in the start-up sequence by referring to line numbers, control valve numbers, equipment item numbers and titles as they appear on the diagram. Normally, all these actions start with introducing the

cold feed. For example, in the case of the thermal cracker project, cold feed from storage would be used for start-up. Sufficient lines and valves would be provided for this purpose and the manual text would be as follows:

'Open block valves at tank TK 101 and at the plant battery limits on line 1234 CS 12″ ST. Open the suction valves on pump P 106 A and start the pump. [Note; Only the term 'Start the pump' need be made as instructions on how to start the pump are included in the manufacturer's documents found in the appendices.] Set the flow control valves 106 FRCV 1 A, B, and C to full open. Set back pressure controller 300 PRCV 1 at the outlet of heater H 301 to 250 psig. Commission the level control indicator 301 LCI 1 and when the desired level in the bottom of fractionator is reached (NLL) open the suction valves to bottoms pump P304 A on line 1235 10″ Cr ST. Control discharge flow from pump P304 by activating control valve 301 LCIV 1 on line 1236 8″ Cr ST.'

- *Normal operation*   This section carries on from the end of 'start-up' with the unit lined out and in stable operation. It then describes a series of procedures during normal operation to 'fine-tune' certain parameters and to maintain the plant on set conditions when minor changes in feed composition, temperature changes, etc. occur. It must be noted here that in modern control systems associated with plant operation these types of adjustments are made automatically with very little need for operator intervention.

  The other adjustments that may still require operator action are those associated with changing the product grade or specification. This section therefore still includes a selected list of operating changes that may be required to be made under normal operating conditions and the procedures used to make these changes.
- *Normal shut-down procedures*   These are usually divided into two forms:

  Short-duration shut-down
  Shut-down for an extended period

  The first type of shut-down is that associated with a minor mechanical difficulty, temporary loss of feed or a minor instrumentation problem, etc. In these cases the feed to the plant is diverted but some or all of the product streams are rerouted back to take the place of the normal feed. The feed heater (if it is a fired heater) may continue to be fired on a lower level of operation but sufficient to enable a quick resumption to normal operation when the fault causing the shut-down has been rectified. The unit under these conditions is said to be 'boxed in'.

  The second type of normal shut-down is the more common and occurs when there is a major fault requiring equipment to be taken out of service for repair. This type of shut-down is also placed on a predefined schedule for routine maintenance. Such a shut-down requires the unit feed to be withdrawn and the unit itself allowed to cool down to ambient temperature, depressured, cleaned out (usually by steam) and drained free of any hazardous material.

  This section of the manual presents both these procedures in the same detail as described earlier for 'normal-start-up'. Again the mechanical flow diagram is used as a reference in these procedures.

- *Emergency procedures*   This section is the most important in the manual and requires careful development and presentation. Its importance is reflected in the fact that very often the preparation of this section becomes a combined effort of both the client process engineer and the contractor's process engineer. Its obvious importance to the safety of all personnel working on the plant is often coupled with the requirements of insurance underwriters and, in the case of a major mishap, its legal significance.

The section should contain at least the following sub-sections:

—Emergency shut-down
—Emergency action by personnel
—Plan showing location of firefighting equipment
—Emergency telephone numbers (by client)
—Location and setting of pressure relief valves

The contractor's process engineer is primarily concerned with the writing of the emergency shutdown procedures. In most process units this follows the same basic principles, which are:

(1)   If there is a fired heater either as a feed preheater or reboiler, shut down the burners immediately and introduce steam into the firebox.

(2)   Again, in the case of fired heaters, take out the fresh feed and recycle prods through the coil until the coil is cool enough not to be damaged. Use vendor data to fix this temperature.

(3)   Cool down fractionating towers by increasing reflux flow and, wherever possible, reducing return temperatures.

(4)   In the case of reactors, remove the feed stream immediately and purge with inert gas such as nitrogen. In reactors in exothermal service and where there is a coolant, increase the coolant duty.

(5)   Divert all product streams to slop but maintain any cooling cycle until safe conditions are reached.

(6)   Depressurize all pressure vessels to flare or to atmosphere at a safe location. (Note: The process engineer must make sure that there exists a depressurizing system on all these vessels and that they are clearly shown on the mechanical flow diagram.)

(7)   As soon as conditions allow, introduce purge steam into all vessels, towers and, where applicable, reactors. In some cases steam would be injurious to catalysts in reactors. Under these circumstances the reactors should be purged with inert gas.

(8)   If the emergency is the result of a fire in a heater firebox due to a fractured coil, shut-down the burners and take out the feed immediately. Introduce snuffing steam into the firebox and do not recycle product or introduce steam into the fractured coil.

As in the case of start-up and normal shut-down procedures the same kind of detail covering the basic actions given above would be written for the emergency shut-down. Again, the mechanical flow diagram would be used as the reference document.

- *Appendices* These should contain as much data as possible to assist safe operation of the plant. As a minimum it should contain:
  —The 'as-built revision' of the mechanical flow diagram
  —An up-to-date equipment list
  —Equipment manufacturer's operating instructions and data
  —A list of hazardous materials used in the process with a summary of recommended handling procedures.

PLANT COMMISSIONING

Process engineers from the contractor's organization are often called upon to assist in supervising plant commissioning. Those from the client's organization are invariably asked to supervise the commissioning. This task begins at the point where the contractor's field organization has completed all their installation activities and have handed the units over to the client's 'care, custody and control'. At this point all contractor's debris has been removed, pressure testing has been completed, and the unit has been reassembled after the initial flushing out.

The commissioning activities fall into the following sequence of events:

- Pre-energizing activities
- Energizing the plant
- Conditioning equipment, calibrating instruments and setting relief valves
- Final check-out and closing up all vessels
- Preparation for 'start-up'
- Start-up
- Lining out
- Performance test runs and guarantee test run

Much of this work is carried out by the future operators of the plant to enhance their familiarity with the process and, as mentioned earlier, the team is supervised and the activities planned by the responsible process engineer(s). A further description of the major commissioning activities now follows.

Pre-energizing activities

When the plant is handed over by the contractor to the client no hazardous material such as fuel oil, fuel gas or permanent electrical power has been introduced into the plant area. Obviously, work to check out and condition equipment utility services must be established, and this will be the objective of this phase. Normally, while the plant is in this 'safe' condition many client companies take advantage of this to carry out their final physical check of the plant on a line-by-line, item-by-item basis.

Although the contractor has flushed the plant out before hand-over a further and more thorough flush-out is advisable before start-up. This is done at this 'safe plant' stage when utility and underground lines can also be flushed out. If the plant is a unit in an existing complex oily or chemical drains would at this point be blocked off from the complex's main drain systems. Immediately prior to the plant being energized these drain systems must be unplugged and checked that they are free from any obstruction.

Energizing the plant

As soon as the responsible process engineer is satisfied that all pre-energizing activities are complete he or she will instruct that the utility systems be commissioned. Normally, this starts with the introduction of permanent power and the checking out of the circuits by the electrical technicians. When this has been completed to the responsible process engineer's satisfaction instructions for commissioning the steam, condensate and fuel systems will follow. Note that the water system is usually commissioned before the final flush-out and the system used for this flushing activity. The commissioning of the fuel systems indicates that the plant is now a 'hazardous area' and all regulations pertaining to this type of area come into effect.

Conditioning equipment

Certain new process equipment will require conditioning before being used in the process. Some of these are:

- *Reactors* In some processes these will have refractory lining which will need to be 'cured'. Curing is the subjection of the refractory to a controlled increasing temperature environment until the curing temperature is reached. The refractory is held at this temperature for a prescribed period of time before cooling back to ambient temperature. The cooling is also undertaken in a controlled fashion. Many reactors and reactor systems also require drying out. This too is accomplished during this period using heated air or inert gas and the recycle gas compressor if there is one in the circuit.
- *Heaters* All heaters will have refractory lining and these will need to be 'cured' as described above. If the heater contains coils that are to be used in steam generation, these need to be treated with hot caustic soda to remove all traces of grease or other undesirable contaminants.
- *Fractionating columns and vessels generally* These usually need to be dried. This is usually done at the same time associated heaters are being conditioned. The hot air or inert gas stream passing through the heater coils to protect them during the refractory curing is routed to the vessels and used for the drying activity.
- *Piping in sour service* Piping and pots, etc. in high H_2S and hydrogen service often need to be acid treated to protect them against local corrosion or embrittlement under these conditions. This is called 'pickling' and is usually accomplished by setting up a temporary system. This contains a reservoir for the fresh acid, one or two small skid-mounted reciprocating pumps and a receptacle for the used acid. The pumps deliver the acid through the piping and equipment for a prescribed period of time.

All these conditioning procedures are provided by the respective equipment manufacturers and would be included as part of the operating manual.

The opportunity is normally taken at this time to calibrate as much instrumentation as possible. Flow meter orifice plates are installed and, where possible, flow meters and control valves used in these procedures. Relief valves are set, and this is always done by the client's organization and the settings certified.

Final check-out and closing up of all vessels

This will be the last opportunity to check such items as the internals of towers, fractionation trays, condition of refractory and other linings, hold-up grids, distributors, and the bottom of the tower baffling system (to and from the reboiler). A final check of the piping layout also needs to be carried out at this point. When all is satisfactory the process engineer will authorize the following final pre-start-up activities to be completed:

- Catalyst loaded into the respective reactors. This is often supervised by the licensor's representative. The reactor is closed up by installing the correct operating gaskets. This point is made because during the pre-start-up activities and conditioning temporary gaskets are used on manway nozzle connections and the like. These are replaced by the correctly specified gaskets for the operating conditions. The temporary gaskets are never used again and are thrown away.
- All towers containing random packing to be loaded with the packing and closed up. Permanent specified gaskets are used to replace temporary gaskets as described above.
- All other vessels and towers are closed up using the permanent specified gaskets.
- In cases where equipment has been subject to a caustic wash the temporary silica level gauges used during the wash are replaced by the specified operational ones.

With the completion of these final checks and vessel close-up the plant is now ready for start-up.

Start-up and lining out

The activities and their sequence for starting up the plant are carried out as described in the operating manual. In the case of an oil refining plant, for example, the first activity is to eliminate air from the plant systems. This is done by using water, steam or inert gas or a combination of all three. The use of steam or water is prohibited for most reactor systems where the catalyst would be irreparably damaged by such contact. Inert gas is circulated using the recycle compressor for this purging. Thus the recycle compressor is commissioned and will then continue to function until plant shut-down. In the case of the fractionating units of the crude distillation and thermal cracking systems, water and steam are used for purging. The water is used to purge the heaters and tower bottom systems and steam for purging the upper sections of the fractionating towers.

After the purging comes the introduction of the cold feeds. In the case of units that contain reactors and use hydrogen under pressure a leak-testing programme is required. First, the plant is subject to leak testing at operating pressure using the inert gas. Then as the inert gas is replaced by hydrogen further leak tests are required. All leaks must be repaired before start-up of these units can begin.

Where water has been used for purging it is replaced by the oil feed. This is termed the 'oil squeeze'. The steam is not replaced by anything. It continues to flow through the tower until after start-up, where it may then be replaced by the vapour phase of fractionation.

Start-up may be defined as beginning when the purge programme shows conditions to be safe to apply heat into the plant. In the case of catalytic plants using a hydrogen stream, the hydrogen stream is circulated and heated to its operating temperature first. When the plant operating conditions are on gas circulation the oil feed is introduced and heated.

In the case of the non-catalytic units where only fractionation is being carried out heat may be applied to the oil feed directly. This is done when the oil squeeze is complete and the water content of the oil flowing out of the plant is at an acceptable level. The plant conditions are obtained in accordance with the procedures described in the operating manual by adjusting heat input, reflux rates, pressures and respective circulating rates in the case of catalytic plants. Up to the point when the operating conditions have been reached the total products have been routed to slop. They may be returned later as feed or used in the refinery fuel system. As soon as the plant has reached its operating conditions product streams are required to be diverted to storage. This is done, however, only after the product quality has been confirmed by laboratory tests.

Performance and guarantee test runs

Usually as soon as possible after the plant has been lined out and is in production a performance test run may be carried out. The purpose of this test run would be to establish as far as possible the limits of the various systems included in the process. Normal feedstock would be used and the performance of the plants under various operating conditions examined. The other objective of this exercise is to familiarize the operators with the plant control responses to the various changes.

The guarantee test run is a more formal requirement. At the end of the test run carried out over a fixed period and under specific conditions, the client will accept the plant completely or require repair and further test runs to meet the guarantees. More details on this test run are given in the next section of this chapter.

Figure 9.3 is an example of the commissioning plan used in the commissioning of a 'grass roots' refinery. This plan reflects the programme for the on-site units

DAY NUMBER

Operation	Unit	Day markers (day no. / date)
Water flushing	Tank farm	17 MAY 1966 (1); 20/5 (3)
	Crude	25/5 (5)
	Visbreaker	24/5 (5)
	Diesel HT	
	Nap HT	10/6 (7)
	Platform	10/5 (7)
	Amine	29/5 (8)
Water circulating	Crude	28/5 (9)
	Visbreaker	28/5 (10); 4/6 (11)
	Diesel HT	28/5 (11)
	Nap HT	
	Platform	13/6 (11)
	Amine	
Drain and inspected	Vessel & Drums as required	
Commissioning fuel oil system		11/6 (13); 28/5 (15)
Dry firing and curing heater refractory (inc. W.H. boiler)	Crude 2F-1	19/6 (18)
	Visb 5F-1	19/6 (18)
	Diesel 6F-1	19/6 (18)
	Nap HT, 3F-1&2	19/6 (20)
	Platformer (4F-1,2,3)	9/7 (13)
Pre-startup line up and steam out	Crude/visb	22/6 (17); 27/6 (22)
	Nap/plat	22/6 (17); 27/6 (24)
	Diesel/amine	22/6 (19); 27/6 (24)
Air dry out and air blow gas systems	Nap HT	28/6 (16)
	Diesel HT	26/6 (16)
	Platformer	18/7 (16)
	Sales gas systems	19/7 (16)
Catalyst loading	Platformer	30/6 (20); 20/7 (21)
	Delsel HT	6/7 (22)
	Nap HT	
Vacuum testing	Platformer	22/7 (22); 22/7 (23)
	Nap HT	22/7 (23)
	Diesel HT	
Caustic then water circulation	Amine	20/6 (21)

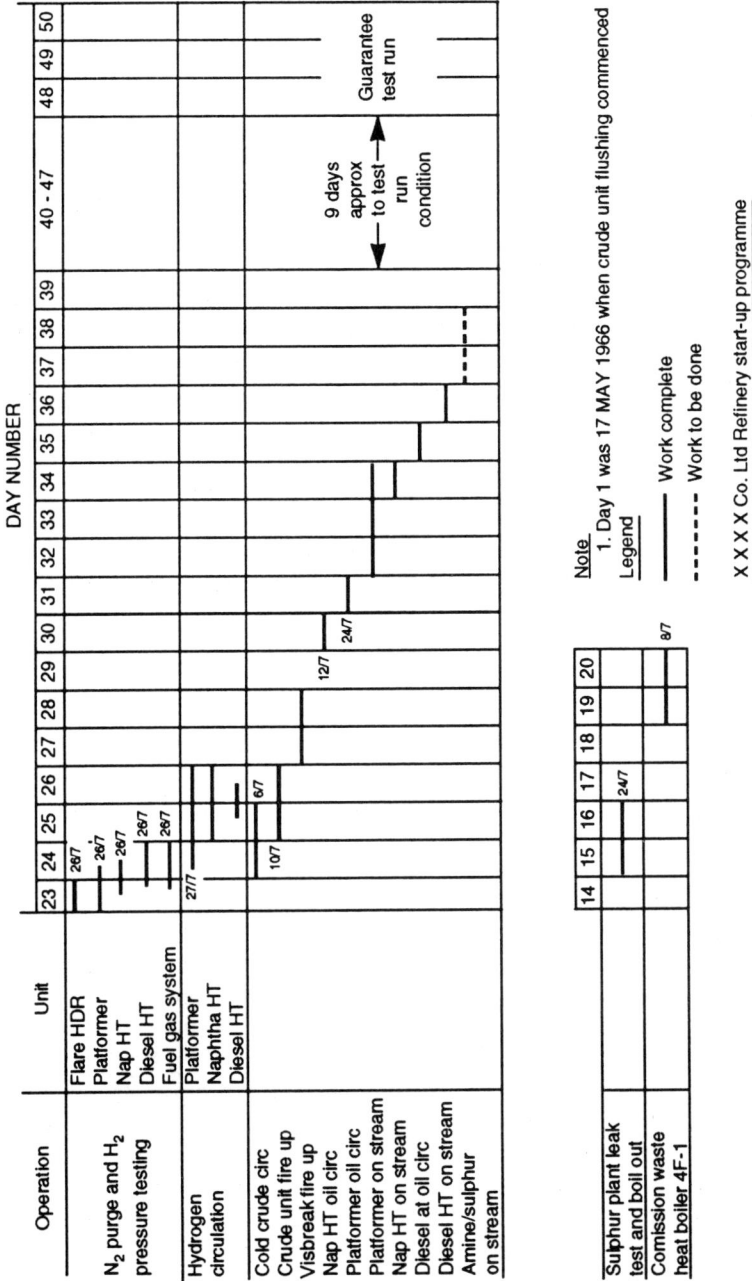

Figure 9.3. Commissioning plan of a refinery

and tank farm only. All the utility plants had already been commissioned and started up before this plan begins. The programme was successful and, apart from some minor problems, the refinery was brought onstream in the time span predicted.

9.4 Process Guarantees and the Guarantee Test Run

One of the last activities that a process engineer performs on a project is the preparation of the process guarantees that are usually required by the client and the procedure for testing the plant to meet the guarantees. The process guarantees may begin to be developed as soon as a firm process has been established and manufacturer's guarantees obtained for the performance of the various manufactured items of equipment. Process engineers may be required to guarantee the performance of any plant that they have calculated and for which they have specified equipment, piping, instruments, etc. They will not be expected to guarantee the performance of any item of plant or piece of equipment they have not specifically designed. They will also not be expected to guarantee any criteria that cannot be measured during a test run. Thus if process engineers calculate a fractionating unit to meet a specified separation they will be expected to guarantee that the design will make the separation at the design feed rate and composition. They will not, however, be expected to guarantee the performance of individual items of equipment contained in the unit and which have been properly specified by them. This guarantee is carried by the respective manufacturers who have designed and fabricated the items.

The process performance is tied also to a guarantee of its efficiency. This will be in terms of a guarantee of the utility consumption in the plant while operating on the design throughput and conditions. Of course, the guarantees as written will differ from process to process but will usually follow a similar pattern or format. This is as follows:

- Description of the feed in terms of throughput, composition or source (in the case of crude oil, for example).
- Design operating conditions and the guarantee of the product specification at these conditions. Alternatively, a guarantee of the fractionation or separation performance of the process in terms of key component separation.
- A guarantee of the hydraulic capacity of the process system. This would only be in the terms that the process system will handle a design quantity per unit time. Note that this hydraulic guarantee would not normally be combined with the stream composition specification guarantee. This is discussed later.
- The utility consumption guarantees. These are usually taken for the total plants in the new or revamped complex. Rarely are utility guarantees written for individual units making up the new or revamped complex.
- A list of the accepted test procedures that will be used.
- The guarantee test run procedures which are written in some detail. Provision is normally made in them for the contractor's process engineer to instruct operating changes to be made as necessary before or during the test run.

● Description of the notices to be given in the event that:

(1) The test run was successful and met all the requirements of the guarantee, or
(2) The test run was not successful and requires to be run again, or
(3) Some or all of the guarantee requirements cannot be met and a limited liability clause is evoked.

The following paragraphs discuss the contents of a performance guarantee in more detail.

Description of the design feedstock

Although the plant may have been designed on the basis of handling more than one feedstock of slightly differing composition, the test run is conducted using one of the design feedstocks as a parameter. This practice is more common in the oil refining industry than in the production of chemicals or even petrochemicals. This is because most refineries handle crude oil from various sources around the world, and these can vary considerably in composition. The intermediate products from the various crude oils therefore can also vary in composition.

The choice and proper definition of the test run feedstock is absolutely basic in developing the guarantees and assessing the result of the guarantee test run(s). This begins by providing a statement of the design feed throughput. This should be as simple as possible and preferably in the units of measurement used in the client's original 'duty specification'. If the measurement units used in the design or the units of measurement used in the operation of the process differ from those in the 'duty specification' then the conversion factors used to convert from one unit of measurement to the others must be noted here. There must be no ambiguity concerning this quantity.

The second item of importance in this section is the composition of the design feedstock. Where possible, this should be a breakdown of the quantity of each component in either weight or molal terms. Occasionally volume composition may be used, but the composition and the units it is quoted *must* be able to be measured using accepted normal test equipment and/or reagents.

When it is not possible to give the feed composition as described above other acceptable criteria for establishing the feed quality may be used. For example, in defining the quality or composition of a crude oil or the products of crude oil distillation boiling points and boiling curves are used (see *Elements of Petroleum Processing* also by this author). Again, however, these characteristics of the feedstock must be measurable by accepted means. The test methods to be used in evaluating the guarantee test run results are listed in this document. This item is discussed later.

Test run conditions and the guarantee of product quality

Following the description covering the design feedstock comes the heart of the performance guarantee—the guarantee of the end product. This guarantee, however, is always tied to the feedstock and to the conditions that the plant is operated on to meet the product specification. Where there is a catalytic or any other type of reactor involved in the process this guarantee may extend to cover the yield of product promised by the contractor or his or her licensor. The pertinent process conditions to obtain the guarantee with respect to the end product are defined here. Obviously these conditions will reflect the basic parameters used in the design of the facility. The process engineer developing the guarantee document must, however, be sure that the conditions described here *are pertinent* to meeting the end product specifications and yield. Incomplete definition of the process conditions may lead to ambiguity and conflict in operating the plant under test run conditions.

The product specification and its yield under certain circumstances should reflect exactly those given in the client's 'duty specification'. Yields and/or product quality which have been left open or to be confirmed by a licensor's pilot plant tests during the compiling of the 'duty specification' must be included here. These should be qualified, however, with reference to the client's acceptance of them and his or her approval for their use in the plant design. They become part of the client's licensor agreement. The contractor may then develop the remaining process guarantees on the design carried out based on this approved licensor data.

Performance of the plant may also be judged by the performance of some engineering principle used in the design. For example, in oil refining the performance of some of the complex fractionators used (e.g. the atmospheric crude distillation unit) is most often defined in terms of fractionating efficiency. This is measured by the difference in temperature reading between a 95 vol% recovery of a light product and the 5 vol% reading of the adjacent heavier product. The test used to determine these readings is the accepted ASTM distillation test carried out routinely in all refineries. The measure of these test temperatures indicates how much of the lighter product is contained in the heavier one and vice versa. This, of course, is also a measure of the respective products' composition and the contaminants they contain.

The hydraulic guarantee

Although this is tied to the plant design in its ability to handle the feed and product streams, as stated earlier, it is guaranteed separately from the performance guarantee described above. This is because very often it is not possible for a contractor to guarantee the quantity of product that the plant will make. An example of this is again seen in the oil industry. When a contractor's process engineer designs the unlicensed units of an oil refinery he or she does so using the design crude's assay data. All the product yields and composition therefore are based on the assay data which the contractor could not have developed—he or she is therefore not expected to guarantee it. From his design calculation, however, he has sized the various systems in terms of piping instruments and, of course, specified the various pumps. He is expected to guarantee these based on the design flows he has used. Thus the hydraulic guarantee should start with the following statement:

When the unit is operated on the design feedstock and at the design feed rate the following systems shall be capable of handling:

- Stream one 450 gpm measured at 60 °F
- Stream two 600 gpm measured at 60 °F

Utility consumption guarantees

Normally contractors offer utility guarantees as a summation of the cost of all utilities consumed by the process. These are also determined as a summary of the utilities consumed by all the plants considered in the test run. The utility guarantee figure is developed from the utility balances compiled during the process design. Many of these figures are based on manufacturers' guaranteed equipment efficiencies for electric motors and turbines. Others are based on the process heat and material balances (see Chapter 2 for a description of utility balances). The calculated utility consumption for each utility is multiplied by the respective utility unit cost as provided by the client in his or her 'duty specification'. The utility guarantee figure is then quoted as the sum of all these cost items as a single cost figure.

In preparing the utility guarantee the following should appear in the statement of the guarantee:

- Unit cost of each utility
- Expected normal consumption of each utility
- Either the *guaranteed* total daily cost of all the utilities as one figure (this would include a contingency factor agreed to by the client) or a guaranteed daily cost of all the utilities as a percentage of the expected normal consumption.

The descriptions of the remaining items that make up the guarantee document are self-explanatory and are illustrated in the example of a typical process guarantee given in Table 9.2. This example is based on the study case for the thermal cracker illustrated in Chapter 8. In the example Table 9.2 only the

Table 9.2. An example of a process guarantee

XYZ REFINERY PROJECT   Contractor Contract 1234

The process performance guarantees are handled on an individual process unit basis as set forth in subsequent sections of this table. The utility guarantee covers all the process units associated with this contract and is lumped into an aggregate utility cost as set forth in section 8.

Where licensed or proprietary processes not owned by contractor are involved or proprietary catalysts are supplied by others, such as in the case of the thermal cracker, naphtha hydrotreater, diesel hydrotreater, and catalytic reformer, contractor offers only the guarantees set forth herein and identified as the responsibility of the contractor. Any other guarantees shall be negotiable between Client and Licensors and are not part of this table.

Table 9.2. (*continued*)

4   THERMAL CRACKER

4.1   Feed

The thermal cracking unit shall be designed to process 34320 BPSD of Kuwait reduced crude to produce gas to C_5, gasoline, gas-oil, and fuel oil as residue. The feedstock shall be material boiling above 700 °F TBP cut point from an atmospheric crude distillation unit. The feedstock shall be substantially in accordance with the Kuwait assay titled 'Assay of Kuwait 31.2 °API crude dated May 1976' and shall have the following properties:

Gravity °API	15.8
ASTM D1160	
	°F
IBP	650
5%	750
10%	775
20%	800
30%	850
Pour point (max) °F	85
Visc. (cSt) at 122 °F	55
Flash point (Pensky–Marten)	> 250

4.2   Performance guarantees

4.2.1   When Owner operates the thermal cracker at conditions defined by the Contractor with the feedstock and rates given in 4.1 above, the estimated (but not guaranteed) yield structure will be as set forth in Table 1 attached.

4.2.2   The thermal cracker will be designed so that it will have the hydraulic capacity to process 34 320 BPSD of the feedstock defined in 4.1 above, with the design yield structure defined in 4.2.1 above.

4.2.3   Based on the yield structure defined in 4.2.1 above, the Contractor guarantees that the fractionation section of the unit shall be designed so that the numerical difference between the 5% point of the thermal cracker gas oil and the 95% point of the thermal cracked naphtha, expressed as ASTM gap, shall be no less than 10 °F, as measured by ASTM D-86.

4.2.4   If the feedstock does not meet the assay defined in section 4.1 or if the thermal cracker is not operated in accordance with Contractor's instructions, then the performance guarantees set forth herein will be modified in accordance with sound engineering practice, as appropriate, based on actual conditions during the test run and using the same data sources and calculation methods as used in the design so as to obtain a true measure of the unit's performance.

4.2.5   The test methods that shall be used in evaluating all aspects of this guarantee shall be as listed below:

Feed flash point — ASTM D-93
Feed distillation — ASTM D-1160
Product distillation — ASTM D-86.
Feed viscosity — ASTM D-445
Feed pour point — —
Gravity — ASTM D-287

Table 9.2. (*continued*)

Table 1. Design yield structure

	wt% on feed
Conversion to 340 °F TBP cut point in cracker furnace and transfer line	25.0
Products:	
Gas to C_5	9.0
Naphtha distillate to 390 °F cut point	20.3
Gas-oil 390 °F to 622 °F TBP cut	24.7
Fuel oil +622 °F cut point	46.0

Process performance and utilities guarantees

8 UTILITIES

8.1   The process units will be designed so that the daily aggregate cost of utilities consumed for all facilities covered in section 8.2 herein shall not exceed $xxxxxxxxx when the process units are operated at the rates and conditions summarized in section 8.2 herein and with the individual unit utility costs as summarized in section 8.3 herein.

8.2   The utilities guarantee is based on operating the process units at the rate and conditions summarized below and defined in more detail in sections 2 to 6 of this table:

(Sec 2) New naphtha hydrotreater	13 500 bpsd fresh feed
(Sec 3) New naphtha splitter	13 500 bpsd fresh feed
(Sec 4) New thermal cracker	34 320 bpsd fresh feed
(Sec 5) New light end unit	3 830 bpsd C_4s, C_3, C_2 mixed feed
(Sec 6) New diesel hydrotreater	9 000 bpsd fresh feed
(Sec 7) New catalytic reformer	9 000 bpsd fresh feed

This utility guarantee covers only those units listed above and does not include the utilities of any revamped units (such as the existing debutanizer). All conditions regarding feedstock composition and operating conditions defined in section 2 through 7 of this table shall extend to the utilities guarantee.

8.3   The following unit costs for the individual utilities shall be used in computing the daily aggregate cost of utilities:

Fuel gas	$0.47 per mmBTU, LHV
Power	$0.042 per kWh
Water	$0.5/1000 US gal
Steam	$0.83/1000 lb

8.4   In the event that the daily aggregate cost quoted in section 8.1 herein is exceeded, Contractor shall have the right to make any alterations it deems necessary in order that the utilities guarantee can be met in a subsequent test run. Contractor shall make alterations or pay a penalty as defined in section 8.5 herein at his sole option.

8.5   The penalty that the Contractor shall pay in the event that the utilities guarantee is not met and the Contractor elects not to modify the plant, shall be the difference between the average daily aggregate cost of utilities for the best test run made under section 10, and the guaranteed daily cost of utilities in section 8.1 herein multiplied by 700.

Table 9.2. (*continued*)

Process performance and utilities guarantees

9 QUALIFICATIONS FOR GUARANTEES

9.1 Notwithstanding any other sections or statements in this exhibit, the guarantees in sections 2 to 8 inclusive above are subject to change if Licensor information, data, or designs for the licensed units should be revised at any time so as to differ from the Licensor's information, data, or designs in the Contractor's possession on [–date–], or if licensed units require more utilities than specified by the aforesaid Licensor information, data or designs.

9.2 In no event shall the Contractor be liable for contingent or consequential damages, including damages for loss of products or profit or for plant downtime.

Process performance and utilities guarantees

10 PERFORMANCE TEST

To determine whether the guarantees defined in sections 2 to 8 are met, the following test run procedure will be used:

10.1 The Contractor will notify the Owner in writing when the plant or any portion thereof is ready for initial operation. Within thirty (30) days thereafter the Owner will perform a series of test runs unless delays are caused by deficiencies that are the Contractor's responsibility. If the performance test run for the plant or any portion thereof is not conducted by the Owner within 30 days after notification by the Contractor to Owner of availability for test run, it shall be conclusively presumed that the performance guarantees have been met and that Contractor's obligations covered herein have been satisfied and Owner agrees to pay any sums due to the Contractor as if the test runs had been successfully met. The time of these test runs may be initiated at any hour of day or night. Test periods shall be of 72-hour duration or less as mutually agreed upon by Owner and Contractor and may be interrupted as follows:

10.1.1 For minor alterations, repairs, failure of feedstock, utility supply or other condition beyond the Contractor's reasonable control, each of which do not exceed 24 hours. The test run shall proceed promptly after the interruption and as soon as the Contractor deems the plant operation has levelled out. The sum of normal operating periods before, between, and after such interruption shall be considered the required test period when it totals 72 hours.

10.2 Owner shall be responsible for supplying all the necessary operating labour, feedstocks, utilities, catalysts, chemicals, sampling laboratory analyses, and other supplies for operating and testing of the process units. Owner shall maintain the process units in accordance with good practice, and all catalysts shall be in essentially new condition to the end that the process units will be in proper condition for the performance tests provided for herein.

10.3 The Contractor shall furnish observers and test engineers, excluding Licensor personnel, to technically advise Owner during the performance tests. Contractor's observers shall have the right to issue instructions regarding the manner in which the plant is to be operated during the test run. Owner shall comply with these instructions unless the instructions are contrary to generally accepted safety practices or expose equipment conditions of temperature, pressure, or stress greater than their maximum allowable operating conditions.

Table 9.2. (*continued*)

10.4 Analyses of the streams and products for these tests shall be determined by methods mutually agreed to by Owner and Contractor. Analysis will be conducted by Owner and may be observed or witnessed by Contractor. In case of disagreement, a referee laboratory may be selected by approval of both parties and paid for by the Owner.

10.5 Samples shall be spot and/or composite and sampling procedure shall be by methods as mutually agree upon by Owner and Contractor. Samples shall be taken at uniform intervals. Elapsed time between samples or sample increments shall not exceed 4 (four) hours, and composite sample increments shall be of equal volume when flow rate is essentially uniform. All flow rates, product rates, and analysis shall be averaged over the test period. Where possible, tanks shall be gauged at frequent intervals to substantiate meter readings.

10.6 The performance guarantees shall be considered satisfactorily met when the average of the performance results during the period meets or exceeds the performance specified in sections 2 to 8 of this table. Owner shall be responsible for the security of the unit operating log sheets, charts, laboratory test results, gauging records, and other pertinent information for the test period. Within ten (10) days after the completion of each performance test, Owner shall submit a written statement for each unit indicating whether the guarantee has been met. If the test is not acceptable to the Owner, then the Owner shall specify in writing to the Contractor in what respect the performance has not been met. On request the Owner shall submit to contractor all records and calculations for review.

10.7 A performance test run shall be stopped when in the judgement of the Contractor alterations, adjustments, repairs, and/or replacements which cannot be made safely with the equipment in operation must be made to enable the plant to meet and fulfil the performance guarantees, or the data obtained during the test will not be sufficient to establish the actual performance of the unit within desired limits of error, or it becomes obvious that such performance test cannot be satisfactorily concluded in the current attempt.

10.8 If a performance test is stopped as provided above, or if the unit or units do not meet their guaranteed performance during the test run, Owner, when requested, shall make the unit or units available to the Contractor as soon as possible to make such alterations or additions thereto as in Contractor's judgement are required to enable the plant to fulfil the aforesaid guarantees. If the Owner does not make the unit or units available within three months, the performance guarantee will be considered to have been met in its entirety. Contractor will make such alterations as it shall deem necessary to make the unit or units perform as guaranteed and a further performance test shall be conducted in accordance with the above procedure.

10.9 Process performance tests for the units described in sections 2 to 7 along with the appropriate pro rata of the utilities guarantee as set forth in section 8, all of this table, may be carried out collectively and/or on an individual basis as may be mutually agreed to by Owner and Contractor. If a test run involving more than one process unit or section of a process unit indicates that any unit or section meets its guarantees as set forth in this table, then that unit or section shall be accepted.

10.10 Notwithstanding any other provision hereof, Contractor's liability for making the changes units to make them perform as guaranteed in sections 2 to 7 of this table shall be limited to $xxxxxx. In determining the total cost expended by Contractor there shall be included the engineering costs to the Contractor of making the necessary design changes, and the labour and material costs of implementing these changes.

performance guarantee for the thermal cracker (section 4) itself is given in full. In an actual process guarantee all the other units such as the hydrotreaters, cat reformer, etc. would be shown in the same detail as the thermal cracker in section 4.

In this example it is assumed that the contract has been awarded on a lump-sum basis for engineering, all other home office functions and the contractor's profit fee. The other costs such as equipment, materials and the installation of the plant is on a basis of cost payable directly by the client.

Appendix 1
GENERAL DATA AND CORRELATIONS

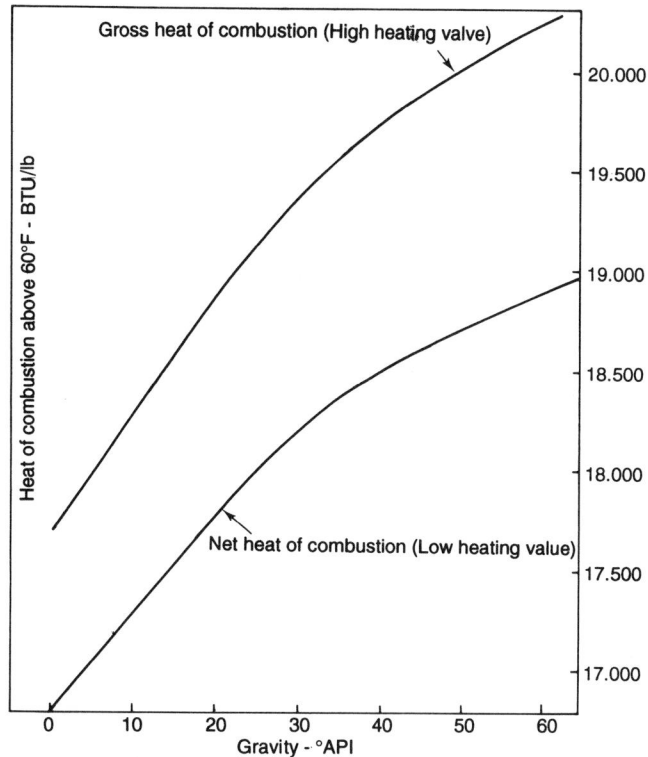

Impurities in average fuels			
°API	%S	%inerts	Total impurity
Residual fuel oil and crudes			
0	2.95	1.15	4.10
5	2.35	1.00	3.35
10	1.80	0.95	2.75
15	1.35	0.65	2.20
20	1.00	0.75	1.75
Crude oils			
25	0.70	0.70	1.40
30	0.40	0.65	1.10
35	0.30	0.60	0.90

Figure A1.1. Heat of combustion of fuel oils. These values represent an average of cracked and virgin fuel oil data up to 20 °API and the correlation allows for average sulphur and inerts (excluding water) found in average fuels. Above 40 °API the correction for impurities is negligible and the curves represent pure petroleum liquids (Reproduced by permission of Kreiger Publishing Company, Malabar, Florida, 1950, Maxwell, Data Book on Hydrocarbons)

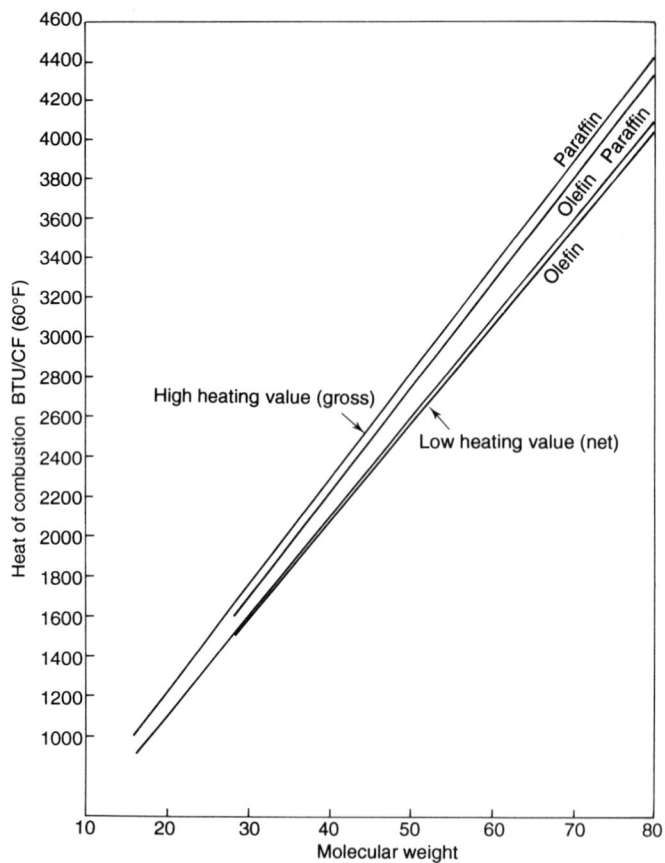

Figure A1.2. Heat of combustion of fuel gases (Reproduced by permission of Kreiger Publishing Company, Malabar, Florida, 1950, Maxwell, Data Book on Hydrocarbons)

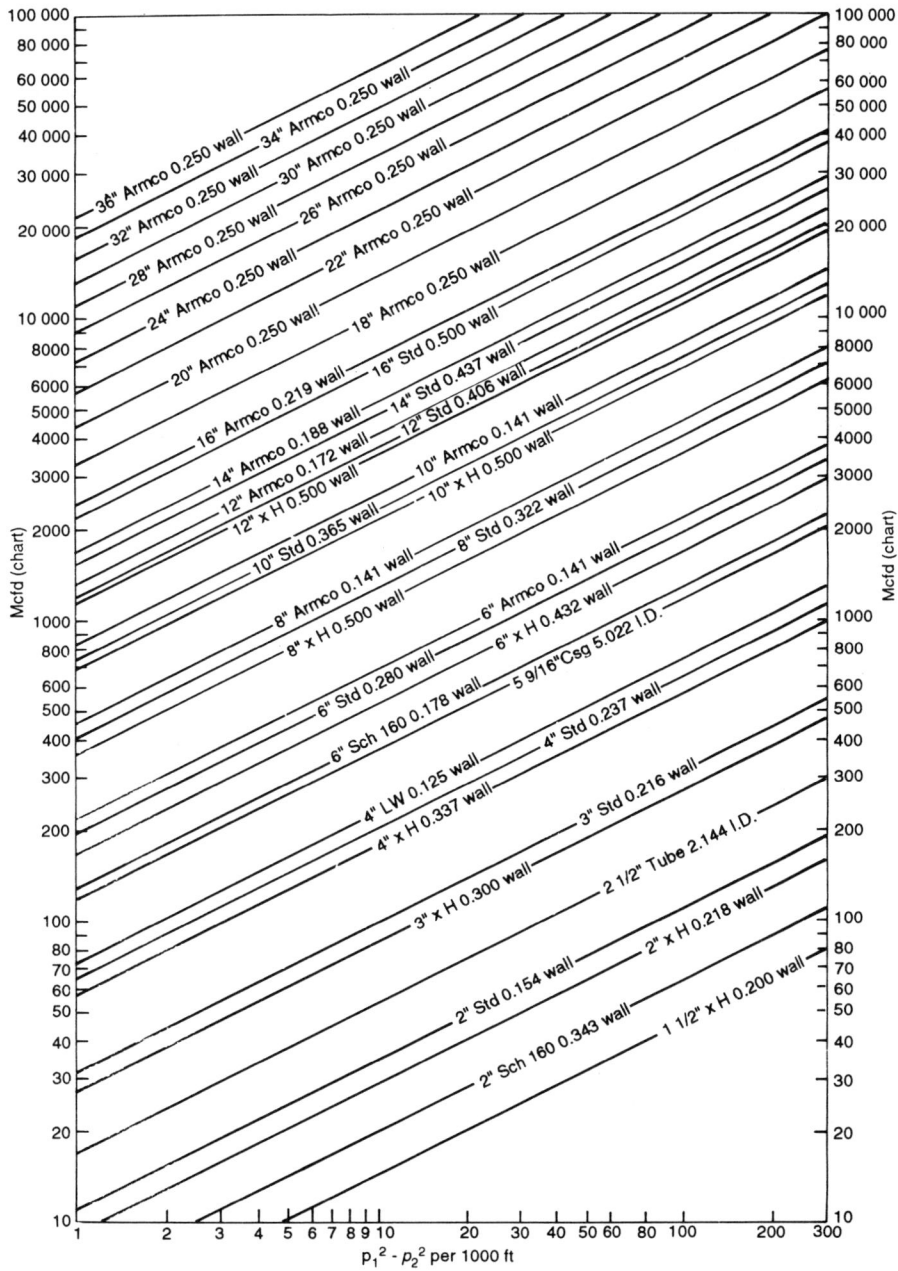

1. Chart—Mcfd—based on 'Weymouth Formule' where specific
gravity = 0.9 (air = 1), flowing temperature = 90°F and
pressure base = 14.65 psia

2. Simplified Mcfd = 1.59 $d^{2/3}$ $(P_1^2 - P_2^2/1000 \text{ ft})^{1/2}$

Figure A1.3. Friction loss for gas flowing in pipes (Reproduced by permission of Gas Processors Suppliers Association)

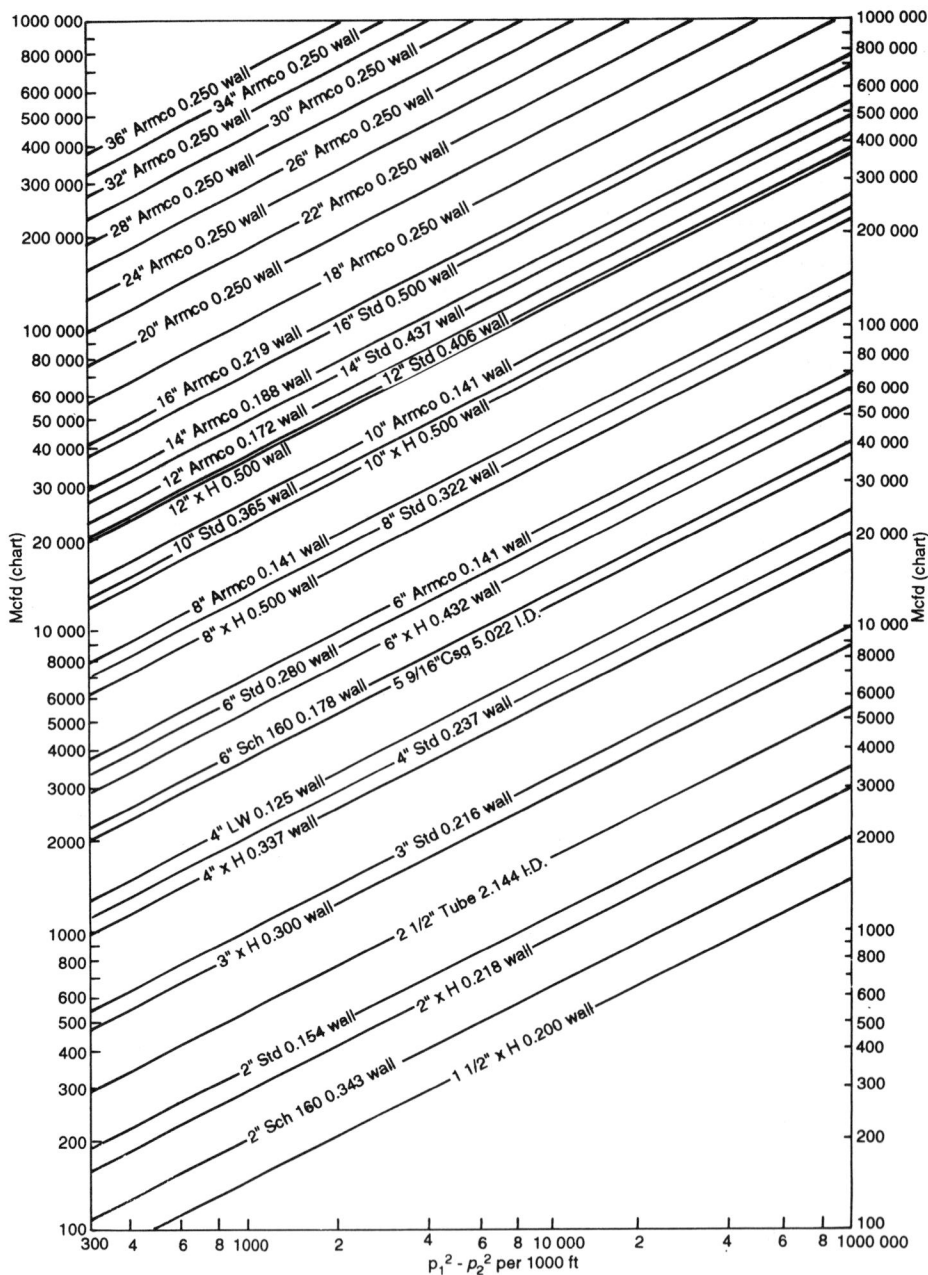

3. For conditions other than in (1) correct Mcfd for chart use as follows:

$$\text{Mcfd(chart)} = \frac{P}{14.65} \left(\frac{SG}{0.9} \times \frac{T}{550} \times \frac{Z}{1} \right)^{1/2} \times \text{Mcfd}$$

Z = Compressibility factor as determined from 'Natural Gas Under Pressure' in the GPSA Engineering Data Book
P = Pressure base other than 14.65 psia
T = Flowing temperature in degrees Rankine (460 + °F)

Figure A1.3. (*continued*)

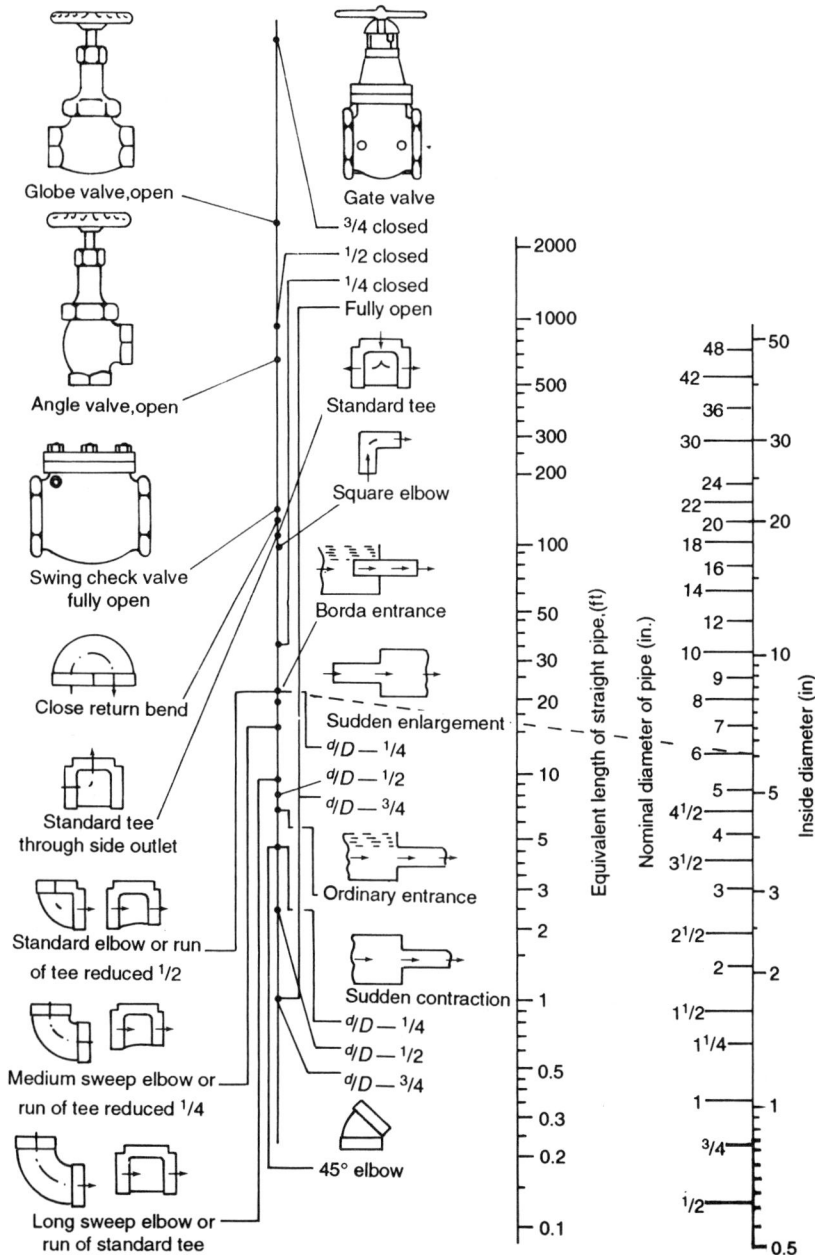

Figure A1.4. Equivalent pipe lengths for fittings. The dotted line shows that the resistance of a 6-inch Standard Elbow is equivalent to approximately 16 feet of 6-inch Standard Pipe. For sudden enlargements or contractions use the smaller diameter, *d*, on the pipe size scale (Reproduced by permission of Ingersoll-Dresser Pump Company)

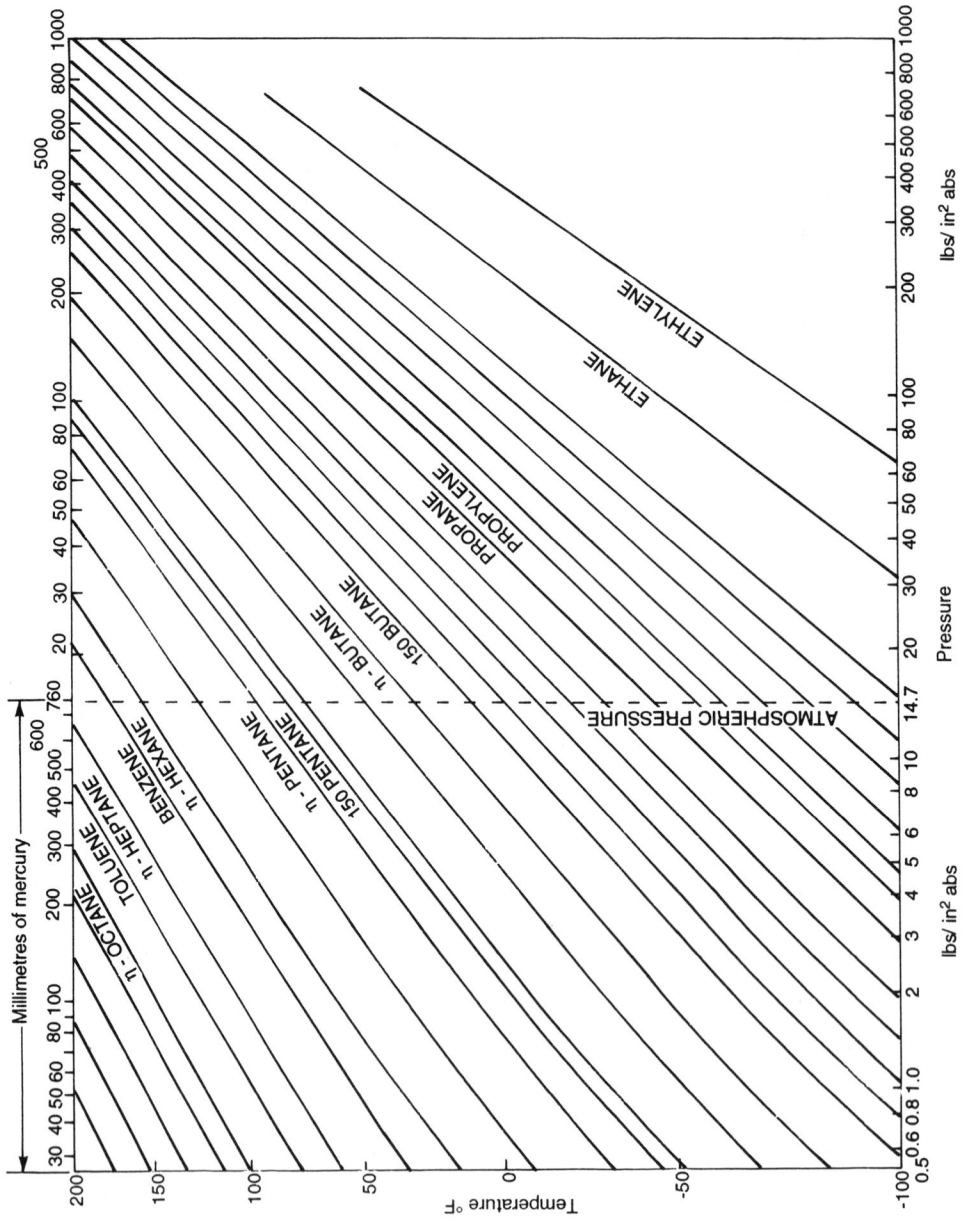

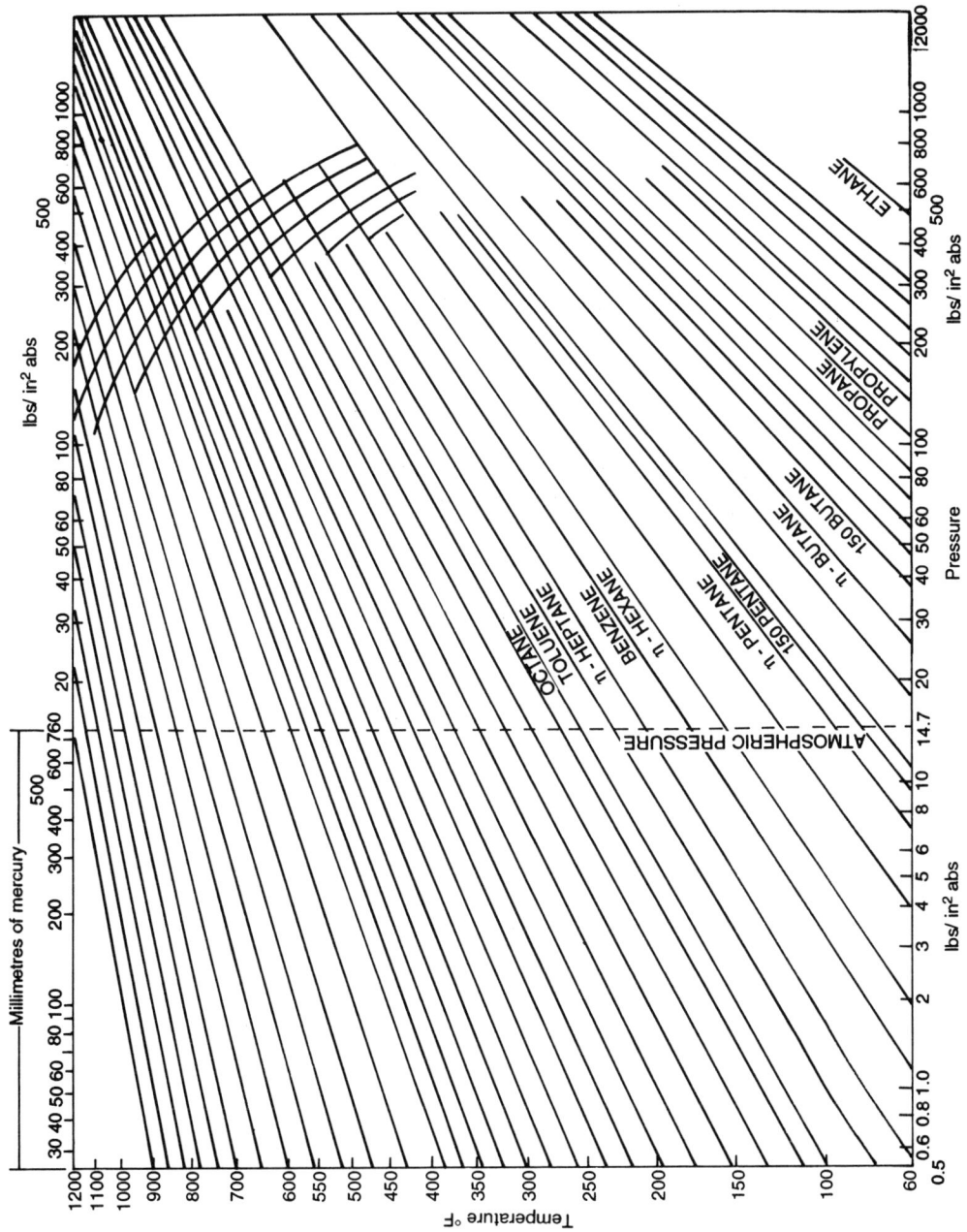

Figure A1.5. Vapour pressure curves for hydrocarbons

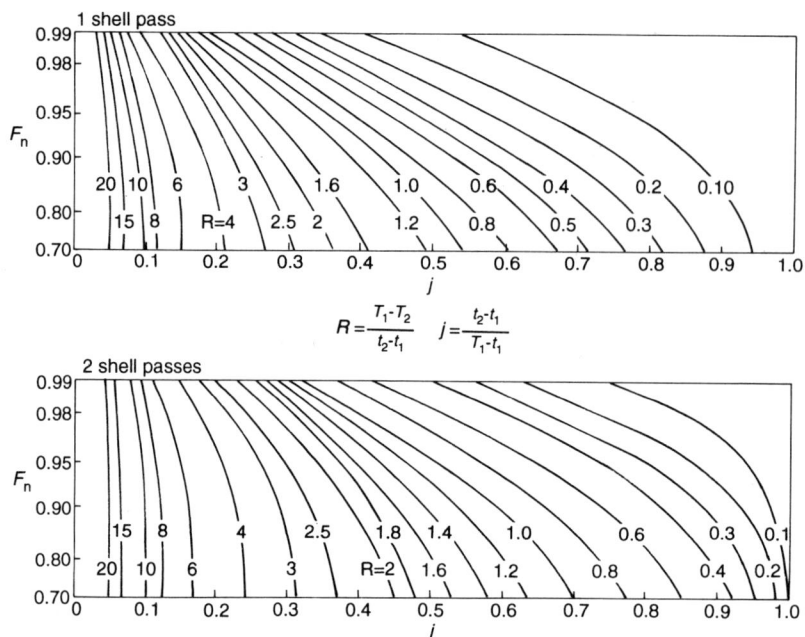

$$R = \frac{T_1 - T_2}{t_2 - t_1} \qquad j = \frac{t_2 - t_1}{T_1 - t_1}$$

Figure A1.6. LMTD correction factors

Table A1.1. Some common overall heat transfer coefficients U_0.

Fluid being cooled	Fluid being heated	U_0 (BTU/h.f^2.°F)
Exchangers		
C$_4$s and lighter	Water	75–110
C$_4$s and Lighter	LPG	75
Naphtha	Naphtha	75
Naphtha	Water	80–100
BPT 450 (kero)	Hy oil (crude)	70–75
Gas oils	Crude	40–50
Gas oils	Water	40–70
Light fuel oil	Crude	20–30
Waxy distillates	Hy oils	30–40
Slurries	Waxy distillate	40
MEA or DEA	Water	140
MEA or DEA	MEA or DEA	120–130
Water	Water	180–200
Air	Water	20–30
Lt HC vapour	H$_2$-rich stream	35–40
Lt HC vapour	Naphtha	38
Condensers		
Full-range naphtha	Water	70–80
Amine stripper O/heads	Water	100
C$_4$s and lighter	Water	90
Reformer effluent (Lt HC)	Water	65

Table A1.2. Friction loss for liquids flowing in pipes (loss in feet of liquid per 1000 feet of pipe) (Reproduced by permission of Ingersoll-Dresser Pump Company)

1-inch (1.049 inch inside dia.) schedule 40 new steel pipe

Flow		Kinematic viscosity (cSt)									
		0.6	1.1	2.1	2.7	4.3	7.4	10.3	13.1	15.7	20.6
US gal per min	Bbl per h (42 gal)	Approx SSU viscosity									
			31.5	33	35	40	50	60	70	80	100
0.5	0.71	0.29	0.28	0.55	0.70	1.12	1.93	2.68	3.41	4.08	5.35
1	1.4	0.96	1.13	1.09	1.41	2.24	3.86	5.36	6.82	8.16	10.7
2	2.9	3.23	3.72	4.41	4.80	4.48	7.72	10.7	13.6	16.3	21.4
3	4.3	6.84	7.63	9.04	9.48	10.8	11.6	16.1	20.5	24.5	32.1
4	5.7	11.4	12.2	14.9	15.9	17.6	15.4	21.5	27.3	32.6	42.8
5	7.1	17.2	19.2	22.1	23.4	26.3	19.3	26.8	34.1	40.8	53.5
6	8.6	24.2	26.8	30.5	32.1	36.8	41.3	32.2	40.9	49.0	64.2
7	10.0	32.3	35.4	40.8	42.6	47.4	53.8	37.5	47.7	57.2	74.9
8	11.4	41.6	45.5	51.1	54.3	59.9	58.2	75.3	54.5	65.2	85.6
9	12.9	51.8	56.2	63.5	66.3	73.4	84.9	91.7	61.4	73.4	96.3
10	14.3	62.7	68.1	76.2	80.1	88.2	101	111	115	81.6	107
12	17.1	89.3	95.3	106	111	122	142	151	162	167	129
14	20.0	120	129	140	147	160	185	198	212	221	150
16	22.8	155	164	181	188	205	234	257	268	279	295
18	25.7	194	202	224	233	254	286	305	326	341	365
20	28.6	237	250	272	281	308	342	372	394	405	437
25	35.7	368	383	410	429	464	501	551	583	599	651
30	42.9	523	545	582	600	640	712	759	803	842	904
35	50.0	708	735	780	795	852	933				

Flow		Kinematic viscosity (cSt)									
		26.4	32.0	43.2	65.0	108.4	162.3	216.5	325	435	650
US gal per min	Bbl per h (42 gal)	Approx SSU viscosity									
		125	150	200	300	500	750	1000	1500	2000	3000
0.1	0.14	1.37	1.66	2.25	3.38	5.65	8.45	11.3	16.9	22.6	33.8
0.3	0.43	4.12	4.98	6.75	10.2	17.0	25.3	33.8	50.7	67.8	102
0.5	0.71	6.86	8.32	11.3	16.9	28.3	42.3	56.4	85	113	189
1	1.4	13.7	18.6	22.5	33.8	56.5	84.5	113	169	226	338
2	2.9	27.5	33.2	45.0	87.8	113	169	226	338	452	676
3	4.3	41.2	49.8	67.5	102	170	253	338	507	678	
4	5.7	55.0	66.5	90.0	136	226	338	452	677	904	
5	7.1	68.7	83.2	113	189	283	423	564	846		
6	8.6	82.4	99.7	135	203	339	607	677			
7	10	96.2	117	158	237	395	591	790			
8	11.4	110	133	180	271	452	676	903			
9	12.9	124	150	203	303	508	760				
10	14.3	137	167	225	338	565	845				
12	17.1	185	200	270	408	678					
14	20.0	192	233	315	474	792					
16	22.8	220	266	360	541	904					
18	25.7	248	299	405	609						
20	28.6	470	332	450	677						

Table A1.2. (*continued*)

1½-inch (1.610 inch inside dia.) schedule 40 new steel pipe

Flow		Kinematic viscosity (cSt)									
		0.6	1.1	2.1	2.7	4.3	7.4	10.3	13.1	15.7	20.6
US gal per min	Bbl per h (42 gal)	Approx SSU viscosity									
			31.5	33	35	40	50	60	70	80	100
1	1.4	0.13	0.10	0.20	0.25	0.41	0.69	0.97	1.23	1.47	1.93
2	2.9	0.42	0.49	0.39	0.51	0.83	1.39	1.93	2.46	2.94	3.86
3	4.3	0.86	0.98	1.17	1.25	1.24	2.08	2.89	3.68	4.41	5.79
4	5.7	1.43	1.63	1.92	2.07	1.65	2.78	3.86	4.91	5.88	7.72
5	7.1	2.11	2.42	2.83	3.05	2.06	3.47	4.82	6.14	7.35	9.65
6	8.6	2.90	3.36	3.89	4.18	4.69	4.17	5.79	7.37	8.82	11.6
8	11.4	4.97	5.60	6.44	6.87	7.77	9.02	7.72	9.83	11.8	15.5
10	14.3	7.51	8.34	9.58	10.1	11.6	13.2	9.65	12.3	14.7	19.3
12	17.1	10.4	11.6	13.4	14.0	15.6	17.9	19.8	14.7	17.6	23.2
15	21.4	16.0	17.4	19.6	20.7	23.2	26.5	29.1	31.1	21.0	29.0
20	28.6	27.2	29.5	32.9	34.6	38.2	43.9	47.6	51.1	53.8	38.6
25	35.7	41.4	44.8	49.5	51.8	57.5	64.6	70.1	75.2	78.9	84.7
30	42.9	58.8	63.0	69.1	72.0	79.0	89.3	97.1	103	109	116.5
40	57.1	102	107	117	122	132	150	160	170	178	191
50	71.4	157	164	178	183	198	222	237	251	263	281
60	85.7	224	233	249	259	279	306	330	347	362	388
70	100	300	312	333	343	369	402	436	457	477	508
80	114	389	403	427	440	470	516	551	580	602	643
90	129	498	508	535	550	585	634	681	715	746	792
100	143	601	624	656	670	714	774	820	863	898	949

Flow		Kinematic viscosity (cSt)									
		26.4	32.0	43.2	65.0	108.4	162.3	216.5	325	435	650
US gal per min	Bbl per h (42 gal)	Approx SSU viscosity									
		125	150	200	300	500	750	1000	1500	2000	3000
1	1.4	2.41	3.00	4.14	6.09	10.2	15.2	20.3	30.4	40.8	69.0
2	2.9	4.95	6.00	8.28	12.2	20.3	30.4	40.6	60.8	81.5	122
3	4.3	7.42	9.00	12.4	18.3	30.4	45.8	60.9	91.3	122	183
4	5.7	9.90	12.0	16.6	24.4	40.6	60.8	81.2	122	163	244
5	7.1	12.4	15.0	20.7	30.4	50.7	76.0	102	152	204	304
6	8.6	14.9	18.0	24.8	36.5	60.8	91.2	122	163	244	365
8	11.4	19.8	24.0	33.1	48.7	81.2	122	163	243	326	487
10	14.3	24.7	30.0	41.4	60.9	102	152	203	304	406	609
12	17.1	29.7	36.0	49.7	73.2	122	182	244	365	490	732
15	21.4	37.1	45.0	62.2	91.4	152	228	304	457	612	914
20	28.6	49.5	60.0	82.8	122	203	304	406	608	815	
25	35.7	61.9	75.0	103	152	254	380	507	760		
30	42.9	124	90.0	124	183	304	456	609	913		
40	57.1	204	216	166	244	406	608	812			
50	71.4	302	317	342	304	501	760				

2-inch (2.067 inches inside dia.) schedule 40 new steel pipe

Flow		Kinematic viscosity (cSt)									
		0.6	1.1	2.1	2.7	4.3	7.4	10.3	13.1	15.7	20.6
US gal per min	Bbl per h (42 gal)	Approx SSU viscosity									
			31.5	33	35	40	50	60	70	80	100
1	1.4	0.04	0.04	0.07	0.09	0.15	0.26	0.36	0.45	0.54	0.71
2	2.9	0.13	0.15	0.15	0.19	0.30	0.51	0.71	0.90	1.06	1.42
4	5.7	0.43	0.50	0.58	0.63	0.59	1.02	1.42	1.81	2.17	2.84
6	8.6	0.87	1.00	1.20	1.27	1.46	1.53	2.13	2.71	3.25	4.26
8	11.4	1.47	1.68	1.97	2.13	2.38	2.04	2.84	3.61	4.33	5.68
10	14.3	2.20	2.55	2.90	3.09	3.52	2.58	3.58	4.52	5.42	7.11
12	17.1	3.06	3.46	3.97	4.23	4.78	5.57	4.27	5.43	6.51	8.53
14	20.0	4.07	4.51	5.22	5.51	6.26	7.28	4.98	6.33	7.59	9.96
16	22.8	5.17	5.79	6.65	7.01	7.92	9.16	9.95	7.23	8.67	11.4
18	25.7	6.44	7.16	8.18	8.63	9.67	11.2	12.6	13.1	9.78	12.8
20	28.6	7.82	8.64	9.77	10.4	11.6	13.5	14.8	15.5	10.8	14.2
25	35.7	11.9	13.0	14.7	15.4	17.2	19.9	21.6	22.8	24.1	17.8
30	42.9	17.0	18.2	20.4	21.5	23.8	27.2	29.9	31.2	33.0	35.6
35	50.0	22.3	24.1	27.0	28.2	31.2	35.6	38.6	40.4	43.4	47.3
40	57.1	28.8	31.0	34.2	36.0	39.6	44.6	48.8	52.9	54.1	60.7
50	71.4	44.1	47.2	52.0	54.0	59.2	66.4	72.0	76.9	78.4	86.2
60	85.7	62.7	66.5	72.2	74.2	82.3	91.8	98.6	105	111	119
70	100	84.1	83.5	95.8	99.4	108	120	130	137	145	156
80	114	109	114	123	127	138	154	166	174	182	195
90	129	137	143	154	158	171	192	204	215	225	239
100	143	167	176	188	193	208	230	244	260	269	289
110	157	202	211	225	231	246	275	290	307	319	335
120	171	238	249	265	273	290	321	341	358	375	396
130	186	277	290	307	316	335	372	392	411	432	459
140	200	320	347	352	364	383	424	449	472	491	521
150	214	366	382	403	415	437	479	510	529	553	586
160	228	414	431	457	469	494	536	572	595	621	659
170	243	467	485	513	522	553	601	639	665	694	729
180	257	524	543	572	583	619	665	714	743	767	804
190	271	584	602	634	649	688	733	792	825	846	884
200	286	643	666	699	716	756	808	851	901	927	977
210	300	709	731	768	786	828	880	935	975		
220	314	778	798	838	858	902	958				
230	328	851	873	912	934	982					
240	343	922	945	988							

Table A1.2. (*continued*)

Flow		Kinematic viscosity (cSt)									
US gal per min	Bbl per h (42 gal)	26.4	32.0	43.2	85.0	108.4	162.3	216.5	325	435	650
		Approx SSU viscosity									
		125	150	200	300	500	750	1000	1500	2000	3000
1	1.4	0.91	1.10	1.49	2.24	3.74	5.60	7.48	11.2	15.0	22.4
2	2.9	1.82	2.21	2.98	4.48	7.49	11.2	15.0	22.4	30.0	44.9
3	4.3	2.73	3.31	4.47	6.73	11.2	16.8	22.4	33.6	45.0	67.4
4	5.7	3.64	4.42	5.98	8.98	15.4	22.4	29.9	44.8	60.0	89.9
5	7.1	4.56	5.52	7.45	11.2	18.7	28.0	37.4	56.0	75.0	112
6	8.6	5.47	6.63	8.95	13.5	22.5	33.6	44.8	67.2	90.0	135
7	10.0	6.38	7.73	10.4	15.7	26.2	39.2	52.3	78.4	105	157
8	11.4	7.29	8.84	11.9	18.0	30.0	34.8	59.8	89.6	120	180
9	12.9	8.20	9.94	13.4	20.2	33.7	50.4	67.3	101	135	202
10	14.3	9.11	11.0	14.9	22.4	37.4	56.0	74.8	112	150	224
12	17.1	10.9	13.3	18.9	26.9	44.9	67.3	89.7	135	180	269
14	20.0	12.7	15.5	20.9	31.4	52.4	78.4	105	157	210	314
16	22.8	14.6	17.7	23.9	35.9	59.9	89.6	120	179	240	359
18	25.7	16.4	19.9	26.6	40.3	67.4	101	135	202	270	404
20	28.6	18.2	22.1	29.6	44.9	74.9	112	150	224	300	449
25	35.7	22.6	27.6	37.3	56.1	93.6	140	187	280	375	562
30	42.9	27.3	33.1	44.7	67.3	112	168	224	336	450	674
35	50.0	31.9	38.7	52.2	78.5	131	196	262	392	525	786
40	57.1	63.0	44.2	59.6	89.8	150	224	299	448	600	899
45	64.3	70.8	80.2	67.1	101	168	252	336	503	675	
50	71.4	92.8	97.1	74.5	112	187	280	374	560	750	
60	85.7	127	134	146	135	225	336	448	672	900	
70	100	162	176	189	157	262	392	523	784		
80	114	208	219	238	180	300	448	596	896		
90	129	257	270	293	327	337	504	673			
100	143	309	322	352	388	374	580	746			
110	157	364	379	412	465	412	617	823			
120	171	423	445	463	537	449	673	898			
130	186	487	510	549	618	487	728				
140	200	548	580	627	703	524	784				
150	214	622	655	705	792	909	840				
160	228	697	737	792	887		896				
170	243	775	808	882			952				
180	257	858	871	962							
190	271	947	986								

$2\frac{1}{2}$-inch (2.469 inches inside dia.) schedule 40 new steel pipe

Flow		Kinematic viscosity (cSt)									
		0.6	1.1	2.1	2.7	4.3	7.4	10.3	13.1	15.7	20.6
US gal per min	Bbl per h (42 gal)	Approx SSU viscosity									
			31.5	33	35	40	50	60	70	80	100
10	14.3	0.92	1.05	1.23	1.31	1.48	*1.26*	*1.75*	2.22	*2.66*	*3.50*
12	17.1	1.28	1.47	1.70	1.80	2.04	*1.51*	*2.10*	2.67	3.19	4.19
14	20.0	1.68	1.92	2.23	2.37	2.65	3.09	*2.45*	*3.11*	3.73	4.89
16	22.8	2.15	2.43	2.81	2.99	3.34	3.86	*2.80*	*3.56*	4.26	5.59
18	25.7	2.68	2.99	3.46	3.68	4.21	4.77	5.26	*4.00*	*4.80*	6.28
20	28.6	3.23	3.61	4.27	4.42	4.94	5.73	6.26	*4.44*	*5.33*	6.96
25	35.7	4.88	5.39	5.17	6.55	7.31	8.42	9.20	9.79	*6.66*	*8.73*
30	42.9	6.67	7.57	8.55	9.07	10.1	11.5	12.6	13.5	14.1	*10.5*
35	50.0	9.18	9.97	11.3	11.8	13.3	15.1	16.4	17.7	18.5	20.0
40	57.1	11.8	12.8	14.3	15.0	16.8	19.0	20.7	22.1	23.2	25.0
45	64.3	14.8	15.9	17.7	18.7	20.8	23.5	25.6	27.1	28.4	30.6
50	71.4	18.1	19.3	21.4	22.4	24.9	28.3	30.6	32.5	34.1	36.8
60	85.7	25.6	27.2	29.9	31.1	34.2	39.1	42.1	44.7	46.7	50.4
70	100	34.2	36.2	39.6	41.4	45.2	51.4	55.4	58.5	61.5	66.0
80	114	44.1	46.7	50.6	53.0	57.3	64.5	69.7	74.2	77.5	82.7
90	129	55.2	58.8	63.2	65.4	70.9	80.0	86.6	91.2	95.3	103
100	143	67.2	72.3	76.7	79.4	85.8	96.3	104	109	115	123
110	157	80.9	86.2	92.4	94.8	103	113	121	130	135	146
120	171	95.7	102	108	112	121	133	143	151	158	169
130	186	112	118	125	130	139	155	166	176	181	194
140	200	129	136	144	150	160	176	188	198	209	221
150	214	147	155	165	170	181	196	212	223	234	250
160	228	167	175	187	191	203	224	238	253	262	281
170	243	188	196	210	214	226	248	267	283	290	312
180	257	210	219	234	239	253	277	297	311	322	345
190	271	233	243	260	265	280	306	328	339	356	378
200	286	258	269	286	292	308	334	357	373	391	417
220	314	310	322	343	351	369	400	427	448	461	486
240	343	367	381	404	416	436	469	494	522	539	573
260	371	429	445	470	482	505	543	575	599	621	660
280	400	497	513	540	556	580	630	657	686	710	758
300	429	568	586	617	632	659	705	748	775	803	849
320	457	643	663	695	716	747	799	837	875	903	952
340	486	725	745	776	800	839	894	933	980		
360	514	809	835	866	892	938	994				

Table A1.2. (*continued*)

Flow US gal per min	Flow Bbl per h (42 gal)	26.4 / 125	32.0 / 150	43.2 / 200	65.0 / 300	108.4 / 500	162.3 / 750	216.5 / 1000	325 / 1500	435 / 2000	650 / 3000
1	1.4	0.45	0.54	0.73	1.10	1.84	2.75	3.67	5.52	7.38	11.0
2	2.9	0.90	1.09	1.47	2.20	3.68	5.50	7.35	11.0	14.8	22.0
4	5.7	1.79	2.17	2.93	4.41	7.36	11.0	14.7	22.1	29.5	44.1
6	8.6	2.69	3.28	4.40	6.62	11.0	16.5	22.0	33.1	44.3	66.2
8	11.4	3.58	4.34	5.87	8.82	14.7	22.0	29.4	44.1	59.1	88.2
10	14.3	4.48	5.43	7.33	11.0	18.4	27.5	36.7	55.2	73.8	110
12	17.1	5.38	6.51	8.80	13.2	22.1	33.0	44.1	66.2	88.6	132
14	20.0	6.27	7.60	10.3	15.4	25.7	38.5	51.4	77.2	103	154
16	22.8	7.16	8.68	11.7	17.6	29.4	44.0	58.8	88.2	118	176
18	25.7	8.06	9.77	13.2	19.8	33.1	49.5	66.1	99.3	133	198
20	28.6	8.96	10.9	14.7	22.0	36.8	55.0	73.4	110	148	220
25	35.7	11.2	13.6	18.3	27.6	46.0	68.8	91.8	138	185	276
30	42.9	13.4	16.3	22.0	33.1	55.2	82.5	110	185	222	331
35	50.0	15.7	19.0	25.6	38.6	64.4	96.3	129	193	258	386
40	57.1	17.9	21.7	29.3	44.2	73.6	110	147	221	295	441
45	64.3	33.0	24.4	33.0	49.6	82.8	124	165	248	332	496
50	71.4	39.2	27.2	36.6	55.2	92.0	138	184	276	369	551
60	85.7	54.0	56.5	44.0	66.2	110	165	220	331	443	682
70	100	70.0	73.5	51.3	77.2	129	193	257	386	517	772
80	114	87.7	93.4	101	88.3	147	220	294	441	591	882
90	129	110	115	125	99.3	166	248	330	497	665	993
100	143	130	137	148	110	184	275	367	552	738	
110	157	154	164	176	197	202	303	403	607	812	
120	171	180	166	205	226	221	330	441	662	886	
130	186	206	216	232	263	239	358	477	717	960	
140	200	234	247	267	299	257	385	514	772		
150	214	265	279	305	333	276	413	551	827		
160	228	296	312	336	374	294	440	588	882		
170	243	328	345	373	415	312	468	624	937		
180	257	364	384	412	461	530	495	661	993		
190	271	403	420	454	514	587	523	698			
200	286	438	457	493	550	628	550	734			
220	314	522	540	586	658	752	605	806			
240	343	612	633	682	760	866	660	881			
260	371	711	732	782	867		715	955			

3-inch (3.068 inches inside dia.) schedule 40 new steel pipe

Flow		Kinematic viscosity (cSt)									
		0.6	1.1	2.1	2.7	4.3	7.4	10.3	13.1	15.7	20.6
US gal per min	Bbl per h (42 gal)		Approx SSU viscosity								
			31.5	33	35	40	50	60	70	80	100
8	11.4	0.22	0.25	0.29	0.32	0.24	0.42	0.59	0.74	0.89	1.18
10	14.3	0.32	0.37	0.43	0.47	0.54	0.53	0.73	0.93	1.11	1.47
15	21.4	0.70	0.76	0.89	0.94	1.07	0.79	1.10	1.40	1.67	2.20
20	28.6	1.12	1.27	1.47	1.57	1.78	2.07	1.46	1.86	2.23	2.93
25	35.7	1.69	1.93	2.23	2.31	2.61	3.01	3.29	2.33	2.79	3.66
30	42.9	2.36	2.64	2.99	3.22	3.60	4.12	4.50	4.83	3.35	4.40
35	50.0	3.13	3.48	3.97	4.21	4.66	5.41	5.89	6.35	6.61	5.13
40	57.1	4.03	4.42	5.02	5.29	5.90	6.60	7.46	7.93	8.37	5.87
50	71.4	6.10	6.70	7.50	7.93	8.76	10.1	10.9	11.7	12.3	13.2
60	85.7	8.57	9.32	10.4	11.0	12.0	13.7	15.0	16.0	16.8	18.0
70	100	11.5	12.4	13.8	14.5	15.9	18.0	19.6	20.9	21.9	23.6
80	114	14.7	15.9	17.5	18.4	20.3	22.9	24.6	26.4	27.7	29.8
90	129	18.4	19.9	21.8	22.8	25.0	28.0	30.4	32.4	33.8	36.3
100	143	22.4	24.2	26.3	27.5	30.2	33.7	36.4	39.0	40.8	43.6
120	171	31.8	34.1	36.9	38.6	41.9	46.8	50.5	53.4	58.4	60.0
140	200	42.4	45.6	49.4	50.9	55.4	65.5	66.0	70.0	73.2	78.6
160	228	54.8	58.0	63.3	65.4	70.4	79.1	83.8	87.9	92.3	98.2
180	257	69.0	72.7	78.7	81.6	87.2	97.2	104	109	114	122
200	266	84.7	88.9	95.7	99.4	106	117	125	131	137	146
225	322	107	112	120	124	132	145	155	164	169	180
250	357	131	137	147	151	160	175	188	195	204	218
275	393	158	164	175	180	191	208	226	233	243	258
300	429	187	193	204	212	225	244	260	273	281	298
325	464	218	225	238	247	261	283	300	316	325	345
350	500	253	260	275	283	300	324	344	361	373	396
375	536	288	298	314	322	341	367	388	407	424	448
400	571	328	339	354	363	385	414	436	458	476	498
425	607	368	381	397	407	432	463	488	511	529	550
450	643	410	427	443	455	460	515	543	568	587	619
475	679	457	473	493	504	532	571	599	625	646	681
500	714	504	524	544	555	589	627	658	684	707	750
525	750	555	574	597	609	644	688	720	748	770	821
550	786	606	627	651	665	703	748	783	814	838	890
575	822	663	685	708	723	761	814	852	886	912	962
600	857	721	742	767	783	820	882	919	960	989	

Table A1.2. (*continued*)

US gal per min	Bbl per h (42 gal)	26.4 / 125	32.0 / 150	43.2 / 200	65.0 / 300	108.4 / 500	162.3 / 750	216.5 / 1000	325 / 1500	435 / 2000	650 / 3000
4	5.7	0.75	0.91	1.23	1.85	3.08	4.62	6.16	9.25	12.4	18.5
6	8.6	1.13	1.37	1.84	2.77	4.62	6.92	9.24	13.9	18.5	27.7
8	11.4	1.50	1.82	2.45	3.70	6.16	9.23	12.3	18.5	24.7	36.9
10	14.3	1.88	2.28	3.06	4.62	7.70	11.5	15.4	23.1	30.9	46.2
12	17.1	2.25	2.73	3.68	5.55	9.24	13.6	18.5	27.7	37.1	55.5
14	20.0	2.83	3.16	4.29	6.47	10.8	16.2	21.5	32.3	43.3	64.7
16	22.8	3.00	3.64	4.90	7.39	12.3	18.5	24.6	37.0	49.5	73.9
18	25.7	3.38	4.09	5.52	8.31	13.9	20.8	27.7	41.6	55.6	83.2
20	28.6	3.76	4.55	6.13	9.24	15.4	23.1	30.8	46.2	61.8	92.4
25	35.7	4.69	5.69	7.67	11.5	19.3	28.6	28.5	57.7	77.3	115
30	42.9	5.63	6.83	9.20	13.9	23.1	34.6	46.2	69.3	92.7	139
35	50.0	6.57	7.97	10.7	16.2	27.0	40.3	53.8	80.9	108	162
40	51.1	7.51	9.10	12.3	18.5	30.8	46.2	61.6	92.5	124	185
50	71.4	9.39	11.4	15.3	23.1	38.5	57.7	77.0	115	154	231
60	85.7	19.2	13.7	18.4	27.7	46.2	69.2	92.4	139	185	277
70	100	25.3	26.6	21.5	32.3	53.9	80.8	108	162	216	323
80	114	31.6	33.6	24.7	37.0	61.6	92.3	123	185	247	369
90	129	38.9	40.9	44.6	41.6	69.3	104	139	208	278	416
100	143	46.1	49.5	53.0	46.2	77.0	115	154	231	309	462
120	171	64.0	67.6	72.9	55.5	92.4	138	185	277	371	555
140	200	83.9	89.1	94.9	108	108	162	215	323	433	647
160	228	106	111	120	135	123	185	246	370	495	739
180	257	131	137	148	164	139	208	277	416	556	832
200	288	157	163	179	198	154	231	308	462	618	924
225	322	191	204	223	242	279	280	346	520	696	
250	357	229	242	261	291	332	288	385	577	773	
275	393	271	285	311	343	396	317	423	635	850	
300	429	316	331	361	398	456	346	462	693	927	
325	464	364	381	416	458	527	375	500	751		
350	500	415	436	467	523	593	672	538	809		
375	536	469	493	528	586	672	746	577	867		
400	571	526	550	592	656	751	843	816	925		
425	607	587	612	656	728	834	937	654	982		
450	643	652	675	728	802	928					
475	679	718	744	801	889						

$3\frac{1}{2}$-inch (3.548 inches inside dia.) schedule 40 new steel pipe

Flow		Kinematic viscosity (cSt)									
		0.6	1.1	2.1	2.7	4.3	7.4	10.3	13.1	15.7	20.6
US gal per min	Bbl per h (42 gal)	Approx SSU viscosity									
			31.5	33	35	40	50	60	70	80	100
20	28.6	0.56	0.63	0.72	0.78	0.88	1.03	0.82	1.04	1.25	1.64
25	35.7	0.82	0.93	1.08	1.14	1.29	1.52	1.67	1.30	1.56	2.05
30	42.9	1.15	1.29	1.50	1.57	1.78	2.09	2.30	1.56	1.87	2.46
35	50.0	1.53	1.70	1.94	2.08	2.31	2.68	2.97	3.16	2.18	2.87
40	57.1	1.95	2.17	2.49	2.68	2.91	3.35	3.70	4.00	4.21	3.28
45	64.3	2.46	2.65	3.03	3.22	3.59	4.11	4.57	4.91	5.19	3.69
50	71.4	2.95	3.25	3.68	3.90	4.32	4.98	5.42	5.87	6.20	6.64
60	85.7	4.17	4.54	5.12	5.38	6.00	6.82	7.41	7.97	8.41	9.21
70	100	6.78	6.02	6.78	7.11	7.84	8.87	9.76	10.4	10.9	11.9
80	114	7.19	7.72	8.57	8.79	9.81	11.2	12.3	13.0	13.7	14.9
90	129	8.92	9.59	10.6	11.2	12.5	13.9	15.0	16.1	16.8	18.3
100	143	10.8	11.7	12.9	13.6	14.8	16.8	18.0	19.3	20.2	21.7
120	171	15.4	16.4	18.1	16.7	20.5	22.9	25.0	26.3	27.7	29.7
140	200	20.5	22.0	23.9	24.9	27.3	30.5	33.1	34.6	36.0	39.1
160	228	26.4	28.3	30.9	31.9	35.0	38.4	41.8	43.9	45.6	49.0
180	257	33.0	35.0	38.0	39.6	42.5	47.7	50.8	54.0	56.3	60.1
200	286	40.3	43.0	46.3	48.0	51.4	57.1	61.8	65.5	68.0	72.2
225	322	50.7	53.2	57.7	59.7	64.7	71.2	75.9	80.1	84.3	88.8
250	357	62.6	65.0	70.9	72.7	78.9	86.5	92.7	96.9	101	107
275	393	75.4	77.9	85.6	87.1	93.3	102	110	115	120	127
300	429	89.2	92.2	99.6	103	109	119	128	135	140	148
325	464	104	108	116	120	127	138	147	154	161	171
350	500	121	124	133	138	146	159	169	178	183	194
375	536	138	142	152	157	167	181	191	200	207	220
400	571	156	161	171	178	188	203	213	225	233	248
425	607	176	181	192	200	211	225	238	252	259	279
450	643	196	203	213	221	234	250	265	280	287	309
475	679	219	225	235	244	257	276	292	310	319	336
500	714	241	249	259	268	282	304	321	340	350	368
550	786	290	300	311	323	340	365	385	407	415	440
600	857	343	355	367	377	399	426	452	466	480	510
650	929	400	414	428	440	461	498	522	540	557	587
700	1000	464	480	494	505	532	572	597	621	637	675
750	1070	532	548	567	576	604	651	682	704	725	769
800	1140	606	624	641	652	684	730	765	794	815	861

Table A1.2. (*continued*)

Flow US gal per min	Flow Bbl per h (42 gal)	26.4	32.0	43.2	65.0	108.4	162.3	216.5	325	435	650
		125	150	200	300	500	750	1000	1500	2000	3000
10	14.3	1.05	1.27	1.72	2.58	4.32	6.48	8.62	12.9	16.9	25.8
15	21.4	1.57	1.91	2.58	3.88	6.47	9.68	12.9	19.4	25.4	38.8
20	28.6	2.10	2.54	3.44	5.17	8.63	12.9	17.3	25.8	33.8	51.7
25	35.7	2.62	3.18	4.29	6.47	10.8	16.1	21.6	32.3	42.3	64.7
30	42.9	3.15	3.82	5.15	7.76	13.0	19.4	25.9	38.8	50.7	77.6
35	50.0	3.67	4.45	6.01	9.05	15.1	22.6	30.2	45.3	59.2	90.6
40	57.1	4.20	5.09	6.87	10.3	17.3	25.8	34.5	51.7	67.6	103
45	64.3	4.72	5.73	7.73	11.6	19.4	29.0	38.8	58.2	76.1	116
50	71.4	5.25	6.36	8.59	12.9	21.6	32.3	43.1	64.7	84.5	129
60	85.7	6.30	7.64	10.3	15.5	25.9	38.8	51.8	77.6	101	155
70	100	12.8	8.91	12.0	18.1	30.2	45.2	60.4	90.6	118	181
80	114	16.1	16.9	13.7	20.7	34.5	51.6	69.0	103	135	207
90	129	19.6	20.8	15.5	23.3	38.8	58.1	77.6	118	152	233
100	143	23.6	24.8	17.2	25.9	43.2	64.6	86.2	129	169	258
120	171	31.9	34.0	37.1	31.0	51.8	77.5	104	155	203	310
140	200	41.5	43.8	48.2	36.2	60.4	90.4	121	181	236	362
160	228	52.3	55.3	60.5	67.7	69.1	103	138	207	270	413
180	257	64.6	67.3	73.9	83.4	77.7	116	155	233	304	466
200	286	77.3	81.5	87.9	99.9	86.3	129	173	258	338	517
225	322	94.4	99.8	108	122	97.1	145	194	291	380	528
250	357	114	120	129	146	108	161	216	323	422	647
275	393	134	141	153	172	199	177	237	358	465	711
300	429	157	164	178	198	233	194	259	388	507	776
325	464	161	188	205	227	266	210	280	420	549	640
350	500	206	215	233	258	301	226	302	452	592	906
375	536	234	245	262	291	339	242	323	485	634	970
400	571	261	274	291	326	379	423	345	517	676	
425	607	291	303	325	363	419	473	387	550	718	
450	643	322	337	359	402	462	525	388	582	761	
475	679	353	368	395	440	505	572	410	614	803	
500	714	389	403	433	482	551	626	431	647	846	
550	786	459	481	514	566	648	733	795	712	930	
600	857	538	566	600	658	749	851	923	778		
650	929	620	648	690	755	865	978		841		
700	1000	708	735	789	863	967			908		

4-inch (4.026 inches inside dia.) schedule 40 new steel pipe

Flow		Kinematic viscosity (cSt)									
		0.6	1.1	2.1	2.7	4.3	7.4	10.3	13.1	15.7	20.6
US gal per min	Bbl per h (42 gal)				Approx SSU viscosity						
			31.5	33	35	40	50	60	70	80	100
20	26.6	0.30	0.34	0.40	0.43	0.49	0.57	*0.50*	*0.83*	*0.75*	*0.99*
30	42.9	0.62	0.70	0.82	0.87	0.98	1.14	1.25	*0.95*	*1.13*	*1.48*
40	57.1	1.05	1.18	1.35	1.44	1.62	1.86	2.04	2.20	*1.51*	*1.96*
50	71.4	1.58	1.76	2.02	2.13	2.37	2.75	3.00	3.21	3.40	*2.47*
60	85.7	2.22	2.44	2.80	2.93	3.27	3.77	4.12	3.29	4.62	*5.01*
70	100	2.96	3.24	3.69	3.88	4.31	4.93	5.39	5.72	6.03	6.53
80	114	3.79	4.16	4.67	4.93	5.44	6.20	6.78	7.23	7.55	8.17
90	129	4.72	5.15	5.77	6.06	6.72	7.63	8.32	8.87	9.29	10.0
100	143	5.77	6.27	6.91	7.33	8.12	9.15	9.97	10.6	11.2	12.0
120	171	8.09	8.61	9.66	10.2	11.2	12.7	13.6	14.6	15.3	16.5
140	200	10.8	11.7	12.9	13.4	14.8	16.6	17.9	19.0	20.0	21.6
160	228	13.9	15.0	16.4	17.1	18.8	21.1	22.7	24.0	25.2	27.2
180	257	17.4	18.7	20.4	21.5	23.2	26.0	28.1	29.6	30.8	33.3
200	286	21.4	22.7	24.9	25.9	28.0	31.4	33.7	35.7	36.9	40.0
220	314	25.6	27.2	29.8	30.8	33.2	37.3	40.2	42.3	44.0	47.0
240	343	30.3	32.0	34.9	36.1	38.8	43.4	46.8	49.1	51.4	54.6
260	371	35.4	37.2	40.4	42.0	45.0	50.1	53.9	56.7	59.3	62.7
280	400	40.8	42.7	46.4	48.2	51.7	57.4	61.6	65.0	67.4	71.3
300	429	48.6	48.7	53.0	54.8	58.8	64.9	69.6	73.3	76.0	81.0
350	500	62.7	65.6	70.6	72.8	77.9	85.4	91.2	96.2	101	107
400	571	61.4	84.7	90.4	93.7	99.8	109	116	122	127	135
450	643	102	106	113	117	124	135	144	151	157	167
500	714	125	130	137	142	151	164	174	182	189	201
550	786	151	157	165	170	180	195	206	216	224	239
600	857	179	185	195	200	212	229	242	253	263	278
650	929	209	216	228	231	246	266	280	291	303	319
700	1000	242	249	260	267	283	306	322	333	346	365
750	1070	276	285	296	305	321	348	366	377	391	414
800	1140	314	324	337	345	362	392	411	426	439	465
850	1215	355	384	378	387	406	438	459	476	489	519
900	1285	396	408	424	434	453	486	510	531	543	574
950	1360	441	451	470	481	502	536	563	584	600	632
1000	1430	488	500	521	527	550	591	621	641	662	694
1100	1570	587	602	627	634	659	708	740	765	790	822
1200	1715	699	712	741	754	780	835	869	898	924	966

Table A1.2. (*continued*)

Flow		Kinematic viscosity (cSt)									
		26.4	32.0	43.2	65.0	108.4	162.3	216.5	325	435	650
US gal per min	Bbl per h (42 gal)	Approx SSU viscosity									
		125	150	200	300	500	750	1000	1500	2000	3000
15	21.4	0.95	1.15	1.55	2.34	3.91	5.85	7.80	11.7	15.7	23.4
20	28.6	1.27	1.54	2.07	3.12	5.21	7.80	10.4	15.6	20.9	31.2
30	42.9	1.90	2.30	3.11	4.68	7.82	11.7	15.8	23.4	31.3	46.8
40	57.1	2.54	3.08	4.15	6.25	10.4	15.6	20.8	31.2	41.8	62.5
50	71.4	3.17	3.84	5.18	7.81	13.0	19.5	26.0	39.0	52.2	78.1
60	85.7	3.80	4.61	6.22	9.37	15.6	23.4	31.2	46.8	62.7	93.7
70	100	4.44	5.38	7.25	10.9	18.2	27.3	36.4	54.6	73.2	109
80	114	8.81	6.15	8.29	12.5	20.8	31.2	41.6	62.4	83.6	125
90	129	10.8	11.3	9.33	14.1	23.4	35.1	46.8	70.2	94.1	141
100	143	12.9	13.7	10.4	15.6	26.0	39.0	52.0	78.0	105	156
120	171	17.6	18.6	20.3	18.8	31.2	46.8	62.4	93.7	125	187
140	200	22.9	24.3	26.5	21.9	36.4	54.6	72.8	109	146	218
160	228	29.0	30.3	33.2	25.0	41.7	62.4	83.2	125	167	250
180	257	35.5	37.4	40.7	45.7	46.9	70.2	93.6	140	188	281
200	286	42.6	45.0	48.7	54.8	52.1	78.0	104	158	209	312
220	314	50.3	53.0	57.1	64.7	57.3	85.8	114	172	230	343
240	343	58.5	61.5	65.1	74.7	62.5	93.6	125	187	251	375
260	371	67.2	70.8	76.8	85.7	67.7	101	135	203	272	406
280	400	76.4	80.5	87.2	97.3	73.0	109	146	218	292	437
300	429	85.8	90.8	98.5	110	127	117	156	234	313	468
325	464	98.5	104	113	125	146	127	169	254	340	508
350	500	112	118	128	143	166	136	182	273	366	547
375	536	127	133	145	161	187	146	195	293	932	585
400	571	143	149	162	180	208	156	208	312	418	625
450	643	178	184	198	222	254	285	234	351	470	703
500	714	213	221	237	265	305	343	260	390	523	781
550	786	252	263	280	313	360	404	286	429	575	860
600	857	296	305	328	364	417	467	507	468	627	937
650	929	338	353	378	419	480	528	583	507	680	
700	1000	386	402	433	474	546	608	663	546	732	
750	1070	437	455	488	533	616	685	745	585	784	
800	1140	490	510	546	570	687	764	830	624	836	
850	1215	544	570	608	663	763	848	920	663	889	
900	1285	603	629	674	739	844	939			941	
950	1360	666	696	743	813	927				993	

6-inch (6.065 inches inside dia.) schedule 40 new steel pipe

Flow		Kinematic viscosity (cSt)									
		0.6	1.1	2.1	2.7	4.3	7.4	10.3	13.1	15.7	20.6
US gal per min	Bbl per h (42 gal)	Approx SSU viscosity									
			31.5	33	35	40	50	60	70	80	100
75	107	0.45	0.49	0.58	0.61	0.68	0.80	0.86	0.93	0.98	0.72
100	143	0.77	0.85	0.96	1.01	1.14	1.30	1.42	1.52	1.62	1.74
125	178	1.14	1.27	1.43	1.51	1.68	1.95	2.10	2.23	2.35	2.57
150	214	1.61	1.78	2.01	2.09	2.32	2.66	2.86	3.08	3.20	3.46
175	250	2.13	2.37	2.63	2.79	3.04	3.52	3.74	3.97	4.24	4.51
200	286	2.75	3.00	3.34	3.55	3.85	4.41	4.76	5.02	5.31	5.69
225	322	3.42	3.74	4.17	4.38	4.78	5.39	5.89	5.16	6.45	7.05
250	357	4.15	4.55	5.07	5.21	5.76	6.94	7.11	7.47	7.77	8.41
275	393	4.99	5.42	6.02	6.28	6.88	7.60	8.35	8.86	9.18	9.81
300	429	5.87	6.38	7.06	7.37	8.09	8.94	9.69	10.3	10.8	11.4
350	500	7.90	8.45	9.38	9.80	10.5	11.8	12.6	13.5	14.2	15.0
400	571	10.2	10.9	11.8	12.5	13.4	15.0	16.0	17.0	18.0	19.1
450	643	12.8	13.6	14.8	15.5	16.7	18.5	19.7	20.1	21.8	23.7
500	714	15.6	16.6	18.0	18.7	20.4	22.6	24.0	25.1	26.4	28.5
550	786	18.8	19.8	21.5	22.3	24.3	26.8	28.5	29.6	29.9	33.4
600	857	22.1	23.3	25.1	26.2	28.4	31.1	33.2	35.0	36.2	38.8
650	929	25.8	27.2	29.2	30.4	32.8	36.1	38.5	40.4	41.8	44.7
700	1000	29.7	31.2	33.5	34.9	37.5	41.1	44.2	46.3	47.9	51.3
750	1070	33.9	35.6	38.2	39.7	42.5	46.7	49.9	51.8	54.0	57.2
800	1140	38.3	40.5	43.2	44.4	47.8	52.7	56.1	58.5	60.9	64.3
900	1285	48.5	50.7	54.4	55.6	59.3	65.4	69.1	72.6	74.6	79.9
1000	1430	59.5	62.2	66.4	67.5	72.4	79.3	83.4	87.6	91.0	95.9
1100	1570	71.6	74.8	79.4	80.8	86.7	94.5	99.6	104	109	114
1200	1715	84.6	87.9	93.4	95.6	102	111	117	121	126	133
1400	2000	115	118	126	128	135	146	155	161	167	177
1600	2285	150	153	162	164	173	187	199	207	213	224
1800	2570	188	193	203	206	216	232	246	256	264	278
2000	2860	231	237	247	253	264	284	296	311	320	334
2200	3140	277	288	297	303	316	338	354	371	382	398
2400	3430	330	341	352	358	374	395	417	430	448	470
2600	3710	387	395	408	418	433	461	485	500	520	543
2800	4000	449	458	470	482	497	526	553	574	595	621
3000	4285	515	526	536	550	567	597	628	655	666	708
3250	4640	605	613	629	641	665	701	729	757	777	817
3500	5000	697	711	729	739	771	808	841	869	897	938

Table A1.2. (*continued*)

Flow		Kinematic viscosity (cSt)									
		26.4	32.0	43.2	65.0	108.4	162.3	216.5	325	435	650
US gal per min	Bbl per h (42 gal)	Approx SSU viscosity									
		125	150	200	300	500	750	1000	1500	2000	3000
50	71.4	0.62	0.74	1.00	1.51	2.52	3.78	5.04	7.57	10.1	15.1
75	107	0.92	1.12	1.51	2.27	3.78	5.66	7.56	11.4	15.2	22.7
100	143	1.23	1.49	2.01	3.03	5.05	7.55	10.1	15.1	20.3	30.2
125	178	2.75	1.86	2.51	3.79	6.31	9.45	12.6	18.9	25.3	37.8
150	214	3.75	3.96	3.01	4.54	7.58	11.3	15.1	22.7	30.4	45.4
175	250	4.90	5.17	5.62	5.30	8.84	13.2	17.6	26.5	35.5	53.0
200	286	6.10	6.51	7.07	6.06	10.1	15.1	20.2	30.3	40.5	60.6
225	322	7.43	7.93	8.66	6.82	11.4	17.0	22.7	34.1	45.8	68.1
250	357	8.91	9.43	10.4	7.57	12.6	18.9	25.2	37.8	50.7	75.7
275	393	10.6	11.1	12.2	13.7	13.9	20.8	27.7	41.7	55.8	83.2
300	429	12.3	12.9	14.2	15.9	15.1	22.6	30.2	45.4	60.9	90.8
350	500	15.9	17.1	18.3	20.8	17.7	26.4	35.3	63.0	71.0	106
400	571	20.1	21.3	23.1	26.2	20.2	30.2	40.3	60.6	81.1	121
450	643	24.7	26.0	28.6	31.9	36.9	34.0	45.3	68.2	91.3	136
500	714	30.0	31.3	34.1	28.0	44.2	37.8	50.4	75.7	101	151
550	786	35.6	36.9	40.2	44.6	52.1	41.6	55.4	83.3	112	166
600	857	41.5	43.1	46.4	51.7	59.1	45.3	80.5	90.9	122	182
650	929	47.7	50.0	53.4	59.6	69.4	49.1	65.5	98.5	132	197
700	1000	54.1	57.0	60.8	68.6	78.8	88.3	70.6	106	142	212
750	1070	60.8	64.4	68.5	76.8	88.5	99.4	75.6	114	152	227
800	1140	68.0	72.1	76.9	85.7	97.8	111	80.6	121	162	242
900	1285	83.9	88.5	95.2	105	120	136	148	136	183	272
1000	1430	101	106	115	126	144	163	177	151	203	302
1100	1570	120	125	136	148	171	192	208	167	223	333
1200	1715	140	146	158	173	200	220	242	182	243	363
1400	2000	184	193	206	230	258	267	316	353	284	424
1600	2285	234	244	260	288	323	363	393	445	324	484
1800	2570	292	299	322	350	399	452	480	543	591	545
2000	2860	350	364	387	425	481	535	576	652	707	605
2200	3140	417	435	459	510	573	628	683	771	833	666
2400	3403	487	507	535	585	668	730	799	885	968	726
2600	3710	564	587	620	677	769	841	913			787
2800	4000	645	669	714	773	874	954				
3000	4285	734	751	805	867	993					
3200	4570	827	850	909	982						

8-inch (7.981 inches inside dia.) schedule 40 new steel pipe

Flow		Kinematic viscosity (cSt)									
		0.6	1.1	2.1	2.7	4.3	7.4	10.3	13.1	15.7	20.6
US gal per min	Bbl per h (42 gal)	Approx SSU viscosity									
			31.5	33	35	40	50	60	70	80	100
150	214	0.42	0.47	0.53	0.56	0.63	0.72	0.79	0.84	0.88	0.96
200	286	0.71	0.78	0.89	0.94	1.05	1.15	1.30	1.38	1.45	1.57
250	357	1.07	1.18	1.33	1.40	1.56	1.77	1.92	2.04	2.14	2.30
300	429	1.50	1.65	1.85	1.94	2.15	2.43	2.65	2.80	2.93	3.15
350	500	2.01	2.19	2.45	2.57	2.81	3.20	3.46	3.69	3.85	4.13
400	571	2.58	2.78	3.12	3.26	3.58	4.04	4.37	4.64	4.83	5.21
450	643	3.21	3.48	3.85	4.05	4.42	5.08	5.39	5.70	5.96	6.38
500	714	3.94	4.23	4.69	4.90	5.33	5.98	6.49	6.82	7.18	7.91
600	857	5.54	5.95	6.61	6.82	7.44	8.27	8.89	9.48	9.83	10.6
700	1000	7.44	7.96	8.71	9.04	9.84	10.9	11.9	12.4	13.0	13.9
800	1140	9.66	10.2	11.1	11.7	12.6	13.9	14.8	15.8	16.4	17.4
900	1285	12.1	12.8	13.8	14.4	15.5	17.1	18.4	19.4	20.2	21.5
1000	1430	14.6	15.6	16.8	17.4	18.7	20.8	22.2	23.3	24.4	26.0
1200	1715	21.0	22.0	23.7	24.5	26.4	28.9	30.7	32.4	33.5	35.9
1400	2000	28.3	29.6	31.8	32.6	35.6	38.2	40.7	42.6	44.3	47.5
1600	2285	36.7	38.4	40.6	42.1	44.5	48.7	51.9	54.1	56.1	59.3
1800	2570	46.1	48.0	50.8	52.3	55.4	60.4	64.5	67.0	69.4	73.5
2000	2860	56.5	58.8	61.9	63.8	67.3	73.4	77.7	81.1	83.8	88.8
2200	3140	67.9	70.2	74.4	76.3	80.5	87.5	92.1	96.8	99.6	105
2400	3430	80.8	83.0	88.0	90.2	94.7	103	108	114	117	123
2600	3710	94.2	97.5	103	105	110	119	125	131	135	142
2800	4000	109	112	118	121	127	136	144	149	155	163
3000	4285	125	129	135	138	145	155	164	170	176	184
3200	4570	142	146	153	156	162	174	184	191	197	208
3400	4860	160	164	172	174	182	196	206	213	220	232
3600	5140	178	183	192	196	204	217	228	237	244	256
3800	5425	199	204	212	217	226	240	251	262	269	285
4000	5715	220	225	234	236	249	265	277	289	295	311
4500	6425	276	284	294	300	311	331	345	358	368	385
5000	7145	341	348	360	365	380	404	418	433	447	466
5500	7855	410	419	433	439	457	480	500	518	532	555
6000	8570	488	498	512	519	540	567	592	609	623	654
6500	9280	573	581	601	609	630	662	686	707	723	755
7000	10000	664	673	692	702	725	763	791	810	829	867
7500	10700	760	773	789	806	827	865	897	925	946	984

Table A1.2. (*continued*)

Flow		Kinematic viscosity (cSt)									
US gal per min	Bbl per h (42 gal)	26.4	32.0	43.2	65.0	108.4	162.3	216.5	325	435	650
		125	150	200	300	500	750	1000	1500	2000	3000
						Approx SSU viscosity					
50	71.4	0.21	0.25	0.34	0.50	0.84	1.26	1.68	2.52	3.38	5.05
100	143	0.41	0.50	0.67	1.01	1.68	2.52	3.36	5.04	6.76	10.1
150	214	1.03	0.75	1.01	1.51	2.52	3.78	5.04	7.56	10.1	15.1
200	286	1.67	1.78	1.34	2.02	3.37	5.04	6.72	10.1	13.5	20.2
250	357	2.46	2.60	2.85	2.52	4.21	6.30	8.40	12.6	16.9	25.3
300	429	3.37	3.56	3.89	3.03	5.05	7.56	10.1	15.1	20.3	30.3
350	500	4.39	4.63	5.04	5.69	5.89	8.82	11.8	17.7	23.6	35.3
400	571	5.54	5.83	6.35	7.18	6.73	10.1	13.5	20.2	27.0	40.4
450	643	6.79	7.16	7.75	8.76	7.58	11.3	15.1	22.7	30.4	45.4
500	714	8.17	8.58	9.32	10.4	8.42	12.6	16.8	25.2	33.8	50.5
550	786	9.65	10.1	11.0	12.3	9.26	13.9	18.5	27.8	37.2	55.5
600	857	11.2	11.8	12.8	14.3	16.6	15.1	20.2	30.3	40.5	60.6
700	1000	14.7	15.5	16.7	18.6	21.7	17.6	23.5	35.3	47.3	70.6
800	1140	18.6	19.4	21.0	23.4	27.1	20.1	28.9	40.3	54.0	80.7
900	1285	22.9	24.0	25.9	28.7	33.1	37.4	30.2	45.4	60.8	90.8
1000	1430	27.4	28.8	31.0	34.4	39.7	44.9	33.6	50.4	67.6	100
1200	1715	37.8	39.4	42.8	47.3	54.3	60.9	66.4	60.5	81.0	121
1400	2000	49.7	52.0	56.1	62.0	70.8	79.3	86.7	70.6	94.5	141
1600	2285	63.2	65.7	70.6	78.0	89.3	99.4	108	80.7	108	161
1800	2570	77.9	81.3	86.9	96.2	110	122	132	150	122	182
2000	2860	93.4	98.0	105	116	132	147	159	180	135	202
2200	3140	111	116	124	137	155	173	186	211	149	222
2400	3430	130	135	145	159	181	201	217	244	266	242
2600	3710	149	155	168	183	208	231	250	279	306	262
2800	4000	170	178	191	210	236	262	283	317	347	282
3000	4285	193	201	215	236	267	296	319	357	389	303
3200	4570	217	225	240	265	299	332	357	396	433	323
3400	4860	242	251	268	294	333	369	397	441	480	343
3600	5140	268	284	296	326	369	407	438	488	529	598
3800	5425	296	307	328	359	404	447	482	536	582	656
4000	5715	325	337	358	394	441	488	526	586	635	718
4500	6425	405	417	442	485	543	601	646	718	777	876
5000	7145	488	505	536	582	656	726	776	860	932	
5500	7855	579	605	634	689	773	855	913			
6000	8570	678	706	644	810	910	993				

10-inch (10.02 inches inside dia.) schedule 40 new steel pipe

Flow		Kinematic viscosity (cSt)									
		0.6	1.1	2.1	2.7	4.3	7.4	10.3	13.1	15.7	20.6
US gal per min	Bbl per h (42 gal)	Approx SSU viscosity									
			31.5	33	35	40	50	60	70	80	100
400	571	0.83	0.92	1.03	1.09	1.19	1.35	1.46	1.55	1.63	1.75
500	714	1.27	1.38	1.53	1.60	1.78	2.00	2.16	2.30	2.40	2.59
600	857	1.78	1.91	2.14	2.24	2.47	2.77	2.96	3.15	3.31	3.54
700	1000	2.39	2.55	2.84	2.97	3.26	3.62	3.93	4.14	4.32	4.67
800	1140	3.06	3.29	3.63	3.79	4.12	4.63	4.99	5.25	5.46	5.86
900	1285	3.84	4.12	4.49	4.72	5.09	5.72	6.14	6.46	6.74	7.19
1000	1430	4.68	4.99	5.42	5.70	6.13	6.90	7.36	7.83	8.10	8.63
1100	1570	5.63	5.97	6.49	6.82	7.34	8.20	8.76	9.25	9.62	10.3
1200	1715	6.61	7.05	7.63	7.85	8.65	9.58	10.3	10.8	11.3	11.9
1300	1855	7.71	8.18	8.85	9.16	9.95	11.0	11.8	12.4	13.0	13.7
1400	2000	8.88	9.42	10.2	10.6	11.4	12.6	13.5	14.2	14.7	15.7
1500	2140	10.1	10.8	11.7	12.0	12.9	14.3	15.3	16.1	16.6	17.8
1600	2285	11.5	12.2	13.2	13.6	14.6	16.0	17.2	18.1	18.7	20.0
1800	2570	14.3	15.1	16.2	16.7	17.9	19.7	20.9	22.1	22.9	24.3
2000	2860	17.6	18.6	19.8	20.6	21.8	24.0	25.5	27.0	28.0	29.5
2200	3140	21.3	22.2	23.7	24.6	26.1	28.6	30.3	32.1	31.8	35.0
2400	3430	25.2	26.3	28.0	28.9	30.7	33.4	35.5	37.3	38.9	41.0
2600	3710	29.6	30.6	32.5	33.5	35.6	38.7	41.0	42.9	44.8	47.3
2800	4000	34.1	35.3	37.4	38.4	40.8	44.5	47.1	49.0	51.0	54.1
3000	4285	39.1	40.2	42.7	43.5	46.6	50.7	53.2	55.7	57.7	61.3
3500	5000	52.5	54.4	57.4	58.9	62.3	66.4	70.6	73.6	76.2	80.8
4000	5715	68.0	70.5	73.9	75.9	79.9	85.8	90.2	94.2	97.1	102
4500	6430	86.1	88.6	92.3	94.8	99.2	107	112	117	120	127
5000	7145	106	109	113	116	122	130	136	142	146	153
5500	7855	128	131	136	139	145	156	162	169	173	182
6000	8570	152	154	161	164	172	183	191	197	204	213
6500	9280	177	180	187	191	201	212	221	228	236	246
7000	10000	205	208	217	220	231	243	255	263	369	282
7500	10700	236	239	246	251	262	277	291	298	303	321
8000	11400	266	272	282	286	296	314	329	337	345	360
8500	12100	301	307	318	321	334	352	367	378	387	403
9000	12900	337	341	354	359	372	392	407	422	429	447
10000	14300	416	422	434	441	453	478	492	511	524	542
11000	15700	503	511	522	533	544	574	593	611	626	649
12000	17150	599	603	617	630	643	679	701	719	737	763

Table A1.2. (*continued*)

Flow		Kinematic viscosity (cSt)									
US gal per min	Bbl per h (42 gal)	26.4	32.0	43.2	65.0	108.4	162.3	216.5	325	435	650
		Approx SSU viscosity									
		125	150	200	300	500	750	1000	1500	2000	3000
150	214	0.25	0.30	0.40	0.61	1.02	1.52	2.03	3.04	4.08	6.09
200	286	0.58	0.40	0.54	0.81	1.35	2.03	2.71	4.06	5.43	8.12
300	429	1.15	1.22	1.33	1.22	2.03	3.04	4.06	6.09	8.15	12.2
400	571	1.83	1.95	2.17	1.62	2.71	4.06	5.41	8.12	10.9	16.2
500	714	2.75	2.91	3.18	3.60	3.39	5.07	6.77	10.1	13.6	20.3
600	857	3.78	3.97	4.34	4.89	4.08	6.08	8.12	12.2	16.3	24.4
700	1000	4.94	5.19	5.66	6.37	4.74	7.10	9.47	14.2	19.0	28.4
800	1140	6.21	6.55	7.10	7.97	9.31	8.12	10.8	16.2	21.7	32.5
900	1285	7.66	8.04	8.71	9.76	11.4	9.13	12.2	18.3	24.5	36.5
1000	1430	9.21	9.61	10.5	11.7	13.6	10.1	13.5	20.3	27.2	40.6
1100	1570	10.9	11.4	12.3	13.8	16.0	18.0	14.9	22.3	29.9	44.6
1200	1715	12.6	13.4	14.4	16.0	18.6	21.0	16.2	24.4	32.6	48.7
1300	1855	14.5	15.4	16.4	18.4	21.2	24.0	17.6	26.4	35.3	52.8
1400	2000	16.6	17.5	18.7	20.9	24.1	27.2	18.9	28.4	38.0	56.8
1500	2140	18.7	19.6	21.2	23.6	27.2	30.6	33.3	30.4	40.8	80.9
1600	2285	21.0	21.9	23.7	26.3	30.4	34.1	37.2	32.4	43.5	64.9
1800	2570	26.0	27.3	29.2	32.2	37.1	41.7	45.5	36.5	48.9	73.1
2000	2860	31.3	32.7	35.0	38.6	44.4	49.8	54.4	40.6	54.3	81.2
2200	3140	37.0	38.6	41.3	45.8	52.3	58.8	64.0	71.9	59.8	89.3
2400	3430	43.1	45.1	48.3	53.4	60.9	68.3	74.2	83.8	55.2	97.4
2600	3710	49.8	52.1	55.4	61.4	70.0	78.3	85.0	96.2	70.7	105
2800	4000	56.8	59.2	63.5	70.1	79.7	88.9	96.1	109	76.1	114
3000	4285	64.3	66.8	71.8	76.6	89.8	99.8	108	122	133	122
3500	5000	84.9	88.3	94.4	103	117	131	142	159	180	142
4000	5715	108	112	119	131	149	165	176	199	217	162
4500	6430	133	139	147	162	182	202	218	244	266	300
5000	7145	160	168	179	195	218	241	262	293	318	360
5500	7855	191	199	212	231	258	286	309	345	373	423
6000	8570	223	232	247	268	302	334	359	399	435	489
6500	9280	258	267	286	310	348	384	411	460	499	560
7000	10000	296	305	326	355	396	436	468	522	566	637
7500	10700	335	347	369	402	447	492	529	589	638	766
8000	11400	377	389	414	452	505	550	594	659	710	797
9000	12900	469	482	512	557	624	679	729	809	869	976
10000	14300	567	582	619	666	743	817	872	964		

12-inch (11.938 inches inside dia.) schedule 40 new steel pipe

Flow		Kinematic viscosity (cSt)									
		0.6	1.13	2.1	2.7	4.3	7.4	10.3	13.1	15.7	20.6
US gal per min	Bbl per h (42 gal)	Approx SSU viscosity									
			31.5	33	35	40	50	60	70	80	100
300	429	0.21	0.24	0.27	0.28	0.31	0.36	0.38	0.41	0.43	0.47
400	571	0.36	0.40	0.45	0.47	0.51	0.59	0.64	0.67	0.70	0.77
500	714	0.54	0.60	0.67	0.70	0.75	0.87	0.95	0.99	1.03	1.12
600	857	0.76	0.83	0.93	0.98	1.04	1.19	1.30	1.39	1.41	1.53
700	1000	0.98	1.11	1.23	1.29	1.37	1.56	1.70	1.82	1.84	2.00
800	1140	1.26	1.42	1.57	1.64	1.74	1.98	2.15	2.30	2.36	2.51
900	1285	1.57	1.76	1.94	1.96	2.15	2.44	2.65	2.82	2.94	3.08
1000	1430	1.92	2.07	2.36	2.38	2.61	2.94	3.19	3.40	3.57	3.70
1200	1715	2.73	2.91	3.18	3.32	3.62	4.07	4.41	4.68	4.91	5.08
1400	2000	3.67	3.90	4.24	4.41	4.80	5.37	5.79	6.14	6.43	6.65
1600	2285	4.75	5.02	5.43	5.64	6.12	6.83	7.35	7.78	8.14	8.51
1800	2570	5.96	6.29	6.77	7.02	7.59	8.44	9.07	9.59	10.0	10.6
2000	2860	7.32	7.69	8.25	8.54	9.21	10.2	11.0	11.6	12.1	12.9
2500	3570	11.3	11.8	12.6	13.0	13.9	15.3	16.4	17.3	18.0	19.2
3000	4285	16.1	16.8	17.7	18.3	19.5	21.4	22.8	23.9	24.9	26.5
3500	5000	21.8	22.6	23.8	24.4	26.0	28.3	30.1	31.6	32.9	34.9
4000	5715	28.3	29.2	30.7	31.5	33.3	36.2	38.4	40.3	41.8	44.3
4500	6430	35.7	36.8	38.5	39.4	41.6	45.0	47.7	49.9	51.7	54.8
5000	7145	44.0	45.2	47.1	48.2	50.7	54.7	57.8	60.4	62.6	66.2
5500	7855	53.1	54.4	56.6	57.8	60.7	65.3	68.9	71.9	74.4	78.7
6000	8570	63.0	64.5	66.9	68.3	71.6	76.8	80.9	84.3	87.2	92.1
6500	9280	73.8	75.4	78.1	79.6	83.3	89.2	93.8	97.7	101	106
7000	10000	85.4	87.2	90.1	91.8	95.9	102	108	112	116	122
7500	10700	97.9	99.8	103	105	109	117	122	127	131	138
8000	11400	111	113	117	119	124	132	138	143	148	155
9000	12850	141	143	147	149	155	164	172	178	183	192
10000	14300	173	176	180	183	190	200	209	217	223	233
11000	15700	209	212	217	220	226	240	250	269	266	278
12000	17150	249	252	258	261	269	283	294	304	312	326
13000	18550	291	295	301	305	314	330	342	353	363	373
14000	20000	336	342	348	353	363	380	394	406	418	434
15000	21400	387	392	399	403	414	433	449	462	473	493
16000	22850	440	445	453	457	469	490	507	522	534	556
18000	25700	557	561	571	577	590	614	634	651	666	692
20000	28600	687	692	703	709	725	752	775	795	812	842

Table A1.2. (*continued*)

Flow		Kinematic viscosity (cSt)									
		26.4	32.0	43.2	65.0	108.4	162.3	216.5	325	435	650
US gal per min	Bbl per h (42 gal)	Approx SSU viscosity									
		125	150	200	300	500	750	1000	1500	2000	3000
100	143	0.08	0.10	0.13	0.20	0.34	0.51	0.68	1.00	1.35	2.00
200	286	0.16	0.19	0.27	0.40	0.67	1.00	1.37	2.05	2.74	4.01
300	429	0.49	0.53	0.41	0.62	1.01	1.51	2.00	3.08	4.11	6.16
400	571	0.81	0.86	0.94	0.82	1.34	2.02	2.68	3.98	5.46	8.21
500	714	1.22	1.25	1.37	1.02	1.71	2.50	3.37	5.01	6.84	10.3
600	857	1.66	1.71	1.87	2.12	1.97	3.02	4.05	6.03	7.99	12.3
700	1000	2.15	2.30	2.43	2.75	2.37	3.66	4.68	7.05	9.38	13.9
800	1140	2.70	2.88	3.05	3.45	2.63	4.04	5.36	8.09	10.7	15.9
900	1285	3.31	3.52	3.74	4.21	4.93	4.44	6.05	9.30	12.1	18.0
1 000	1430	3.97	4.22	4.48	5.04	5.89	5.13	6.84	10.0	13.5	20.0
1 200	1715	5.43	5.77	6.33	6.88	8.01	5.91	7.89	12.1	16.2	24.1
1 400	2000	7.10	7.53	8.24	8.96	10.4	11.8	9.47	14.6	18.9	28.2
1 600	2285	8.96	9.48	10.4	11.3	13.1	14.8	10.5	16.2	21.7	32.4
1 800	2570	11.0	11.6	12.7	13.8	16.0	18.0	19.7	17.7	24.2	36.5
2 000	2860	13.2	14.0	15.2	16.6	19.1	21.5	23.6	20.5	27.4	40.1
2 500	3570	19.6	20.6	22.4	25.3	28.0	31.5	34.3	25.7	34.2	51.3
3 000	4285	27.2	28.4	30.8	34.6	38.4	43.0	46.8	53.1	41.1	61.6
3 500	5000	36.2	37.3	40.3	45.2	50.2	56.0	60.9	68.8	47.9	71.9
4 000	5715	43.4	47.3	51.0	57.0	63.2	70.5	76.5	86.3	94.5	82.1
4 500	6430	57.6	58.3	62.8	69.9	77.6	86.4	93.6	105	115	92.4
5 000	7145	69.8	70.3	75.6	84.1	93.3	104	112	126	138	103
5 500	7855	82.8	83.6	89.5	99.4	110	122	132	148	162	183
6 000	8570	96.8	98.2	104	116	128	142	154	172	188	212
6 500	9280	112	114	120	133	148	164	176	197	215	243
7 000	10000	128	131	137	152	174	186	201	224	244	275
7 500	10700	145	148	155	172	196	210	226	253	274	310
8 000	11400	163	167	174	192	220	235	253	282	306	345
9 000	12850	202	208	215	237	270	289	311	346	375	422
10 000	14300	204	253	260	286	325	347	373	415	450	505
11 000	15700	290	301	309	338	384	411	441	490	530	594
12 000	17150	341	354	361	395	448	479	514	569	616	689
13 000	18550	394	409	417	456	516	551	591	655	707	790
14 000	20000	452	469	477	521	588	628	673	745	804	898
15 000	21400	513	532	541	589	664	710	760	840	906	
16 000	22850	578	598	608	662	745	796	851	940		

14-inch (13.124 inches inside dia.) schedule 40 new steel pipe

Flow		Kinematic viscosity (cSt)									
		0.6	1.13	2.1	2.7	4.3	7.4	10.3	13.1	15.7	20.6
US gal per min	Bbl per h (42 gal)	Approx SSU viscosity									
			31.5	33	35	40	50	60	70	80	100
	571	0.22	0.25	0.28	0.30	0.32	0.37	0.41	0.43	0.45	0.49
	714	0.34	0.38	0.42	0.45	0.48	0.55	0.61	0.62	0.66	0.72
	857	0.48	0.52	0.59	0.62	0.66	0.76	0.83	0.87	0.90	0.98
	1000	0.63	0.70	0.78	0.81	0.87	0.99	1.09	1.16	1.18	1.28
	1140	0.81	0.89	0.99	1.04	1.10	1.26	1.37	1.47	1.49	1.61
	1285	0.98	1.11	1.23	1.28	1.36	1.55	1.69	1.80	1.85	1.97
	1430	1.20	1.34	1.49	1.55	1.65	1.87	2.03	2.16	2.25	2.36
	1715	1.69	1.82	2.07	2.09	2.29	2.58	2.80	2.98	3.13	3.24
	2000	2.27	2.43	2.66	2.77	3.03	3.40	3.68	3.91	4.09	4.23
	2285	2.94	3.13	3.40	3.55	3.86	4.32	4.66	4.94	5.18	5.37
	2570	3.69	3.91	4.24	4.41	4.78	5.34	5.75	6.09	6.37	6.67
	2860	4.52	4.78	5.16	5.36	5.80	6.48	6.94	7.35	7.68	8.11
	3570	6.97	7.32	7.84	8.11	8.73	9.67	10.4	10.9	11.4	12.2
	4285	9.94	10.4	11.1	11.4	12.2	13.5	14.4	15.2	15.8	16.8
	5000	13.4	14.0	14.8	15.3	16.3	17.8	19.0	20.0	20.8	22.2
	5715	17.5	18.1	19.1	19.6	20.9	22.8	24.2	25.5	26.5	28.1
	6430	22.0	22.7	23.9	24.5	26.0	28.3	30.0	31.5	32.7	34.7
	7145	27.0	27.9	29.2	30.0	31.7	34.4	36.4	38.1	39.6	41.9
	7855	32.6	33.6	35.1	35.9	37.9	41.0	43.4	45.4	47.0	49.8
	8570	38.7	40.0	41.5	42.4	44.6	46.2	50.9	53.2	55.0	58.2
	9280	45.3	46.5	48.4	49.4	51.9	55.9	58.9	61.5	63.7	67.3
	10000	52.5	53.7	55.8	57.0	59.7	64.1	67.6	70.5	72.9	76.9
	10700	60.1	61.5	63.7	65.0	68.1	72.9	76.8	80.0	82.7	87.1
	11400	68.3	69.8	72.2	73.6	76.9	82.3	86.5	90.0	93.0	98.0
	12850	86.2	87.9	90.7	92.4	96.3	103	108	112	115	121
	14300	106	108	111	113	118	125	131	136	140	147
	15700	128	130	134	138	141	150	156	162	167	175
	17150	152	155	159	161	167	176	184	191	196	205
	18550	179	181	186	188	195	205	214	221	227	238
	20000	207	210	215	217	224	236	246	254	261	273
	21400	237	240	246	249	256	269	280	289	297	310
	22850	270	273	279	282	290	304	316	326	334	349
	25700	341	345	351	355	365	381	395	407	417	434
	28600	420	425	432	436	447	466	482	496	508	528
	35700	655	661	671	676	691	716	737	756	772	800

Table A1.2. (*continued*)

Flow		Kinematic viscosity (cSt)									
		26.4	32.0	43.2	65.0	108.4	162.3	216.5	325	435	650
US gal per min	Bbl per h (42 gal)	Approx SSU viscosity									
		125	150	200	300	500	750	1000	1500	2000	3000
200	286	0.11	0.14	0.19	0.28	0.47	0.68	0.93	1.36	1.86	2.77
300	429	0.32	0.34	0.27	0.42	0.69	1.02	1.40	2.09	2.71	4.19
400	571	0.52	0.55	0.60	0.55	0.92	1.39	1.82	2.73	3.72	5.46
500	714	0.78	0.80	0.88	0.67	1.16	1.73	2.28	3.46	4.57	6.73
600	857	1.06	1.09	1.20	1.36	1.42	2.07	2.75	4.19	5.58	8.37
700	1000	1.38	1.47	1.56	1.76	1.57	2.51	3.21	4.82	6.43	9.64
800	1140	1.73	1.85	1.95	2.21	1.82	2.73	3.68	5.55	7.44	10.9
900	1285	2.12	2.26	2.39	2.70	2.07	3.19	4.14	6.18	8.29	12.6
1 000	1430	2.54	2.70	2.86	3.22	3.78	3.41	4.65	8.91	9.30	13.8
1 200	1715	3.47	3.69	3.92	4.40	5.14	4.09	5.67	8.28	11.2	16.7
1 400	2000	4.53	4.81	5.27	5.73	6.67	7.57	6.27	10.0	12.9	19.3
1 600	2285	5.71	6.05	6.63	7.21	8.36	9.48	7.28	10.9	14.7	22.2
1 800	2570	7.01	7.42	8.11	8.83	10.2	11.6	8.29	12.7	16.6	24.7
2 000	2860	8.43	8.91	9.72	10.6	12.2	13.8	15.1	13.8	18.8	27.7
2 500	3570	12.5	13.2	14.3	15.6	17.9	20.2	22.0	16.8	22.8	34.4
3 000	4285	17.2	18.1	19.6	22.1	24.5	27.5	30.0	34.1	27.1	41.9
3 500	5000	22.8	23.7	25.7	28.8	32.0	35.8	39.0	44.1	33.0	48.2
4 000	5715	29.2	30.0	32.5	36.3	40.4	45.1	48.9	55.3	60.6	54.8
4 500	6430	36.4	37.0	39.9	44.6	49.5	55.2	59.9	67.5	73.9	60.9
5 000	7145	44.2	44.6	48.1	53.6	59.5	66.2	71.7	80.7	88.3	67.3
5 500	7855	52.5	52.8	58.9	63.3	70.2	78.1	84.5	95.0	104	77.4
6 000	8570	61.3	61.9	66.3	73.3	81.8	90.8	98.1	110	120	136
6 500	9280	70.8	71.8	76.4	84.8	94.1	104	113	126	138	158
7 000	10000	80.9	82.3	87.2	96.6	107	119	128	143	156	177
8 000	11400	103	105	111	122	140	150	161	180	196	221
9 000	12850	127	131	136	150	172	184	198	221	240	270
10 000	14300	154	159	165	181	207	221	238	265	287	323
11 000	15700	183	190	195	215	264	261	281	312	339	380
12 000	17150	215	223	228	251	285	305	327	363	393	441
13 000	18550	248	258	264	289	326	351	276	418	451	506
14 000	20000	284	295	302	330	374	400	428	475	513	574
15 000	21400	323	335	342	373	422	451	484	536	578	646
16 000	22850	363	377	384	419	473	506	542	599	647	722
18 000	25700	451	468	476	518	583	623	667	736	793	884
20 000	28600	548	567	576	626	703	777	803	885	953	

16-inch (15.000 inches inside dia.) schedule 40 new steel pipe

Flow		Kinematic viscosity (cSt)									
		0.6	1.13	2.1	2.7	4.3	7.4	10.3	13.1	15.7	20.6
US gal per min	Bbl per h (42 gal)	Approx SSU viscosity									
			31.5	33	35	40	50	60	70	80	100
600	857	0.25	0.27	0.31	0.32	0.35	0.40	0.44	0.46	0.48	0.52
700	1000	0.33	0.36	0.41	0.43	0.46	0.53	0.58	0.61	0.63	0.68
800	1140	0.42	0.46	0.52	0.54	0.58	0.66	0.73	0.78	0.79	0.86
900	1285	0.52	0.58	0.64	0.67	0.72	0.82	0.89	0.95	0.97	1.05
1 000	1430	0.54	0.70	0.78	0.81	0.87	0.99	1.07	1.15	1.17	1.25
1 200	1715	0.87	0.98	1.08	1.13	1.20	1.36	1.48	1.58	1.65	1.72
1 400	2000	1.16	1.30	1.43	1.50	1.59	1.79	1.94	2.06	2.17	2.25
1 600	2285	1.50	1.61	1.79	1.84	2.02	2.27	2.46	2.61	2.74	2.83
1 800	2570	1.88	2.01	2.20	2.29	2.50	2.80	3.03	3.21	3.37	3.48
2 000	2860	2.30	2.45	2.67	2.78	3.03	3.39	3.65	3.87	4.05	4.23
2 200	3140	2.77	2.94	3.19	3.32	3.60	4.02	4.33	4.59	4.80	5.05
2 400	3430	3.27	3.46	3.75	3.89	4.22	4.70	5.06	5.35	5.59	5.93
2 600	3710	3.82	4.03	4.35	4.52	4.89	5.43	5.84	6.17	6.45	6.88
2 800	4000	4.41	4.65	5.00	5.18	5.60	6.21	6.67	7.04	7.35	7.87
3 000	4285	5.04	5.30	5.69	5.90	6.35	7.04	7.55	7.97	8.31	8.88
3 500	5000	6.80	7.12	7.61	7.86	8.44	9.31	9.96	10.5	10.9	11.7
4 000	5715	8.82	9.20	9.79	10.1	10.6	11.9	12.7	13.3	13.9	14.8
4 500	6430	11.1	11.5	12.2	12.6	13.5	14.7	15.7	16.5	17.2	18.3
5 000	7145	13.6	14.2	14.9	15.4	16.4	17.9	19.0	20.0	20.7	22.0
6 000	8570	19.5	20.1	21.2	21.7	23.0	25.0	26.5	27.8	26.8	30.5
7 000	10000	26.4	27.2	28.4	29.1	30.7	33.2	35.2	36.8	38.1	40.3
8 000	11400	34.4	35.3	36.7	37.6	39.5	42.6	44.9	46.9	48.5	51.3
9 000	12850	43.3	44.4	46.1	47.1	49.4	53.0	55.8	58.2	60.2	63.5
10 000	14300	53.4	54.5	56.5	57.6	60.3	64.5	67.8	70.6	72.9	76.8
12 000	17150	76.5	78.0	80.5	81.9	85.3	90.8	95.1	98.8	102	107
14 000	20000	104	106	109	110	114	121	127	131	135	142
16 000	22850	135	137	141	143	146	156	163	168	173	181
18 000	25700	171	173	177	180	185	195	203	210	216	225
20 000	28600	211	213	218	221	227	239	248	256	262	274
22 000	31400	255	258	263	266	273	286	297	306	313	326
24 000	34300	303	306	312	315	323	338	350	360	369	384
26 000	37100	355	359	365	369	378	394	407	418	428	445
28 000	40000	411	415	422	426	436	454	468	481	492	511
30 000	42850	472	476	484	488	499	518	534	548	560	581
32 000	45700	537	541	549	554	566	587	604	620	633	656

Table A1.2. (*continued*)

Flow		Kinematic viscosity (cSt)									
		26.4	32.0	43.2	65.0	108.4	162.3	216.5	325	435	650
US gal per min	Bbl per h (42 gal)	Approx SSU viscosity									
		125	150	200	300	500	750	1000	1500	2000	3000
400	571	0.28	0.29	0.21	0.32	0.54	0.81	1.06	1.61	2.21	3.23
500	714	0.40	0.43	0.47	0.41	0.68	1.01	1.34	2.05	2.73	4.10
600	857	0.57	0.58	0.64	0.50	0.79	1.21	1.63	2.42	3.26	4.97
700	1000	0.74	0.76	0.83	0.94	0.92	1.41	1.89	2.86	3.78	5.59
800	1140	0.92	0.99	1.04	1.18	1.05	1.68	2.15	3.23	4.30	6.46
900	1285	1.13	1.20	1.27	1.44	1.18	1.77	2.41	3.68	4.83	7.33
1000	1430	1.35	1.44	1.52	1.72	1.38	2.02	2.70	4.04	5.46	8.20
1200	1715	1.85	1.96	2.08	2.35	2.75	2.36	3.15	4.84	6.51	9.68
1400	2000	2.41	2.56	2.81	3.05	3.56	2.86	3.67	5.65	7.58	11.4
1600	2285	3.03	3.22	3.53	3.84	4.46	5.07	4.20	6.71	8.61	12.9
1800	2570	3.72	3.94	4.32	4.70	5.45	6.18	4.72	7.08	9.77	14.7
2000	2860	4.47	4.73	5.17	5.63	6.52	7.38	5.52	8.07	10.8	16.1
2200	3140	5.27	5.58	6.09	6.63	7.67	8.67	9.50	9.06	11.5	17.9
2400	3430	6.14	6.49	7.08	7.71	8.90	10.0	11.0	9.44	12.6	19.4
2600	3710	7.06	7.46	8.13	8.86	10.2	11.5	12.6	10.4	13.6	21.1
2800	4000	8.04	8.49	9.24	10.1	11.6	13.0	14.3	11.4	14.7	22.6
3000	4285	9.08	9.58	10.4	11.4	13.1	14.7	16.0	12.4	15.7	24.3
3500	5000	11.9	12.6	13.6	15.3	17.0	19.1	20.8	23.6	18.9	29.2
4000	5715	15.3	15.9	17.2	19.3	21.4	24.0	26.1	29.5	22.1	32.2
4500	6430	19.0	19.5	21.1	23.7	26.3	29.4	31.9	36.0	39.5	35.4
5000	7145	23.1	23.5	25.4	28.4	31.6	35.2	38.2	43.1	47.2	41.0
6000	8570	32.2	32.5	35.0	39.0	43.3	48.2	52.2	58.7	64.2	49.7
7000	10000	42.5	43.0	46.0	51.1	56.7	63.0	68.1	76.4	83.3	94.5
8000	11400	54.0	55.0	58.3	64.6	71.7	79.4	85.7	96.0	105	118
9000	12850	66.7	68.4	71.8	79.5	91.2	97.5	105	118	128	144
10000	14300	80.7	83.2	86.6	95.7	110	117	125	141	153	172
12000	17150	112	117	120	132	151	161	173	193	209	235
14000	20000	148	155	158	174	197	211	227	252	273	306
16000	22850	189	197	201	220	250	267	287	318	343	384
18000	25700	235	244	249	272	308	329	352	390	421	470
20000	28600	285	296	301	329	371	396	424	469	506	564
22000	31400	340	352	358	390	439	486	502	554	597	664
24000	34300	399	413	420	456	512	566	585	645	694	772
26000	37100	462	478	487	527	591	652	674	742	798	886
28000	40000	530	548	559	602	674	743	768	846	909	

Table A1.3. Values for coefficient C

Values of coefficient $C = 520\sqrt{k\left(\dfrac{2}{k+1}\right)^{\frac{k+1}{k-1}}}$; $k = \dfrac{C_p}{C_v}$

k	C	k	C	k	C	k	C	k	C	k	C
0.41	219.28	0.71	276.09	1.01	316.56*	1.31	347.91	1.61	373.32	1.91	394.56
0.42	221.59	0.72	277.64	1.02	317.74	1.32	348.84	1.62	374.09	1.92	395.21
0.43	223.86	0.73	279.18	1.03	318.90	1.33	349.77	1.63	374.85	1.93	395.86
0.44	226.10	0.74	280.70	1.04	320.05	1.34	350.68	1.64	375.61	1.94	396.50
0.45	228.30	0.75	282.20	1.05	321.19	1.35	351.60	1.65	376.37	1.95	397.14
0.46	230.47	0.76	283.69	1.06	322.32	1.36	352.50	1.66	377.12	1.96	397.78
0.47	232.61	0.77	285.16	1.07	323.44	1.37	353.40	1.67	377.86	1.97	398.41
0.48	234.71	0.78	286.62	1.08	324.55	1.38	354.29	1.68	378.61	1.98	399.05
0.49	236.78	0.79	288.07	1.09	325.65	1.39	355.18	1.69	379.34	1.99	399.67
0.50	238.83	0.80	289.49	1.10	326.75	1.40	356.06	1.70-	380.08	2.00	400.30
0.51	240.84	0.81	290.91	1.11	327.83	1.41	356.94	1.71	380.80	2.01	400.92
0.52	242.82	0.82	292.31	1.12	328.91	1.42	357.81	1.72	381.53	2.02	401.53
0.53	244.78	0.83	293.70	1.13	329.98	1.43	358.67	1.73	382.25	2.03	402.15
0.54	246.72	0.84	295.07	1.14	331.04	1.44	359.53	1.74	382.97	2.04	402.76
0.55	248.62	0.85	296.43	1.15	332.09	1.45	360.38	1.75	383.68	2.05	403.37
0.56	250.50	0.86	297.78	1.16	333.14	1.46	361.23	1.76	384.39	2.06	403.97
0.57	252.36	0.87	299.11	1.17	334.17	1.47	362.07	1.77	385.09	2.07	404.58
0.58	254.19	0.88	300.43	1.18	335.20	1.48	362.91	1.78	385.79	2.08	405.18
0.59	256.00	0.89	301.74	1.19	336.22	1.49	363.74	1.79	386.49	2.09	405.77
0.60	257.79	0.90	303.04	1.20	337.24	1.50	364.56	1.80	387.18	2.10	406.37
0.61	259.55	0.91	304.33	1.21	338.24	1.51	365.39	1.81	387.87	2.11	406.96
0.62	261.29	0.92	305.60	1.22	339.24	1.52	366.20	1.82	388.56	2.12	407.55
0.63	263.01	0.93	306.86	1.23	340.23	1.53	367.01	1.83	389.24	2.13	408.13
0.64	264.72	0.94	308.11	1.24	341.22	1.54	367.82	1.84	389.92	2.14	408.71
0.65	266.40	0.95	309.35	1.25	342.19	1.55	368.62	1.85	390.59	2.15	409.29
0.66	268.06	0.96	310.58	1.26	343.16	1.56	369.41	1.86	391.26	2.16	409.87
0.67	269.70	0.97	311.80	1.27	344.13	1.57	370.21	1.87	391.93	2.17	410.44
0.68	271.33	0.98	313.01	1.28	345.08	1.58	370.99	1.88	392.59	2.18	411.01
0.69	272.93	0.99	314.19*	1.29	346.03	1.59	371.77	1.89	393.25	2.19	411.58
0.70	274.52	1.00	315.38*	1.30	346.98	1.60	372.55	1.90	393.91	2.20	412.15

Values of C for gases

	Mol wt	$k = C_p/C_v$	C	$C/356$		Mol wt	$k = C_p/C_v$	C	$C/356$
Acetylene	26	1.28	345	0.969	Hydrochloric acid	36.5	1.40	356	1.000
Air	29	1.40	356	1.000	Hydrogen	2	1.40	356	1.000
Ammonia	17	1.33	351	0.986	Hydrogen sulphide	34	1.32	348	0.978
Argon	40	1.66	377	1.059	Iso-butane	58	1.11	328	0.921
Benzene	78	1.10	327	0.919	Methane	16	1.30	346	0.972
Carbon disulphide	76	1.21	338	0.949	Methyl alcohol	32	1.20	337	0.947
Carbon dioxide	44	1.28	345	0.969	Methyl chloride	50.5	1.20	337	0.947
Carbon monoxide	28	1.40	356	1.000	N-butane	58	1.11	328	0.921
Chlorine	71	1.36	352	0.989	Natural gas	19	1.27	345	0.969
Cyclohexane	84	1.08	324	0.910	Nitrogen	28	1.40	356	1.000
Ethane	30	1.22	339	0.952	Oxygen	32	1.40	356	1.000
Ethylene	28	1.20	337	0.947	Pentane	72	1.09	325	0.913
Helium	4	1.66	377	1.059	Propane	44	1.14	331	0.930
Hexane	86	1.08	324	0.910	Sulphur dioxide	64	1.26	342	0.961

* Interpolated values, since C becomes indeterminate as k approaches 1.00. (Reproduced by permission of Gas Processors Suppliers Association)

Table A1.4. Relief valve selection table

| Orifice & area | Standard connections ASA flanges | | Max. back press. (psig) | | 450 °F max. Carbon steel spring & body | | | | 800 °F max. Alloy steel spring, carbon steel body | | | |
| | | | | | | | Pressure limits, psig Inlet | | | | Pressure limits (psig) Inlet | |
	Inlet	Outlet	Conventional	Bellows	Crosby style	Farris style	−20 to 100 °F	450 F	Crosby style	Farris style	450 F	800 F
D E 0.110 0.196 in²	1–150	2–150	230	–	JO-25	A-10	275	165	JO-26	A-20	165	92
	1–300	2–150	230	–	JO-25-3	A-11	275	275	JO-26-3	A-21	275	275
	1–300	2–150	230	–	JO-35	A-12	720	650	JO-36	A-22	650	365
	1–600	2–150	230	–	JO-45	A-13	1440	1305	JO-46	A-23	1305	730
	1½–900	2–300	600	–	JO-55-9	A-14	2160	1955	JO-56-9	A-24	1955	1100
	1½–1500	2–300	600	–	JO-55	A-15	3600	3255	JO-56	A-25	3255	1830
	1½–2500	2½–300	600	–	JO-65	A-16	6000	5430	JO-66	A-26	5430	3050
F 0.307 in²	1½–150	2–150	230	230								
	1½–300	2–150	230	230								
	1½–300	2–150	230	230			Same as above for D and E orifices					
	1½–600	2–150	230	230								
	1½–900	2½–300	600	500								
	1½–1500	2½–300	600	500								
	1½–2500	2½–300	600	500	JO-65	A-16	5000	5000	JO-66	A-26	5000	3050
G 0.503 in²	1½–150	2½–150	230	230								
	1⅝1–300	2½–150	230	230			Same as above for D and E orifices					
	1½–300	2½–150	230	230								
	1½1–600	2½–150	230	230								

→	1½-900	2½-300	600	470	JO-55-9	A-14	2160	1955	JO-56-9	A-24	1955	1100
	2-1500	3-300	600	470	JO-65	A-15	3600	3255	JO-66-15	A-25	3255	1830
	2-2500	3-300	600	470	JO-75	A-16	3600	3600	JO-66	A-26	3600	3050
H 0.785 in²	1½-150	3-150	230	230	JO-25	A-10	275	165	JO-26	A-20	165	92
	1½-300	3-150	230	230	JO-25-3	A-11	275	275	JO-26-3	A-21	275	275
	2-300	3-150	230	230	JO-35	A-12	720	650	JO-36	A-22	650	365
	2-600	3-150	230	230	JO-55	A-13	1440	1305	JO-56	A-23	1305	730
	2-900	3-150	230	230	JO-55-9	A-14	2160	1955	JO-56-9	A-24	1955	1100
	2-1500	3-300	600	415	JO-75	A-15	2750	2750	JO-66	A-25	2750	1830
J 1.287 in²	2-150	3-150	230	230	JO-25	A-10	275	165	JO-26	A-20	165	92
	2-300	3-150	230	230	JO-25-3	A-11	275	275	JO-26-3	A-21	275	275
	2½-300	4-150	230	230	JO-35	A-12	720	650	JO-36	A-22	650	365
	2½-600	4-150	230	230	JO-55	A-13	1440	1305	JO-46	A-23	1305	730
	3-900	4-150	230	230	JO-65	A-14	2160	1955	JO-66-9	A-24	1955	1100
	3-1500	4-300	600	230	JO-75	A-15	2700	2700	JO-66	A-25	2700	1830
K 1.838 in²	3-150	4-150	230	150	JO-25	A-10	275	165	JO-26	A-20	165	92
	3-300	4-150	230	150	JO-25-3	A-11	275	275	JO-26-3	A-21	275	275
	3-300	4-150	230	150	JO-35	A-12	720	650	JO-36	A-22	650	365
	3-600	4-150	230	200	JO-55	A-13	1440	1305	JO-46	A-23	1305	730
	3-900	6-150	230	200	JO-65	A-14	2160	1955	JO-66-9	A-24	1955	1100
	3-1500	6-300	600	200	JO-75	A-15	2160	2160	JO-66	A-25	2160	1830
L 2.853 in²	3-150	4-150	230	100	JO-25	A-10	275	165	JO-26	A-20	165	92
	3-300	4-150	230	100	JO-25-3	A-11	275	275	JO-26-3	A-21	275	275
	4-300	6-150	230	170	JO-35	A-12	720	650	JO-36	A-22	650	365
	4-600	6-150	230	170	JO-45	A-13	1000	1000	JO-46	A-23	1000	730
	4-900	6-150	230	170	JO-55	A-14	1500	1500	JO-66	A-24	1500	1100
→	4-1500	6-150	230	170	-	-	-	-	JO-76	A-25	1500	1500

continued overleaf

Table A1.4. (*continued*)

Orifice & area	Standard connections ASA flanges — Inlet	Standard connections ASA flanges — Outlet	Max. back press. (psig) Conventional	Max. back press. (psig) Bellows	450 °F max. Carbon steel spring & body — Crosby style	Farris style	Pressure limits, psig Inlet −20 to 100 °F	Inlet 450 F	800 °F max. Alloy steel spring, carbon steel body — Crosby style	Farris style	Pressure limits (psig) Inlet 450 F	Inlet 800 F
M 3.60 in² →	4–150	6–150	230	80	JO-25	A-10	275	165	JO-26	A-20	165	92
	4–300	6–150	230	80	JO-25-3	A-11	275	275	JO-26-3	A-21	275	275
	4–300	6–150	230	160	JO-35	A-12	720	650	JO-36	A-22	650	365
	4–600	6–150	230	160	JO-55	A-13	1100	1100	JO-56-6	A-23	1100	730
	4–900	6–150	230	160	–	–	–	–	JO-56	A-24	1100	1100
N 4.34 in² →	4–150	6–150	230	80	JO-25	A-10	275	165	JO-26	A-20	165	92
	4–300	6–150	230	80	JO-25-3	A-11	275	275	JO-26-3	A-21	275	275
	4–300	6–150	230	160	JO-35	A-12	720	650	JO-36	A-22	650	365
	4–600	6–150	230	160	JO-45	A-13	1000	1000	JO-56	A-23	1000	730
	4–900	6–150	230	160	–	–	–	–	JO-66	A-24	1000	1000
P 6.38 in² →	4–150	6–150	230	80	JO-25	A-10	275	165	JO-26	A-20	165	92
	4–300	6–150	230	80	JO-25-3	A-11	275	275	JO-26-3	A-21	275	275
	4–400	6–150	230	150	JO-35	A-12	525	525	JO-36	A-22	525	365
	4–600	6–150	230	–	JO-45	A-13	1000	1000	JO-56	A-23	1000	730
	4–900	6–150	230	–	–	–	–	–	JO-66	A-24	1000	1000
Q 11.05 in²	6–150	8–150	115	70	JO-25	A-10	165	165	JO-26	A-20	165	92
	6–300	8–150	115	70	JO-25-3	A-11	165	165	JO-26-3	A-21	165	165
	6–300	8–150	115	115	JO-35	A-12	300	300	JO-36	A-22	300	300
	6–600	8–150	115	115	JO-55	A-13	600	600	JO-56	A-23	600	600

R 16.0 in²	6-150	8-150	60	60	JO-25	A-10	100	100	JO-26	A-20	100	92
	6-300	8-300	60	60	JO-25-3	A-11	100	100	JO-26-3	A-21	100	100
	6-300	10-150	100	100	JO-35	A-12	230	230	JO-36	A-22	230	230
	6-600	10-150	100	100	JO-45	A-13	300	300	JO-46	A-23	300	300
T 26.0 in²	8-150	10-150	30	30	JO-25	A-10	65	65	JO-26	A-20	65	65
	8-300	10-150	30	30	JO-25-3	A-11	65	65	JO-26-3	A-21	65	65
	8-300	10-150	60	60	JO-35	A-12	120	120	JO-36	A-22	120	120

Table A1.5. Temperature/viscosity relationship in fuel oils

Viscosity of oil		Temperature (°F) to obtain viscosity		
In SSU at 100°F	In SSF at 122°F	100 SSU	150 SSU	200 SSU
100		100	80	70
150		120	100	90
200		135	110	100
300	21	150	128	115
400	26	160	138	128
500	30	170	145	133
750	39	185	160	147
1000	50	190	168	155
1200	60	200	175	160
1500	75	205	180	168
2000	100	215	190	175
2500	115	220	195	182
3000	135	225	200	187
3500	145	229	205	190
4000	160	235	208	194
4500	170	237	212	197
5000	190	240	215	200
5500	200	245	218	204
6000	220	248	220	208
7000	260	250	225	210
8000	285	254	227	214
9000	320	258	230	216
10000	342	260	235	218

Table A1.6. Relationship of chords, diameters and areas

R*	L*	A*	R*	L*	A*	R*	L*	A*	R*	L*	A*	R*	L*	A*	R*	L*	A*
0.070	0.511	0.0308	0.120	0.650	0.0680	0.170	0.751	0.113	0.220	0.828	0.163	0.280	0.898	0.230	0.390	0.977	0.361
1	0.514	0.0315	1	0.652	0.0688	1	0.753	0.114	1	0.829	0.164	5	0.903	0.236	5	0.979	0.367
2	0.517	0.0321	2	0.654	0.0697	2	0.755	0.115	2	0.831	0.165	0.290	0.908	0.241	0.400	0.980	0.374
3	0.521	0.0328	3	0.657	0.0705	3	0.756	0.116	3	0.832	0.166	5	0.913	0.247	5	0.982	0.380
4	0.524	0.0335	4	0.659	0.0714	4	0.758	0.117	4	0.834	0.167	0.300	0.917	0.252	0.410	0.984	0.386
5	0.527	0.0342	5	0.661	0.0722	5	0.760	0.117	5	0.835	0.169	5	0.921	0.258	5	0.986	0.392
6	0.530	0.0348	6	0.663	0.0731	6	0.762	0.118	6	0.836	0.170	0.310	0.925	0.264	0.420	0.987	0.398
7	0.533	0.0355	7	0.665	0.0739	7	0.763	0.119	7	0.838	0.171	5	0.930	0.270	5	0.989	0.405
8	0.536	0.0362	8	0.668	0.0748	8	0.765	0.120	8	0.839	0.172	0.320	0.933	0.276	0.430	0.991	0.412
9	0.539	0.0368	9	0.670	0.0756	9	0.786	0.121	9	0.841	0.173	5	0.937	0.282	5	0.993	0.418
0.080	0.542	0.0375	0.130	0.672	0.0765	0.180	0.768	0.122	0.230	0.842	0.174	0.330	0.941	0.288	0.440	0.994	0.424
1	0.545	0.0382	1	0.674	0.0774	1	0.770	0.123	1	0.843	0.175	5	0.945	0.294	5	0.995	0.430
2	0.548	0.0389	2	0.677	0.0782	2	0.772	0.124	2	0.845	0.176	0.340	0.948	0.300	0.450	0.996	0.437
3	0.552	0.0396	3	0.679	0.0791	3	0.773	0.125	3	0.846	0.177	5	0.951	0.306	0.460	0.997	0.450
4	0.555	0.0403	4	0.682	0.0799	4	0.775	0.126	4	0.848	0.178	0.350	0.955	0.312	0.470	0.998	0.462
5	0.558	0.0410	5	0.684	0.0808	5	0.777	0.127	5	0.849	0.179	5	0.958	0.318	0.480	0.998	0.475
6	0.561	0.0418	6	0.686	0.0817	6	0.778	0.128	6	0.850	0.180	0.360	0.961	0.324	0.490	0.999	0.488
7	0.564	0.0425	7	0.688	0.825	7	0.780	0.129	7	0.851	0.181	5	0.964	0.330	0.500	1.0	0.50
8	0.567	0.0432	8	0.691	0.0834	8	0.781	0.130	8	0.853	0.182	0.370	0.967	0.337			
9	0.570	0.0439	9	0.693	0.0842	9	0.783	0.131	9	0.854	0.183	5	0.969	0.343			
0.090	0.573	0.0446	0.140	0.695	0.0851	0.190	0.784	0.132	0.240	0.855	0.184	0.380	0.971	0.348			
1	0.576	0.0454	1	0.697	0.0860	1	0.786	0.133	5	0.860	0.190	5	0.977	0.354			
2	0.578	0.0461	2	0.699	0.0869	2	0.787	0.134	0.250	0.866	0.196						
3	0.581	0.0469	3	0.700	0.0878	3	0.789	0.135	5	0.872	0.202						
4	0.583	0.0476	4	0.702	0.0887	4	0.790	0.136	0.260	0.878	0.207						
5	0.586	0.0484	5	0.704	0.0896	5	0.792	0.137	5	0.883	0.213						
6	0.589	0.0491	6	0.706	0.0905	6	0.794	0.138	0.270	0.888	0.218						
7	0.592	0.0499	7	0.708	0.0914	7	0.795	0.139	5	0.893	0.224						
8	0.594	0.0506	8	0.710	0.0923	8	0.797	0.140									
9	0.597	0.0514	9	0.712	0.0932	9	0.798	0.141									

* This table relates the downcomer area the weir length, and the height of the circular segment formed by the weir

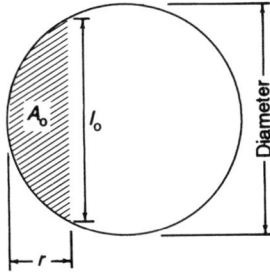

Figure labels: A_o, l_o, r, Diameter

R	L	A	R	L	A	R	L	A
0.100	0.600	0.0521	0.150	0.714	0.0941	0.200	0.800	0.142
0.101	0.603	0.0529	0.151	0.716	0.0950	0.201	0.802	0.143
0.102	0.605	0.0537	0.152	0.718	0.0959	0.202	0.803	0.144
0.103	0.608	0.0545	0.153	0.720	0.0969	0.203	0.805	0.145
0.104	0.610	0.0555	0.154	0.722	0.0978	0.204	0.806	0.146
0.105	0.613	0.0561	0.155	0.724	0.0987	0.205	0.808	0.148
0.106	0.615	0.0568	0.156	0.726	0.0996	0.206	0.809	0.149
0.107	0.618	0.0576	0.157	0.728	0.1005	0.207	0.810	0.150
0.108	0.620	0.0584	0.158	0.729	0.1015	0.208	0.812	0.151
0.109	0.623	0.0592	0.159	0.731	0.102	0.209	0.813	0.152
0.110	0.625	0.0600	0.160	0.733	0.103	0.210	0.814	0.153
0.111	0.628	0.0608	0.161	0.735	0.104	0.211	0.816	0.154
0.112	0.630	0.0616	0.162	0.737	0.105	0.212	0.817	0.155
0.113	0.633	0.0624	0.163	0.738	0.106	0.213	0.819	0.156
0.114	0.635	0.0632	0.164	0.740	0.107	0.214	0.820	0.157
0.115	0.638	0.0640	0.165	0.742	0.108	0.215	0.822	0.158
0.116	0.640	0.0648	0.166	0.744	0.109	0.216	0.823	0.159
0.117	0.643	0.0656	0.167	0.746	0.110	0.217	0.824	0.160
0.118	0.645	0.0664	0.168	0.747	0.111	0.218	0.826	0.161
0.119	0.648	0.0672	0.169	0.749	0.112	0.219	0.827	0.162

$$R = \frac{^*\text{Downcomer rise}}{\text{Diameter}} = \frac{r}{\text{Dia.}}$$

$$L = \frac{^*\text{Weir length}}{\text{Diameter}} = \frac{l_O}{\text{Dia.}}$$

$$A = \frac{^*\text{Downcomer area}}{\text{Tower area}} = \frac{A_D}{A_S}$$

Table A1.7. Standard exchanger tube sheet data

d_o = o.d. of tubing (in.)	BWG gauge	l = thickness (ft)	d_i = i.d. of tubing (in.)	Internal area (in.²)	External surface per foot length (ft²)
$\frac{3}{4}$	10	0.0112	0.482	0.1822	0.1963
$\frac{3}{4}$	12	0.00908	0.532	0.223	0.1963
$\frac{3}{4}$	14	0.00691	0.584	0.268	0.1963
$\frac{3}{4}$	16	0.00542	0.620	0.302	0.1963
$\frac{3}{4}$	18	0.00408	0.652	0.334	0.1963
1	8	0.0137	0.670	0.355	0.2618
1	10	0.0112	0.732	0.421	0.2618
1	12	0.00908	0.782	0.479	0.2618
1	14	0.00691	0.834	0.546	0.2618
1	16	0.00542	0.870	0.594	0.2618
1	18	0.00408	0.902	0.639	0.2618
$1\frac{1}{2}$	10	0.0112	1.232	1.192	0.3927
$1\frac{1}{2}$	12	0.00908	1.282	1.291	0.3927
$1\frac{1}{2}$	14	0.00691	1.334	1.397	0.3927
$1\frac{1}{2}$	16	0.00542	1.37	1.474	0.3927

Table A1.8. Typical factors used in capacity factored estimates

The factors presented here are those for the value of b in the capacity cost equation:

$$c = (k/B)^b$$

Type of plant	b
Atmospheric or vacuum crude distillation	0.6
Catalytic reforming	0.6
Fluid catalytic cracking	0.7
SO₂ extraction	0.6
Naphtha splitter	0.7
Thermal cracking, visbreaking	0.7
Delayed coking	0.6
Fluid coking	0.65
Gas compression—reciprocating	0.9
Gas compression—centrifugal	0.75
Light ends absorption	0.6
Polymerization (catalytic)	0.65
Alkylation (sulphuric acid and HF)	0.6
Isomerization units	0.6
Caustic wash	0.6
Amine units (MEA or Dea)	0.5
Hydrofining	0.6
Steam cracking	0.6

Table A1.8. (*continued*)

Type of plant	b
Ethylene production	0.6
Olefin extraction	0.4
Dehydrogenation plants	0.6
Hydrogen production	0.55
Sulphur production	0.55
Concentrators	0.7
Tankage	0.6
Steam generation	0.55
Feed and product handling	0.6
Offsites (general)	0.6
Utilities (general)	0.55

Appendix 2
LINEAR PROGRAMMING AIDS DECISIONS ON REFINERY CONFIGURATIONS (Reprinted from *Oil & Gas International*, June/July 1969)

D. S. J. Jones
Fluor (England) Ltd
and
J. N. Fisher
Bonner & Moore Associates, Inc., Houston, Texas

It is becoming more and more evident that there is a definite economic incentive to studying problems associated with refinery planning by mathematical modeling using the linear programming approach.

Ever-increasing investment costs associated with the tightening of product specifications, changing crude slates, alterations in the energy pattern in marketing countries, and expanding petrochemical requirements, all make an error in decision judgment of refinery processing increasingly costly. Wherever this basic risk can be reduced by modern mathematical techniques, the potential saving in capital or investment could make the financial outlay on such a study insignificant.

What follows is a description of a typical refinery simulation study but this is only one of an increasing number of problems that can now be solved by mathematical modeling using linear programming.

Wider Impetus

The application of linear programming techniques using electronic computers has long been used in the oil refining industry for the development of planning and operating policies. With the production of modern high-speed, large-capacity computers, this technique is having a considerably wider impetus within the industry.

Briefly, linear programming is the developing of linear sub models which mathematically describe many of the various operations within the industry, such as refinery processing, crude and product flows, marketing demands, etc. These linear sub models are inter-related to build up a complete mathematical model of the specific operation. By the use of the computer, the equations within the model can be solved to optimize, on a selective basis, the operation under study.

The growth of mathematical models using this technique provides management with a means of making an increasing number of decisions which do have a calculable basis. Thus, in many cases, the need for decisions based only on individual experience or 'feel', with its obvious inherent dangers, is being eliminated.

One such study carried out by Bonner & Moore Inc. and Fluor Corp. for a client illustrates a recent application of mathematical modeling and linear programming techniques.

Definition of Problem

The client, an American company, wished to build a refinery in Europe. It had already executed a maketing survey in the area and could specify quantity and quality, together with prices, of the products which would meet its market requirements. Its management now had to decide the economic optimum refinery configuration that would satisfy its crude and product slate. At this stage only one type of cude was intended for the refinery, and to some extent this simplified the problem. However, to satisfy other considerations, management required the solutions to the following premises:

- The refinery configurations, which would satisfy a *minimum* investment, when producing a high volume of gasoline with and without a low sulfur content limitation of the fuel oil. All other products were to meet quality and quantity requirements.
- The refinery configurations which would give the *maximum* return on investment to satisfy a fixed crude throughput with no quantity restriction on the product slate, and then to satisfy a limited restriction on the product slate with no limit to the crude throughput.

Such a problem lends itself readily to linear programming and thus Fluor together with Bonner & Moore built and developed a refinery simulation model to solve these two premises.

Process Consideration

The first step in constructing the model was to establish as many processing units as could conceivably contribute to the solution of the problem. For instance, with such large requirements of gasoline there would obviously be required a cracking unit of some kind. Thus, the model included a fluid cracker, hydrocracker, coker and visbreaker. Some combination or a single one of these processes must satisfy the premises of the problem.

Similarly, the lower sulfur content of the fuel oil would probably require some form of residue treating. Thus two severity desulfurizers for both short and long residue respectively were included together with a process for hydrocracking these residues.

The many process units now included were then defined in terms of feed streams, product yields and quality, and operating costs, all based as a percentage on the feed streams. This part of the study was the first important step which required the expertise of specialists. This data forms the basis for the rational solution to the problem, and therefore it was necessary to be accurate and to augment prediction and theory with realism and technical experience.

For instance, in arriving at the yields from the crude and vacuum units the effect of fractionation on the product yield was considered. Realistic ASTM distillation gaps were used that could be met by a commercial distillation unit.

In the fluid catalytic cracking units a more sophisticated approach was needed to correlate the yields from the many feedstocks which would be independent of thermodynamic considerations. Here a base case feed yield data (in this case a straight run waxy distillate) at a conversion of 75% using zeolite catalyst was used.

Yields from all other feedstocks (including those which had been hydrotreated) were related by first principle kenetic and thermodynamic considerations to the base case. A short and simple computer program was used for this purpose, and it was also possible to simulate the effect of changing the quantity of zeolite catalyst by this means. The results of these computerized calculations were checked against existing plant data before being incorporated into the study.

In other processes such as hydrocracking, hydrotreating, visbreaking, etc., care was taken that only proven yield data or correlations were used.

Catalytic reforming yield data was obtained from a correlation which related yield to severity for a basic naphthene and aromatic content of the feed-stock. A whole range of severity operations from 95 to 105 O.N. (Research) clear was encompassed in the study. Spot checks of the predicted yield by this method against actual yield from an operating unit showed that the method was viable and acceptably accurate.

Basic Economic Data

Having developed the physical yield structure of the 'model', the next step was to complete the basic data by providing investment and maintenance costs.

There is, of course, a considerable wealth of plant cost data available to a contractor from the projects he has completed over the years. However, there is always the need to analyze these costs, and to review them in terms of up to data material and labor cost changes.

For this study, a large amount of cost data was statistically analyzed for each type of plant. From this analysis, a base cost and an empirical expotential factor was developed in order to relate a total investment cost to capacity in as realistic a way as possible. This relationship can be expressed mathematically in a non-linear form.

$$C = C_0(T/T_0)^K$$

where

C = investment cost
C_0 = base cost at a base throughput T_0
T = new throughput
K = an empirical constant

The inclusion of a non-linear form for inventment costs in a linear program required special consideration, and we shall see later how this was utilized.

In the model, many of the units considered were licensed processes, for which a royalty would be paid. A value in terms of a paid up royalty in dollars per barrel of throughput was included in the investment. Where chemicals and catalysts were used, the first inventory of these was also included as part of the investment.

Chemicals and loss of catalyst was considered as an operating cost based on usage as were utilities. Labor, a fixed operating cost, was included with the return on investment. Maintenance cost was fixed as a percentage of the total maintenance cost.

Model Development

At the same time as the process and cost data were being generated, the basic form that the model would finally take was also being developed. This consisted of defining the various optional routes of each stream within the simulated refinery model.

The optional routing of the streams was carefully selected. This selection had to satisfy at least one of two requirements. Firstly, would such a routing actually contribute to satisfying the product slate and the premises of the problem? Secondly, would such a routing be feasible under actual operating conditions?

Just as a refinery is described by the units of which it is comprised, so also was the refinery linear model described. Here, each processing unit was considered a submodel in itself and these submodels were defined by their process and economic data.

These data were arranged in tabular form which were easily accessed and listed in recognizable terminology. An example of a submodel tabulation as used in this study is shown in Table 1. From this tabulation, a matrix generator, called GAMMA[1,2], was used to assemble the many submodels into a complete LP matrix. The matrix was solvable by a linear programming system called OMEGA[3]. These tabular input arrays were also used by the solution report writers as we shall see later.

This complete mathematical model of the refinery was displayed by an equation listing of the entire contents of the data. The equations showed the inter-relationships of the many variables, including the refinery streams, the blending constraints, the unit to investment ties, etc. These were also in a form which was recognizable to the engineers working on the project.

Having now assembled all the data in a manner usable for linear programming, it was necessary to check it for errors. Various computer techniques had been developed for this purpose, and these, together with a secondary check by the process engineers, substantially eliminated the possibilities of obvious error and invalid data.

However, as an added safeguard, a final checkout was carried out by actually solving a test case. These results were scrutinized to ensure that the output gave a realistic refinery configuration and that all was in balance.

Optimizing and Other Techniques

A major value of linear programming is that once an environment is reflected within a LP framework, this environment can be optimized. In this study, optimization could be

Table A2.1. An extract from a typical base data file

TABLE TEF—H2 TREAT OF CRACKED GAS OILS			
	KGO COKER GAS OIL	VBO VIS BR GAS OL	CYO CAT CY LE OIL	*
REFORMER H2 FOEB	0.0244	0.0209	0.0096	XX + H2F
GAS, FOEB	—	—	—	XX + GS1
PROPANE	—	—	—	XX + C3S
ISOBUTANE	—	—	—	XX + IC4
N-BUTANE	—	—	—	XX + NC4
C5-380 HYDRO GASL	−0.0916	−0.0550	−0.0527	XX + TNP
LOSS OR GAIN	−0.0169	−0.0109	0.0034	XX + LOS
DESULF COKER GAS O	−0.9159	—	—	XX + SKO
LT COKER GAS OIL	1.0000	—	—	XX + KGO
LT VIS BR GAS OIL	—	1.0000	—	XX + VBO
H2 TREAT LT VIS GO	—	−0.9550	—	XX + SVO
H2 TRTED CYCLE OI	—	—	−0.9603	XX + SCY
CYCLE OIL	—	—	1.0000	XX + CYO
FUEL MMBTU/UNIT	0.1113	0.1113	0.1113	XX + FUL
ELEC KWH/UNIT	3.0764	3.0764	3.0764	XX + KWH
STEAM MLB/UNIT	0.0029	0.0029	0.0029	XX + STM
CHEM ROYALTY CATAL	0.0023	0.0023	0.0023	XX + CRC
REPT FEED COLLECTO	1.0000	1.0000	1.0000	XX + FOR
* * * * * *	—	—	—	STOPIT
REPORT WRITER AID	1.0000	1.0000	1.0000	UNPACK
SULFUR TO RECOVER	−8.8500	−5.1400	−5.9800	XX + SUA
H2 TRT LT GO CAP	1.0000	1.0000	1.0000	HTXCPE
GAS PLANT MAX CAP	—	—	—	GPXCAP

accomplished either by maximizing profit or minimizing expense, and in this specific case, the former was selected. It should be emphasized that optimization can only be achieved for the environment reflected in the model. Great care was taken therefore to reflect all the meaningful, worthwhile options known to be available.

There are many refinery variables that are of a non-lienar nature. Among these non-linearities are the effects of TEL on motor gasoline, the capital expenditures in relation to size of the units, and many severity effects within the various processes. When these effects could be described on a cost basis by a *convex* curve, they were generally included in the model as linear segments of a curve (see Fig. 1).

If the severity to value of product relationship was a concave curve, only one variable could be used to reflect changing severity. This by nature has to be an estimate with a review of the estimate upon solution.

The TEL to Octane relationship is highly non-linear and a new approach was used to reflect this in the model. This approach had considerable advantage over the older techniques in that it was relatively easy to understand and use.

It was capable of reflecting the value of lead susceptibility to the individual components available for the blends. It also had the capability to represent accurately more than one type of octane (i.e., research, motor, road, etc.) with these effects also reflected back to the various components.

This technqiue required the aid of a recursive routine to update various model coefficients that reflected the actual susceptibilities at the solution point. The model then re-optimized with the recalculated octane to TEL slopes and again they were checked. This was repeated until no further change was required.

The non-linearities of the capital expenditures are concave in nature. Thus, the initial investment cost estimate was updated by means of a recursive routine which calculated the investment cost per unit of activity at the solution level. This recursive routine is described in detail next. Among the various factors that could be considered in investment cost to size relationships, offsites, insurance, taxes (both income and property), overhead, maintenance, labor, royalties, escalation, plant or service factors, depreciation, and the expected economic life of the various facilities.

Solution Approach

In a study of capital expenditure (commonly called facilities planning) such as this, there are many possible mathematical techniques which can be used to obtain a solution.

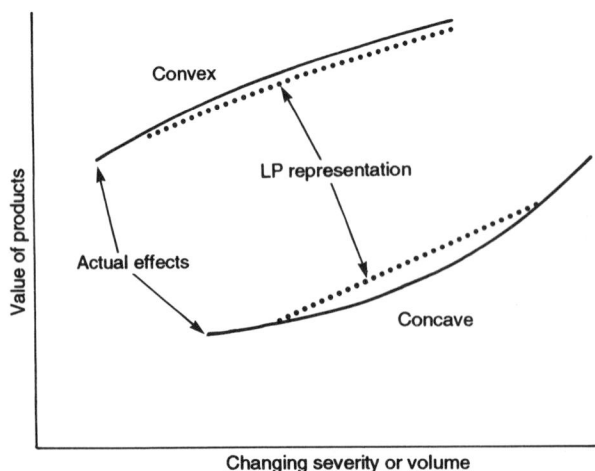

Figure A2.1. LP representation of the convex and concave curves

Experience has shown, however, that linear programming on a computer is by far the most economic approach to solving this type of problem. By this method, most of the many possible solutions can be examined quickly and effectively. Before discussing the solution approach used in this study, let us quickly review the investment environment for any typical refinery unit. This is illustrated in a simplified form in Fig. 2.

The total cost curve, on linear graph paper, shows that as a volume throughput is increased, the total cost of a unit will increase. However, the plot of unit volume cost ($/bpcd) against throughput shows the reverse; that is, the cost per unit of throughput will decrease as the volume of throughput increases. It is this type of cost that is reflected in an LP model.

Fig. 2 also shows a plot of the average cost slope and the incremental cost slope at a given throughput level. This *average cost* slope includes all the costs associated with a particular unit. The slope is linear and must pass through the origin. The *incremental cost* slope reflects the change in cost per unit throughput over a short range. This slope is always tangential to the total cost curve at any throughput under consideration.

It can be seen that the *average cost* gives the model 'greater than to be expected' incentive for changing the size of the unit, while the incremental cost curve on the other hand gives the expected incentive. The incremental cost ignores all fixed costs at the solution level, and consists mainly of the expected return on investment.

The technique used in this study was to begin the solution by establishing a very nominal cost on all units. This allowed any unit to be chosen in the solution. Once the LP had selected an optimal unit configuration with those nominal costs, a recursive program for investment cost estimating was used to determine the *average cost* of the units at the solution throughputs. These average costs were based on a minimum return of investment, and an expected economic life calculated on a discounted cash flow basis.

These new calculated costs were then substituted for the original arbitrary cost in the model. This average costing tended to delete the very unrealistically small units that may have been chosen in the unrestricted configuration. Again, the model was optimized with these new costs. The recursions and solutions were repeated until no further cost, configuration or size changes were required. All the items described above were carried out in a single computer run, and this solution was saved on tape and reported. Close scrutiny of the results then followed to make sure that they were reasonable and that there was no automatically restricted unit that might have been selected had a different solution path been used.

A second step was then commenced which restored the solution of the first step based on

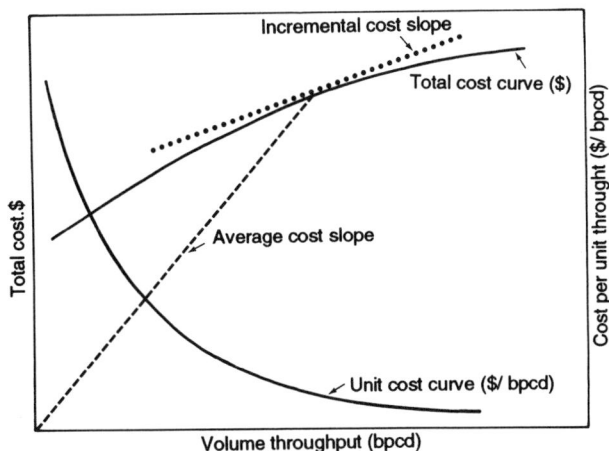

Figure A2.2. Cost parameters for any typical refinery unit

the *average cost* to the computer. The investment costs for the configuration were then recalculated using the *incremental cost* concept. This incremental cost required no return on investment and an infinite economic life was assumed for the units. The only costs that were recognized in this step were the incremental maintenance for each unit. Therefore, the building costs did not economically suppress the size of units or the total investment. The unit sizes and total investment were optimized on the premised product slate and available raw materials. A similar recursive step was again used, but this time it included only the incremental maintenance cost.

The model contains a variable that carries the sum total of the investment. This variable is updated at the same time as the unit costs are updated by the recursive operations described. Using this total investment variable it was possible to step down (paramatrize) on the total investment, until a feasible solution was no longer possible. An infeasible solution in this context is that in which the case flow for the configuration becomes less than zero.

As the total investments are reduced, the unit sizes, the product slates and raw materials are changed within the framework of the overall premises. Therefore, by this parametric sweep, the refinery configurations (both in size and form) could now be found which would satisfy the following requirements:

- Maximum investment
- Maximum return on investment
- Maximum expected cash flow
- Minimum investment.

The results of each parametric step were reported and documented.

Solution Analysis

Besides the specially designed report writing technique already discussed, the LP also has a standard number of solution reports and these are generally of a more technical nature. Although of great value, they do not readily lend themselves to immediate and apparent interpretation. It is imperative, however, that some members of a team using an LP can read and interpret these outputs, particularly as one major use of these reports is to highlight any obvious errors that may have been overlooked. Some of these reports used in this study were called the BI/DJ and Range output. They warrant a brief explanation.

The BI/DJ output (see Table 2) gives the solution level activities (BI) for all the variables that were selected in the optimum solution. For those variables that were not selected, the cost or decrease in profit that would occur were they forcibly included in the configuration (or basis) is given by the DJ value.

The Range information is a complimentary report to the BI/DJ. The ranges give the incremental volume associated with the DJ's and the cost ranges associated with the basis variables. The valuable use of the BI/DJ and the range files can best be expressed by an example, using the extracts shown in Table 2, 3 and 4.

Consider the component CYO that has been selected in the configuration. (The code FU/CYO in this case indicates that the refinery stream CYO is routed to fuel.) The quantity of the stream CYO that enters the fuel blend is 1,457 bpcd. Now consider the component 'HBM'. This has not been selected in the basis, and this stream has a DJ value (i.e., no prefix). To route the 'HBM' stream to fuel forcibly would cost $0.6095/bbl. This information is interesting, but has no real value unless the ranges for their streams are known.

Because the FU/CYO item is selected in the basis, the range data for this is found in the primal range output (Table 4). Interpreting the statement for this stream in this output means that optimum volume fuels of the solution would not have changed even if the variable had a very small *negative* incentive (less than $0.000001). Further, the situation would not have changed even if this variable had a *positive* cost incentive of up to $0.02807/bbl.

Table A2.2. An extract of a typical BI/DJ output

Status	Label	Cost	BI/DJ	Status	Label	Cost	BI/DJ
BI	FU + VIS	0.000	12.9443	BI	FU/VBR	0.000	0.0000
BI	FU/CYO	0.000	1.4567		GO-DIN	0.000	0.0574
BI	FU/HBA	0.000	0.0000	BI	GO − SPG	0.000	0.0029
	FU/HBB	0.000	0.0591		GO − XXX	−10.000	14.1623
BI	FU/HBE	0.000	0.0000	BI	GO − 446	0.000	8.7949
	FU/HBF	0.000	0.0000	BI	GO − 675	0.000	12.5429
	FU/HBG	0.000	0.1084	BI	GO + POR	0.000	1.9087
	FU/HBH	0.000	1.3164	BI	GO + SPG	0.000	0.0097
	FU/HBM	0.000	0.6095	BI	GO + SUL	0.000	0.0754

Table A2.3. An extract of the corresponding primal range output

		Limits of range			
		Negative		Positive	
		Variable	Cost	Variable	Cost
LP label	LP cost	affected	increment	affected	increment
MD/KER	0.000000	MD/HDF	−0.005291	MD + SUL	0.131973
FU/CYO	0.000000	HF/CYO	−0.000000	XX + VBR	0.028074
FU/HBA	0.000000	HF/HBA	−0.000000	GO/HDA	1.293871
FU/HBE	0.000000	HF/HBE	−0.000000	H3FHCE	1.085441
FU/HRB	0.000000	HF + POR	−0.000000	HF + VIS	0.000000
FU/HRE	0.000000	HF/HRE	−0.000000	HVFHBE	0.164348
FU/HRG	0.000000	HF/HRG	−0.000000	GO/HDG	1.243582
FU/HSB	0.000000	HF/HSB	−0.000003	GO/HDB	1.510931
FU/KEX	0.000000	XX + KEX	−2.541661	SLYKER	19.910502

Table A2.4. An extract of the corresponding dual range output

		Limits of range	
	Original	Variable	Positive volume
LP label	activity	affected	increment
FU/HBB	0.000000	FU/HRB	0.000000
FU/HBF	0.000000	HF/HBF	0.000000
FU/HBG	0.000000	CCFHSG	0.000000
FU/HBH	0.000000	TEFCYO	0.109287
FU/HBM	0.000000	FU/CYO	0.549070
FU/HRA	0.000000	FU/VBR	0.000000
FU/HRF	0.000000	FU/VBR	0.000000
FU/HRH	0.000000	FU/VBR	0.000000
FU/HSA	0.000000	CCFHSA	0.000000

Let us now look at the 'HBM' variable. This was not chosen in the basis and it appears in the dual range output (Table 4). Interpreting the data for this item shows that to route 'HBM' to fuel would cost 60 cents/bbl for the first 549 bpcd. All that is known thereafter is that the cost per barrel over 549 bpcd would increase.

The example chosen here describes the economic analysis of two optional streams which can be logically routed ao a fuel blend. It is emphasized that the BI/DJ and range outputs, however, contain similar information for *all* variables whether process units, refinery streams, product specifications, etc. contained in the model.

Computer Report Writer

The data generated by the computer contains all the facts relevant to the solution. However, to all but a few highly trained people the data in this form would be meaningless and of no practical use. The LP system used in this study contained a specially designed report writer, coded in a language called DART[4,5]. This converted and assembled the computer LP output into management orientated reports that could easily be read and understood without sacrificing the relevant technical content.

In this particular case, too, much of the data, as produced by the report writing sequence, was in such a form as to be reproducible and able to be included in the final documentation. Table 5 shows an example of such a report. (Note: the actual calculated data in this example has been deleted).

For the parametric series discussed above, a special report writing technique was developed which allowed each succeeding parametric step to be repeated in a case stacking fashion. This type of report was considerably condensed from the reports described earlier.

Table A2.5. Material and economic balance

Product or feed	Price	M B/CD	M $/CD	MM $/YR
Premium gasoline	—	—	—	—
Inter. gasoline	—	—	—	—
Regular gasoline	—	—	—	—
Hi. vis. hvy fuel	—	—	—	—
Kerosene, regular	—	—	—	—
Propane LPG	—	—	—	—
Marine diesel	—	—	—	—
Sulfur MM LBS		(*Actual data has been deleted*)		
Shortage and fuel	—	—	—	—
Total production	—	—	—	—
Crude	—	—	—	—
Total feedstocks	—	—	—	—
Tel. in liters	—	—	—	—
Production margin	—	—	—	—
Expenses				
Utilities chem. and royalties	—	—	−6.064	—
Operating labor, super, and lab.	—	—	−2.930	—
Maint., ins., tax, and overhead	—	—	−3.061	—
Capital recovery	—	—	−12.817	4.678
Total expenses	—	—	−24.872	—
Earnings (loss)	—	—	−4.202	−1.534
Cash flow, earnings plus capital recovery	—	—	—	3.144

Investment	*MM$*
Plant	27.929
Offsite	—
Catalyst and royalties	—
Added offsites, wharfs, etc.	—
Total	27.929

Years to payout, inv./cash flow*   8.882
ROI 7 years, percent 3.0
ROI 10 years, percent 10.5
ROI 16 years, percent 15.1

* Figures reported before income tax withdrawn

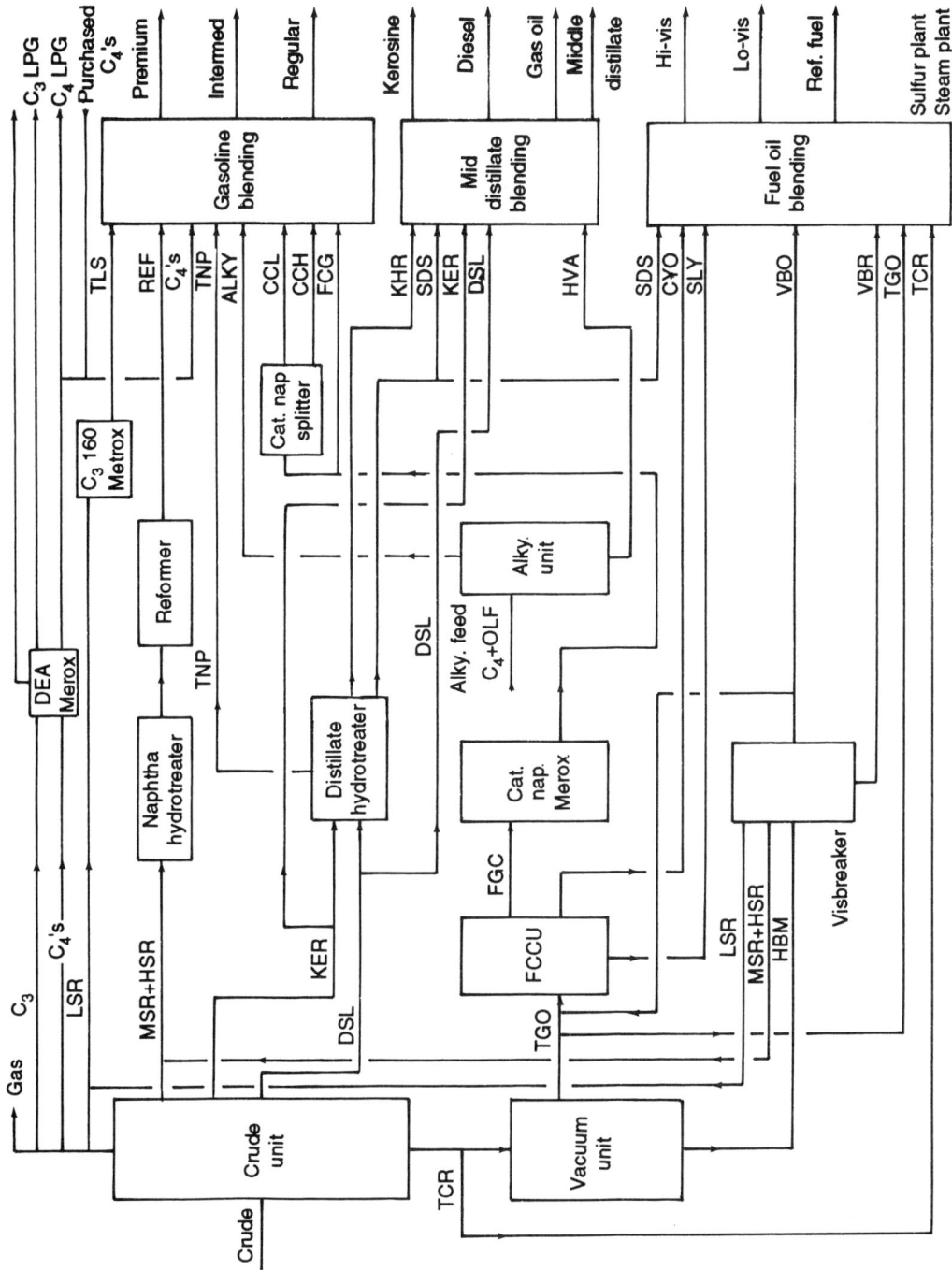

Figure A2.3. Typical final optimum configuration

Final Documentation

When the solutions to the four premises of the problem had been determined, using the techniques described, they existed, hidden among the mass of tabulated data that formed the computer output. It remained now to extract the pertinent section of the output and to present it so that the many objective of the study, which was to provide management with information to make a good decision, could be achieved. The most common means of doing this—and the one chosen on this occasion—is by a written report in which the data is summarized, discussed and the conclusion stated.

Although it is not proposed to discuss the general techniques of technical report writing here, some fundamental requirements of a complex presentation such as this are worth highlighting. This report had to satisfy two principal functions. The first, to present as succinctly as possible the conclusions, and the interpretation of those conclusions, for the convenience of the client's management. Secondly, it had to present all the back-up data in as short a form as possible that would be necessary to enable the client's own staff to check and confirm the conclusions reached.

This second function was satisfied in this report in the form of an appendix. This included copies of the actual pertinent computer printouts complete with tabular listing of the submodels, economic balances, etc. These data were further augmented by the summary of the economic and yield output for the respective parametric runs.

The main body of the report consisted of a short description of the study, together with the discussion of the result. The results were however succinctly described by two illustrations for each of the four cases of the problem. The first illustration showed the ultimate refinery complex which satisfied the premise of the case studied, and a typical example is shown in Fig. 3. The second illustration, typified by Fig. 4, gave the basic economic trend for this configuration and also described the yields of major products for each parametric case.

The charts shown are meant only as an example and the figures are fictitious or have been purposely deleted. However, it can be seen that Fig. 3 describes the result in terms of the processes that must be built to satisfy the premise. Fig. 4 shows why such a configuration is the optimum, and what the ultimate product slate would look like. Such a chart also gives the client's management an opportunity to assess quickly the effect of changing a basic premise such as maximum return on investment or minimum capital investment.

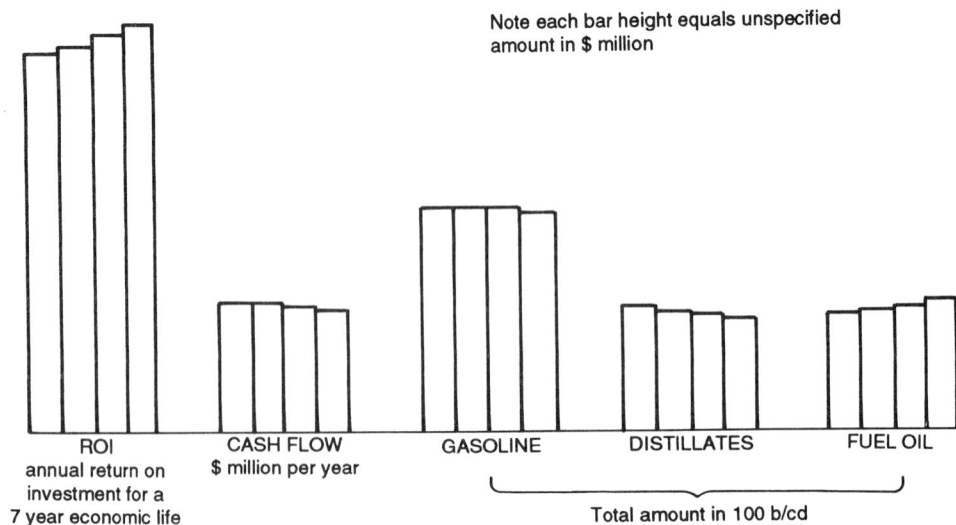

Figure A2.4. Bar chart summarising economic studies for configuration in fig A2.3

Bibliography

1. Hearn, J.: 'GAMMA—A General Description', Bonner and Moore Publication. Document C 139-1, March 10, 1965.
2. Hearn, J.: 'GAMMA Primer', Bonner and Moore Publication, Document M G1-1, May 10, 1965.
3. Romberg, F. A.: 'OMEGA—A General Description', Bonner and Moore Publication, Document C 74-2, January 1965.
4. Romberg, F. A.: 'Dart Systems for Larger Computers', Bonner and Moore Publication, Document C 55-2, May 31, 1965.
5. Romberg, F. A.: 'Dart 2—A General Description', Bonner and Moore Publication. Document C 53-2, March 31, 1967.

REFERENCES

1. Fenske, *Ind. Eng. Chem.*, **24** (1932).
2. Underwood, *Chem. Eng. Progress*, **44**, 603 (1948).
3. Gilliland, *Ind. Eng. Chem.*, **32** (1940).
4. *GPSA Engineering Data Book*, 9th edition.
5. D'Arcy and Weisbach, *Hydraulic Institute Engineering Data Book*.
6. L. F. Moody, *ASME* (1944).
7. Maxwell, *Data Book on Hydrocarbons*.
8. Heilman, *Trans. Am. Soc. Mech. Eng.* **51**, 287 (1929).
9. Brown and Souders, *Ind. Eng. Chem.*, **52**, No. 8 (1960).
10. Sherwood, Shipley and Holloway, *Ind. Eng. Chem.*, **30**, 765 (1930).
11. Sieder and Tate, *Ind. Eng. Chem.*, **28** 1429 (1935).
12. Chilton and Colburn, *Ind. Eng. Chem.*, **26**, 1183 (1934).

INDEX